INTRODUCTION TO Magnetic Resonance Spectroscopy ESR, NMR, NQR

INTRODUCTION TO
Magnetic Resonance Spectroscopy
ESR, NMR, NQR

Third Edition

D.N. SATHYANARAYANA
Formerly, Chairman,
Department of Inorganic and Physical Chemistry
Indian Institute of Science, Bangalore

ISBN: 978-81-94778-76-9

Edition: 2022

Printed at: Shree Maitrey Printech Pvt. Ltd., Noida

To my wife, Vijayalakshmi,
to whom I dedicate this book.

Preface to the Third Edition

The main objective of 'Introduction to Magnetic Resonance Spectroscopy' is to acquaint readers with more details than that found in basic spectroscopy books. What has not changed in this edition is the breadth of coverage, one of the features that has led to its wide use as a graduate level text and reference material. In Magnetic Resonance Spectroscopy wide range of structural techniques are discussed in depth. These techniques are used in organic chemistry as much as in inorganic chemistry.

Several new problems have been added and detailed solutions are provided for all the end-of-chapter problems. This is deliberate to make the students aware of the problem solving strategies. Solving problems is a very real part of the curriculum. I have attempted to correct as many as possible of the errors that appeared in the earlier edition.

It is a pleasure to thank Professor E. Arunan, the chairman of the Department for extending all the facilities. I thank my wife Vijayalakshmi and our sons Supradeep and Sushirdeep for their cooperation and support.

D.N. SATHYANARAYANA

Preface to the Second Edition

I am happy to note that my book has been quite favourably accepted by students and teachers of various universities and for this I am grateful. Readers of this book will, if they are familiar with the first edition, immediately recognize that there are no major changes.

I have continued to handle other aspects of the book in the same way as in the earlier edition. It has always been my desire to keep the physical size of the book to about that of the first edition.

This book is intended for graduate students and research workers of chemistry, physics and biological sciences who need a text that is pitched at a modest relatively non-mathematical level and that can be applied in solving problems related to structure and behaviour of organic and inorganic molecules. The theoretical basis of the more important effects are given where it seemed appropriate.

I wish to record my sincere thanks to Professor A.G. Samuelson, the Chairman of the Department for providing the facilities. My thanks are due to my wife, Vijayalakshmi and sons, Supradeep and Sushirdeep for their forbearance and encouragement.

Care has been taken to make the book error and obscurities free. Feedback from teachers and students especially concerning errors and deficiencies in this edition are welcome. Suggestions for further improvement of the text will be greatly appreciated. Finally I am deeply grateful to I.K. Publishers at New Delhi and Bangalore for believing in this project from its onset and overseeing various aspects of its production.

Bangalore
July 18, 2013

D.N. SATHYANARAYANA

Preface to the First Edition

The contents of this book are concerned solely with spectra arising from the interaction of magnetic field component of the electromagnetic radiation with nuclear and electron spins. The chapters deal successively with the fundamentals of electron spin resonance, nuclear magnetic resonance and nuclear quadrupole resonance spectroscopy. This book has the aim of providing an introduction to magnetic resonance spectroscopy while making use of illustrations. The book is primarily aimed at students taking an introductory course in magnetic resonance spectroscopy at the Masters Degree level and those entering research in chemistry, biochemistry, pharmacy and physics. It should also serve as a source book for self study. They hopefully will find the book easy to read and understand.

The past three decades have witnessed the development of nuclear magnetic resonance spectroscopy into one of the most powerful analytical tools. An indepth understanding of this advanced technology is necessary for fruitful applications of magnetic resonance spectroscopy. In a subject as large as magnetic resonance spectroscopy, a book of this length presents the author with a difficult dilemma; one can either attempt to treat all areas equally, albeit at a superficial level, or one is forced to restrict the scope so as to achieve a reasonable coverage of selected topics. The second approach has been adopted here for two reasons. First, it provides for greater coherence and a more satisfying level of exposition. Secondly, with the tremendous strides, magnetic resonance spectroscopy is now the norm rather than the exception.

The first part covers the basic material necessary for understanding the topic of ESR spectroscopy. The subsequent parts, part II and part III, have dealt with NMR and NQR spectroscopies.

Many of my students have contributed to the development of the book – in particular I wish to thank Dr. Priyadarsi De, Dr. S. Subrahmanya and Dr. B. Veerendra. I am profoundly grateful to Professor S.N Bhat, Jawaharlal Nehru Centre for Advanced Scientific Research, Bangalore, Professor Arabinda Ray, Sardar Patel University, V. V. Nagar, Gujarat and Professor S. Manogaran, Indian Institute of Technology, Kanpur, who have carefully and perceptively read the whole manuscript and have made constructive criticisms. I am most thankful to the various authors and publishers who have given their permission to reproduce the respective copyrighted material.

I would like to thank Professor P. Balaram, the Director, Indian Institute of Science, Bangalore for general help. My special thanks are due to Professor A.R Chakravarty, the former Chairman, Professor K.L. Sebastian, the former

Chairman, and Professor A.G. Samuelson, the Chairman, Department of Inorganic and Physical Chemistry of the Indian Institute of Science, for their full support by extending all the facilities to me. I am greatly thankful to my esteemed colleague Professor V. Krishnan for his help which has enabled me to complete this work. The Financial assistance received from the Department of Science and Technology, New Delhi, towards the writing of the book under the scheme "USERS" is gratefully acknowledged. I gratefully acknowledge the financial assistance received towards the preparation of the manuscript by the Curriculum Development Cell of the Centre for Continuing Education of the Indian Institute of Science. I would like to record my appreciation of the assistance of Mr. C.Sridhar for his expert typing of the manuscript, Mr. M.B. Madhusudhana Reddy, Chemistry Department, Bangalore University and Mr. M. Srinivasa Murthy for the preparation of some of the drawings. I would like to thank my publishers for their full support throughout the production process. It is inevitable that in a book of this size and complexity that there will be occassional errors. I shall be grateful to the readers who point out any errors and/or make other constructive suggestions.

It is always a joy and a privilege to express my gratitude to my wife, Vijayalakshmi and our children, Supradeep and Sushirdeep for their love, support and emotional sustenance during the writing of this book.

D.N. SATHYANARAYANA

Acknowledgements

It is not feasible to list all the sources which have helped me in writing this text. However, I would like to specially acknowledge the following works from which some of the illustrations have been adopted, I am most thankful to these authors and publishers for the permission to reproduce the copyright material.

The permission of the following individuals, institutions, and journals for reproduction of illustrations is gratefully acknowledged.

Fig. 3.28	A. Carrington and J.dos Santos–Veiga, Mol. Phys. **5,** 21 (1962).
Figs. 3.30 and 6.1	J.A. Weil, J.R. Bolton and J.E. Wertz, Electron Paramagnetic Resonance, Wiley–Interscience, New York (1994).
Fig. 3.34	D.H. Levy and R.J. Meyer, J. Chem. Phys. **41**, 1062 (1964).
Fig. 4.3	Professor K.R. Nagasundara and Dr. S. Rekha, Department of Chemistry, Bangalore University, Bangalore.
Fig. 6.2	R.S. Drago, Physical Methods in Chemistry, W.B. Saunders & Co., Philadelphia (1977).
Fig. 7.6	C.A. Hutchison, Jr. and B.W. Mangum, J. Chem. Phys. **34,** 908 (1961).
Fig. 7.8	R.E. Jesse, P. Biloen, R. Prins, J.D.W. van Voorst and G.J. Hoijtink, Mol. Phys. **6,** 633 (1963).
Fig. 8.12	A.H. Maki and B.R. McGarvey, J. Chem. Phys. **29,** 35 (1958).
Fig. 8.14	B.J. Hathaway and D.E. Billing, Coord. Chem. Revs. **5,** 143 (1970).
Fig. 9.3	G. Labrauze, J.B. Raynoor and E. Samuel, J. Chem. Soc. Dalton Trans. 2425 (1980).
Figs. 12.7 and 19.17	P.J. Hore, Nuclear Magnetic Resonance, Oxford University Press, Oxford (1995).
Fig. 14.20	Ramey, O'Brien, Hasegawa and Borchert, J.Chem. Phys. **69**, 3418 (1965).
Fig. 14.22	F.A. Cotton and D.L. Hunter, J. Amer. Chem. Soc. **98**, 1413 (1976).
Fig. 15.7	K.G.R. Pachler and P.L. Wessels, J. Magn. Reson. **12**, 337 (1973).
Figs. 15.9, 15.15 and 19.22	A–U Rahman and M.I. Choudhary, Solving Problems with NMR Spectroscopy, Academic Press, San Diego, California (1996).

Fig. 15.12	H. Gunther, NMR Spectroscopy, Basic Principles, Concepts and Applications in Chemistry, 2nd edition, John Wiley, New York (1995).
Fig. 16.6	M.L. Martin, J.J. Delpuech and G.J. Martin, Practical NMR Spectroscopy, Wiley, New York (1986).
Fig. 16.10	J.W. Akitt, NMR and Chemistry, An Introduction to the Fourier Transform Multinuclear Era, 2nd edition, Chapman and Hall, London (1983).
Fig. 17.2	Farrar, Johannensen and Coyle, J. Chem. Phys. **49**, 281 (1968).
Fig. 18.4	E. de Boer and C. McLean, Mol. Phys. **9**, 191 (1965).
Fig. 18.6	S.A. Richards, Laboratory Guide to Proton NMR Spectroscopy, Blackwell Scientific Publications, Oxford (1988).
Fig. 18.8	From Bruker CXP Application Notes.
Fig. 18.9	R.K. Harris, Nuclear Magnetic Resonance Spectroscopy, Pitman, London (1983).
Figs. 19.7 and 19.21	J.K.M. Sanders and B.K. Hunter, Modern NMR Spectroscopy, Oxford University Press, Oxford (1987).
Fig. 19.12	G. Solomons and C. Fryhle, Organic Chemistry, John Wiley and Sons Inc., New York (2000).
Fig. 20.4	E.A. Kravchenko, N.V. Timofeeva and G.Z. Vinogradova, J. Mol. Struct. **58**, 253 (1980).

Contents

GENERAL INTRODUCTION 1

INTRODUCTION

The particular spectroscopic methods selected are those which deal with nuclear spin and electron spin. They are:

NMR nuclear magnetic resonance,

NQR nuclear quadrupole resonance and

ESR electron spin resonance.

This particular selection has been made on the basis that they have much in common and that all are particularly useful in chemistry. This chapter provides an introductory survey of the techniques discussed later and treats matters which are common to them all.

All electrons and some nuclei possess a property conveniently called 'spin'. The spectra they can give rise to come under the purview of magnetic resonance spectroscopy. Magnetic resonance is one of the few techniques available for obtaining structural and dynamic information of the molecules at the atomic level. Other spectroscopic techniques usually give information at the molecular level. The magnetic resonance spectroscopy gives highly specific information. The theory behind magnetic resonance spectroscopy is common to both electron spin resonance (ESR) and nuclear magnetic resonance (NMR), but there are differences in the signs and magnitudes of the magnetic interactions involved. Major differences arise from the size of the magnetic moments and the frequencies of the radiation used to bring resonance. There are many studies specifically designed to utilize the differences between ESR and NMR phenomena. For example, electron spins often lead to line broadening in NMR spectra and nuclear spins cause splitting in ESR spectra. These interactions provide information on molecular structure. ESR is a useful complement to NMR, which cannot normally be used for paramagnetic substances. Both depend on the magnetic moment associated with a spinning particle either an electron or a nucleus. The splitting of the energy levels by a magnetic field is customarily referred to as the Zeeman effect, after the Dutch Physicist Pieter Zeeman who discovered the effect. Accordingly, ESR is the study of transitions between electron Zeeman levels while NMR is the study of transitions between nuclear

Zeeman levels. An interesting aspect of the ESR as well as NMR spectroscopy is that the separation between the spin energy levels can be varied by varying the applied magnetic field. In other branches of spectroscopy, the energy levels are fixed for a particular system. The requirement of a static magnetic field is an unique aspect of magnetic resonance spectroscopy.

All the three methods have some limitations on the type of system which can be examined. ESR is restricted to samples containing unpaired electrons, principally the compounds of *d* and *f*-block elements and free radicals. The NMR and NQR techniques require particular isotopes containing the nuclei with suitable spins and certain other properties. NQR can only be used with solid samples whereas NMR is simple with liquids and difficult for solids. Some systems can only be studied at low temperatures. There is a wide variety of materials which can be studied easily by each technique and in cases where more than one technique can be applied the information obtained is usually complementary. The following chapters deal with each technique in turn.

The objective of this book is to develop a qualitative treatment of the theory of magnetic resonance spectroscopy (ESR, NMR and NQR) relevant to the applications, describe the instrumentation for measurement of spectra and discuss some representative applications to chemical problems.

ELECTROMAGNETIC RADIATION

According to wave theory, apparently different forms of electromagnetic radiation, e.g. visible light, infrared rays, radiowaves, etc. consist of oscillating electric and magnetic fields perpendicular to each other and to the direction of propagation. The waves travel with the velocity of light c (3×10^8 m s^{-1}). For simplicity, a plane polarized radiation is shown in Fig. 1.1 where the component of the electric field is present only in the *yz*–plane and the magnetic field component lies in the *xz*–plane. The intensity of the radiation is proportional to the square of the amplitude of the wave. At any given time, the wave possesses different electric and magnetic field strengths at different points along the direction of propagation. The variation in the magnitude of wavelength λ (Fig. 1.1) results in different forms of radiation. Radiation characterized by a single wavelength is said to be monochromatic, while a polychromatic radiation consists of radiations of different wavelengths, which can be separated into monochromatic radiations.

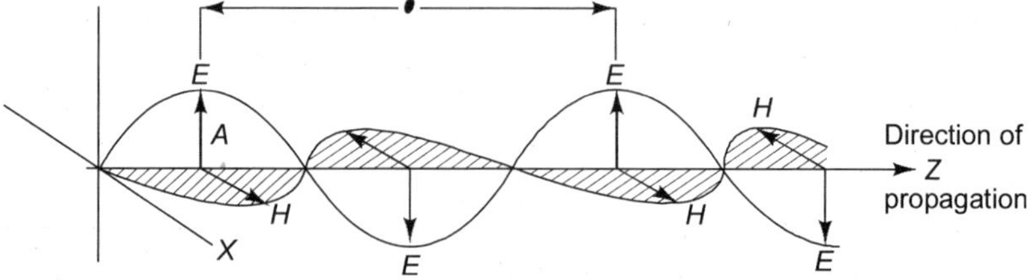

Fig. 1.1 Representation of a plane polarized electromagnetic radiation propagating in the Z-direction. A is the amplitude of the wave.

Radiation is characterized by its wavelength λ or its frequency (ν). The frequency ν is expressed in Hz. The relationship between these two quantities is given by

$$\nu\,(\text{Hz}) = \frac{c(\text{cm s}^{-1})}{\lambda(\text{cm})} \tag{1.1}$$

Figure 1.2 shows the electromagnetic spectrum covering an astonishingly wide range of values for the frequency and wavelength. It ranges from gamma rays to radiofrequencies, a span of over 20 orders of magnitude in frequency. In the electromagnetic spectrum and in the series of well-known spectrometric methods, NMR, NQR and ESR spectroscopic techniques take their place at the longer wavelength end.

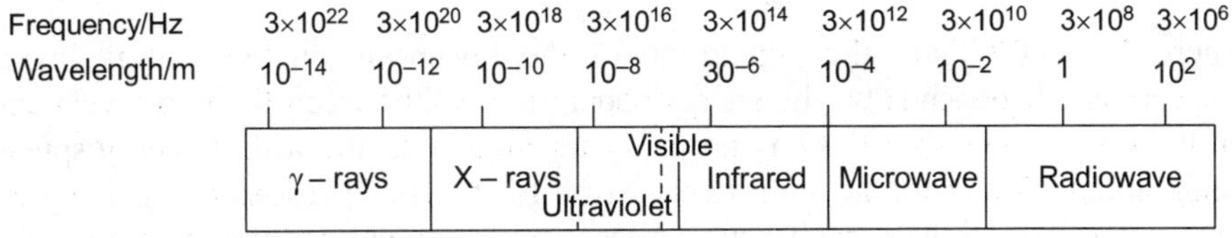

Fig. 1.2 The electromagnetic spectrum.

Electromagnetic radiation can be regarded as a stream of elementary particles known as photons. Different forms of radiation have photons of different energy, which travel with the velocity of light. In magnetic resonance spectroscopy, our concern will be the interaction of the magnetic field component of the microwave or radiofrequency radiation with the molecules. This interaction results in resonance absorption of the radiation by the molecules. Magnetic dipole transitions are inherently much weaker in intensity than the electric dipole transitions. In other forms of spectroscopy, namely, rotational, vibrational and electronic spectroscopy, the interaction of molecules with the electric field component of the electromagnetic radiation occurs.

All spectroscopic techniques are essentially concerned with measuring energies. The three methods of magnetic resonance differ in their interactions with the molecules which give rise to the spectral transitions and therefore in energy ranges (Fig. 1.2). The energy range involved is of the order of 0.001–10 Joules mole^{-1}. The present techniques span 10 kHz to 1000 MHz and 9 to 36 GHz. The relatively small magnitudes of these ranges have two important implications which may affect the number of lines in the spectrum and the total intensity of the spectrum.

The frequency of the radiation must match with the energy difference (ΔE) between any two quantized energy levels of the molecule for the resonance absorption of the radiation to take place. The frequency of radiation ν required for matching is given by the relation

$$\Delta E = hc/\lambda \tag{1.2}$$

where h is the Planck's constant. Absorption of one quantum of energy $h\nu$ will rise one molecule from the lower energy state to the next higher energy state.

NUMBER OF SPECTRAL LINES

Generally, the sample absorbs electromagnetic radiation and the electrons (nuclei) are excited from the ground state to various higher energy states. The number of absorption lines observed depends on the number of excited states and the selection rules. However, if the ground state is a set of closely spaced sublevels, transitions from one or more of the lower-lying levels are possible provided that they are populated. The population of a set of energy levels follows the Boltzmann distribution.

$$N_2/N_1 = \exp(-\Delta E/kT) \quad (1.3)$$

where N_1 and N_2 are the population of the lower and higher energy levels respectively. It depends on the energy separation ΔE between the two levels and on the thermal energy kT. kT is normally much greater, the widest energy spread being about 25 J mol^{-1} at room temperature for magnetic resonance techniques. This means that all the substates have almost equal populations and that transitions are possible from each of them. For example, in elemental iodine, the ^{127}I nuclei have two excited states lying at 344 and 1032 MHz above the ground state (Fig. 1.3) but the NQR selection rule allows transitions only between adjacent states. However, since the first excited state has a population almost equal to that of the ground state, two lines appear in the spectrum, at 344 and 688 MHz.

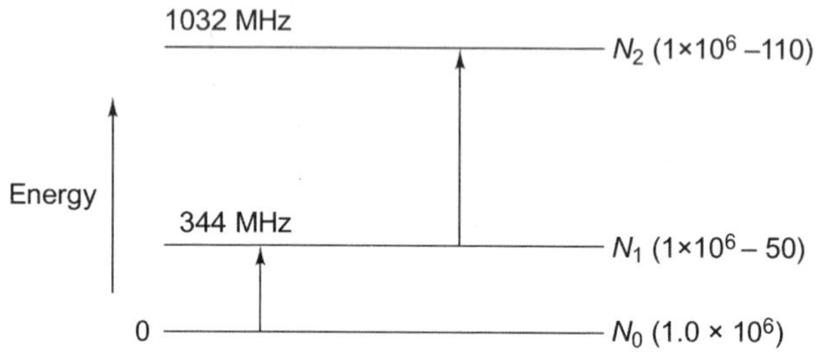

Fig. 1.3 Energy levels, populations and NQR transitions for I_2 at 298 K. Values in the parenthesis indicate the relative populations of the energy levels.

SPECTRAL LINE WIDTH

All spectra display absorption bands rather than discrete lines. The spectral band from each technique has a characteristic minimum width which is determined by the nature of the de-excitation processes involved. This is because the energies of the excited states actually cover a range ΔE which depends on the time scale of the deactivation process. The time scales are usually expressed as half lives $\tau_{1/2}$ i.e. the time during which a 50% reduction in population occurs or as the mean lifetimes, $\tau = \tau_{1/2} \ln 2$. The energy range ΔE and lifetime τ are related by the uncertainty principle,

$$\Delta E.\ \tau \geq h / 4\pi (= 3.18 \times 10^{-11}\ \text{J mol}^{-1}) \quad (1.4)$$

Thus longer the lifetime of the excited state, the more sharply defined is the energy and narrower the absorption line. Lifetimes involved vary from technique to technique and sometimes more than one deactivation process may occur. The two most common deactivation mechanisms possible in NMR and ESR are the spin-lattice and spin-spin relaxation discussed later.

When two or more chemically distinct species coexist in dynamic equilibrium e.g. two conformers, one often sees spectral peaks corresponding to the individual species in some form of spectroscopy while other spectroscopic methods may give a single average peak. Each spectroscopic technique is thus sensitive to a particular time scale. A process may cause broadening of the spectral line in some spectral region while the same process in some other spectral region may have no effect. This feature can again be interpreted from the Heisenberg's uncertainty principle, equation (1.4). Since $\Delta E = h\Delta\nu$, we have,

$$\Delta \nu . \tau \geq 1/4\pi \quad (1.5)$$

If a molecule undergoes a transition from the ground state to an excited state having a lifetime of 10^{-5} s, the uncertainty in the excited state energy and hence the width of the associated spectral line are

$$\Delta E = \frac{h}{4\pi \times 10^{-5}} = \frac{6.626 \times 10^{-34}}{4 \times 3.143 \times 10^{-5}} = 0.527 \times 10^{-29}\ \text{J}$$

$$\Delta \nu = 1/(4\pi\ \tau) = 1/(4 \times 3.143 \times 10^{-5}) - 0.796 \times 10^{4}\ \text{Hz}$$

Consider a system which gives rise to two distinct peaks due to different sites as in an exchange process. When the rate of the chemical exchange comes closer to the frequency difference $\Delta\nu$, the two peaks begin to broaden. As the rate of exchange becomes faster, they move closer, merge and then a sharp single peak results.

If there is a process, chemical or physical, that is fast, the excited state lifetime, τ is shortened by this process and the spectral line will be broadened. In the infrared, for example, it is possible to resolve (that is, detect the presence of) two bands corresponding to two different sites that are separated by 0.1 cm^{-1}. Substituting this value in equation (1.5), we get

$$\Delta \nu = 0.1\ \text{cm}^{-1} \times 3 \times 10^{10}\ \text{cm/s} = 3 \times 10^{9}\ \text{Hz}$$

$$\tau = \frac{1}{4\pi(3 \times 10^{9}\ \text{Hz})} = 2.65 \times 10^{-11}\ \text{s}$$

Therefore, a process that will give rise to lifetimes of ~ 10^{-11} s or less would cause broadening. Lifetimes are the reciprocals of the first order rate constants and hence a process whose rate constant is at least 10^{11} s^{-1} is expected to cause broadening in the infrared band. Thus it is possible to obtain information about the molecular dynamics from the line broadening. For ESR transitions, typical lifetime of the excited spin state is 10^{-8}s which gives roughly a line width of ~10^7 Hz.

If the species interconvert so fast that the lifetime in either one of the states is less than τ, then only an averaged line results. This is calculated by substituting the difference between the frequencies $\Delta\nu$ of the two states in equation (1.5). In NMR, for a system in which, for example, the two proton peaks are separated by 100 Hz, we get

$$\tau = 1/4\pi(100\ \mathrm{s}^{-1}) = 0.8 \times 10^{-3}\ \mathrm{s}$$

Thus an exchange process involving a proton that has a rate constant of ~5 × 10^2 s^{-1} would cause these two resonances to appear as one broad line. As the rate constant becomes much higher than 5 × 10^2 s^{-1} (e.g. by rising the temperature for a positive activation enthalpy process), the broadened single resonance begins to sharpen. When the process becomes very fast, the line width is no longer influenced by the chemical process and a sharp line results. In NMR and ESR, rate processes occurring in the range of 10^{-1} to 10^{-5}s and 10^{-4} to 10^{-8}s, respectively can be investigated. Thus the relaxation times and chemical exchange which are widely studied in NMR are difficult to study by ESR. The ESR line widths and splittings correspond to a much shorter time scale than the NMR line widths and splittings.

BAND INTENSITIES

When the system is irradiated at the resonance frequency, not only the nuclei (electrons) in the ground state are excited but those in the excited state undergo emission to the lower state on a similar scale. Stimulated absorption and emission processes are simply absorption and emission occurring in the presence of radiation. The number of molecules undergoing transition in either direction is directly proportional to the population of the state from which it occurs. The intensity of the absorption line depends on the difference in population of nuclei (electrons) undergoing the two types of transitions. These differences are very small for most of the energy ranges involved presently. ESR gives the greatest spread of energies and therefore the most intense spectra and it means that it is the most sensitive technique. While NMR samples must contain at least ~ 10^{-6} mol of the nuclei (i.e. millimolar solutions) to give a reasonable spectrum, ESR can detect as few as 10^{11} spins i.e. 10^{-12} mol and will routinely give good spectra on samples containing far less than a micromole. For all the techniques, sensitivity is improved by cooling the samples so that the population differences become greater (kT becomes smaller).

The spacing of energy levels in NMR and ESR depends directly on the strength of the external magnetic field. The sensitivity can therefore be increased by increasing the magnetic field. In NQR, the spacing of the energy levels is governed entirely by the electronic contributions within the sample. The NQR spectrum shows transitions between closely spaced levels (typically 10 – 100 MHz). Consequently it is difficult to obtain an intense NQR spectrum.

A problem that may arise from the small difference in the population of the energy states is that of saturation. In order to achieve net absorption of radiation, the rate of excitations from the ground state must exceed that of emission from the upper states. If the rate of excitation becomes higher, the population of the two states may rapidly become identical. The intensity of the spectrum diminishes and may become zero. There is usually an optimum power level of the radiation for each type of spectroscopic technique and for each sample.

Once the radiation has been absorbed, the atom (nucleus) does not exist in the excited state for long. The excitation process is faster, energy of the excited state in solution in NMR is $\sim 10^8$ Hz. Spontaneous emission to the lower state occurs after a delay which may be a few microseconds. The energy accumulated by the sample on excitation can be dissipated by other means e.g. by transfer to the lattice. Such relaxation processes occur considerably more rapidly than spontaneous emission. Useful information can be gained by studying these processes, which are manifested as changes in the line widths.

SPIN TRANSITIONS

Schematic representation of absorption, emission and stimulated emission are shown in Fig. 1.4. Stimulated absorption results in the transition of a molecule from a lower energy state to one of higher energy that is driven by the electromagnetic radiation at the transition frequency ν. The transition rate from the lower level 1 to the upper level 2 is given by $B_{12}\,\rho(\nu)$ where B_{12} is known as the Einstein's coefficient for stimulated absorption and $\rho(\nu)d\nu$ is the energy density of the radiation in the frequency range $\nu + d\nu$. Molecules or atoms in the upper state can return to the lower state by spontaneous or stimulated emission of radiation. The rate of stimulated emission is given by $B_{21}\,\rho(\nu)$ where B_{21} is the Einstein's coefficient for stimulated emission. The excited state could also spontaneously emit at the rate A_{21}, where A_{21} is Einstein's coefficent for spontaneous emission. The total rate of the upward and downward transitions is proportional to the number of molecules in the lower and higher energy levels respectively as discussed before. Let A_{21} be the transition probability per unit time for spontaneous emission from level 2 to level 1. Then the number of spontaneous emissions per second is $N_2\,A_{21}$ where N_2 denotes the number of molecules in the level 2. The stimulated emission from level 2 to level 1 is proportional to the energy density $\rho(\nu)$ of the radiation of frequency ν. It also depends on N_2 and B_{21} and the number of stimulated emissions per second will be $N_2\,B_{21}\,\rho(\nu)$. The total number of transitions from level 2 to level 1 $= N_2\,A_{21} + N_2\,B_{21}\,\rho(\nu)$. Similarly, the rate of stimulated absorption from level 1 to level 2 is $N_1\,B_{12}\,\rho(\nu)$ where N_1 is the number of molecules in the level 1.

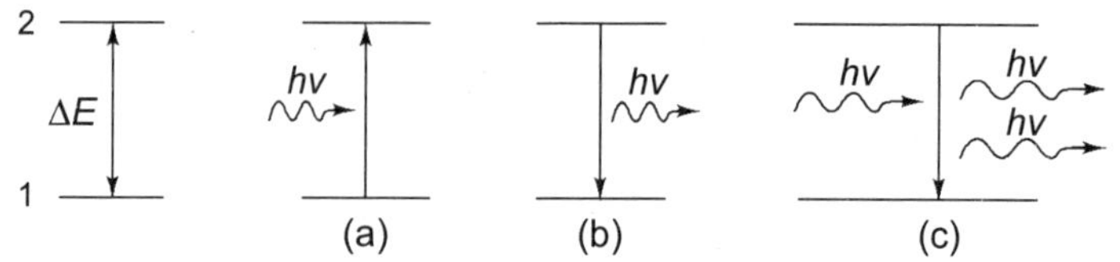

Fig. 1.4 Schematic representation of (a) absorption, (b) spontaneous emission and (c) stimulated emission of radiation.

Under thermal equilibrium, the net rate of the downward transitions is equal to that of the upward transitions. That is,

$$N_2 A_{21} + N_2 B_{21}\, \rho(\nu) = N_1 B_{12}\, \rho(\nu) \tag{1.6}$$

$$\rho(\nu)\,(N_1 B_{12} - N_2 B_{21}) = N_2 A_{21} \tag{1.7}$$

Einstein showed that the two coefficients for stimulated absorption and emission are equal, $B_{12} = B_{21} = B$ and the coefficient for spontaneous emission is relate to them by

$$A_{21} = \left(\frac{8\pi h\nu^3}{c^3}\right) B \tag{1.8}$$

Thus spontaneous emission is predominant in the visible and ultraviolet region and much less in the infrared and negligible in the microwave and radiofrequency region.

The transition probability varies directly as the radiation density. Only low powers of microwave or radiofrequency radiation should be used since radiation tends to equalize the population of the states it links. The intensity of the radiation should be set at a level high enough to allow accurate detection, but low enough to avoid equalizing of the population between the two levels.

EXERCISES

1. Calculate the frequency and the wavenumber (in cm^{-1}) of the electromagnetic radiation of wavelength $\lambda = 15$ μm.
2. Convert the electromagnetic radiation of 1750 cm^{-1} into frequency and wavelength, and energy in Joules.
3. For light of wavelength 254 nm, calculate the frequency and the amount of energy absorbed by a single molecule when it absorbs a quantum of this light and the energy absorbed by one mole of the molecules.
4. Find the wavelength of the electromagnetic radiation required to cause rupture of a C—C bond with dissociation energy of 380 kJ mol^{-1}.
5. A transition takes place from the ground state to an upper state having a lifetime of 10^{-4}s. Find the uncertainty in the excited state energy level and the width of the associated spectral line.

6. Calculate the ratio of the rate of stimulated emission to that of spontaneous emission at 900 K, when the wavelength of the radiation is 5000 Å.
7. What is the wavelength (in cm) and energy (in kJ mol^{-1}) associated with the nuclear magnetic resonances of a proton at 100 MHz?
8. What type of fine structure one expects to observe on an ESR spectrum?
9. Which of the common spectroscopic transition occurring in the time scale 10^{-15} to 10^{-8} Hz will possess the broadest bands due to uncertainty?
10. Find the value of kT at 298 K and 4.2 K in MHz and kJ mol^{-1}. By comparing with vibrational (~ 1000 cm^{-1}) and rotational (~ 10 cm^{-1}) energies, what inference would you draw?
11. The ^{127}I nuclei have an excited state at 1032 MHz from the ground state. Calculate the Boltzmann population ratio between the ground and excited states.
12. A gamma-ray photon emitted during the decay of ^{60}Co nucleus has an energy of 2.14×10^{-16}J. Find the frequency and the wavelength in metres of this radiation.
13. Indicate the spectral region you would expect the absorption of radiation to occur and the wavenumber of the radiation. What sort of transition each region corresponds to?

    (a) 30.0 kcal mol^{-1} (b) 0.15 nm (c) 9 GHz and (d) 10 eV.
14. The energy of a microwave photon absorbed giving rotational transitions of a molecule is 1.98×10^{-24} J. Calculate wavelength and energy absorbed by 1 mol of the molecule in Joules.
15. Indicate the spectroscopic techniques associated with the following time scales.

    (a) 10^{-19} s^{-1} (b) 10^{-15} s^{-1} (c) 10^{-11} s^{-1} (d) 10^{-1} s^{-1}
16. Calculate the ratio of the Einstein coefficients for spontaneous and stimulated emission A and B for the following transitions occurring at (a) 500 nm, (b) 3000 cm^{-1}, (c) 500 MHz and (d) 3.0 cm. What do the results signify?

Part One

Electron Spin Resonance

BASIC THEORY 2

GENERAL REMARKS

Magnetic interactions were detected as fine structure in the atomic spectra long before they were observed directly as in magnetic resonance spectroscopy. ESR spectroscopy is a technique in absorption spectroscopy for detecting paramagnetic materials and paramagnetism arises from the presence of unpaired electrons. The ESR was first observed by Soviet Physicist Zavoisky in 1944 when he detected a peak for the paramagnetic resonance absorption from $CuCl_2.2H_2O$. Other names for ESR spectroscopy include electron paramagnetic resonance (EPR) and electron magnetic resonance (EMR).

Chemical bonding in most substances results from pairing of electrons. Electron(s) are either transferred from one atom to another to form an ionic bond or electrons are shared between two atoms to form a covalent bond. Substances having paired electrons are repelled slightly in a magnetic field and they are termed as diamagnetic. Substances having unpaired electrons are slightly attracted in a magnetic field and they are termed as paramagnetic. Substances with unpaired electrons either occur naturally or could be produced artificially. Examples of naturally occurring paramagnetic substances are oxygen, nitric oxide, nitrogen dioxide, transition metal ions and their complexes. A molecule containing an unpaired electron is known as a free radical. Free radicals such as methyl radicals ($^{\bullet}CH_3$) may be formed as intermediates in a chemical reaction or by irradiation of a suitable substance by X-rays and other high energy radiations. Free radicals have been observed in gaseous, liquid and solid samples. They are usually short-lived.

The technique of ESR spectroscopy has been applied in many fields and the chief of these are the transition metal complexes, unpaired electrons in semiconductors, organic free radicals, color centres, ferro and antiferromagnetic materials and radiation damage centres. A principal feature of ESR spectra of interest in reaction mechanism is its ability to detect short-lived paramagnetic intermediates. Emphasis will be given here to studies relating to the electronic structure of free radicals. Free radicals can be studied by ESR spectroscopy if the lifetimes of the radicals are greater than 10^{-6}s. If the lifetimes of free radicals are shorter than 10^{-6}s, they may be studied at low temperature in a matrix which leads to lengthening of their lifetimes.

The theory of ESR spectroscopy in liquid phase is much simpler than that in the solid state. We shall confine our discussion here mostly to solutions of organic free radicals containing one unpaired electron and we shall then briefly touch upon the spectra of transition metal complexes.

ELECTRON SPIN AND MAGNETIC MOMENT

The fundamental properties of an electron are its mass, charge, orbital angular momentum and intrinsic angular momentum, which is known as the spin. The angular momentum is a fixed property of bodies such as atoms, nuclei and electrons when they are isolated. Stern and Gerlach experiments showed that electrons have intrinsic angular momentum, in addition to any angular momentum they may possess as a result of their orbital motion. In the Stern-Gerlach experiment, a beam of silver atoms travel from a source through an inhomogeneous magnetic field. The detector records that the silver atoms traverse in two distinctive paths as illustrated in Fig. 2.1. The intrinsic spin angular momentum has a fixed value of $(\sqrt{3}/2)h/2\pi$ and the Z-components of $\pm$ $(1/2)h/2\pi$ where h is the Planck's constant.

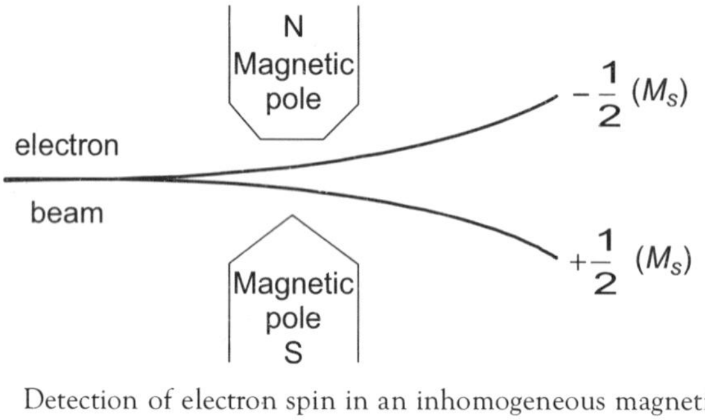

Fig. 2.1 Detection of electron spin in an inhomogeneous magnetic field.

Two types of magnetic dipoles are possible—one arising from the motion of the electron about the nucleus of the atom and the other from the spinning of the electron about an axis through its centre. The magnetic moment that results from the latter is directly proportional to the magnitude of angular momentum. The strength of the magnetic dipole is expressed in terms of its moment. One component of the magnetic moment can be measured, for example, by observing the amount of deflection of the particle in an inhomogeneous magnetic field (Stern-Gerlach experiment). In most cases (over 99 per cent), the total electron magnetic dipole is due to the spin with only a small orbital contribution. To a first approximation, the orbital and spin angular momentum may be treated independently and later account for the spin-orbit interaction as a correction term, when required. The convention as to the direction of the spin and orbital angular momentum and the resulting magnetic moment is indicated in Fig. 2.2.

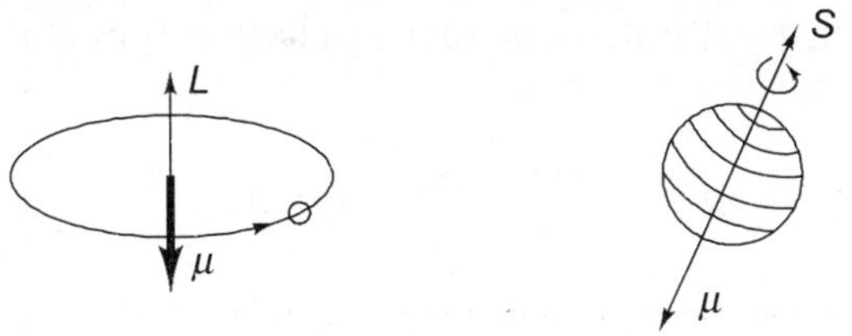

Fig. 2.2 The convention as to the direction of rotation and the relative directions of the spin (S), orbital angular momentum (L) and magnetic moment (μ).

As a consequence of Heisenberg's Uncertainty principle, in an atomic system, all that can be measured with certainty with regard to the spin angular momentum is its square (S^2) and a component which conventionally is taken as the Z-component, S_z.

A system with more than one electron can be described using the total spin quantum number $S = \sum s_i$, which is the vectorial sum of the individual spins i. Depending on the system under consideration, it can take on integral or half-integral values 0, $\frac{1}{2}$, 1, $\frac{3}{2}$ ... The magnitude of the total spin angular momentum is given by

$$S = \sqrt{S(S+1)}\, h/2\pi \tag{2.1}$$

Both μ and S lie along the same line perpendicular to the rotation of the electron as shown in Fig. 2.2. Usually the Z-axis is selected as the direction of the external magnetic field. The magnetic spin quantum number M_s can take ($2S$+1) values, varying from S, S–1,... – S. As noted before, the component of S for an unpaired electron along the Z-direction is S_z, that is,

$$S_z = m_s\, h/2\pi \tag{2.2}$$

where the quantum number m_s (for a single electron) can only have the values of ±1/2. The value of 1/2 is justified on the basis of advanced theoretical treatments as well as from experimental observations. The components of the spin angular momentum along the Z-direction for S = 1/2 and S = 1 are shown in Fig. 2.3.

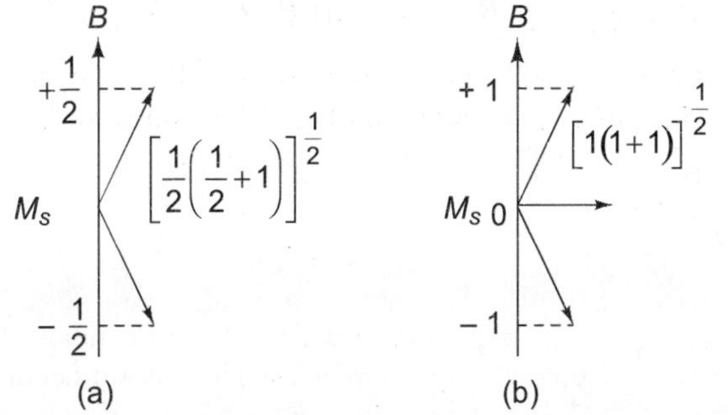

Fig. 2.3 Possible values of the spin angular momentum $[S(S+1)]^{\frac{1}{2}}$ and its components M_S in units of $h/2\pi$ in the direction of magnetic field of strength B for (a) S = 1/2 and (b) S = 1.

The magnetic moment generated from an electron spin is related to the spin angular momentum by the equation,

$$\mu = g\left(\frac{eh}{4\pi m}\sqrt{S(S+1)}\right) h/2\pi \tag{2.3}$$

The component of the electron spin magnetic moment along the *Z*-direction is given by

$$\mu_z = -g\frac{eh}{4\pi m} M_s = -g\,\beta_e M_s \tag{2.4}$$

where g is a dimensionless constant which is usually known as the Landé splitting factor or simply as the g-factor. It is a physical property of the electron. For a free electron g has a value of 2.00232. The negative sign in equation (2.4) arises because of the negative charge of the electron and it shows that the magnetic moment vector μ_z of the electron spin is opposite in direction to the spin angular momentum (see Fig. 2.2). For the electron, the ratio $eh/4\pi m$ denoted by β_e (the Bohr magneton) has a value of 9.2741×10^{-24} J T^{-1} and it is the unit of magnetic moment of the spinning electron.

Equation (2.4) can also be written as

$$\mu_z = \gamma_e\left(\frac{h}{2\pi}\right) M_s$$

where γ_e is the magnetogyric ratio of the electron.

ESR TRANSITIONS

In the absence of an external magnetic field, the magnetic moments of the unpaired electrons in a sample are randomly oriented. But in the presence of a magnetic field, they tend to be aligned, in the direction of the applied field, as shown in Fig. 2.4. It is necessary to begin with a study of non-interacting magnetic dipoles in a constant applied magnetic field. The potential energy of a classical magnetic dipole of moment μ of the electron in a magnetic field of strength B is given by

$$E = -\boldsymbol{\mu}.\boldsymbol{B} = -\mu B \cos\theta \tag{2.5}$$

where θ denotes the angle between the direction of the magnetic moment μ and the direction of the applied magnetic field *B*. The minus sign in equation (2.5)

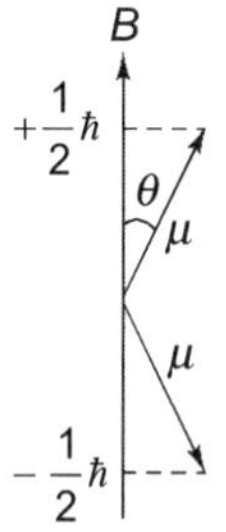

Fig. 2.4 Two possible orientations of an electron spin magnetic moment in an external magnetic field.

shows that the energy is minimum when the magnetic moment is oriented antiparallel to the external magnetic field.

From the equations (2.4) and (2.5), we obtain a set of energies

$$E = g\,\beta_e\,B\,M_s \tag{2.6}$$

The separation between the two spin energy levels varies directly with the applied field for the unpaired electron with $S = 1/2$ as shown in Fig. 2.5. The transition between the two levels $M_s = +1/2$ and $M_s = -1/2$ can be induced by an electromagnetic radiation of appropriate frequency ν if the energy of the photon $h\nu$ matches with the energy separation ΔE, that is

$$\Delta E = h\nu = g\,\beta_e\,B \tag{2.7}$$

The equation (2.7) represents the resonance condition.

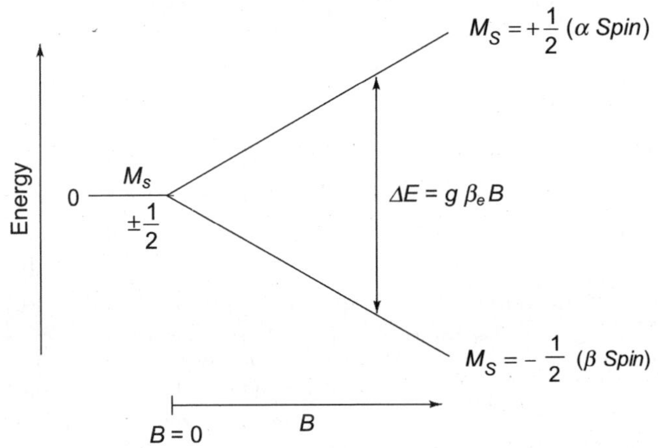

Fig. 2.5 Energy level diagram for the simplest system S=1/2 in a varying magnetic field showing an ESR transition.

The ESR experiment consists of the measurement of such energy differences. The magnetic field generally employed is about 0.34 *T* (3,400 gauss). The ESR frequency for an unpaired electron in a field of 0.34 *T* assuming $g = 2.0$ would be

$$\nu = \frac{g\beta_e B}{h}$$

$$= \frac{(2.0)\,(9.274\times10^{-24}\ \mathrm{J\,T^{-1}})\,(0.34\,\mathrm{T})}{6.626\times10^{-34}\ \mathrm{Js}}$$

$$= 9.518\times10^{9}\ \mathrm{Hz} = 9.518\ \mathrm{GHz}$$

Thus the ESR transitions fall in the microwave region of the electromagnetic spectrum. For a typical ESR experiment, $h\nu$ corresponds to an energy of 10^{-3} kJ mole^{-1} (or about 1 cm^{-1}).

SELECTION RULES

For molecules to absorb electromagnetic radiation, two conditions must be met. (i) The energy quantum of the electromagnetic radiation must correspond to the separation between the energy levels in the molecule. (ii) The oscillating magnetic dipole component of the electromagnetic radiation must be able to interact with the molecule by stimulating an oscillating magnetic dipole in the molecule. ESR transitions are caused by the interaction of the electron spin magnetic moment with the oscillating magnetic field of the incident microwave radiation.

To carry out an ESR experiment, the sample under study is placed between the poles of an electromagnet and it is irradiated with an appropriate microwave radiation. The magnetic field is then varied through the resonant condition. When the incident microwave radiation frequency matches with the separation between the electron spin energy levels, the microwave radiation absorbed by the sample is monitored by means of a detector. The ESR spectra are plotted as the intensity of absorption versus the applied magnetic field (See chapter 4).

g-FACTOR

Different microwave frequencies may be used in different spectrometers and the frequency (or the applied magnetic field) of each spectrometer also needs to be varied to some extent depending on the sample to obtain resonance signals. Consequently neither the magnetic field nor the frequency of the microwave radiation can be utilized to denote the spectral positions. It is therefore convenient to refer to the observed absorption in terms of its *g*-value.

Molecules of different types come into resonance at different applied magnetic fields. This is because the application of an external magnetic field may generate an internal magnetic field in the sample by inducing the electrons to circulate through the molecular framework. This orbital motion sets up a small additional magnetic field. The magnitude of the additional field would be proportional to the applied magnetic field itself. The electron spin magnetic moment interacts with the local magnetic field. There are also local fields that are permanent and do not depend on the presence of an external magnetic field. These permanent local fields are generated more commonly by the presence of magnetic nuclei in the molecule. The internal magnetic fields may reinforce or oppose the external magnetic field. Such local fields are accounted for by allowing the *g*-factor to vary, that is,

$$g_{eff} = \frac{\Delta E}{\beta_e B_{res}} = \frac{h\nu}{\beta_e B_{res}} \qquad (2.8)$$

where B_{res} is the magnetic field at resonance. For example, an irradiated specimen of MgO has an intense signal at 9.418 GHz in a magnetic field of 0.1629 T. Then the *g*-value is given by

$$g = \frac{h\nu}{\beta_e B_o} = \frac{(6.626\times10^{-34}\ \text{Js})(9.418\times10^{9}\ \text{s}^{-1})}{(9.274\times10^{-24}\ \text{JT}^{-1})(0.1629\ \text{T})} = 4.13$$

For a free electron in vacuum, the g-value is 2.00232. In chemical systems, the unpaired electron occupies an orbital which may be localized on a single atom or extensively delocalized across the molecule or the radical. The g-value now reflects the nature of the orbital. The value of g may then be quite different from 2.0; it seldom falls much below this figure, but it may reach 9.0 or more. The g-values for the unpaired electrons in the sample are sensitive to the chemical environment of the paramagnetic ion.

The g-factor can thus be considered as a quantity characteristic of the molecule in which the unpaired electrons are located. The g-value deviations from the free electron value are much like the chemical shifts of NMR discussed later. Magnetic interactions of the electron spin with the orbital angular momentum of the unpaired electron (that is, the spin-orbit coupling) are the sole cause for the deviation of g-values. Most free radicals have very low molecular symmetry and hence the orbital angular momentum will be generally quenched leading to very small spin-orbit coupling. As a result, the g-values are nearly equal to the free electron g-value. The small deviations ($\pm$ 0.05 or smaller) often observed are accounted for by mixing of the low-lying excited electronic states with the ground electronic state. The g-values for free radicals are thus of rather limited use.

PRESENTATION OF SPECTRA

A major disadvantage of the broad signals of the ESR spectrum is that these are more difficult to observe and measure than sharp signals. For this reason, ESR spectra are almost always obtained as the first derivative (the slope) of the absorption curve plotted against the strength of the magnetic field. Much higher sensitivity can be achieved by this method if the line width is very broad. The derivative curve in Fig. 2.6(b) corresponds to a single absorption peak of Fig. 2.6(a). The derivative curve crosses the abscissa at the maximum in the absorption curve. The curve(c) of Fig. 2.6 is the absorption counterpart of the derivative curve (d). The shoulders in Fig. 2.6(c) do not pass through a maximum and consequently the derivative curves in Fig. 2.6(d) do not cross the abscissa. They represent overlapping bands.

A number of factors affect the shape, position and width of an ESR signal. The intensity of an ESR absorption is the total area under the resonance curve and it is proportional to the concentration of the paramagnetic species present. This relation could be used for estimating the amount of the paramagnetic substance by measuring the ratio of the area for a known amount of the sample to that of the unknown sample. This procedure is valid if the known and unknown samples are similar type of spin systems, such as free radicals. With all the parameters which affect the intensity taken properly into account, the area under the signal could be used to determine the concentration of the paramagnetic species in the sample.

The method is very sensitive and under favourable conditions, concentrations as low as 10^{-13} mole of the free radicals could be detected. Integrated absorption intensities can, therefore, be used for the estimation of concentrations and for kinetic studies of the decaying unstable radicals. Although absolute intensities cannot be measured reliably, relative intensities of the signals are in principle proportional to the relative number of electrons in the system. The relative intensities can be approximated by comparing the peak to peak (maximum to minimum) height on the first derivative curve when the spectra are relatively simple.

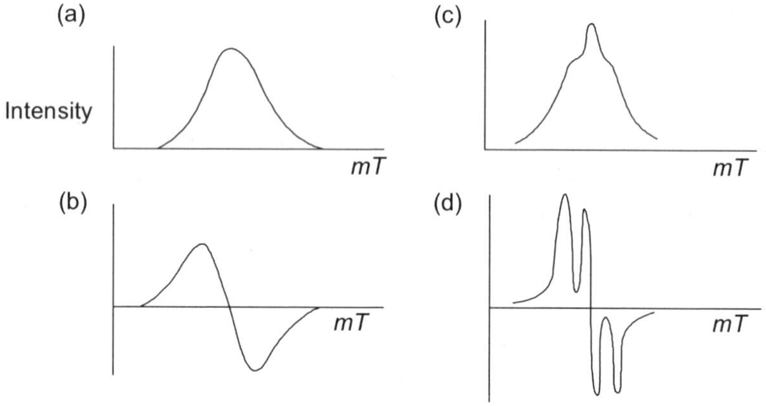

Fig. 2.6 Typical ESR absorption curves (a) and (c) and their respective first derivative curves (b) and (d).

INTERACTION OF MAGNETIC DIPOLES WITH MICROWAVE RADIATION

As stated before, in a magnetic field, the electrons distribute themselves in the two energy levels $M_s = +1/2$ (α) and $M_s = -1/2$ (β) for the simplest system with $S = 1/2$. The incident radiation $h\nu$ absorbed by electrons in the lower energy (N_β) state will induce the electrons to undergo a transition to the higher energy state (N_α). The electrons in the higher energy state can lose energy by stimulated emission and return to the lower level. The net absorption of the microwave radiation is thus proportional to the difference in the population between the ground state and that of the excited state.

The ratio of population of these two states separated by an energy difference ΔE can be described by the Boltzmann distribution,

$$N_\alpha/N_\beta = \exp(-\Delta E/kT) \qquad (2.9)$$

where k is the Boltzmann constant and T is the absolute temperature in degrees Kelvin, and it depends on the establishment of a thermal equilibrium. Strictly speaking, electrons obey Fermi-Dirac statistics but when the interactions between the electron spins are weak, it can be shown that Boltzmann statistics apply. Substituting for ΔE from equation (2.7) gives

$$N_\alpha/N_\beta = \exp(-g\beta_e B/kT) \qquad (2.10)$$

Since $g\beta_e B$ (~ 0.3 cm^{-1} for B = 0.3 T) is about three orders of magnitude lower than kT (~ 200 cm^{-1}) at room temperature, the exponential may be expanded ($e^{-x} = 1 - x$ if $x << 1$) to obtain

$$\frac{N_\alpha}{N_\beta} = 1 - \frac{g\beta_e B}{kT} = 1 - \frac{h\nu}{kT} \qquad (2.11)$$

At room temperature (T = 300 K), for g = 2.0 and B = 0.34 T, since ν is of the order ~ 10^{10} Hz, equation (2.11) gives the ratio $N_\alpha/N_\beta = 0.9984$ or $N_\beta/N_\alpha = 1.0015$. Thus, in ESR experiments, the lower state is slightly more populated than the upper state and thus only a small fraction of the electron spins actually contribute to the absorption. The population difference varies inversely as the absolute temperature and directly to the magnetic field applied. The sensitivity of the ESR technique can therefore be increased by employing higher external magnetic fields and lower temperatures which increase the population difference.

LARMOR PRECESSION

According to both classical and quantum mechanics, a magnetic dipole which is not pointing exactly along or against the direction of the applied magnetic field (Fig. 2.4) behaves in the same way as a spinning top tilted from the vertical. The interaction between the magnetic moment μ and the applied magnetic field B generates a torque tending to align the magnetic moment towards B. The direction of the torque is at right angles to the spin angular momentum vector in a plane parallel to the horizontal plane as shown in Fig. 2.7. Since the electron is spinning, the effect of the torque is to cause a type of motion called precession as shown in Fig. 2.7. It precesses about the magnetic field direction at a rate determined by the strength of the magnetic field and the rate of spin, like a spinning gyroscope precessing about the gravitational field. The resulting motion takes successive positions on a cone whose axis coincides with the external magnetic field direction (taken as the Z-axis). The spin angular momentum vector can lie on either one of the two cones (the other cone is not shown in Fig. 2.7). The angular velocity ω of the precessional motion known as Larmor precession in radians per second is given by

$$\omega = \gamma B \qquad (2.12)$$

where γ is the magnetogyric ratio of the electron. The rotational frequency of precession of the magnetic moment is known as the Larmor frequency.

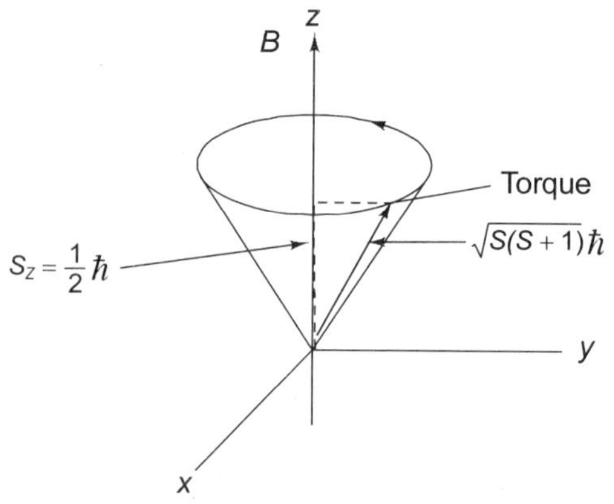

Fig. 2.7 The spin angular momentum in an external magnetic field. The torque exerted on the magnetic moment produces a precessional motion around the magnetic field direction.

The Larmor frequency is directly proportional to the applied magnetic field and it also depends on μ. If the equation (2.12) is expressed in Hz, then the following equation for the frequency of the precessional motion is obtained.

$$\nu = \frac{\gamma B}{2\pi} = \frac{g\,\beta_e B}{h} \tag{2.13}$$

The precessing magnetic moment μ shown in Fig. 2.7 can be resolved into two components: μ_z which is parallel to the axis of the applied magnetic field and μ_x which is perpendicular to the direction of the applied magnetic field. While μ_z is static, μ_x rotates at the same angular frequency as the precessing magnetic moment. The net magnetic moment pointing in the Z-direction is proportional to the population difference between the two energy levels.

RESONANCE PHENOMENA

In the ESR experiment, an oscillating magnetic field B_1 is supplied along the X-direction from a microwave radiation using a coil along an axis perpendicular to the main applied field as shown in Fig. 2.8.

Mathematically, the resolution of a plane (linearly) polarized wave into two counter rotating components could be written as

$$\begin{aligned} B_r &= (B_1)_x \cos 2\pi\nu t + (B_1)_y \sin 2\pi\nu t \\ B_l &= (B_1)_x \cos 2\pi\nu t - (B_1)_y \sin 2\pi\nu t \end{aligned} \tag{2.14}$$

where B_r and B_l are the right and left circularly polarized electromagnetic fields, respectively. The sum of B_r and B_l components that is, $2(B_1 \cos 2\pi\nu t)$ has only an x component with respect to the precessing electron magnet and the counteracting field is at frequency ν_o away from it and may be ignored. The component rotating in the direction opposite to the Larmor precession will have very little effect on the

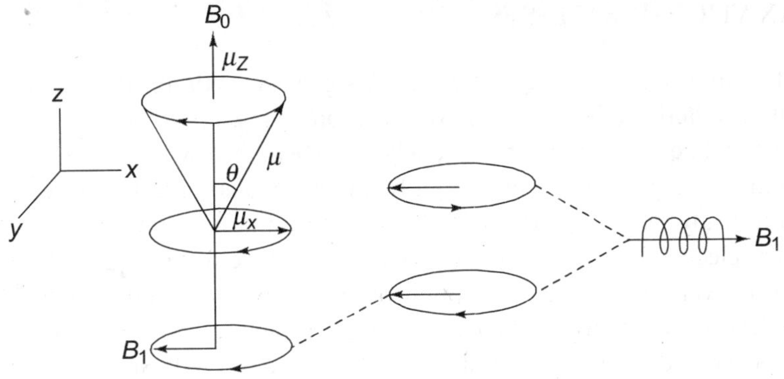

Fig. 2.8 Precessional motion of the electron magnetic moment μ around the applied field B_0 and representation of a linear oscillating magnetic field as addition of two circularly polarized components.

resonance. The electron spin magnet will also attempt to precess about the direction of the smaller magnetic field B_1 in addition to precessing about the main field B_0. The magnetic field B_1 will be relatively ineffective, because (i) it is smaller and (ii) the orientation of the magnet relative to it will continuously be changing as a result of the precessional motion of μ about B_0. However, if B_1 is the magnetic field vector of a circularly polarized electromagnetic field rotating with an angular velocity ω in the x, y–plane, the system can be viewed in a frame of reference rotating with B_1. When rotating with the angular velocity $\omega = \omega_1$, the electron magnet will appear stationary and maintain a constant orientation relative to B_1. That is, we have now set up a coordinate system rotating with the Larmor frequency and if there are no other magnetic fields acting, the magnetic moment vector μ would appear stationary in the new frame. In other words, in the rotating frame, the effect of the static magnetic field is effectively reduced to zero. If the angular velocity of the second magnetic field B_1 is different from that of the Larmor frequency, B_1 will also be rotating in the rotating coordinate system. Its effect will be to exert a torque on the electron magnet tending to align it towards a plane perpendicular to B_0. If the field B_1 rotates at the Larmor frequency itself, then in the rotating frame, it will behave like a constant magnetic field and the torque being always in the same direction will cause large oscillations. It can thus precess freely about B_1 in addition to B_0 and the energy can be transferred from the field B_1 to the electron magnet and vice versa. This is the classical description of the magnetic resonance phenomenon. The microwave radiation field thus induces transitions between the two allowed orientations in the field B_1 rather than causing continuous precession about B_1. The field of the rotating electron moment can be detected by a coil whose axis is in the X-direction and therefore unaffected by the driving frequency. One uses electronic means to detect the phase lag of the electron spin rotation relative to the driving field. Magnetic resonance spectrometers detect only the component of magnetization induced in the sample in the plane of the field M_{xy} and not M_z.

RELAXATION PROCESSES

As noted above, the intensity of an ESR signal depends on the existence of a population difference between the two electron spin energy levels. At resonance, if the thermal equilibrium is not seriously disturbed, there will be a net absorption of radiation energy because of a slightly larger population in the lower state. To maintain steady state conditions (thermal equilibrium), there must be a mechanism by which electrons excited to the upper level can lose their energy and return to the lower level. Otherwise, absorption of energy would continue until both the energy levels are equally populated when no further net absorption of radiation would occur and no resonance signal would be observed. Such a condition is referred to as "saturation" and it is normally avoided in ESR measurements by working with a low power microwave radiation to induce the electron spin transitions. The rate of the transitions is directly dependent on the microwave radiation density.

The onset of saturation is accompanied by the gradual broadening and distortion of the resonance signal. By switching off the incident microwave radiation, the system will return to the equilibrium because of thermal effects. The processes which restore the equilibrium are known as the relaxation processes, whereby the spin populations return to their equilibrium value. The return to the equilibrium is an exponential process and is characterized by a relaxation time.

Two relaxation processes occur, namely, the spin-lattice and spin-spin relaxation. The spin-lattice relaxation involves the thermal equilibration of the spin system with its environment usually referred to as the "lattice" and the decay process occurs with a time constant T_1. When the spin population ratio is disturbed, to restore the equilibrium, the energy is transferred from the spin system to the lattice. Generally, the term lattice refers to the degrees of freedom of the system other than those directly concerned with the spin. Spin relaxation is possible because the spin system is coupled to the thermal motions of the lattice, be it a gas, liquid or solid and the lattice is at thermal equilibrium. This process of non-radiative transfer of excess magnetic energy is called the spin-lattice relaxation. It is also known as the longitudinal relaxation because spin energy is defined along the direction of the external magnetic field. The magnitude of T_1 varies with temperature and it increases as the temperature decreases. The variation in T_1 for different systems is quite large. A large value of T_1 indicates very slow relaxation.

The spin-spin relaxation or transverse relaxation is denoted by a characteristic time constant T_2. It can be explained on the basis that different paramagnetic ions in the sample are in slightly different magnetic environments. Although the anisotropic interactions such as the dipole-dipole interactions average to zero in solution, if the paramagnetic ions are rotating slowly some remain in one orientation for a significantly longer time and resonate at the appropriate value of the applied magnetic field; other paramagnetic ions stay in other orientations and resonate at slightly different applied magnetic fields. Thus, the paramagnetic species of the sample resonate over a small range of magnetic fields. The individual spins lose phase coherence with each other in the process of decaying into random orientations. The spin-spin interaction results from the small magnetic fields existing at the neighbouring paramagnetic species. In other words, the static field variations cause the spins to have slightly different precessional frequencies and thus provide a means for the loss of phase coherence. The spin-spin relaxation does not alleviate saturation, since it is not concerned with net energy changes. The magnitude of T_2 can be reduced by increasing the distance between the paramagnetic ions by diluting the sample with an isomorphous diamagnetic compound (for example, a small amount of $CuSO_4$ in $ZnSO_4$ crystal). This is usually done for the study of transition metal complexes.

The spin lattice relaxation is mainly caused by the 'magnetic noise' or time dependent fluctuations in the local magnetic field. The noise is mainly due to neighbouring magnetic dipoles undergoing relaxation processes or rotational motions or exchange of electrons between different chemical environments. Each process has a characteristic time scale and the noise has a well-defined frequency distribution. Any component of the noise close to the Larmor frequency will tend to induce transitions between the energy levels similar to an applied microwave radiation field inducing magnetic resonance. The components of the noise much above and below the Larmor frequency will be ineffective. The effect of inducing transitions between the energy states is to restore thermal equilibrium. Thus, magnetic noise is a major factor contributing to the spin-lattice relaxation rate, that is $1/T_1$. The transitions also reduce the lifetimes of the energy states and cause loss of phase coherence between the precessing spins. In the limit where the molecular motions are very slow, there are still local variations in the magnetic field throughout the sample due to other magnetic dipoles but they do not vary with time or at least vary only very slowly compared to the Larmor frequency. Consequently, there is no mechanism for inducing transitions between energy states and hence they have a longer life time.

Many mechanisms contribute to spin-lattice relaxation time T_1. One of the most important of these is the interaction between the spin and orbital motion of the electron. If the spin-orbit coupling is strong, T_1 may be so short that the spectral line is extremely wide and virtually undetectable. For many transition metal ions, this is the case. Lowering of the temperature lengthens T_1 and sometimes temperatures as low as 1 K may be required to record the spectrum.

Spin-lattice relaxation is usually efficient at room temperature ($T_1 \sim 10^{-6}$ s) but becomes gradually less efficient at lower temperatures. The spin-spin relaxation is generally efficient and gives a relaxation time of 10^{-6} to 10^{-8} s. Thus, ESR relaxation times are often too short to be accurately measurable by pulsed techniques which are common in NMR. The spin-spin relaxation may be the dominant form of relaxation in the solid state ($T_1 >> T_2$) and at low temperatures but is often relatively unimportant in liquids having a good deal of thermal motions ($T_1 \sim T_2$).

TRANSITION PROBABILITY

As noted earlier, the difference in population between the two electron spin energy levels in a static magnetic field is achieved through Boltzmann distribution. For a system of two energy levels ($S = 1/2$), any change in the population difference say n, where $n = N_\beta - N_\alpha$, where N_β is the population of $M_s = -1/2$ level and N_α that of $M_s = +1/2$ level can be achieved only by a net loss or gain in the total energy of the spin system. The spin system has its maximum energy when $n = 0$. As n increases, the spin system loses energy. From thermodynamic consideration, this implies that the spin system must be in thermal contact with its surroundings as depicted in Fig. 2.9.

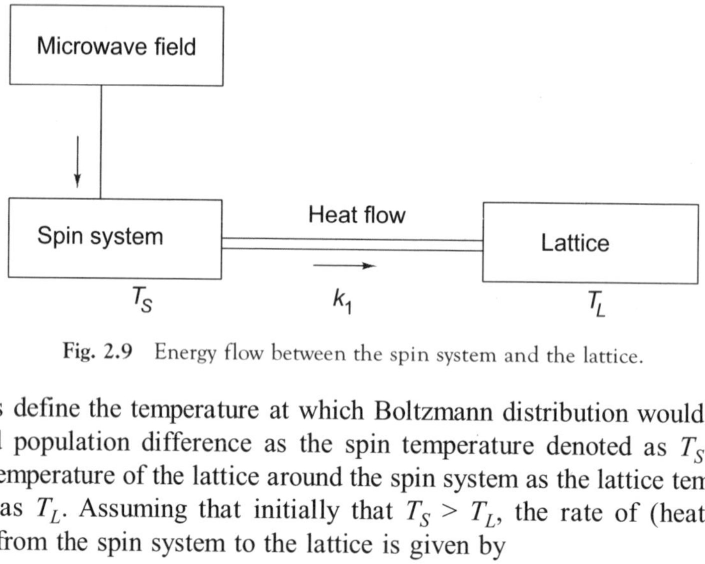

Fig. 2.9 Energy flow between the spin system and the lattice.

Let us define the temperature at which Boltzmann distribution would give the observed population difference as the spin temperature denoted as T_S and the normal temperature of the lattice around the spin system as the lattice temperature denoted as T_L. Assuming that initially that $T_S > T_L$, the rate of (heat) energy transfer from the spin system to the lattice is given by

$$\frac{dE}{dt} = k_1 k (T_S - T_L) \tag{2.15}$$

where k is the Boltzmann constant and k_1 is the first order rate constant. Since the unit of k_1 is s^{-1}, k_1 can be represented as the inverse of time that is, $k_1 = 1/T_1$, where T_1 is the spin-lattice relaxation time, which characterizes the efficiency of the thermal contact between the spin system and the lattice. Energy will flow from the spin system to the lattice until $T_S = T_L$. The total system will then be in thermodynamic equilibrium.

If the system is irradiated with a microwave radiation of appropriate frequency to cause resonance absorption, the energy will flow into the spin system, due to the absorption of energy from the microwave radiation. The rate of radiant energy absorption is characterized by the rate constant k_2, which will be proportional to the microwave power density, $\rho(\nu)$. If $k_2 << k_1$, the spin system can efficiently release this extra energy to the lattice until $T_S \approx T_L$ and little change in n from the unperturbed value would occur. However, when $k_2 \approx k_1$, the spin system can no longer release its extra energy to the lattice sufficiently quickly leading then to $T_S > T_L$ resulting in a decrease in n. This forms the basis for the occurrence of saturation. Spin systems with short T_1s will not readily go to saturation and vice versa.

As discussed earlier, there are relaxation mechanisms through which transfer of excess spin energy occurs so that a constant excess spin of the lower energy is maintained. Since in the absence of an external magnetic field, the two levels are equally populated, the establishment of thermal equilibrium between the two spin states of the system, after the application of the external magnetic field clearly shows that interactions between the spin system and its surroundings exist and they cause the spin orientation to change while the excess magnetic energy is transferred to other degrees of freedom.

As noted above, spin-lattice relaxation is possible because the spin system is coupled to the lattice and the lattice is at thermal equilibrium. It implies that the probabilities for the stimulated upward and downward spin transitions are not equal.

The probability of ESR transitions taking place per unit time between the levels m and m' as derived from the first order perturbation theory is given by

$$P_{mm'} = \gamma^2 B_1^2 |< m | S_x | m' >|^2 \, \delta(\nu_{mm'} - \nu) \tag{2.16}$$

Here B_1 is the magnitude of the microwave radiation field applied in the X-direction. The transition probability is proportional to the square of the amplitude of the microwave field and hence low power microwave radiation is used to avoid saturation. In equation (2.16), $< m | S_x | m' >$ is the quantum mechanical matrix element of the X-component of the electron spin operator and it is zero unless $m = m' + 1$, that is, $\Delta M_S = \pm 1$, and $\delta(\nu_{mm'} - \nu)$ is known as the Dirac delta function which is zero unless $\nu_{mm'} = \nu$, since $\delta(\nu_{mm'} - \nu) = 1$ if $\nu_{mm'} = \nu$ and $\delta(\nu_{mm'} - \nu) = 0$ if $\nu_{mm'} \neq \nu$, where $\nu_{mm'}$ is the frequency corresponding to the energy difference between the states m and m'. The delta function predicts an infinitely sharp line which is unrealistic. It is, therefore, replaced by a line space function $g(\nu)$. The equation (2.16) then becomes,

$$P_{mm'} = \gamma^2 B_1^2 |< m | S_x | m' >|^2 g(\nu) \tag{2.17}$$

For a single electron, $S = ½$, and hence the equation (2.17) can be rewritten as

$$P_{mm'} = \gamma^2 B_1^2 |< \alpha | S_x | \beta >|^2 g(\nu) \tag{2.18}$$

Since there is only one allowed transition, the equation (2.18) becomes

$$P = \gamma^2 B_1^2 g(\nu) \tag{2.19}$$

The ESR spectral lines in solution generally show the Lorentzian line shape which is given by the expression,

$$g(\omega) = \left(\frac{T_2}{\pi}\right) \frac{1}{1 + T_2^2 (\omega - \omega_0)^2} \tag{2.20}$$

The line shape function is given in angular frequency scale $g(\omega) = g(\nu)/2\pi$ and T_2 is the relaxation time (see chapter 5). The resonance has a sharp peak at $\omega = \omega_0$ and the full width between the points where the absorption has half-its maximum height is $2/T_2$. A characteristic form of the Lorentzian line is shown in Fig. 2.10. The two peaks of the derivative curve correspond to the points of maximum slope in the absorption curve and the separation between them is $2/T_2\sqrt{3}$.

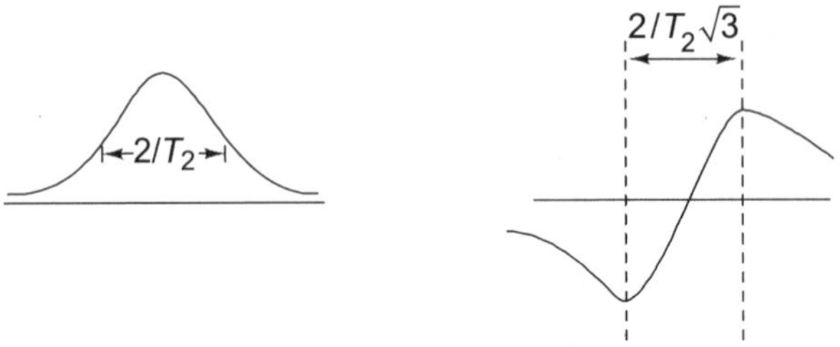

(a) (b)

Fig. 2.10 Line shape of ESR absorption curve: (a) Lorentzian curve and (b) its first derivative curve.

HYPERFINE STRUCTURE

3

NUCLEAR HYPERFINE SPLITTING

The magnetic interactions between the electron spin and the nuclear spin cause the ESR lines of a sample to split into a simple or a more complex pattern of lines. It is referred to as the nuclear hyperfine interaction and this interaction energy is very small, about 10^{-2} cm^{-1}. The concomitant splitting of the electron spin energy levels is called hyperfine splitting and the resulting group of lines in the ESR spectrum is called the hyperfine structure. The spacing between the lines of a hyperfine multiplet provides the hyperfine coupling constant which is denoted by the symbol 'a'. Important applications of ESR spectroscopy to chemical and physical problems stem from the occurrence of hyperfine structure. Analysis of the hyperfine structure can lead to the identity of the free radical and provide an insight into the electronic structure of the paramagnetic species.

Under normal experimental conditions, the hyperfine interaction is much smaller than the Zeeman splittings. The magnitude of the hyperfine coupling constant 'a' depends on the following factors: (i) the ratio of the nuclear magnetic moment to the nuclear spin (that is, the nuclear *g*-factor) and (ii) the electron spin density in the immediate vicinity of the nucleus.

RADICALS CONTAINING ONE PROTON

The simplest case is that of the interaction of an unpaired electron with a nucleus such as a proton, which has a spin (I) of 1/2. The classic example of this is the hydrogen atom. The hydrogen atom in free space is a simple system to discuss since due to its spherical symmetry anisotropic effects are absent. The ESR spectrum of hydrogen atom, exhibits instead of a single line as shown in Fig. 3.1, a pair of lines as shown in Fig. 3.2 separated by the hyperfine coupling constant

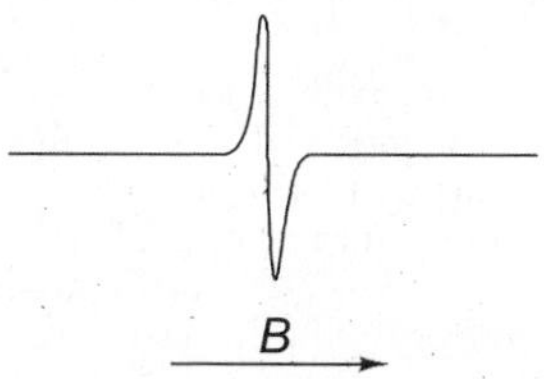

Fig. 3.1 The ESR spectrum of hydrogen atom if $a = 0$. The spectrum is plotted as microwave radiation absorbed (ordinate) against the magnetic field (abscissa) in the first derivative mode.

$a = 50.7$ mT. The g–value is measured midway between the two absorption maxima. The average position of the two lines (0.33535 T for a spectrometer operating at 9.5 GHz) corresponds to $g = 2.0023$.

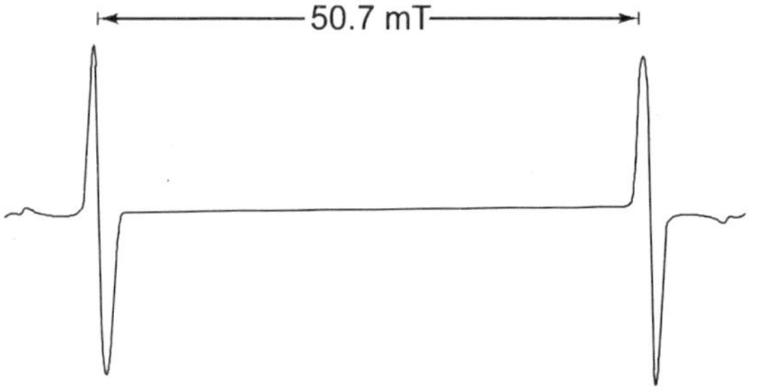

Fig. 3.2 The ESR spectrum of hydrogen atom showing hyperfine splitting.

When the nuclear spin I is non-zero, in the presence of a strong magnetic field, the nuclear spin magnetic quantum number m_I can take $(2I + 1)$ values ranging from $+I$ to $-I$ that is, $I, I - 1, I - 2,..0, -I + 1, -I$. The sum of the m_I values for all the nuclei gives the total spin magnetic quantum number M_I. For an individual nucleus, M_I is the same as m_I. For a nuclear spin $I = 1/2$, m_I can have two values, $+1/2$ and $-1/2$. Since the proton has a spin $I = 1/2$, the energy of the electron will now be modified by the orientation of the magnetic moment of the nuclear spin, which can also be aligned parallel or antiparallel to the external magnetic field.

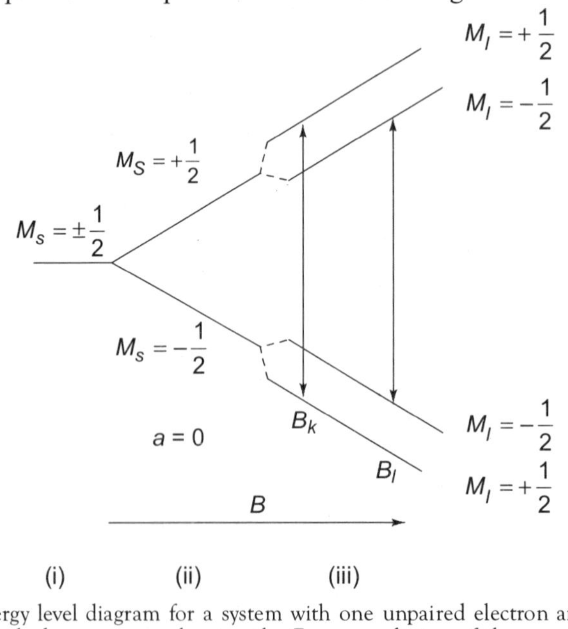

Fig. 3.3 The energy level diagram for a system with one unpaired electron and a nucleus with $I=\frac{1}{2}$, e.g. hydrogen atom, showing the Zeeman splitting of the unpaired electron split further by the nuclear spin $I = \frac{1}{2}$ in a varying external magnetic field: (i) in the absence of magnetic field, (ii) electron spin in the presence of a magnetic field and (iii) electron bound to a proton in an external magnetic field. Allowed transitions are shown by the arrows. The hyperfine splitting constant is given by the difference in magnetic fields $B_l - B_k$.

The electron spin senses a slightly different magnetic field depending on whether the spin of the nucleus is parallel or antiparallel to the applied magnetic field. Each orientation of the electron spin can thus be associated with either of the two proton spin orientations. Thus, the nuclear hyperfine interaction splits each of the electron Zeeman levels into two levels as shown in Fig. 3.3. The probability that any one of the nuclei will have $m_I = +1/2$ is almost the same as that it will have $m_I = -1/2$. Hence, the ESR spectral line of the hydrogen atom will be split into two components of equal intensity as shown in Fig. 3.2. The hyperfine splittings are independent of the applied magnetic field.

SPIN HAMILTONIAN

The complete energy term for electron spin resonance is the sum of the expressions for different interactions and is known as 'spin Hamiltonian'. The interpretation of the ESR spectrum is made in terms of an 'effective spin Hamiltonian', which contains only those effects of the many which are considered to interpret the spectrum of the compound studied. The importance of spin Hamiltonian is that it provides a standard phenomenological way in which the ESR spectrum can be described in terms of a small number of constants. It becomes possible to characterize the spectrum by specifying the effective spin and a small number of parameters (such as g, a, etc.). These parameters measure the magnitudes of various terms in the spin Hamiltonian.

For a sytem with $s = 1/2$, the two spin states are characterized by the quantum number $m_s = \pm 1/2$ as noted before. Thus, if $S\hbar$ is the angular momentm operator, then

$$\hat{S}_z \psi_\alpha = +(1/2)\,\psi_\alpha$$

$$\hat{S}_z \psi_\beta = -(1/2)\,\psi_\beta \tag{3.1}$$

The symbols α and β represent the states with $m_s = +1/2$ and $-1/2$, respectively and $\hat{S}_z$ is the spin operator. The interaction between the electron magnetic moment μ_e and an applied magnetic field B is represented by the Hamiltonian,

$$H = -\mu_e B \tag{3.2}$$

which becomes

$$H = g\,\beta_e B \hat{S}_z \tag{3.3}$$

if the applied field is in the Z-direction. The Hamiltonian operating on the electron spin functions α and β produces the result as shown in Fig. 3.4 (ii). The spin Hamiltonian for the hydrogen atom is very simple because the hydrogen atom is spherically symmetrical. The nucleus and the electron both interact with the applied magnetic field giving an energy

$$H = g\beta_e BS_z - g_N \beta_N BI_z \tag{3.4}$$

Here, g and g_N are the g-factors for the electron and the nucleus, respectively. The nuclear spin energy is much smaller than the electron spin energy.

The two possible spin functions for the electron are denoted by the symbols $|\alpha_e>$ and $|\beta_e>$ for the spin quantum numbers $m_s = +1/2$ and $-1/2$, respectively. Thus,

$$S_Z|\alpha_e> = (+1/2)\,|\alpha_e> \tag{3.5}$$

$$S_Z|\beta_e> = (-1/2)\,|\beta_e>$$

Similarly, for the proton spin we write,

$$I_Z|\alpha_N> = (+1/2)\,|\alpha_N>$$

$$I_Z|\alpha_N> = (-1/2)\,|\beta_N> \tag{3.6}$$

Each orientation of the electron spin can be associated with either of the two proton spin orientations as in an hydrogen atom. The above four states are degenerate but in the presence of an external magnetic field the degeneracy of the energy levels is removed. $\hat{S}_z$ does not operate on the nuclear spin function and $\hat{I}_z$ does not operate on the electron spin function.

The zero order energy values for the four states are

$$E_1 = (1/2)\,g\beta_e B - (1/2)\,g_N \beta_N B\,|\alpha_e\,\alpha_N>$$

$$E_2 = (1/2)\,g\beta_e B + (1/2)\,g_N \beta_N B\,|\alpha_e\,\beta_N> \tag{3.7}$$

$$E_3 = (-1/2)\,g\beta_e B - (1/2)\,g_N \beta_N B\,|\beta_e\,\alpha_N>$$

$$E_4 = (-1/2)\,g\beta_e B + (1/2)\,g_N \beta_N B\,|\beta_e\,\beta_N>$$

The complete Hamiltonian for the hydrogen atom (and other isotropic systems) in free space is

$$H = g\beta_e BS_z - g_N \beta_N BI + aI.S \tag{3.8}$$

where a is the isotropic hyperfine coupling constant.

For a spherical system in an applied magnetic field along the z-axis, the equation (3.8) modifies to

$$H = g\beta_e BS_z - g_N \beta_N BI_z + aI.S \tag{3.9}$$

The first term of the Hamiltonian (3.9) describes the electron Zeeman interactions and the second term describes the interaction of the nuclear magnetic moment of the hydrogen atom with the applied magnetic field. It is of opposite sign and smaller in magnitude than the first term. The combined effect of the first two terms of the equation (3.9) upon the energies of the spin states of the hydrogen atom in a magnetic field is shown in Fig. 3.4 (iii). There are four possible electron and nuclear spin functions and four energies are obtained. The two transitions ($\Delta M_S = \pm 1$) would now have the same energy (see Fig. 3.4 (iii)). Thus, considering only the first two terms of the Hamiltonian, the ESR spectrum of the hydrogen atom would be the same as that of a free electron, i.e. one signal at $g = 2.0023$ is expected.

The term $aI.S$ in the Hamiltonian describes the coupling of the electron and nuclear spin moments, which classically corresponds to the scalar product of these two vectors. The term $aI.S$ does not however indicate anything about the mechanism of coupling. The splitting constant a gives the magnitude of the interaction and has the dimensions of energy. This is the Fermi contact interaction to be discussed later. The result of this interaction is added to the energies in Fig. 3.4 (iii), which leads to the modification of the energies as shown in Fig. 3.4 (iv). Now the two transitions are not of equal energy and the separation between the two transitions is 'a'.

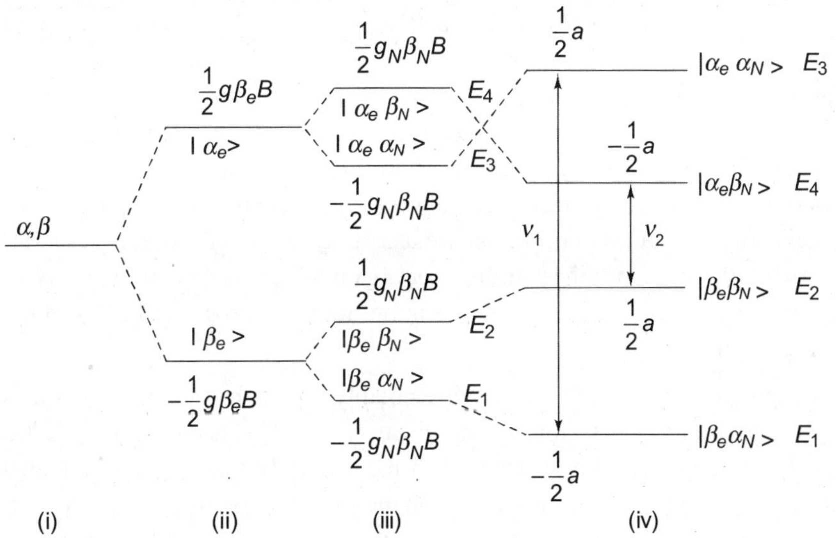

Fig. 3.4 Spin energy levels and the allowed ESR transitions for hydrogen atom: (i) free electron, (ii) electron Zeeman levels, (iii) electron plus nuclear Zeeman levels and (iv) with hyperfine interaction.

We can neglect the term describing the smaller nuclear Zeeman interactions since they do not affect the liquid phase spectrum and write a simpler Hamiltonian,

$$H = g\beta_e BS_z + aS_z I_z \tag{3.10}$$

The first order energies in the hydrogen atom are then given by

$$E = g\,\beta_e Bm_s + a\,m_s m_I \tag{3.11}$$

Figure 3.4 (iv) shows that one transition is lower by 1/2 a and the other is higher by 1/2 a and the difference between the two transitions is a. Substituting the values of m_s and m_I into the equation (3.11) enables us to reproduce the energies given in Fig.3.4 (iv).

SELECTION RULES

Transitions where only the electron spin changes are called ESR transitions while those which involve the nuclear spin alone are called NMR transitions. The transitions where both the electron spin and the nuclear spin change are forbidden and have a very low probability. The selection rules which govern the ESR transitions are

$$\Delta M_S = \pm 1 \text{ and } \Delta M_I = 0 \qquad (3.12)$$

It is seen from Fig. 3.4 that there are two possible transitions which are shown by the arrows satisfying the selection rules. As the magnetic field is varied, there are two values of the magnetic field at which resonance takes place. The energies for the two transitions are $g\beta_e B \pm (1/2)a$ and the separation between the two peaks is simply a, the hyperfine coupling constant. The splitting indicated in Fig. 3.4 implies a positive value for a. The lower energy state would have been $m_S = -1/2$ and $m_I = -1/2$ (that is $\beta_e\beta_N$) if 'a' were negative. In general, the sign of a cannot be determined from the spectrum. The coupling constant may carry a subscript such as a_H, a_N, etc. depending on the nucleus involved. The coupling constants of free radicals in solution are usually given in units of millitesla, mT or Gauss (G). Proton hyperfine splittings for organic free radicals are typically in the range 0–3 mT. The hyperfine coupling constant for the hydrogen atom (50.68 mT) is the largest known.

The hyperfine coupling constants are notably larger than the nuclear spin-spin coupling constants by a factor of approximately 10^6. This is due to the fact that an electron can approach the nucleus much more closely than can another nucleus and hence will interact with it more strongly. The position of the main line or the centre of gravity of the hyperfine pattern provides the g-factor. The interaction of the electron spin with the external magnetic field and the hyperfine interaction do not change the magnitude of the g-value.

The interaction between the electron spin and the nuclear spin is expressed by the term aM_SM_I, where a is the hyperfine splitting constant ignoring the small nuclear Zeeman interaction. The equation (2.7) (chapter 2) now modifies to

$$E = g\beta_e M_S B + aM_S M_I \qquad (3.13)$$

Thus for an electron spin interacting with a single nucleus such as in the case of hydrogen atom, the energies of the four spin states are as follows.

$$E_{1/2,1/2} = (1/2)\, g\beta_e B + (1/4)a$$
$$E_{1/2,-1/2} = (1/2)\, g\beta_e B - (1/4)a$$
$$E_{-1/2,-1/2} = (-1/2)\, g\beta_e B + (1/4)a$$
$$E_{-1/2,1/2} = (-1/2)\, g\beta_e B - (1/4)a \quad (3.14)$$

The two allowed ESR transitions (Fig. 3.4 (iv)) are

$$h\nu_1 = g\beta_e B + (1/2)a$$
$$h\nu_2 = g\beta_e B - (1/2)a \quad (3.15)$$

The energy difference between the two transitions gives the hyperfine splitting constant.

It is of interest to calculate the magnetic field at resonance for ^{23}Na as an example. For Na ($I = 3/2$), the ESR resonance at 9,300 MHz ($g = 2.0022$) is split into a quartet by the hyperfine coupling, $a = 885$ MHz. To calculate the magnetic field at resonance, we need both g and a values. From equation (3.13), we have,

$$h\nu_0 = g\beta_e B_0 + h a m_I$$

Hence, $$B_0 = h\,(g\beta_e B)^{-1}(\nu_0 - a m_I)$$

On substituting the values, we get

$$B_0 = \frac{6.626\times10^{-34}\,Js\,(9.3\times10^{9}\,s^{-1} - 8.85\times10^{8}\,s^{-1} m_I)}{(2.00\,22)(9.274\times10^{-24}\,JT^{-1})}$$

On simplification after inserting the values of $m_I = \frac{3}{2}, \frac{1}{2}, -\frac{3}{2}$ and $-\frac{1}{2}$, we obtain B values of 0.2844, 0.3160, 0.3476 and 0.3792 T respectively.

RADICALS CONTAINING A SET OF EQUIVALENT PROTONS

Consider now a system of an electron spin interacting with two equivalent protons. A simple way of describing it is to regard that the proton 1 splits the original single line due to the electron spin transition into a doublet, each line being split into a further two line spectrum by the interaction with proton 2 as shown in the energy level diagram in Fig. 3.5. A set of three energy levels, with doubly degenerate central energy level results. There are now three values of the applied magnetic field at which resonances occur according to the selection rules. A hyperfine structure of three equally spaced lines is observed, the intensity of the central line being twice that of the other two; the central line arises from the transition between the two doubly degenerate energy levels.

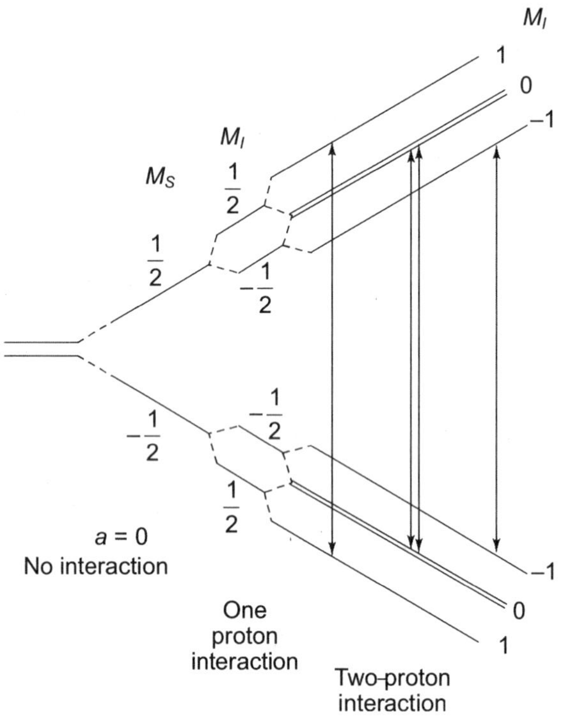

Fig. 3.5 The electron spin energy levels of a free radical (S = ½) interacting with two equivalent protons in a varying external magnetic field.

The $^{\bullet}CH_2OH$ radical is one of the simplest systems which contains one set of two equivalent methylene protons. This free radical may be generated by adding an aqueous H_2O_2 solution to a solution of Ti^{+3} in CH_3OH. When the pH of the solution is sufficiently low (~1.0), the OH proton contributes no detectable hyperfine splitting since it undergoes rapid exchange.

For a free radical with two equally coupled nuclear spins of I = 1/2, we can postulate three different nuclear spin subpopulations as shown in Table 3.1.

Table 3.1 Intensity ratio in the hyperfine structure for two equally coupled nuclear spins, I = 1/2

Nuclear spin arrangement	m_I: m_1	m_I: m_2	$M_I = \Sigma m_I$	B_{local}	***Resonant field,*** B_{res}	***Intensity ratio***
↑↑	1/2	1/2	1	a	$B' - a$	1
↑↓	1/2	−1/2	0	0	B'	2
↓↑	−1/2	1/2				
↓↓	−1/2	−1/2	− 1	a	$B' + a$	1

Since there are two different ways to obtain $M_I = 0$, but only one way to obtain $M_I = 1$ or -1, the observed relative intensities for the corresponding transitions are in the ratio of 1:2:1. Nuclei which occupy symmetrically equivalent positions in the molecule have identical coupling constants. Such nuclei are termed as equivalent nuclei. For the $^{\bullet}CH_2OH$ radical, the two protons of the $-CH_2$ group are equivalent. No hyperfine splitting is observed from carbon and oxygen atoms since the abundant isotopes ^{12}C and ^{16}O have no nuclear spin. The stick diagram given in Fig. 3.6 illustrates the splitting of the resonance as each proton hyperfine interaction is added.

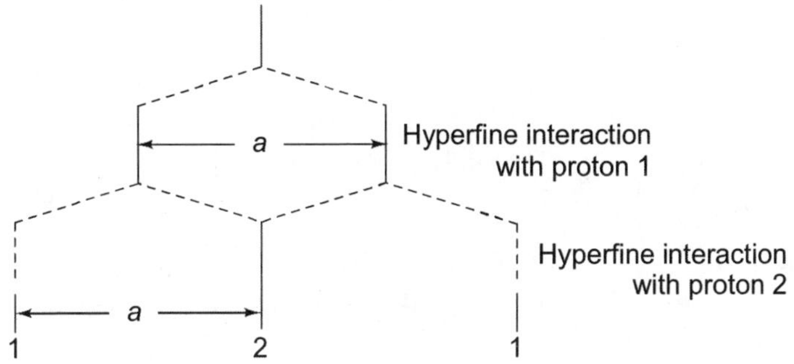

Fig. 3.6 The origin of the three-line hyperfine structure (solid lines) from two equivalent protons, $a_1 = a_2 = a$.

A simple spectrum consisting of a 1 : 2 : 1 triplet arises from the CH_2 protons and the hyperfine splitting constant has been found to be 1.72 mT. The spectrum is shown in Fig. 3.7.

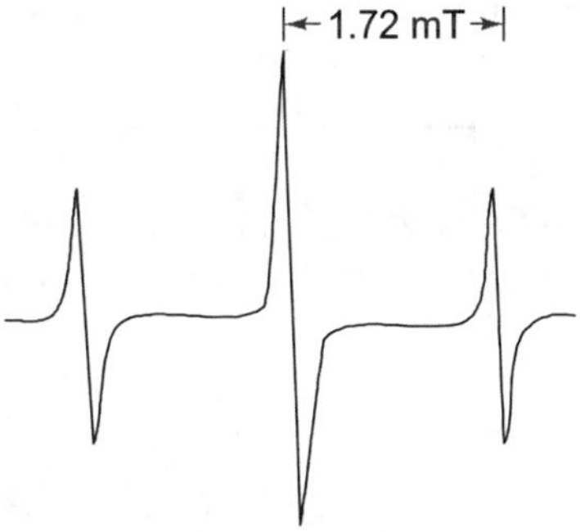

Fig. 3.7 The ESR spectrum of $^{\cdot}CH_2OH$ at pH ~ 1.0.

Consider now a free radical with three equivalent protons e.g. $^{\bullet}CH_3$. We can postulate four different subpopulations of the nuclear spins as shown in Table 3.2 for a free radical with three equally coupled nuclei of $I = 1/2$. The energy level diagram is shown in Fig. 3.8 and the allowed ESR transitions are also shown therein.

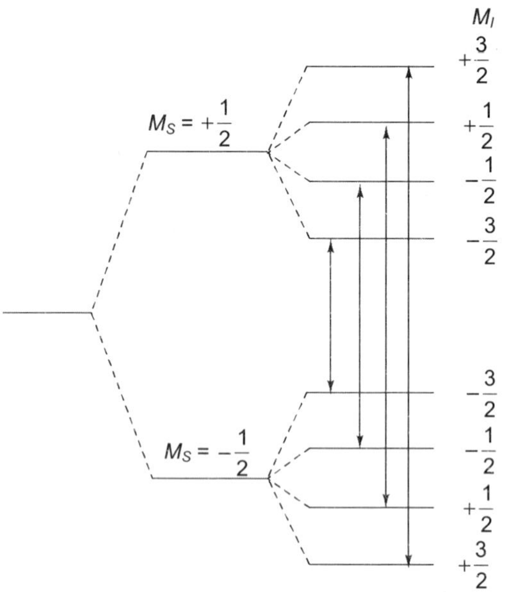

Fig. 3.8 The energy level diagram for $^{\bullet}CH_3$ radical and the allowed ESR transitions in a constant external magnetic field.

There are four transitions for the $^{\bullet}CH_3$ radical. Since four nuclear spin states are possible in the ratio 1 : 3 : 3 : 1 (Table 3.2), in the ESR spectrum of $^{\bullet}CH_3$ four lines in the intensity ratio of 1 : 3 : 3 : 1 are observed. Electron irradiation of liquid ethane in a microcavity produces methyl radicals. The spectrum of the methyl radical is shown in Fig. 3.9. The spectrum exhibits four equidistant lines with a relative intensity ratio of 1 : 3 : 3 : 1 and a separation of 2.30 mT.

Table 3.2 Intensity ratio in the hyperfine structure for three equally coupled nuclei $I = 1/2$

Nuclear spin arrangement	m_I m_1	m_2	m_3	$M_I = \Sigma m_I$	B_{local}	*Resonant field,* B_{res}	*Intensity ratio*
↑↑↑	1/2	1/2	1/2	3/2	$(3/2)a$	$B' - (3/2)a$	1
↑↑↓	1/2	1/2	–1/2				
↑↓↑	1/2	–1/2	1/2	1/2	$(1/2)a$	$B' - (1/2)a$	3
↓↑↑	–1/2	1/2	1/2				
↑↓↓	1/2	–1/2	–1/2				
↓↑↓	–1/2	1/2	–1/2	–1/2	$(-1/2)a$	$B' + (1/2)a$	3
↓↓↑	–1/2	–1/2	1/2				
↓↓↓	–1/2	–1/2	–1/2	–3/2	$(-3/2)a$	$B' + (3/2)a$	1

Extension of the treatment to n equivalent nuclei is straightforward. When several nuclei each with nuclear spin $I = 1/2$ are equally coupled to an electron

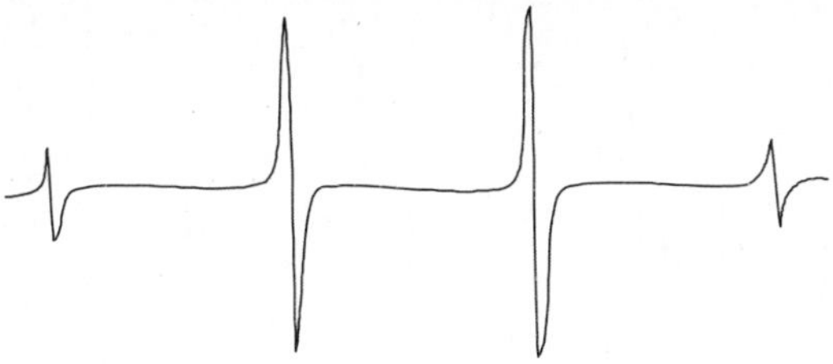

Fig. 3.9 The ESR spectrum of $^{\bullet}CH_3$ radical.

spin, a hyperfine structure with (n+1) (that is, $2nI + 1$, where $I = 1/2$) equally spaced lines are obtained. The hyperfine lines exhibit an intensity ratio that follows the distribution of the binomial coefficients of $(1 + x)^n$. Alternatively, these relative intensities can be read off from the nth row of the Pascal's triangle which is shown below, where n indicates the number of equivalent interacting nuclei.

Pascal's triangle

Nuclei	Hyperfine splitting pattern
$n = 0$	1
$n = 1$	1 1
$n = 2$	1 2 1
$n = 3$	1 3 3 1
$n = 4$	1 4 6 4 1
$n = 5$	1 5 10 10 5 1
- - -	- -

After the second row, each member in a particular row is the sum of the two nearest numbers in the row just above the one under consideration, of course keeping 1 at the terminals. Thus, we can read off the expected pattern for two equivalent protons ($n = 2$) from the third row and the results for three equivalent protons ($n = 3$) from the fourth row of the triangle and so on.

A free radical with four equivalent protons is p-benzosemiquinone radical ion $[O\text{–}C_6H_4\text{–}O]^-$. The oxidation of hydroquinone in alkaline ethanol readily produces p-benzosemiquinone radical ion. As expected, it exhibits a spectrum with five hyperfine lines ($n + 1$, where $n = 4$) with relative intensities of the lines in the ratio of 1 : 4 : 6 : 4 : 1. The hyperfine coupling constant is 0.237 mT. The spectrum of p-benzosemiquinone radical ion is given in Fig. 3.10.

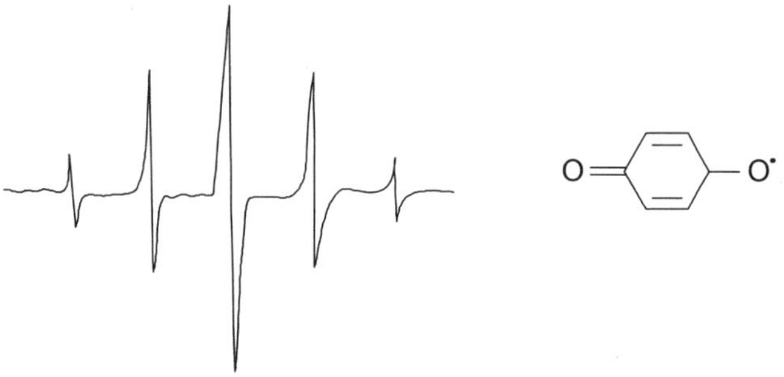

Fig. 3.10 The ESR spectrum of *p*-benzosemiquinone radical.

Finally a free radical with five equivalent protons, for example, cyclopentadienyl radical is considered. All the five protons in the molecular plane are equivalent. A six-line spectrum with their relative intensities in the ratio of 1:5:10:10:5:1 is expected as shown in Fig. 3.11. The hyperfine coupling constant was found to be 0.6 mT.

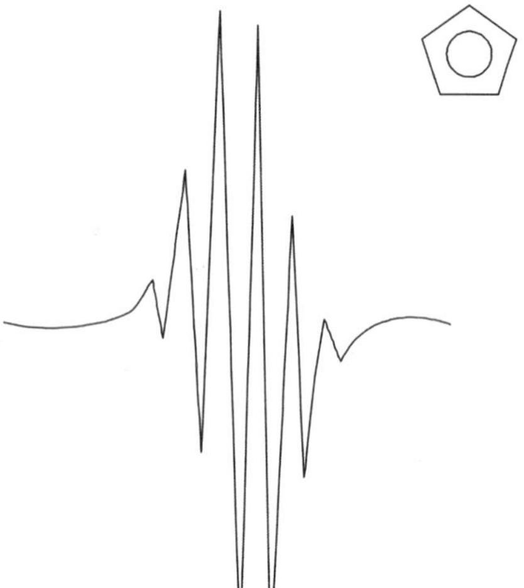

Fig. 3.11 The ESR spectrum of cyclopentadienyl radical.

RADICALS CONTAINING MULTIPLE SETS OF EQUIVALENT PROTONS

The hyperfine structure becomes more complex when non-equivalent nuclei are involved. If the unpaired electron interacts with, for example, two sets of inequivalent nuclei, say m set of nuclei with spin I_i and n set of nuclei with spin I_j, then the number of lines expected in the hyperfine pattern is the product of number of lines expected for each of the sets, that is, $(2mI_i + 1)(2nI_j + 1)$, equivalently $(m + 1)(n + 1)$ lines if the nuclei are protons. In particular, if the hyperfine interaction with one group of nuclei possessing the total nuclear spin I_1 is greater than the hyperfine interaction with the other group of nuclei possessing a total spin I_2, then the hyperfine structure resulting from both the interactions consists of $(2I_1 + 1)$ lines due to the stronger interaction, each of these lines being split into $(2I_2 + 1)$ closely spaced lines due to the weaker interaction. The final set of energy levels is independent of the order in which the splitting of the energies are taken. However, crossing of many lines is avoided if the larger hyperfine coupling is taken first for drawing the splitting diagram. When the hyperfine splitting constants are close enough, an overlapping spectrum different from the 'idealized' spectrum would result making it difficult to recognize the spectral pattern at the first instance. This feature is illustrated for ethyl radicals later.

The simplest case of coupling of two inequivalent nuclei, both with $I = 1/2$, is glycolic acid HO•CHCOOH radical. The energy levels are shown in Fig. 3.12 along with the allowed transitions. The hyperfine splitting pattern for a radical with two non-equivalent protons with the hyperfine coupling constant a_1 greater than a_2 is analysed in Fig. 3.13. The spectrum of HO•CHCOOH in solution given in Fig. 3.14 is consistent with the calculated spectrum. The hyperfine coupling constants have been found to be a(CH) = 1.71 mT and a(COH) = 0.26 mT. It is to be noted that, in general, it is not possible to assign the hyperfine splitting constants solely from an analysis of the spectrum. The assignment can be made by comparison of these hyperfine coupling constants with those of other similar radicals.

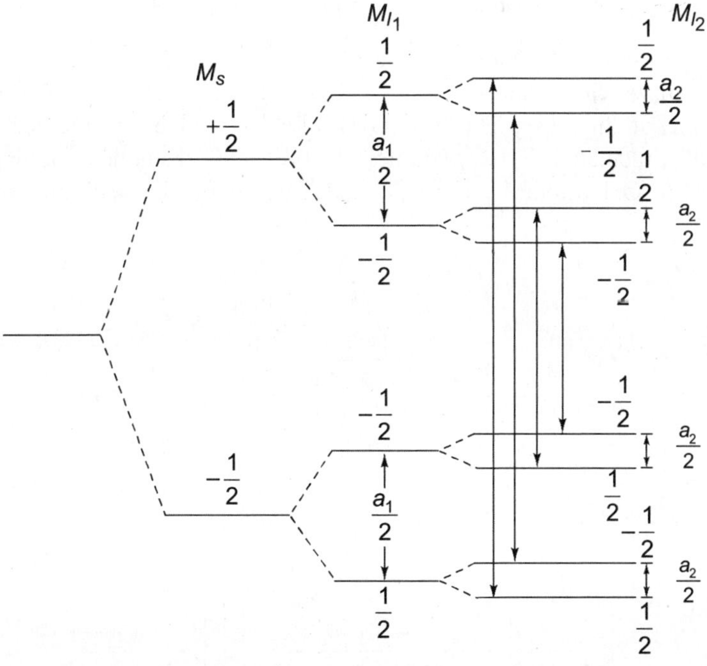

Fig. 3.12 The energy level diagram and the allowed ESR transitions for a free radical with two non-equivalent nuclei, each of $I = 1/2$, in a fixed external magnetic field assuming $a_1 > a_2 > 0$.

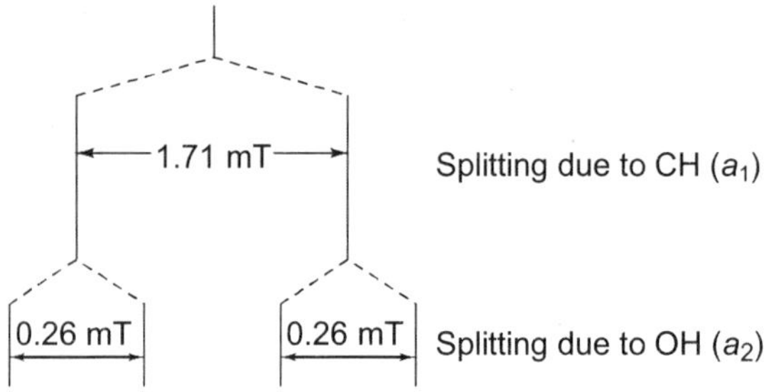

Fig. 3.13 The analysis of the ESR hyperfine structure (solid lines) for two inequivalent protons as in glycolic acid radical, HO$^•$CHCOOH.

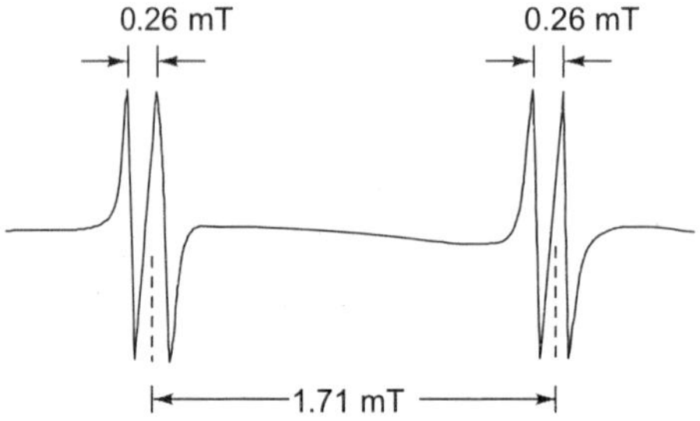

Fig. 3.14 The ESR spectrum of glycolic acid radical HO$^•$CHCOOH in solution.

The next system to be considered is that of a free radical containing three protons, two of which are equivalent. The $^•CH_2OH$ radical considered previously is an example of this case. The $^•CH_2OH$ radical can also be obtained by the photolysis of a methanol – H_2O_2 solution. In less acidic solutions, the hyperfine coupling due to OH proton is also observed. The energy level diagram for the

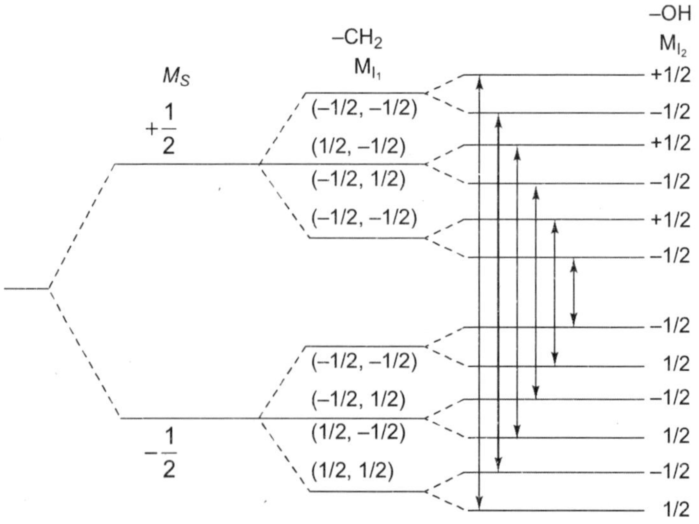

Fig. 3.15 The energy level diagram for $^•CH_2OH$ radical and the allowed ESR transitions.

system is constructed by successive splitting of the energy levels and it is shown in Fig. 3.15. Apparently the smaller splitting is due to that of the OH proton as noted from a comparison with the spectrum of the $HO^{\bullet}CHCOOH$ discussed previously. The calculated ESR spectrum of $^{\bullet}CH_2OH$ using 'sticks' of appropriate length and assuming $a(CH_2)$ to be greater than $a(OH)$ is presented in Fig. 3.16. The spectrum of $^{\bullet}CH_2OH$ given in Fig. 3.17 is consistent with the theoretical spectrum. The hyperfine lines arise from a doubling by the unique proton of the three transitions expected for two equivalent protons. From the spectrum, it was found that $a(CH_2)$ = 1.74 mT and $a(OH)$ = 0.115 mT.

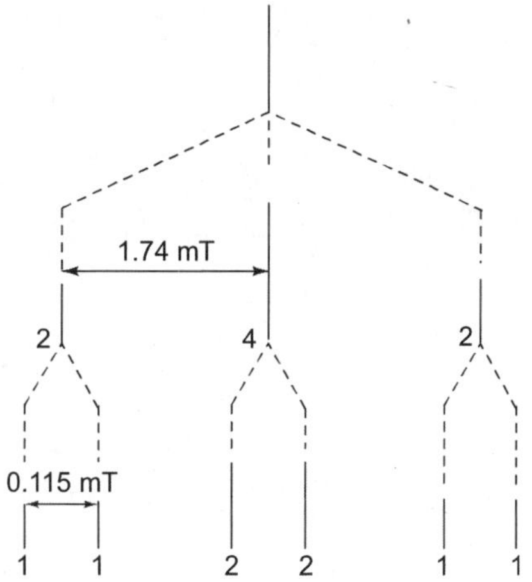

Fig. 3.16 The analysis of the hyperfine structure (solid lines) of $^{\bullet}CH_2OH$ radical in solution, $a(CH_2) > a(OH)$.

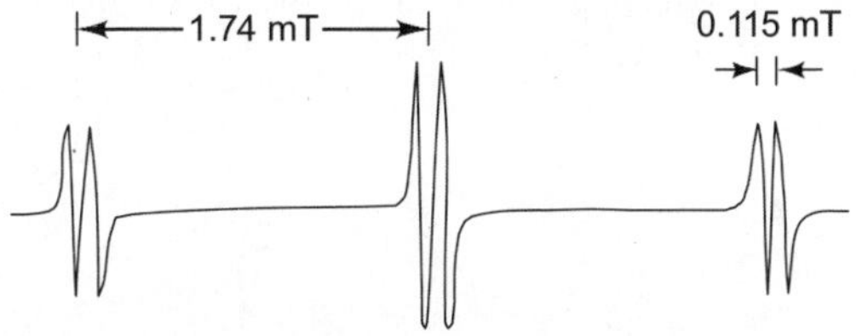

Fig. 3.17 The ESR spectrum of $^{\bullet}CH_2OH$.

Ethyl radicals may be produced in solution by irradiating liquid ethane with high energy electrons. A triplet hyperfine splitting from the two equivalent CH_2 protons and a quartet from the three equivalent CH_3 protons may be predicted. If the $-CH_2$ hyperfine splittings are larger than those of the $-CH_3$ protons, the spectrum shown in Fig. 3.18 (a) is expected. On the other hand, if the $-CH_3$ hyperfine coupling constant is larger than that of the $-CH_2$ proton hyperfine coupling constant, a spectrum similar to that presented in Fig. 3.18 (b) is expected. The CH_2 and CH_3 proton hyperfine coupling constant has been found to be 2.25 and 2.71 mT, respectively. Interestingly, the far away CH_3 group shows a higher coupling constant than the CH_2 group. The analysis of the hyperfine splitting pattern is shown in Fig. 3.19. Four triplets are expected. The components of each triplet are in the intensity ration of 1:2:1 and the four triplets are in the intensity ratio of 1:3:3:1. However, it can be seen from the calculated hyperfine splitting pattern, that the triplets are not found completely together. In the actual spectrum, the lines (1,2,4); (3,5,7), (6,8,10) and (9,11,12) constitute the triplets as seen from Fig. 3.18(c). If the four sets of triplets were 'in order', the intensity ratio would be (1:2:1), (3:6:3), (3:6:3) and (1:2:1). However, the observed ratio of the intensity of the hyperfine lines is 1:2:3:1:6:3:3:6:1:3:2:1, as noted from Figs. 3.18(c) and 3.19.

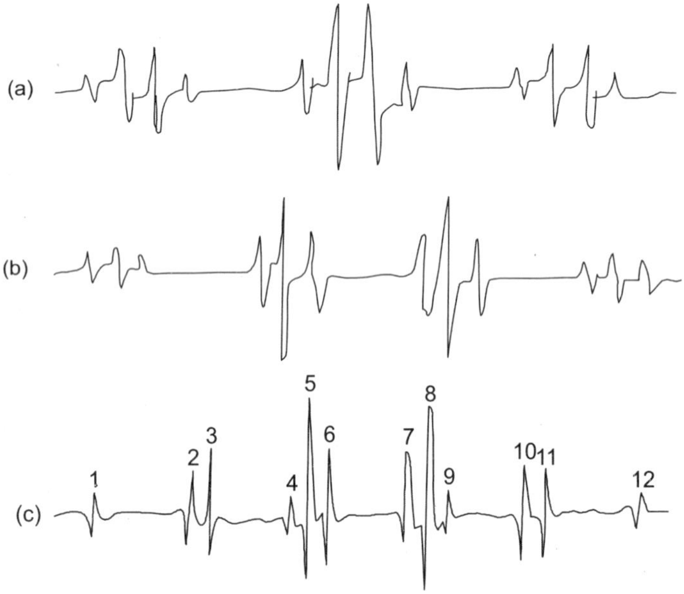

Fig. 3.18 The ESR spectra of $^{\bullet}CH_2CH_3$ radical; predicted spectrum : (a) $a_{CH_2} > a_{CH_3}$, (b) $a_{CH_3} > a_{CH_2}$ and (c) the actual spectrum in solution.

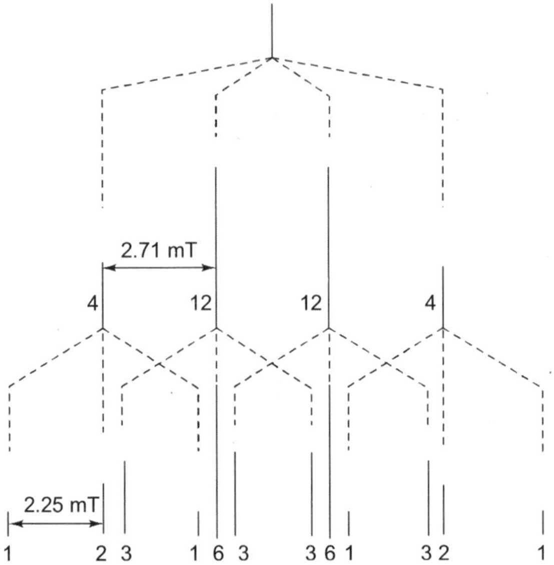

Fig. 3.19 The analysis of the hyperfine structure of the ESR spectrum (solid lines) of $^{\bullet}CH_2CH_3$ radical.

RADICALS CONTAINING OTHER NUCLEI OF SPIN $I = 1/2$

The hyperfine splitting in the ESR spectrum of organic free radicals is most frequently due to electron-proton hyperfine interactions, but other magnetic nuclei besides proton also cause hyperfine splitting. Hyperfine splittings from ^{13}C, ^{19}F or ^{31}P are usually indistinguishable from the proton hyperfine splittings. Additional evidence is needed to identify the interacting nucleus.

We shall now discuss the ESR spectrum of ^{13}C labelled methyl radical. Each of the four nuclear spin energy levels of the $^{\bullet}CH_3$ radical discussed earlier will now be split into two by the ^{13}C nuclear spin ($I = 1/2$). We therefore expect to observe eight lines in the hyperfine structure with the relative intensities 1:1:3:3:3:3:1:1 instead of four lines observed for the normal methyl radical ($^{12}CH_3$). If the larger hyperfine splitting arises from the protons, a spectrum with four sets of doublets as shown in Fig. 3.20 (a) would be expected. However, if the larger hyperfine splitting arises from ^{13}C, two sets of quartets would be expected as depicted in Fig. 3.20 (b). Experimentally, the hyperfine splittings have been found to be 4.10 mT and 2.30 mT for ^{13}C and ^{1}H, respectively. These splittings are somewhat close enough so that neither spectrum (a) nor spectrum (b) of Fig. 3.20 is obtained, but rather an 'overlapping' spectrum depicted in Fig. 3.20 (c) results. The analysis of the hyperfine splitting pattern for $^{13}CH_3$ radical is shown in Fig. 3.21 using the experimental values of the coupling constants. The natural abundance of ^{13}C is only about 1.1% and hence the splitting due to ^{13}C is usually not observed in the spectrum of the normal methyl radical.

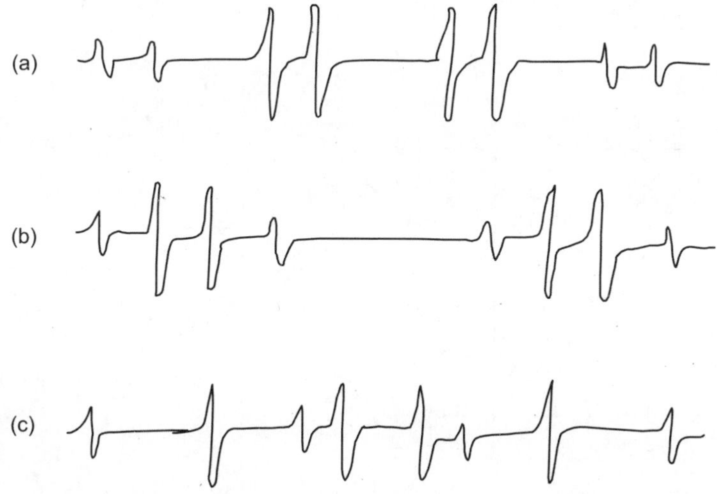

Fig. 3.20 The ESR spectra of $^{13\bullet}CH_3$; expected spectrum: (a) $a_H > a_{^{13}C}$, (b) $a_{^{13}C} > a_H$ and (c) actual spectrum (^{13}C enriched).

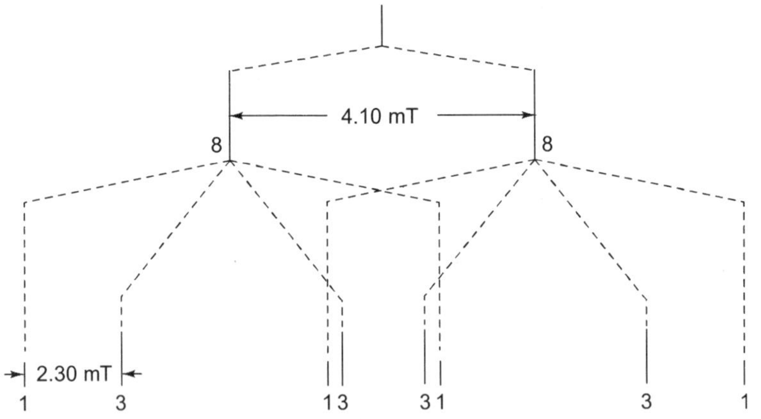

Fig. 3.21 The analysis of the hyperfine splitting of the ^{13}•CH_3 radical.

RADICALS CONTAINING NUCLEI WITH SPIN $I > 1/2$

The above discussion on hyperfine interaction can easily be extended to nuclei with values of spin I other than 1/2. For example, deuterium or nitrogen containing radicals show hyperfine structure due to ^2H or ^{14}N which have a nuclear spin of $I = 1$. The possible values of M_I are 1, 0, –1 so that an electron coupled to a single ^2H or ^{14}N nucleus should show a triplet hyperfine pattern with all lines of equal intensity. For an electron interacting with a single nucleus such as deuterium, there are now six spin energy states represented by (M_S, M_I) values. An energy level diagram for a free radical interacting with a nuclear spin of $I = 1$ is shown in Fig. 3.22 along with the allowed ESR transitions. The spectrum of di-t-butyl nitroxide radical given in Fig. 3.23 is in agreement with the prediction. The hyperfine splitting constant is 1.5 mT.

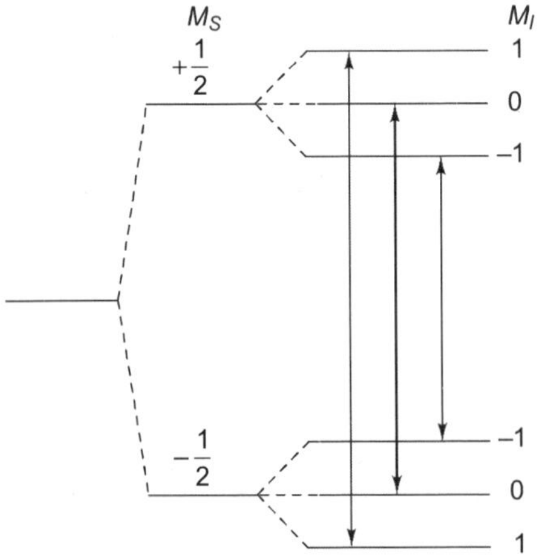

Fig. 3.22 The energy level diagram and allowed ESR transitions for a system with $S = 1/2$ interacting with a nucleus $I = 1$ (e.g. ^2H) at a constant magnetic field.

The NH_3 radicals are produced at room temperature by irradiation of the single crystals of ammonium perchlorate by X-rays. For •NH_3 radical, the expected

Fig. 3.23 ESR spectrum of di-t-butyl nitroxide radical in solution.

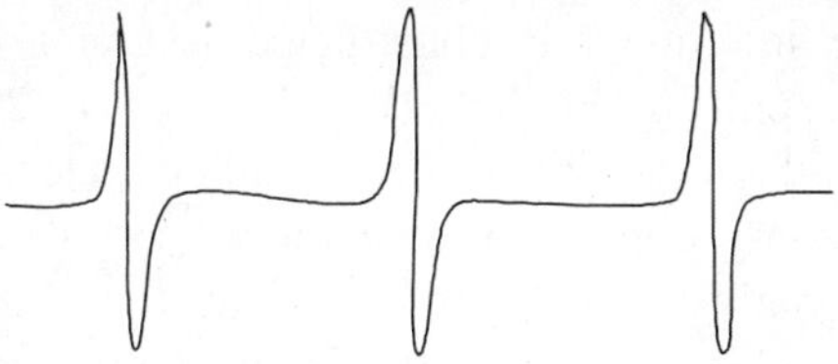

hyperfine interaction involves the nitrogen nucleus and the three equivalent protons. The hyperfine splitting arising from the interaction of the unpaired electron with the nitrogen gives rise to a 1:1:1 triplet. Each of these lines is split into a quartet by the three equivalent protons in the intensity ratio 1:3:3:1 giving a twelve-line pattern. The stick diagram drawn assuming $a_N > a_H$ as given in Fig. 3.24 (a) shows that the lines (1,3,6,9), (2,5,8,11) and (4,7,10,12) constitute the three quartets. On the other

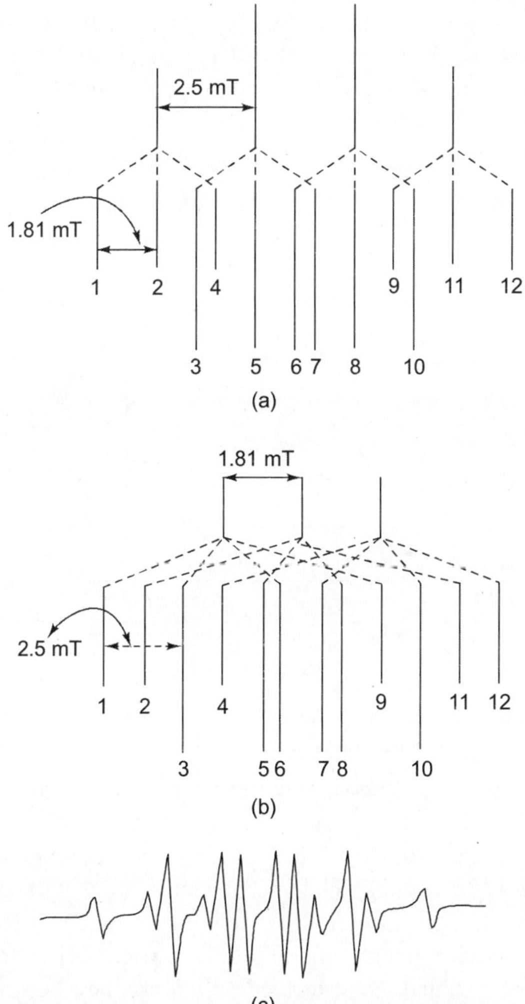

Fig. 3.24 The analysis of the hyperfine structure of $^{\bullet}NH_3$ radical considering hyperfine splitting from (a) protons first, (b) nitrogen first and (c) the ESR spectrum of $^{\bullet}NH_3$ radical.

hand, the stick diagram constructed assuming $a_H > a_N$ also shows a twelve-line hyperfine splitting [Fig.3.24 (b)]. The lines (1,2,4), (3,5,7), (6,8,10) and (9,11,12) constitute the four triplets. The calculated intensity patterns for either of the 'stick diagrams' are consistent with the observed spectrum given in Fig. 3.24 (c). Experimentally it has been found that $a_N = 1.81$ mT and $a_H = 2.5$ mT. The Figs. 3.24 (a) and 3.24 (b) illustrate that the crossing of the lines is minimized in the stick diagram if the larger hyperfine splitting constant is taken first.

For the $^{\bullet}NH_2$ radical, the hyperfine splitting arising from the nitrogen (a_N) is larger so that a triplet spectrum, where each of the lines is again split into a triplet in the intensity ratio of 1:2:1 is obtained.

If an unpaired electron interacting with a ^{14}N nucleus interacts with a second ^{14}N nucleus, each of the lines of the triplet is split further into three lines giving in general nine lines. However, if the two ^{14}N nuclei are equivalent, a five-line spectrum with relative intensities of 1:2:3:2:1 is obtained due to merging of some of the lines, as evident from the transitions shown in Fig. 3.25.

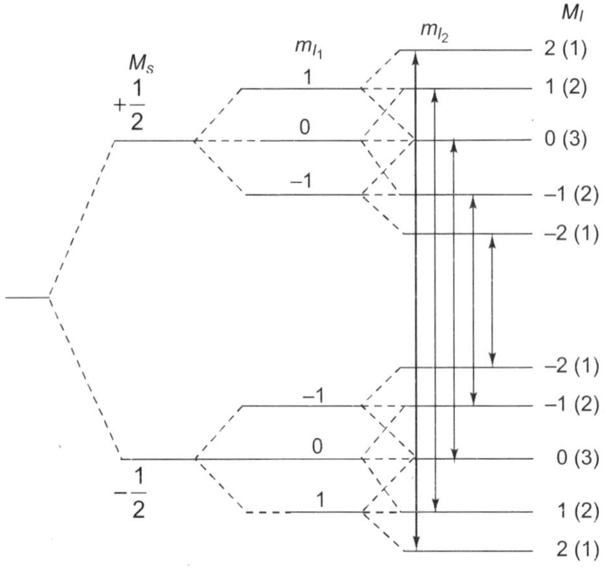

Fig. 3.25 The energy level diagram showing the hyperfine levels for two equivalent nuclei with $I = 1$. Number in the parenthesis indicates the degeneracy of each level.

An example of two equally coupled nitrogen nuclei is found in 1, 1- diphenyl - 2 –picryl hydrazyl (DPPH) radical. The hyperfine coupling constants due to nitrogen are 0.77 (a_1) and 0.92 (a_2) mT. The proton coupling constants for DPPH are small, about 0.1–0.2 mT and the hyperfine structure due to protons is generally not observed in the spectrum of DPPH. It is to be noted that the two hydrazyl nitrogens of DPPH are not magnetically equivalent. However, due to rapid chemical exchange in fairly concentrated solutions, they become equivalent. As powder or in concentrated xylene solution, DPPH exhibits only one band while in dilute solution (10^{-3} M), it shows a five-line spectrum in the intensity ratio of

1:2:3:2:1 (see Chapter 4). This intensity ratio can also be derived from Pascal's triangle. The ESR spectrum of DPPH in solution is presented in Fig. 3.26. The reconstruction of the spectrum of DPPH using the 'stick diagram' is shown in Fig. 3.27.

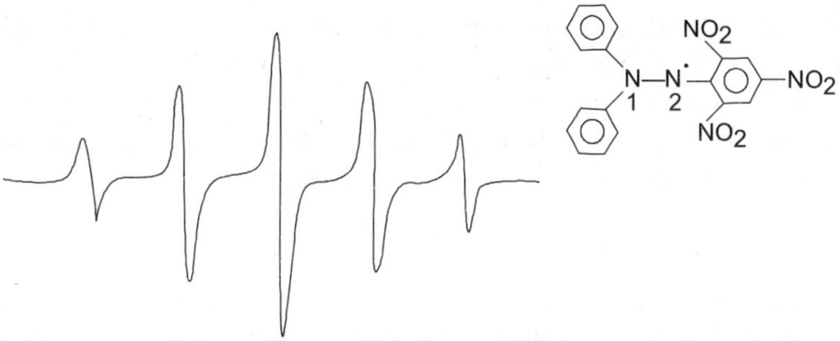

Fig. 3.26 ESR spectrum of DPPH in solution; a_1 = 0.77 and a_2 = 0.92 mT.

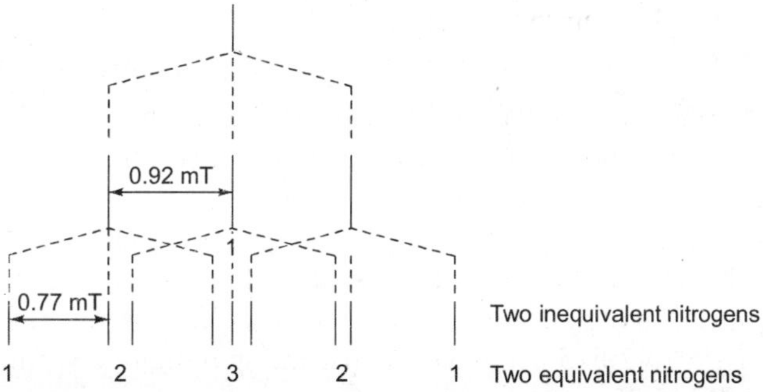

Fig. 3.27 The analysis of the hyperfine structure of DPPH radical.

The spectrum of pyrazine mono-negative radical ion demonstrates the splitting of the ESR resonance by two different types of nuclei. Pyrazine when reacted with an alkali metal such as potassium in tetrahydrofuran yields pyrazine mono-negative radical ion. Interaction of the unpaired electron with two equivalent nitrogens produces five ($2nI + 1$, $n = 2$) lines each of which is split into a quintet by the four equivalent ring protons and a 25-line spectrum should result. The observed spectrum shown in Fig. 3.28 is consistent with this prediction. From the analysis of the

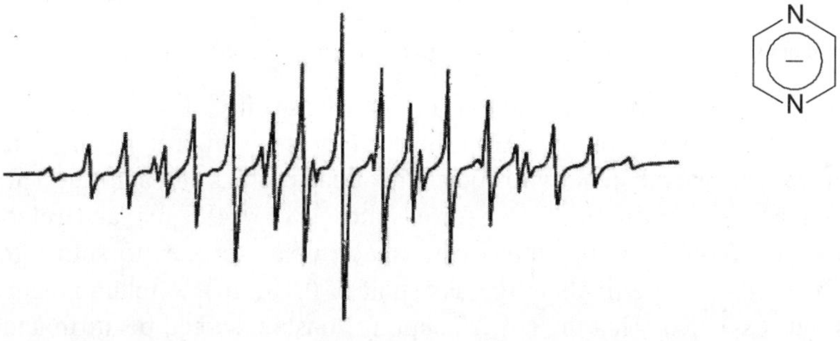

Fig. 3.28 ESR spectrum of pyrazine anion radical in solution. The counter ion is K^+.

spectrum, hyperfine coupling constants $a_N = 0.72$ mT and $a_H = 0.27$ mT have been obtained and the larger coupling constant arises from the nitrogen nuclei. In the spectrum of the sodium salt of pyrazine anion, a further splitting into 1:1:1:1 quartets is observed and it indicates the existence of ion pairs in solution.

Nearly all radicals containing nitrogen have well-resolved ^{14}N hyperfine structure if there is any *pi* electron density on the nitrogen atom. The nuclear quadrupole interaction is always much smaller than the ESR hyperfine splittings and hence there is no line broadening of the type found in the proton NMR spectra of nitrogen compounds due to ^{14}N nuclear quadrupole moment.

AROMATIC RADICALS

In the absence of air, aromatic hydrocarbons react with alkali metals in solvents such as dioxane, tetrahydrofuran, etc., yielding colored solutions. It is known that electron transfer from the alkali metal to the hydrocarbon occurs and negative radical ions such as $C_6H_6^-$, $C_{10}H_8^-$, ... are produced. The positive and negative radical ions of aromatic hydrocarbons are quite stable in solution.

Benzene dissolved in an inert solvent such as tetrahydrofuran will react with sodium metal resulting in benzene negative ion (which is stable below –70 °C) according to the reaction $C_6H_6 + Na \rightarrow C_6H_6^- + Na^+$. The additional electron of the benzene negative ion occupies a *pi* molecular orbital composed of atomic $2p_Z$ orbitals of carbon and interacts with all the six hydrogen atoms. The benzene anion radical shows an ESR spectrum with seven hyperfine lines with relative intensities of 1:6:15:20:15:6:1 and the hyperfine splitting is 0.375 mT. The spectrum is shown in Fig. 3.29.

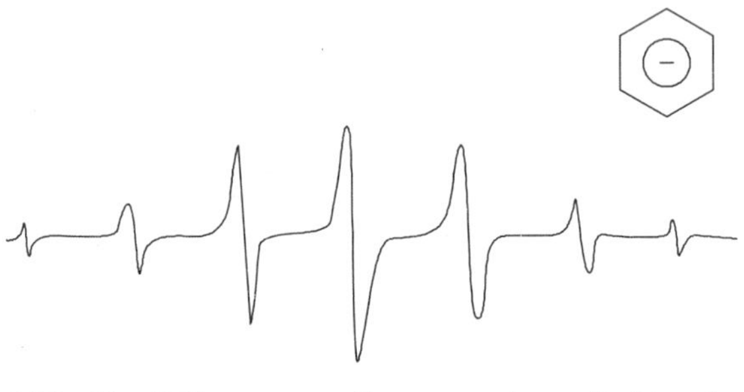

Fig. 3.29 The ESR spectrum of benzene anion radical in solution.

The spectrum of naphthalene anion radical was one of the first to be investigated and it is shown in Fig. 3.30. There are two sets of protons labelled as α and β protons in naphthalene with four equivalent protons in each set. The α protons produce a hyperfine quintet with a splitting of 0.495 mT and relative intensities of 1:4:6:4:1. The β protons cause each of these lines to split into a further quintet with a hyperfine splitting constant of 0.187 mT. A total number of 25 lines is thus expected. That the larger coupling constant should be attributed to the α protons is inferred from the spectrum of β-deuterated naphthalene anion radical. It also finds support from Huckel molecular orbital calculations discussed later. Substitution of deuterium for hydrogen leads to a reduction in the coupling constant according to the relation,

$$a_D = (g_D/g_H)\, a_H = 0.15\, a_H$$

where the nuclear g-factors $g_D = 0.85745$ and $g_H = 5.5857$. As with NMR spectroscopy, deuterium substitution is often used to facilitate the interpretation of the ESR spectra.

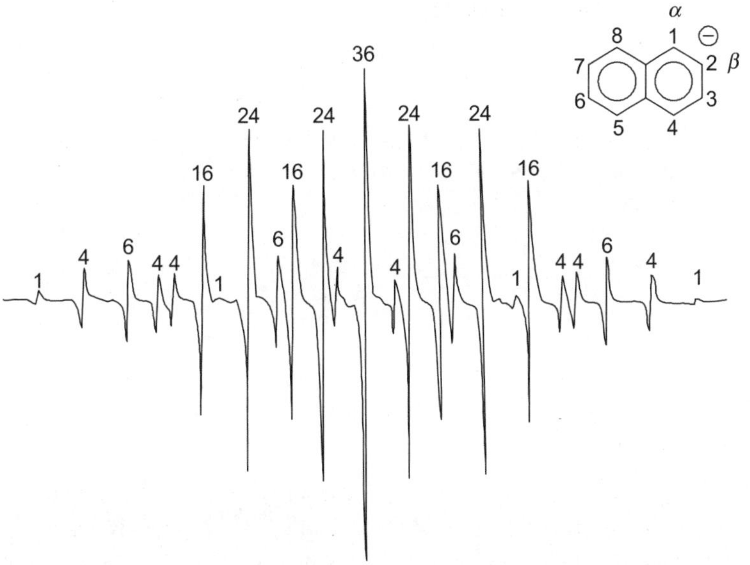

Fig. 3.30 The ESR spectrum of naphthalene anion radical in solution. The counter ion is K^+. The number above each line corresponds approximately to the relative intensity.

ORIGIN OF HYPERFINE INTERACTION

(a) Dipolar Interaction

Free radicals are readily produced and trapped in solids by high energy X-rays or other radiations. The hyperfine lines in the spectra of free radicals as solids are generally very broad compared to those obtained in solution. The principle source of this line broadening lies in the dipolar interactions between the magnetic moments of the unpaired electron and the neighboring magnetic nuclei. This can be large when the separation between the dipoles is of the order of bond lengths. A characteristic feature of this dipole-dipole interaction is that its magnitude depends on the orientation of the radicals with respect to the applied magnetic field, that is, it is not the same in all directions. Interaction between the magnetic moments of the electron and the nucleus is entirely analogous to the dipolar coupling between two bar magnets. The classical expression for the energy of interaction between two dipoles of the electron and the nucleus, μ_e and μ_N, respectively is given by

$$B_{\text{local}} = \mu_e \mu_N \frac{(3\cos^2\theta - 1)}{r^3} \tag{3.16}$$

where θ is the angle between the direction of the magnetic field and the line joining the electron and the nucleus, and r is the distance between the electron and nuclear dipoles as depicted in Fig. 3.31. The energy of interaction, B_{local} parallel to the external magnetic field is given by

$$B_{\text{local}} = \mu_N(z) \frac{(3\cos^2\theta - 1)}{r^3} \tag{3.17}$$

where $\mu_N(z)$ is the component of the nuclear magnetic moment along the direction of the external magnetic field.

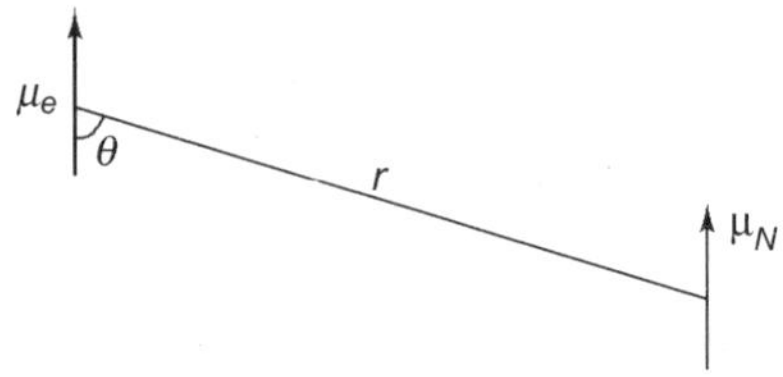

Fig. 3.31 Dipolar interaction between the magnetic moments of the electron (μ_e) and the nucleus (μ_N).

The dipolar interactions will modify the energy levels of the unpaired electron of the free radicals trapped in solids. Evidently, B_{local} depends markedly on the value of θ. The dipolar interaction thus depends on the orientation of the dipoles in the applied magnetic field. The resultant local field is not the same in all direction, that is, it is not isotropic but anisotropic. The g-value and hyperfine splitting would be anisotropic, since they depend on the direction of the orientation of the sample molecules in the applied magnetic field. The situation in solution is simplified since the radicals tumble randomly and rapidly. When all the orientations occur as in solutions (of solvents of low viscosity), the average value of $<\cos^2\theta> = 1/3$. All dipolar interactions are thereby averaged to zero. Most ESR spectra discussed earlier refer to liquid samples in which the radicals are free to orient rapidly. Rapid tumbling over different orientations averages out any anisotropy in the g-factor and in the hyperfine splitting as discussed above. For this reason, narrow hyperfine lines are observed in solution in contrast to broad bands generally observed for solids.

There are two other major factors which control the magnitude of hyperfine coupling constant, a. First factor is the Fermi contact term. Secondly, there are spin-polarization (exchange) effects. They are discussed below.

hydrogen atom in its ground state in which the electron occupies 1*s* orbital, as noted above, is 50.68 mT. The electron density in the 1*s* orbital of hydrogen atom is obviously unity. Thus, if the measured hyperfine splitting from a particular proton in a radical is a_H, the fractional 1*s*-character of the unpaired electron at the nucleus is given by $a_H/50.7$ mT; it is a measure of the contribution of the 1*s* orbital to the wave function of the unpaired electron. For example, for the vinyl radical, the observed hyperfine coupling constant for the *cis*-hydrogen atom is 6.8 mT. It means, broadly speaking, that the unpaired electron density at the hydrogen atom under consideration is about 68/507.

The theory of isotropic hyperfine splitting for other nuclei such as ^{13}C, ^{14}N, ... is more complex than for the protons. The ^{13}C hyperfine splitting is not directly proportional to the *pi* electron spin density at the carbon atom. For example, for the methyl radical it was noted earlier that a_C = 4.1 mT and for the benzene negative ion, a_C = 0.28 mT. If it were the case, for benzene, the proton hyperfine splitting should have been nearly 0.7 mT.

The magnetic field produced by dipole-dipole interaction of *p* electrons and Fermi contact interaction of *s* electrons can be considerably large. The 1*s* electron of the hydrogen atom, as noted before, experiences a nuclear field of 50.7 mT as a result of contact interaction while a 2*s* electron in nitrogen experiences 55.2 mT and a 2*s* electron in fluorine atom experiences a very large field of nearly 1720 mT. An electron in the 2*p* orbital of nitrogen experiences an average field of nearly 3.4 mT from ^{14}N nucleus. The magnitude of the contact interaction can be interpreted in terms of the *s*-character of the unpaired electron. The dipole-dipole interaction can be interpreted in terms of the *p*-character. This gives valuable information on the nature of the molecular orbital and also of the atomic orbital occupied by the unpaired electron. The anisotropic coupling constant for ^{14}N and ^{19}F arising from the 2*p*-character of the orbital in which the unpaired electron resides is respectively 3.4 and 108.4 mT. Similarly, for ^{13}C, the isotropic and anisotropic coupling constant arising from the electron in the 2*s* and 2*p* orbitals, is 113.0 and 6.6 mT respectively. It shows the nature of the hybridization of the orbital occupied by the unpaired electron.

(c) Spin Polarization

The methyl radical discussed previously has a planar structure and the unpaired electron resides in the $2p_Z$ orbital of carbon. Since the hydrogen atoms are located in the nodal plane of the $2p_Z$ orbital, no isotropic proton hyperfine splitting should be observed, at least, to a first approximation. The same situation exists for hydrogen atoms in aromatic *pi* radicals. For example, in benzene anion radical, the unpaired electron occupies a *pi* molecular orbital composed of a linear combination of $2p_Z$ atomic orbitals of the carbon atoms. The molecular orbital formed by the $2p_Z$ atomic orbitals of carbon lying in the plane of the molecule is a nodal plane. We expect, therefore, that the coupling constant between the unpaired electron and the ring protons to be zero. The spectra of benzene and other conjugated molecules clearly demonstrate that proton hyperfine splittings do occur in *pi* radicals. It is therefore necessary to consider how the unpaired spin finds its way to the 1*s* orbital of these hydrogen atoms. The mechanism by which the isotropic hyperfine interaction with the ring protons in aromatic *pi* radicals and other systems such as methyl radical occurs has been attributed by McConnell and others to the interaction between π and σ electrons by an indirect mechanism known as spin polarization. McConnel has shown that the unpaired π-electron can interact through an atomic exchange mechanism with a σ-bond to produce an appreciable resultant electron spin polarization in an s-atomic orbital at the aromatic protons. In this manner, the amplitude of the wave function of the unpaired electron has a finite value at the nucleus of the hydrogen atom. Consequently an electron spin-nuclear spin interaction (i.e. Fermi contact interaction) can occur. It depends on the unpaired electron spin density on the particular carbon and is in general manifested in the magnitude of the proton hyperfine splitting and also proton NMR chemical shift (see chapter 18). A simplified view of this effect is described below.

Consider a C—H fragment of a conjugated system such as benzene. The $2p_x$, $2p_y$ and 2*s* orbitals of the carbon atom form the trigonal sp^2 hybrid orbitals. The hydrogen atom bonds to one of the sp^2 hybrid orbitals. The two possible configurations of the electron spin in the C—H sigma bonding orbital is shown in Fig. 3.33. If there were no electron in the $2p_z$ orbital, both the electronic configurations (a) and (b) of Fig. 3.33 would be equally probable. The presence of an unpaired electron in the π-orbital (that is, the $2p_z$ orbital of carbon perpendicular to the molecular plane) say in the α spin (up spin) polarizes the electron pair of the C—H sigma bond in such a manner that the electron having parallel spin (that is, the α spin) is preferentially located in its proximity resulting in a small unpaired spin density at the proton. This effect is called spin polarization. At the site of the hydrogen nucleus, a spin population of the opposite sign will be induced and the induced spin population will be proportinal to the spin density in the atomic orbital of carbon say ϕ_μ . According to Hund's rule, when electrons occupy singly two different orbitals on the same atom (here carbon), the more stable arrangement of the ground state is that in which the electron spins are parallel. In other words, the arrangement (a) of Fig. 3.33 in which the two electrons on the carbon atom have parallel spins has a slightly greater probability than the arrangement (b) in which the spins are antiparallel resulting in a small net negative spin density at the hydrogen 1*s* orbital (that is, excess of β spin over α spin). There will be a positive spin density at the carbon nucleus. The negative spin density at the proton will produce a negative proton hyperfine splitting. This effect known as the spin polarization leads to a hyperfine splitting of the ESR signal but the splittings are significantly smaller than those due to direct contact of the unpaired electron with a proton.

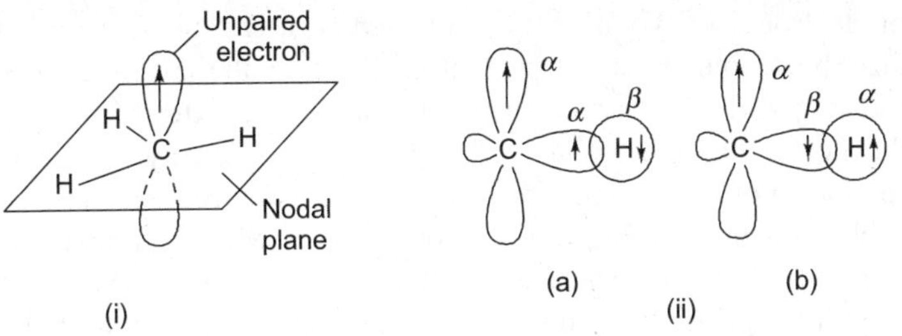

Fig. 3.33 (i) C–H fragment showing the nodal plane of the $2p_z$ atomic orbital of carbon containing the unpaired electron. (ii) Arrangement of electron spins (a) and (b) in the C–H sigma bonding orbital.

The stabilizing feature of the arrangement (a) over that of (b) of Fig. 3.33 is the electron exchange interaction. The two electrons which are mainly localized on carbon can be interchanged in arrangement (a). Such an exchange interaction not only gives rise to spin polarization but is also consistent with the Hund's rule.

The concept of unpaired spin density will now be discussed. It was assumed that when an electron is added on to a conjugated molecule to form a negative ion, the other electrons in the molecule would be unaffected. The density of the electron spin or spin density would then be just that of the added unpaired electron. However, the other electrons in the molecule are affected slightly by the added electron. Consequently, in some regions of the molecule, the 'paired' electron becomes slightly unpaired. Thus, the spin density will not be equal to the unpaired electron density.

Spin density may be defined by the relation,

$$\rho_i = \rho_{i\,(\alpha)} - \rho_{i\,(\beta)} \tag{3.22}$$

where ρ_i is the spin density in the region of i of the molecule and $\rho_{i\,(\alpha)}$ and $\rho_{i\,(\beta)}$ are the total probability densities of the electrons with α and β spins, respectively in the region of i. By adding the densities of all the electrons with α and β spins respectively $\rho_{i\,(\alpha)}$ and $\rho_{i\,(\beta)}$ are obtained. Again, consider the $>$C–H fragment of a conjugated molecule. There will be a net negative electron spin density (that is, excess of β spin over α spin) at the proton. There will be a positive spin density at the carbon nucleus. The negative spin density at the proton will produce a negative proton hyperfine splitting as mentioned before.

In a conjugated molecule, the hyperfine splitting from the delocalization of the unpaired spin density $\rho(r_N)$ on to a hydrogen atom of the molecule is given by

$$a = \frac{8\pi}{3} g\beta_e g_N \beta_N \rho(r_N) \tag{3.23}$$

where $\rho(r_N)$ can roughly be thought of as the difference in the average number of electrons at the nucleus that have been characterized by α and β spins. When there is an excess of α spin density (that is, the electron spin moment parallel to the field) at the nucleus, the nucleus is said to experience a positive spin density. An excess of electron density with β spin is called negative spin density. Positive spin density may be denoted by an arrow that is aligned with the applied magnetic field and negative spin density by an arrow opposed to the external magnetic field. It should be noted that the amount of unpaired spin density on an atom in the molecule does not correspond directly to the unpaired electron density. An unpaired electron in the orbital of an atom in a molecule can polarize the paired spins (in an orthogonal sigma bond) so that one of the electrons is more often in the vicinity of the other. This puts unpaired spin density at the nucleus of the atom even though there is no unpaired electron density delocalized on it.

The calculations for $>$C — H fragment have shown that the proton hyperfine splitting constant a_H is proportional to the unpaired *pi* electron density on the carbon atom, ρ_C to which hydrogen atom in question is bonded. That is,

$$a_H = Q\,\rho_C \tag{3.24}$$

where Q is the value of a_H when there is one full electron on the carbon orbital. Thus, the quantity Q is proportional to the probability of the electron being located at the site of the nucleus. Positive spin density on carbon leads to negative spin density at the hydrogen. Hence Q is a negative quantity. Therefore, the hyperfine parameter for the proton of a –CH fragment 'a' should be accompanied by a negative sign. For aromatic fragments positive carbon spin densities (α i.e. with a magnetic moment vector parallel to B_o) lead to high field proton chemical shifts and negative carbon spin densities (β) lead to low field proton chemical shifts. The sign of the spin density can therefore be obtained directly from the NMR spectrum. The observation of contact shift is of course not restricted to aromatic protons (see chapter 18). The negative sign of Q(CH) denotes that the unpaired electron spin induced in the 1s orbital of the proton is opposite in sign to that induced in the carbon sp^2 orbital. However, the above discussion is valid irrespective of the sign of Q and a. The value of Q varies from one radical to another say from –2 to –3 mT but is often taken as –2.3 mT from the spectrum of benzene negative ion. When two or more lines occur, we call the distance between the centres of the first and the last line as the extent of the spectrum and the extent of the spectrum of benzene anion radical is close to 2.25 mT. A similar value for Q is obtained using a_H of the methyl radical ($\rho_C = 1$). It is usually somewhat larger for radical cations than for the corresponding radical anions. The proton hyperfine splitting in benzene anion radical is 0.375 mT and the electron density by the symmetry of the molecule is 1/6th at each carbon atom.

Employing the value of 50.7 mT for the hyperfine splitting caused by a single electron in the 1s orbital of hydrogen atom, according to the relation

$$a = Q\rho \tag{3.25}$$

the values of ρ_{1s} for each proton in $^{\bullet}CH_3$ and $C_6H_6^-$ are 0.045 (= 23/507) and 0.0075 (= 3.75/507), respectively. In other words, the unpaired electron spends for example, in the methyl radical nearly 5% of its time in the 1s orbital of each hydrogen atom and the rest 85% near the carbon atom.

The alkyl radicals are distinct from the conjugated radicals in that the unpaired electron is localized primarily on a particular carbon atom. The data for $^\bullet CH_3$ and $^\bullet CH_2CH_3$ are given below.

$\rho(^\bullet CH_3) = 1.0$	a_H 2.30 mT	Q 2.30 mT
$\rho(^\bullet CH_2CH_3) = 0.910$	2.24 mT	2.435 mT

For $^\bullet CH_3$ radical, the doublet splitting due to ^{13}C is 4.1 mT which corresponds to a contribution from the carbon 2*s* orbital of about 4% ($C_{2s} = 0.04$). This again is a small factor which is, however, of great significance in the interpretation of the hyperfine structure of the ESR spectra of many organic free radicals.

Analysis of the ESR spectra with the aid of McConnell's relation allows us to measure directly the *pi* electron spin densities. The above relationship permits the unpaired spin density distribution on the carbon skeleton of the free radical to be obtained from the experimental proton coupling constants. Since the charge distribution can also be calculated from the molecular orbital theory, the molecular orbital calculations could be utilized to assign the measured hyperfine splitting constants to particular protons where this is in doubt. The spin distributions in these radicals can be explained successfully by simple Huckel molecular orbital theory. The agreement between the theory and the experiment may confirm the identity of a particular radical. For example, from the calculated spin densities it was found that the larger splitting for naphthalene negative ion arises from the α protons. Thus, it provides a test for the predictions of the molecular orbital theory.

The general problem is one of relating observed coupling constants to spin densities at the nucleus. Another problem is related to the observation of negative spin densities. Both these problems require some consideration of electron correlation. Electron correlation in molecules arises from coulombic repulsion and spin exchange effects.

SIGMA RADICALS

If the unpaired electron in a radical resides in a π-molecular orbital, the radical is called a π-type radical. Most organic free radicals discussed earlier belong to this category. For π-type radicals considered earlier, the unpaired electron is located chiefly in a carbon $2p_z$ (or π) orbital. Small isotropic hyperfine couplings are observed and these arise primarily from the indirect spin polarization mechanism. However, there are a limited number of free radicals in which the unpaired electron occupies a σ orbital. Such radicals are known as σ radicals.

There are a number of radicals known which exhibit proton hyperfine splittings of the order of 6 to 18 mT. These splittings are far too large and evidently the orbital of the unpaired electron has considerable *s* character. In these radicals, the unpaired electron is primarily located in a σ orbital which would form a σ bond with another atom such as hydrogen. Most σ orbitals have considerable *s* orbital component. Hence, large isotropic hyperfine couplings are characteristic of σ

orbitals and the sign of the splitting constant should be positive because of the direct interaction of the proton with the *s*-component of the σ orbital. In ethynyl radical, $^{\bullet}C \equiv CH$, the unpaired electron occupies the orbital otherwise directed to a hydrogen atom in acetylene molecule. Likewise, in the formyl radical, the unpaired electron is primarily in a σ orbital which would bond a hydrogen atom in formaldehyde molecule.

When HI is photolysed in solid CO at 4 K, HCO radical is produced. It shows an ESR spectrum consisting of a doublet separated by 13.6 mT. The HCO radical is known from its gas phase electronic absorption spectrum to have a bent structure. In the corresponding DCO radical, the spectrum consists of a triplet of equally intense lines due to the hyperfine interaction of the electron spin with deuterium nuclear spin. The hyperfine splitting for the DCO radical was found to be 2.0 mT. The ratio $a_D/a_H = 20/136 = 0.15$ is in good agreement with the ratio of the nuclear *g* values, $g_D/g_H = 0.153$. In the HCO radical, the spin density in the 1*s* orbital of hydrogen is approximately 0.27, since the proton hyperfine coupling constant is 13.6 mT. This is the largest known proton coupling constant except for that of the hydrogen atom.

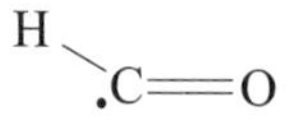

Another example of a σ radical is the vinyl radical ($^{\bullet}CH{=}CH_2$), for which in a rigid system the hyperfine coupling constants are 1.57 mT for H_1, 6.85 for H_2 (*trans*) and 3.42 for H_3(*cis*) hydrogens. In the vinyl radical, the unpaired electron occupies a σ orbital. The hyperfine splitting constant 6.85 mT at the *cis* position indicates that the unpaired spin spends a fraction of roughly 68.5/506.8 ≈ 0.135 of its time in the ls orbital of *cis* hydrogen. This fraction represents the unpaired spin density in the 1*s* orbital.

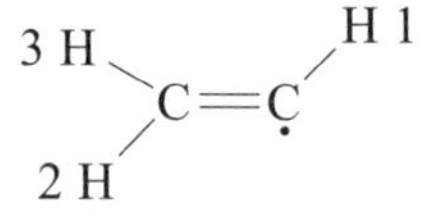

Even the largest value of the hyperfine splitting constant of the vinyl radical is considerably less than that for the HCO radical. The difference is attributable largely as due to the large variation in the splitting constant with the bond angle. Hyperfine splittings in the sigma type radicals may exhibit large anisotropy which is discussed later.

Simple alkyl radicals have been studied in the liquid phase. They are formed by the continuous electron irradiation of liquid hydrocarbons at low temperatures. The hyperfine splitting constants (in mT) in ethyl, isopropyl and isobutyl radicals are cited below. A notable feature of the results is that the methyl or β proton coupling is actually larger than that for α protons. It is attributed as due to hyper conjugation.

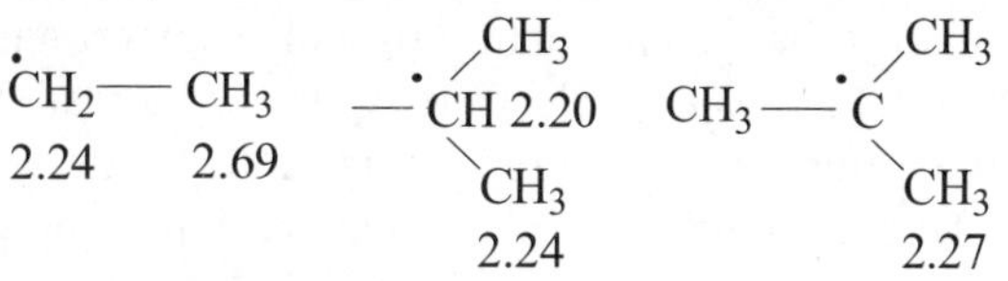

Many *pi* electron radicals containing methylene groups also have larger hyperfine splittings due to hyper conjugation. For instance, $H_2CN^{\bullet}$ has proton hyperfine splitting a_H = 8.7 mT, and the conjugated cyclohexadienyl radical shows large methylene proton splitting even though the π-electron is spread out over a larger ring.

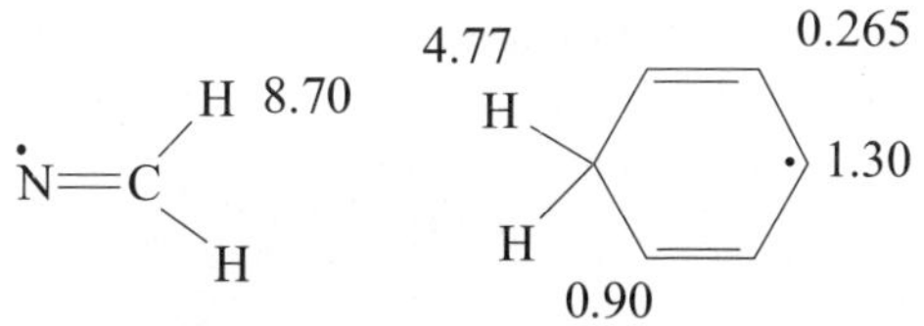

(Hyperfine splitting constant in mT)

It is of interest to compare the ^{13}C coupling constant in $^{\bullet}CH_3$ and $^{\bullet}CF_3$ radicals. For the planar CH_3 radical, a_C = 4.1 mT, whereas for the $^{\bullet}CF_3$ radical, a_C = 27.16 mT. The large increase in the ^{13}C hyperfine splitting for $^{\bullet}CF_3$ radical can be interpreted in terms of a large distortion towards pyramidal structure in $^{\bullet}CF_3$; this distortion leads to some *s* character in the orbital of the unpaired electron.

If the free radical is a sigma radical as, for example, the vinyl radical, the protons in the molecule make a contribution to the sigma molecular orbital containing the unpaired electron. Consequently, the unpaired electron is delocalized directly on to the proton and a_H is proportional to the square of the molecular orbital wave function ψ^2 according to the equation (3.26) and since simple Huckel molecular orbital calculations do not take into account sigma bonding, extended Huckel molecular orbital methods are utilized. Methods have been reported for evaluating ψ^2 at the hydrogen nucleus from the molecular orbital wave function and a_H is calculated using

$$a_H\,(\text{gauss}) = 1887\ \psi_H^{\ 2} \tag{3.26}$$

Generally, for other protons in the molecule (with some exceptions), spin polarization effects make a relatively insignificant contribution when direct delocalization is appreciable. The use of molecular orbital methods for the interpretation of the hyperfine splittings of atoms other than hydrogen is considerably complex.

ASSIGNMENT OF SPECTRA USING HUCKEL MOs

The Huckel molecular orbital (HMO) theory gives valuable assistance to the interpretation of the ESR spectra of π-electron radicals. The HMO theory is also useful for obtaining the electron densities at each atom and thus obtain the nature of the MO occupied by the unpaired electron. For an outline of the Huckel molecular orbital theory for the calculation of molecular orbital energies and the unpaired electron distribution in *pi* electron systems, the reader may consult the books listed in the references. References 11 and 12 describe in detail the procedure for the Huckel molecular orbital calculations. For illustration, 1, 3–butadiene anion is considered first.

1, 3–Butadiene anion radical is obtained by the electrolytic reduction of butadiene in liquid ammonia at 195 K. It contains two sets of protons, one set consists of two CH protons and the other set contains four CH_2 protons. It is rather difficult to predict whether the ESR spectrum of butadiene anion should consists of three quintets or five triplets, which depends on the hyperfine splitting constants. However, it is easily shown from HMOs of butadiene which is discussed below that the latter situation holds. The spectrum of 1, 3–butadiene anion is shown in Fig. 3.34. The hyperfine splitting constants are $a(CH_2) = a_1 = a_4 = -0.762$ mT and $a(CH) = a_2 = a_3 = -0.279$ mT. The analysis of the ESR spectrum of butadiene anion is given in Fig. 3.35.

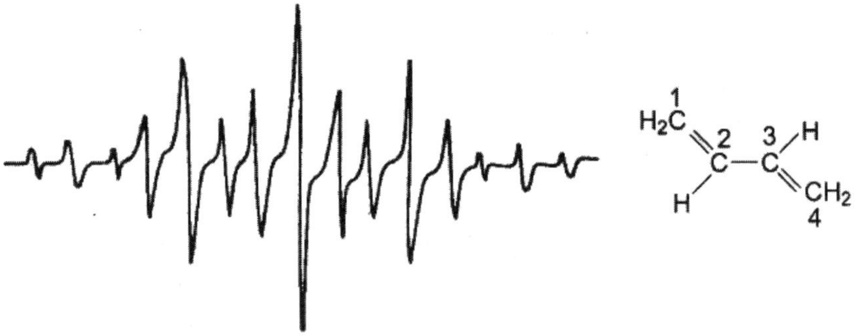

Fig. 3.34 ESR spectrum of 1, 3-butadiene anion radical.

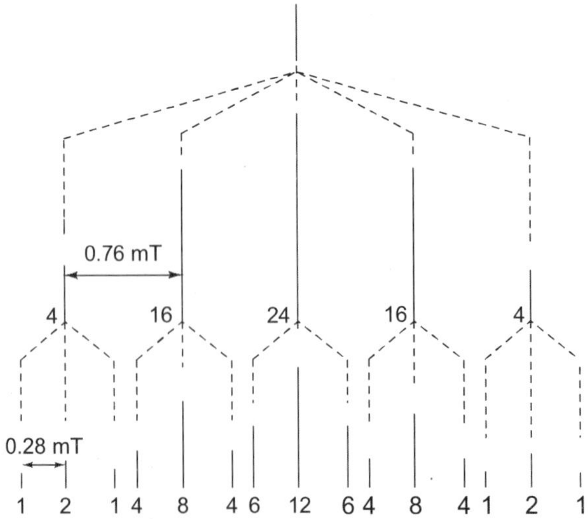

Fig. 3.35 Analysis of the hyperfine structure in the ESR spectrum of 1, 3-butadiene anion radical.

1, 3–Butadiene anion has five *pi* electrons. The lowest unoccupied Huckel MO of butadiene is $\psi_3 = 0.602\ \phi_1 - 0.372\ \phi_2 - 0.372\ \phi_3 + 0.602\ \phi_4$, where ϕ denotes the $2p_z$ carbon atomic orbitals. The unpaired electron resides in the MO ψ_3. From the coefficients of the atomic orbitals in the MO ψ_3, we obtain the electron density at the atoms 1 and 4, that is $\rho_1 = \rho_4 = (0.61)^2 = 0.372$. Similarly, the electron density at the atoms 2 and 3 of butadiene anion is, $\rho_2 = \rho_3 = (0.37)^2 = 0.137$. It is observed that the ratio of the hyperfine splittings $a_1/a_2 = 2.73$ agrees well with the ratio of the *pi* electron densities $\rho_1/\rho_2 = 2.72$. This correspondence is apparently consistent with the linear relationship between the unpaired *pi* electron density and the proton hyperfine splitting as given by the McConnell relation for conjugated organic free radicals.

For naphthalene negative ion, the HMO theory places the unpaired electron in the lowest unoccupied MO of the neutral naphthalene molecule. This MO is

$$\psi_6 = 0.425\ (\phi_1 + \phi_4 - \phi_5 - \phi_8) + 0.263\ (-\ \phi_2 - \phi_3 + \phi_6 + \phi_7) \qquad (3.27)$$

where ϕ *s* are the carbon $2p_\pi$ atomic orbitals. The numbering of atoms of naphthalene is shown on p. 49. Since the unpaired electron occupies the MO ψ_6, the unpaired electron density could be calculated using the coefficients of the atomic orbitals forming the MO ψ_6. We find that for α carbon atoms, $\rho_\alpha = (0.425)^2 = 0.181$ and for β carbon atoms $\rho_\beta = (0.263)^2 = 0.069$. The predicted coupling constants are – 0.42 and 0.16 mT in reasonable agreement with the observed values – 0.495 and – 0.187 mT. The calculated ratio of the α to β coupling constants ($a_\alpha / a_\beta = 2.72$) is in very good agreement with the observed value of 2.70 (that is, 4.95/1.87), confirming the predictions of the HMO theory.

ALTERNANT HYDROCARBONS

The individual carbon atoms of a *pi* electron system can be starred in such a way that two starred or two unstarred atoms are not the nearest neighbours. Such a system is known as an alternant hydrocarbon. They are further divided into odd or even alternant hydrocarbons when they contain odd or even number of carbon atoms, respectively. In the odd alternant hydrocarbon, the number of starred atoms is taken to be larger than the number of unstarred atoms. Some examples of alternant hydrocarbons are given in Fig. 3.36.

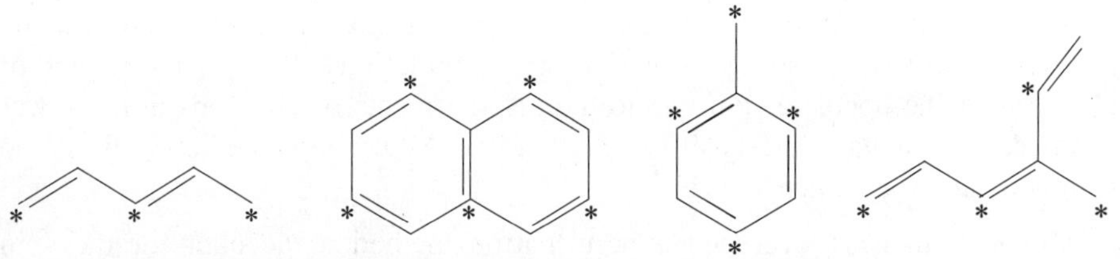

Fig. 3.36 Some examples of alternant hydrocarbons.

One of the serious limitations of the HMO theory in its application to the interpretation of the ESR spectra is that it predicts zero unpaired electron density at positions where small hyperfine splittings are observed, for example, for the central atom in allyl radical and the meta positions in benzyl radical which are odd alternant hydrocarbons. This is a typical case of the failure of the HMO method. Allyl radical shows hyperfine splitting of 0.407 mT for the proton on the central carbon atom, and 1.394 and 1.484 mT for protons on the terminal carbon atoms. The two hydrogen atoms of the terminal carbons are not equivalent. It is not known which of the CH_2 protons has the larger hyperfine splitting. The HMOs of allyl radical formed from the carbon $2p_z$ orbitals ϕ_1, ϕ_2 and ϕ_3, one each from each of the three carbon atoms are given below.

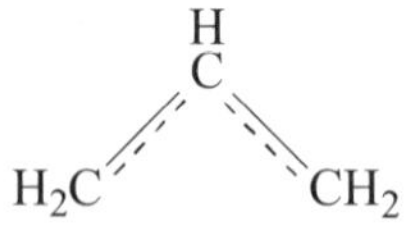

$\psi_1 = 0.5\ \phi_1 + 0.707\ \phi_2 + 0.5\ \phi_3$ (bonding m.o)

$\psi_2 = 0.707\ \phi_1 - 0.707\ \phi_3$ (non-bonding m.o)

$\psi_3 = 0.5\ \phi_1 - 0.707\ \phi_2 + 0.5\ \phi_3$ (anti-bonding m.o)

The unpaired electron in the allyl radical resides in ψ_2 and hence the *pi* electron density should be 1/2 on the end atoms and zero at the central atom, which is contradictory to the observed experimental proton hyperfine splitting of 0.407 mT for the central atom. The MO ψ_2 has a node at the central atom 2.

In the case of odd alternant hydrocarbons, as discussed above, the hyperfine splittings at the unstarred atoms could not be explained by the HMO method. It was hence necessary to introduce the concept of spin densities which unlike the unpaired electron densities have both magnitude and sign. In general, however, when the unpaired electron density is calculated by the HMO method to be zero or very small, a negative spin density must generally be assumed in order to interpret the ESR spectra successfully. To obtain the negative spin densities by the MO method, it is necessary to include electron repulsions which are neglected in the HMO method. The effect of spin correlation in allyl radical is to pile up positive spin (α) on the —CH_2 carbons and leave negative (β) spin at the —CH carbon. These negative spin populations are dealt with by more advanced MO methods.

It should be noted that in the one-electron HMO model, only the unpaired electron in the singly occupied MO, ψ_k will contribute to the spin population ρ_μ at the $2p_z$ orbital of atom μ. The resulting spin population must, therefore, be necessarily always positive. Negative spin populations, that is, an excess of electrons of the opposite spin due to the displacements of electrons in the lower levels can occur particularly at centres where HMO theory predicts little or no spin population as stated above.

McLauchlan has developed a perturbation method to account for the spin densities, where the HMOs give zero electron density. The perturbation calculations have predicted that for allyl radical the spin density on the central carbon atom 2 is negative and the spin densities on the other two carbon atoms 1 and 3 are positive. Similar calculations by McLauchlan's method for the benzyl radical discussed below predict negative spin densities at the meta positions. Since the algebraic sum of the spin densities must be equal to one, the presence of negative spin density at one position means that there must be an extra compensating spin density elsewhere in the molecule. Hence, the overall spread of the hyperfine splitting pattern tends to be considerably larger than the magnitude of Q (see equation (3.24)). Nuclear magnetic resonance experiments have shown that negative *pi* electron spin densities also occur in even alternant hydrocarbons such as pyrene negative ion.

The unpaired electron of the benzyl radical resides in a non-bonding MO. No hyperfine splitting is expected for the protons in 3 and 5 positions since the coefficients in the HMO are zero as noted from the non-bonding MO of the benzyl radical given below.

$$\psi_0 = \psi_4 = -0.378\ \phi_2 + 0.378\ \phi_4 - 0.378\ \phi_6 + 0.756\ \phi_7.$$

Benzyl radical shows appreciable hyperfine splitting in the ESR spectrum from the unstarred meta positions where zero electron density is predicted by the HMO theory. The experimental values of the proton hyperfine splittings for benzyl radical are compared below with the theoretically predicted values.

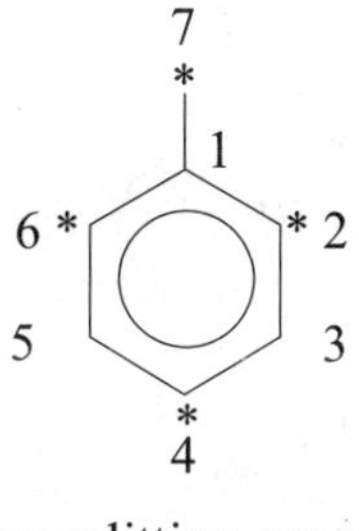

Hyperfine splitting constant, mT

Atom position	**Experiment**	**Theory**
2,6	– 0.49	– 0.40
4	– 0.61	– 0.40
3,5	+ 0.15	0.0
7 (CH_2)	– 1.59	– 1.59

In spite of these differences, one can regard the calculated values as being in good agreement with the experiment considering the approximations made in the HMO method. The MO calculations which include configuration interactions have shown a small negative spin density at the meta carbon atoms of the benzyl radical. It means that a_H is positive for the protons at the meta positions in benzyl radical.

Examination of the ESR spectra of the radical ions provides a test for the theoretical predictions on alternant hydrocarbons. The highest occupied and the lowest unoccupied MOs are paired in alternant hydrocarbons and these are the orbitals occupied by the unpaired electron in the two ions, the cation and the anion. It follows that the spin density distributions should be identical in the cation as well as in the anion, $\psi_r^+ \equiv \psi_r^-$. For benzene anion, the proton hyperfine coupling constant is 0.375 mT while for the cation, it is slightly higher, 0.440 mT. Unfortunately, this prediction cannot be checked for naphthalene since the oxidation of naphthalene using $SbCl_5$ yields a dimeric radical $(C_{10}H_8)_2^+$ and not $C_{10}H_8^+$ monomer. However, from simulating X–band ESR spectrum taken in boric acid glass, the proton hyperfine coupling constants in napthalene cation and anion have been estimated and they are listed below. For anthracene, the radical anion and the cation have been prepared; the hyperfine coupling constants are given below.

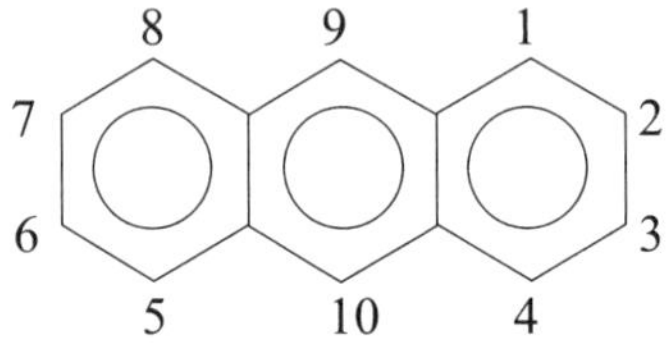

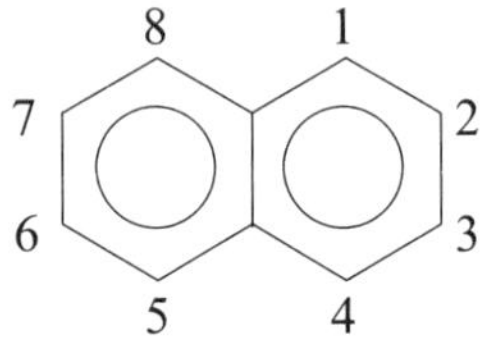

	a_1	a_2	a_9
Cation	0.306	0.140	0.653 mT
Anion	0.274	0.151	0.534 mT

	a_1	a_2
Cation	0.540	0.160 mT
Anion	0.495	0.187 mT

The observed proton hyperfine splittings are very similar but not identical and they tend to be generally higher in the cation as in benzene. The electron spin density distribution and the atomic orbital coefficients for the highest occupied HMO of anthracene cation are given in Fig. 3.37. The HMO theory predicts from electron densities $\rho_2 : \rho_1 : \rho_9$ as 0.048 : 0.096 : 0.194, that is, the ratio of the hyperfine splittings $a_2 : a_1 : a_9$ to be in the ratio of 1 : 2 : 4, which is in satisfactory agreement with the observed ratio.

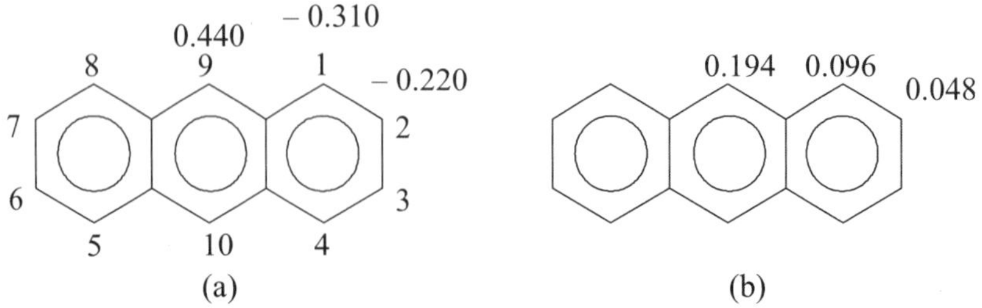

Fig. 3.37 • (a) Atomic orbital coefficients for the highest occupied HMO and (b) the calculated electron densities for anthracene anion.

As an example of a heterocyclic compound, the ESR spectrum of pyrazine which was discussed earlier is considered. For pyrazine negative ion, the HMO theory places the unpaired electron in the lowest unoccupied MO (ψ_4) of the neutral molecule,

$$\psi_4 = 0.542\,(\phi_1 + \phi_4) - 0.321\,(\phi_2 + \phi_3 + \phi_5 + \phi_6)$$

The ratio of the electron density on the nitrogen to carbon atoms is $(0.542)^2 / (0.321)^2 = 2.80$. It is in fair agreement with the ratios of the observed coupling constants $a_N = 0.722$ mT and $a_H = 0.266$ mT giving $a_N / a_H = 2.71$.

The utilization of the results from molecular orbital calculations for the interpretation of the hyperfine splittings from atoms other than hydrogen such as ^{13}C, ^{14}N, ^{17}O, etc. is more complex.

HYPERFINE SPLITTING CONSTANTS

The magnitude of hyperfine splitting constant varies with the nature of the free radical. It is a measure of the strength of the interaction between the nuclear and electron spins. With transition metal ions, where the electron interacts exclusively with a single atom, hyperfine splitting of the order of 10 mT or more may be observed. In organic free radicals, where the electron often interacts with several nuclei (say as in aromatic compounds) and in solvents of low viscosity, the hyperfine splitting between the individual lines may be low with the extent of the spectrum between the terminal lines typically of about 2–3 mT.The hyperfine splitting constants (in mT) for some free radicals are listed in Fig. 3.38. The splitting constants given are chiefly for electron-proton interactions.

The ESR spectra of organic free radicals in solution are easy to analyze provided the lines do not overlap. Frequently, however, the lines are not completely resolved rendering the analysis difficult and subject to error. The hyperfine structure can become exceedingly complex when there are numerous interacting nuclei, but in principle, the spectra of all paramagnetic species in solution can be interpreted by the application of some simple methods.

The assignment of the hyperfine coupling constants for free radicals could be made by comparison with related systems aided by selective deuteration, and by computation of electron densities using molecular orbital methods as discussed above. Difficulties arise when the hyperfine lines do not all have the same width, when they overlap to a significant extent or when the radical is involved in exchange reactions where redistribution of spins may occur. When the resolution of the spectrum is poor or the lines are numerous, it is advantageous to carry out a computer simulation of the spectrum based on assumed hyperfine splittings and line widths. Methods are available for specifically decoupling hyperfine interactions to simplify the spectra and assist the assignment. Under this, the use of double resonance techniques discussed later may be mentioned. The measurement of spectra at different temperatures and in solvents of different pH and dielectric media may prove useful. Some of these aspects are discussed later.

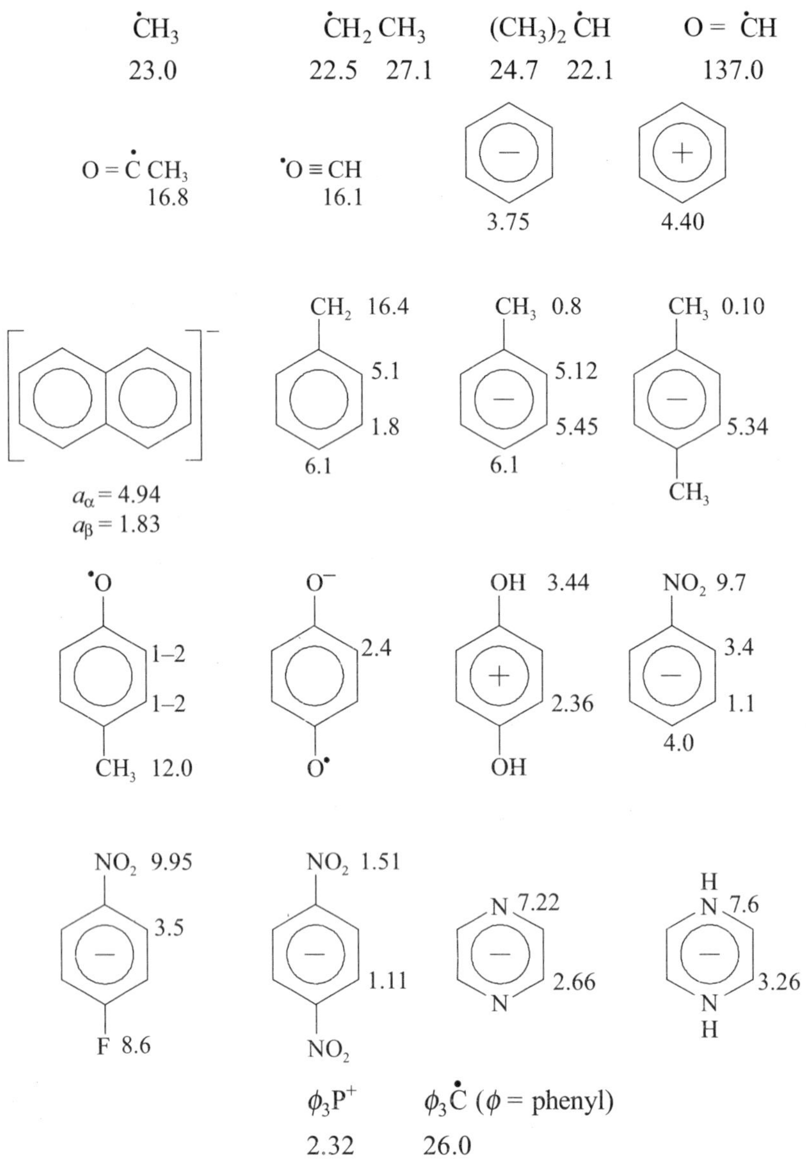

Fig. 3.38 Hyperfine splitting constants (in mT × 10) for some simple radicals.

The following are some important rules which can help in the interpretation of ESR spectra.

(i) The positions of the lines in the spectrum are expected to be symmetric about a centre point.

(ii) A spectrum having no intense central line indicates the presence of an odd number of equivalent nuclei of half integral spins. However, the observation of a centre line does not exclude the presence of an odd number of nuclei.

(iii) The stick-plot reconstruction should match the experimental line positions especially in the wings of the spectrum. Electron densities from Huckel molecular orbital calculations could be employed for assigning the spectrum.

(iv) If the overlap of the lines is serious, or the lines are numerous, it is desirable to perform a computer simulation of the spectrum, based on assumed line widths and hyperfine splitting constants. Such programs have been described in the literature.

(v) The separation between the two outermost lines is always the smallest hyperfine splitting.

SECOND ORDER SPLITTING

The analysis of the hyperfine splittings which has been presented above is valid only in cases where the hyperfine splitting (a) is very much smaller than the electron Zeeman energy ($g\beta H$). The spectra obtained are referred to as the first order spectra. Additional splitting of the lines may occur when the hyperfine couplings are very large or when small magnetic fields are used. These additional splittings are usually referred to as 'second order' splittings since the energies of these levels must be calculated using second order perturbation theory. This analysis is not presented here.

APPLICATIONS

Identification and Structure Elucidation

The object of an ESR spectroscopic work is often the determination of the identity of the paramagnetic species. This is best accomplished by comparing one's spectrum with a known spectrum. The g-factor, hyperfine structure, line width and intensity are characteristic of a particular species.

One of the advantages of ESR spectroscopy is its extreme sensitivity and hence very small amounts of the paramagnetic material can be detected. For example, under favorable conditions, the signal of a radical can be detected from 10^{-12} g of the material. This extreme sensitivity can be utilized in the study of free radicals. Nearly all useful information in the studies of free radicals are obtained from a study of the hyperfine structure. The ESR spectroscopy is suitable for detecting intermediates in chemical reactions and weak magnetism of the phosphorescent triplet states. There have been several studies reported in which ESR spectroscopy has been employed to identify and provide structural information about the free radicals generated on irradiating the material with high energy radiations.

ESR spectroscopy provides information on the extent of electron delocalization. It gives useful information in some cases on the shape of the radicals under study. An interesting example of this is the determination of the structure of the simplest π radical, the methyl radical. The CH_3 radical exhibits a four-line ESR spectrum with a hyperfine splitting of 2.30 mT. According to McConnell's equation, using $\rho^{\pi} = 1$ and $a_H = 2.30$ mT, for $^{\bullet}CH_3$ radical, a value of 2.30 is obtained for Q which is close to the value obtained for most aromatic radicals. This suggests that the methyl radical has a planar structure. However, the observed splitting is not an unambiguous evidence for the planarity of the $^{\bullet}CH_3$ radical, since the value of the proton hyperfine splitting if the $^{\bullet}CH_3$ radical was not planar is not known. The ESR spectrum of $^{13\bullet}CH_3$ radical is useful for this purpose. A hyperfine splitting of 3.85 mT was observed for ^{13}C in the ESR spectrum of ^{13} $^{\bullet}CH_3$. If the methyl radical were tetrahedral, the unpaired electron would be in one of the sp^3 hybrid orbitals. With 25 percent s character in the hybrid orbital, the interaction of the electron spin with the ^{13}C nucleus according to the theoretical estimates should give a splitting of 30.0 mT. A much smaller hyperfine splitting observed for ^{13}C in $^{\bullet}CH_3$ radical suggests a very small amount of s-character for the unpaired electron signifying that the unpaired electron is in a predominantly p-type orbital. The methyl radical is, therefore, most likely to be planar. On the other hand, the ^{13}C hyperfine splitting constant for the CF_3 radical is 27.1 mT. This large increase in ^{13}C hyperfine splitting is attributed to a nearly pyramidal structure of CF_3. Thus CF_3 is a sigma radical.

Study of Transient Paramagnetic Species

Many chemical reactions are known to involve the formation of paramagnetic species as intermediates and hence their identification is very important in mechanistic studies. The ESR spectroscopy is one of the most direct methods for detecting their presence. However, short lifetimes of these species as well as their low concentration may render their detection difficult. The ESR spectra of several free radical reaction intermediates have been observed by allowing the reaction mixture to flow rapidly through the resonant cavity. As long as the flow is continued, the short-lived radicals observed in the cavity are continually replenished and a steady state concentration is maintained. For example, by mixing an acidified solution of $TiCl_3$ with an acidified solution of CH_3OH and H_2O_2, the radical $^{\bullet}CH_2OH$ is generated. The reactions shown below are known to take place. The hydroxyl radical is apparently responsible for the abstraction of the hydrogen atom because of its greater reactivity.

$$Ti^{3+} + H_2O_2 \rightarrow Ti^{4+} + {}^{\bullet}OH + OH^-$$

$$Ti^{3+} + {}^{\bullet}OH \rightarrow Ti^{4+} + OH^-$$

$$^{\bullet}OH + H_2O_2 \rightarrow H_2O + {}^{\bullet}O_2H$$

$$CH_3OH + {}^{\bullet}OH \rightarrow {}^{\bullet}CH_2OH + H_2O$$

The ESR spectrum of CH_3CHOH radical produced as a transient species in the UV photolysis of a solution of H_2O_2 in ethanol has been obtained. The photolysis generates OH radical which then abstracts hydrogen atom from the ethanol molecule. This is an excellent example of the ESR spectroscopy in the identification of radical intermediates in chemical reactions. ESR spectroscopy is thus useful for the studies on reaction kinetics, which are not extremely fast by simply following the signal intensity of the paramagnetic species, either the reactant or the product, as a function of time. In addition, in the study of very rapid electron transfer and other rate processes ESR spectroscopy has been very useful.

Biochemical Applications

ESR spectroscopy has extensive biological applications. The investigations of plant and animal tissues have shown the presence of small but significant concentrations of free radicals. Important information regarding the chemical, structural and kinetic properties of biomolecules can be obtained from ESR spectroscopy by the use of spin-labelling technique. Compounds which do not contain unpaired electrons can be studied when they are chemically bonded to a stable free radical. These radicals or spin labels as they are called are generally nitroxide radicals of the type $R_1R_2N^{\bullet}O$, where N atoms are bonded to the molecule to be studied. The unpaired electron is largely localized on the nitrogen atom. They are stable and show well-resolved spectra that are sensitive to the molecular environment.

Spin-labelling techniques have been used for the study of biological macromolecules such as proteins and nucleic acids. Usually the label is located at or near an active site in them. Following the changes in the ESR spectrum of the radical in the free and bound state, information regarding structural and kinetic properties of the biomolecule close to the binding site or about the macromolecule as a whole can been obtained.

Analytical Applications

Analytical applications of ESR spectroscopy involve the quantitative determination of certain metal ions such as Mn^{2+}, V^{4+}, etc. and polynuclear hydrocarbons. The area under an absorption signal is proportional to the number of unpaired electron spins in the sample. If suitable precautions regarding instrumental conditions, temperature, etc. are taken, studies can be made quantitative. In simple cases, the use of peak to peak heights of the derivative spectra as a direct measure of concentration could be utilized.

The integrated intensity is usually related to the concentration of the paramagnetic species by comparison with a standard containing a known number of unpaired electrons and having the same line shape (Gaussian or Lorentzian, see Chapter 5) as the unknown. A standard frequently used is 1,1-diphenyl-2-picrylhydrazyl (DPPH) which contains 1.53×10^{21} unpaired spins per gram. Secondary standards include charred dextrose. A dual sample cavity cell is used to minimize difficulties with actual physical interchange of the standard and the sample.

EXPERIMENTAL ASPECTS: ESR

4

THE ESR SPECTROMETER

A simple diagram of an ESR spectrometer is given in Fig. 4.1. The main components of the spectrometer are a magnet, power supply, source of microwave radiation, sample cavity and a means of detecting the absorption at resonance after amplifying the signal.

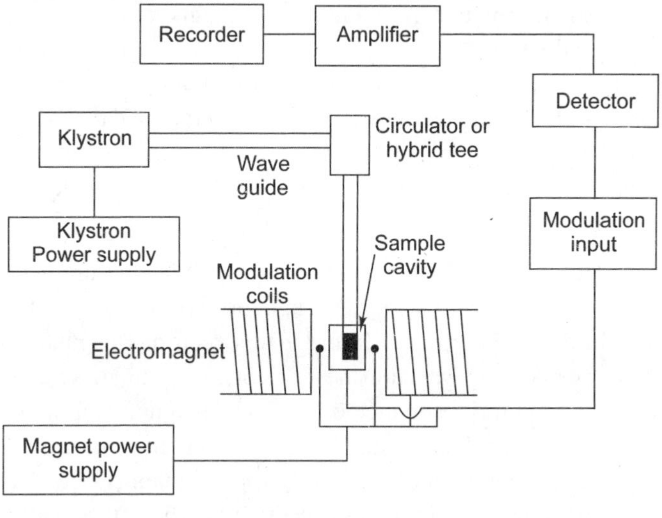

Fig. 4.1 A simple block diagram of an ESR spectrometer.

Source

The source of microwave radiation is a klystron oscillator with a frequency which is customarily operated in the range 9,000 to 10,000 MHz. There are devices to measure the frequency of the microwave radiation produced and to adjust the level of the microwave power incident upon the sample. The microwaves are brought from the klystron to the sample cavity and then to the detector by wave guides. The klystron is connected to a length of the wave guide terminating in a cavity cell containing the sample. The sample is placed in a specially designed cell.

Sample Cavity

The resonant cavity cell which contains the sample is constructed in such a way so as to maximize the applied field along the sample dimension. In most spectrometers, dual cavity cells are generally used. This is done for the simultaneous observation of the sample and the reference material. By the use of a reference compound and by comparing the relative signal heights, the sources of error are compensated. For studying single crystals and solid samples, rotatable cavity cells are generally used.

Magnet

The resonance cavity cell is placed between the poles of an electromagnet. An operating range of 0.20 to 0.60 T is usually used. The magnetic field should be stable and uniform over the sample volume. The spacing between the electron spin energy levels can be varied by changing the magnetic field so that the spin transitions can be detected with a monochromatic radiation source by sweeping the magnetic field through the resonance condition. Thus the ESR spectrometer operates at a fixed microwave frequency. The magnetic field is varied. This method is usually adopted in practice, since it is difficult at high frequencies to sweep the frequency of the microwave radiation with a constant magnetic field at the sample. One can set the field at any position using the field dial and sweep from that position. The field sweep is assumed to be linear and the *g*-values of the other peaks are calculated relative to the standard. A frequency counter should be used to measure the microwave frequency.

The magnetic field homogeneity and stability for ESR measurements are much less acute than for NMR spectroscopy. A magnetic field homogeneity of 1 in 10^5 is adequate for the ESR experiment. In order to sweep the magnetic field over a range, provision is made by varying the current in a pair of sweep coils. The magnetic field is varied until resonance occurs. The magnetic dipole vector of the microwave field, which induces transitions between the energy levels of the unpaired electron is perpendicular to the main applied magnetic field. The sample in the cavity cell is placed in the centre of gravity, where the magnetic field is maximum. The changes in the power level due to absorption of the microwave radiation are detected, amplified and recorded with a pen recorder.

Microwave power should be sufficiently low so that no saturation occurs. The population difference between the spin states is a function of the incident microwave radiation density. A higher microwave power produces spin transitions at a faster rate and hence decreases the lifetime of the spin state. It results in an increased uncertainty in the excited state energy leading to an increased line width and may lead to saturation.

ESR experiments are generally performed at one of the several fixed frequencies referred to as *X*, *Q* and *K*-band from their original use in radar. The *X*-band is the lowest frequency (about 9.4 GHz for a magnetic field 0.34 T; 1 Tesla = 1 kg $s^{-1}C^{-1}$ = 1 J $C^{-1}m^{-2}$ s = 1×10^4 Gauss) and the most commonly used. The frequency of 9 GHz was chosen from the already developed marine radar.

The K (~ 25.0 GHz) and the Q bands (35.0 GHz) operate in magnetic fields of about 0.9 and 1.25 T, respectively. The X-band spectrometer is more convenient and often used for studying samples in liquid phase while the Q band spectrometer is generally used for low temperature work and for studies on single crystals. The sensitivity of the instrument increases roughly as the square of the microwave frequency (ν^2). It results in better spectral resolution and reduction of second order effects. High field-high frequency measurements are sometimes needed to detect subtle spectroscopic details. The D band (140 GHz) spectrometer operates in a magnetic field of 4.90 T. It is rather difficult to attain homogeneity at higher field and for aqueous solutions dielectric absorption by the solvent becomes more serious.

For maximum sensitivity, the spectrum should be obtained at as low a temperature as possible, even when the sample temperature does not govern the line width, since the amplitude of the signal is inversely proportional to the absolute temperature. However, the temperature generally has an effect on the line width. For each system, there is usually an optimum temperature at which the line width is minimum. This temperature is usually below the room temperature. A decrease in the sample temperature therefore often gives an improved spectrum.

So far, we had dealt with continuous wave situations in which B_1 has an amplitude constant with time. A new class of pulsed techniques including one- and two-dimensional Fourier transform ESR spectroscopy is now practised with electron spins. They give large improvement in the signal to noise ratio.

SAMPLING PROCEDURE

ESR spectra can be obtained for gas, solution, powder, single crystal and frozen solution. Commonly the spectra of samples are obtained at room temperature in solution or frozen solutions. The best results are obtained for frozen solution when the solvent freezes to form a glass. Otherwise paramagnetic aggregates are formed which lead to a spectrum with broadened lines. Even in a well-formed glass, one cannot usually detect hyperfine splitting smaller than 0.3 or 0.4 mT. Some of the solvents which form good glasses are toluene, methyl cyclohexane, triethanolamine, etc. Some of the mixtures of solvents which form good glasses are the mixtures of ethanol and methanol, toluene and acetone, diethyl ether and ethanol, etc. In most studies, water, alcohol and other solvents of high dielectric constants are not desirable, since they strongly absorb the microwave radiation. However, when there is no alternative to solvents of high dielectric constants, very narrow cavity tubes are used for samples which have a strong absorption. The sample concentration should be 10^{-6} to 10^{-9} M depending on the nature of the substance. The great advantage of ESR spectroscopy is that there is little interference from the diamagnetic bulk of the sample. Broadening of the spectral line may occur if the concentration of the paramagnetic species is high. For this reason, the ESR spectrum is obtained for such a substance whose trace quantity is incorporated into a diamagnetic compound isomorphous with the compound under study. Solid samples cannot be satisfactorily diluted by just mixing with a diamagnetic powder. Dilution must occur at the molecular level, i.e. by recrystallisation in the presence of a diamagnetic substance which is isomorphous with the sample in order to obtain a uniform distribution of the paramagnetic substance in the diamagnetic host.

Sample tubes usually have diameters of a few millimeters and packed to a depth of a few millimeters to 2-3 cm. A quartz sample tube is preferable because pyrex absorbs more of microwave power and also exhibits an ESR signal. It is necessary to remove oxygen from the solvent, otherwise it can lead to broadening of the resonance. Generation of free radicals *in situ* can be made by photolysis, radiolysis or electrolysis and concentration of the radicals can be adjusted in a flow system.

The ESR spectra in the gas phase are highly complicated because the rotational angular momentum of the molecule, which is quenched in the liquid and solid states strongly couples to the electron spin and orbital angular momentum.

REFERENCE SPECTRA

The ESR spectra are generally recorded as the first derivative spectra rather than as simple absorption peaks. The spectra are thus displayed in first-derivative form dI/dB. The position of the maximum absorption is the point at which the central trace crosses the base line. The signal width is the spacing between the maximum and the minimum (peak to peak). The spectrum of the free radical, 1, 1 -diphenyl-2- picryl hydrazyl (DPPH) is usually employed as the standard. In Fig. 4.2(a), the spectrum of DPPH in benzene is shown. It exhibits a five-line hyperfine pattern with an intensity ratio of 1 : 2 : 3 : 2 : 1. In solutions of high concentration or in solids, rapid electron exchange between adjacent radicals averages out the hyperfine structure to produce a singlet ($\Delta B_{pp} \sim 0.27$ mT), as shown in Fig. 4.2(b). DPPH in benzene has a *g*-factor of 2.0035 and as powder 2.0037 very close to that of the free electron value of 2.0023. It contains 1.53×10^{21} spins per gram of the compound. Charred dextrose ($g = 2.0028$) can be used as a secondary standard. The *g*-values of the sample peaks are calculated relative to the standard.

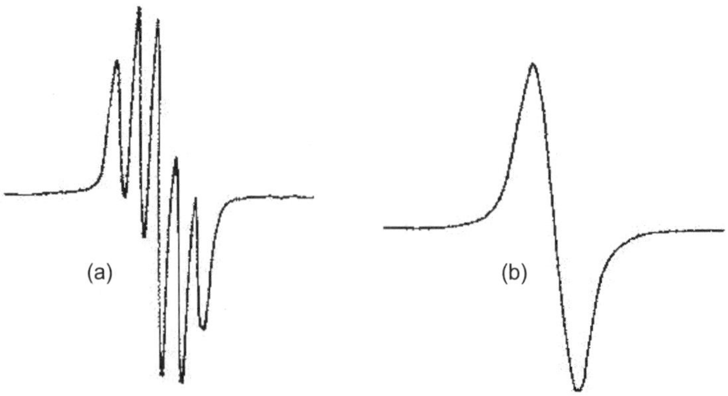

Fig. 4.2 ESR spectrum of DPPH: (a) in benzene and (b) as solid.

A spectrum consists of a field sweep so that if the signal is observed at a lower magnet field than the standard, the corresponding g-value is higher and the higher g-values occur on the low field side of the spectrum.

The use of DPPH as a standard requires an indication of its method of synthesis. The sample should be protected from contact with oxygen. The most suitable sample is DPPH crystallized from benzene in which the effect of oxygen absorption is insignificant. DPPH cannot be employed as an internal standard for other organic free radicals because there is only slight variation in the g-values and the unknown substance cannot be distinguished from the standard substance. In the case of free radicals, a trace of Cr(III) entrapped in a tiny chip of ruby crystal cemented to the sample cell is employed as a reference. This standard shows an intense signal and has a g-value of 1.40.

DETERMINATION OF g-VALUE

The best method to obtain the g-value is to measure the magnetic field separation between the centre of the sample spectrum and that of the reference substance whose g-value is accurately known. In most cases, as mentioned earlier, DPPH is generally used as the standard. In the spectrometer, the standard substance is placed along with the unknown in the dual cavity cell. The ESR spectrum will show two signals with a field separation of ΔB. The g-value for the sample is given by

$$g_{sample} - g_{ref} = \left(\frac{B_{sample} - B_{ref}}{B_{ref}} \right) g_{ref} \tag{4.1}$$

$$g_{sample} = g_{ref} \left(1 - \frac{\Delta B}{B} \right) \tag{4.2}$$

where B is the magnetic field at the resonance frequency of the sample and ΔB is taken as positive if the sample has its resonance at a higher field and g_{ref} is the g-value of the reference.

The g-factor for the sample can also be calculated from the relation,

$$g_{sample} = \frac{g_{ref}\, B_{ref}}{B_{sample}} \tag{4.3}$$

The g-value, A/B ratio, (*i.e.* the ratio of the area of the positive peak to the negative peak), the line width and spin concentration (*i.e.* the number of spins per gram) are all important parameters obtainable from an ESR spectrum to characterize the sample. To obtain the g-value and the spin concentration, the ESR spectra of the sample and the reference, either charred dextrose or DPPH, should be recorded under identical experimental conditions. The spin concentration can approximately be determined by comparing the area under the signal for the sample with that of charred dextrose as the standard for which the spin concentration is known.

The determination of the *g*-value is illustrated taking the spectrum of the copper(II) complex of a Schiff base obtained by the condensation of 2-(4-aminophenyl)benzimidazole and 2, 2′-dehydropyrrollidine *N*-aldehyde given in Fig. 4.3. It is a dimer of composition $Cu_2L_2Cl_4$, where L represents the Schiff base ligand. The resonances were observed for the complex at magnetic fields of $B_\perp$ 321.4 and $B_{||}$ 307.1 mT and for the reference (DPPH) at 328.6 mT, giving $\Delta B_\perp - 7.2$ mT and $\Delta B_{||} - 21.5$ mT. Using the equation (4.2), on substituting the values, we get

$$g_\perp = g_{\text{ref}}\left(1 - \frac{\Delta B_\perp}{B_\perp}\right)$$

$$g_\perp = 2.0037\left(1 - \frac{-7.2}{321.4}\right) = 2.049$$

$$\text{Similarly, } g_{||} = 2.0037\left(1 - \frac{-21.5}{307.1}\right) = 2.144$$

$$g_{\text{av}} = \left(\frac{1}{3}\right)(2\, g_\perp + g_{||}) = 2.081$$

Using the relation (4.3),

$$g_{\text{sample}} = \frac{g_{\text{ref}}\, B_{\text{ref}}}{B_{\text{sample}}}$$

On substituting the values, we obtain,

$$g_\perp = \frac{(2.0037)\,(328.6)}{321.4} = 2.049$$

Similarly, $g_{||}$ can be calculated.

The half-field signal for the dimer appears at 114.3 mT.

Many chemical and physical processes of interest occur only in a definite temperature range and hence temperature control of the sample is essential. In ESR studies, variable temperature facilities are useful and are available with most instruments. The population distribution of the different spin states is a function of the temperature. In general, the *g*-factor and hyperfine interaction constants are independent of the temperature. The line shape, line width and relaxation times T_1 and T_2 are the main factors which are sensitive to the temperature. At subambient temperatures, the sensitivity of the detectable levels is increased. The sensitivity of the ESR spectrometer is also increased by working at higher magnetic fields. For example, the sensitivity of spectrometer working at 35.0 GHz is nearly twenty times greater than that working at 9.5 GHz. Use of higher frequencies improves the spectral resolution. Hyperfine splittings and line widths are independent of the applied magnetic field whereas the *g*-value is proportional to the magnetic field.

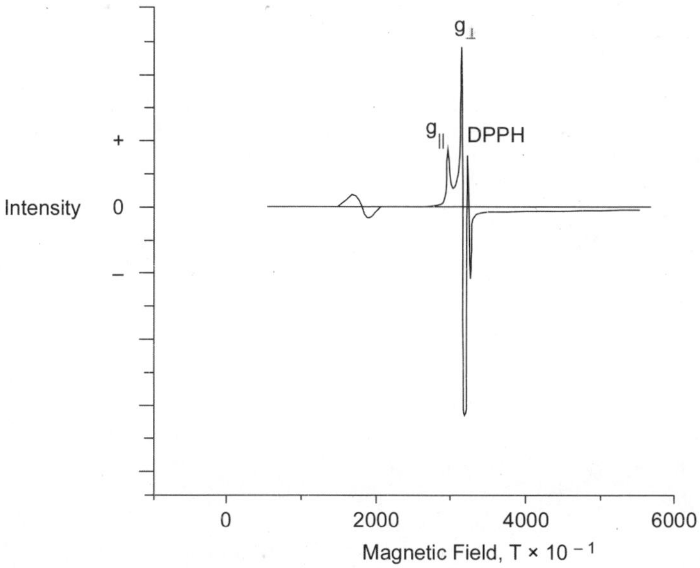

Fig. 4.3 The ESR spectrum of the copper(II) complex, $Cu_2L_2Cl_4$ where L denotes the Schiff base (see text).

Free radicals can be studied by ESR spectroscopy provided the lifetimes of such radicals are greater than 10^{-6}s. If the lifetimes are less than 10^{-6} s, they should be produced at low temperatures and the matrix isolation technique may be used to increase their lifetimes. Special arrangements may be necessary to produce, stabilize and observe the free radicals generated *in situ* with a very short time interval. These may include facilities for (*i*) irradiation of the sample with ultraviolet, gamma or X-rays, (*ii*) electrolytic redox reaction, (*iii*) continuous flow systems allowing radicals to be observed in a short time after mixing the reagents and (*iv*) rapid freezing to arrest the reactions for a known time after mixing in a flow system so that spectra of the unstable radicals can be observed at leisure.

Many paramagnetic species which are too unstable to be observed in solution may be observed in rigid solids. For solid systems, even the qualitative aspects of an ESR spectrum may be markedly dependent upon the orientation of the crystalline sample in a magnetic field. Free radicals oriented in solids and transition metal ions in single crystals may show such an anisotropy. When the resolution of the spectrum is poor or when the spectral lines are too many, it may be advantageous to carry out a computer simulation of the spectrum based on assumed hyperfine splittings and line widths. Such computer programs are available.

SPECTRAL CHARACTERISTICS: LINE WIDTH AND ANISOTROPY

5

LINE WIDTH

INTRODUCTION

As seen from the spectra presented earlier, the ESR spectral lines always have a finite width. Typical line width and other measurable parameters of an ESR peak having Lorentzian line shape are illustrated in Fig. 5.1. The origin of line widths in the ESR spectrum is more complex than that in the NMR spectrum. The theoretical width of an ESR absorption line is very small, about 0.01 mT (10 μT). Lines as narrow as this are often found for solutions of organic free radicals. However, various relaxation processes may shorten the lifetime of the excited state and lead to line broadening. This is particularly important for transition metal complexes. If there are any low-lying excited electronic states, relaxation becomes quite

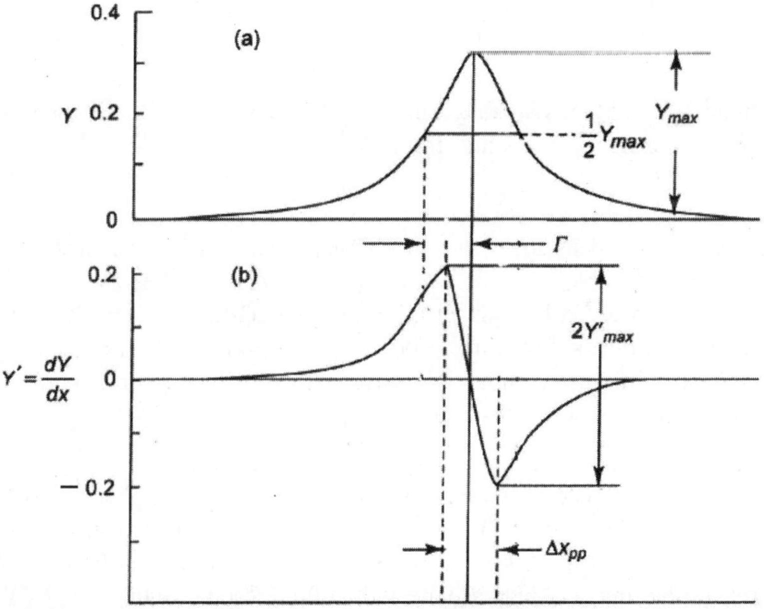

Fig. 5.1 Lorentzian line shape : (a) absorption spectrum and (b) its first derivative spectrum. Y_{max} denotes peak intensity, $2Y'_{max}$ denotes peak to peak amplitude in the first derivative curve. Γ is the half width at half-height, and Δx_{pp} denotes peak to peak width = $\left(\frac{2}{\sqrt{3}}\right)\Gamma$.

efficient and lines are considerably broadened. The line widths may vary from 0.01 mT for radical ions to 0.3 mT for exchange narrowed free radicals to over 100 mT for transition metal complexes. If the signals are too broad, it can interfere with the interpretation of the spectrum. We shall briefly discuss the factors that influence the line widths in ESR spectroscopy; these factors can possibly be varied to narrow the observed signals. The microwave radiation power level should be kept low so that broadening due to saturation effects do not occur.

LIFETIME BROADENING

All quantum mechanical transitions have a limiting nonzero spectral width called the lifetime broadening which arises from the finite lifetime of the excited state. Thus all ESR spectral lines have a finite inherent line width apart from any instrumental broadening that may occur. If, on an average, a system exists in a state for a time τ, the mean lifetime of the state, then the uncertainty in the energy δE of the system is given by

$$\delta E \approx h/2\pi\tau \tag{5.1}$$

This equation reminds us of the Heisenberg's uncertainty principle and hence this lifetime broadening is often known as the uncertainty broadening as discussed earlier. When the uncertainty in energy is expressed in frequency unit through $\delta E = h\delta v$, the above relation becomes

$$\delta v \approx \frac{1}{2\pi\tau} \tag{5.2}$$

The line width may be calculated in terms of the spread in the magnetic field, δB. Since $\delta E = g\,\beta_e \delta B$, we obtain from equation (5.1)

$$\delta B \approx h/2\pi g\beta_e\tau \tag{5.3}$$

Here δB is defined as the width at half-maximum on the absorption curve. Transitions thus occur over a range of applied fields and the resonance line is broadened. The observed ESR spectrum represents a statistical average over the ensemble of the radicals. Since this process applies equally to all members of the spin system, lifetime broadening is a case of homogeneous broadening and it defines the minimum line width for a given spectral line.

If T_1 is the spin lattice relaxation time, then a small value of T_1 will lead to a large value of δE and hence a smearing out of the energy levels leading to broadening of the ESR signal. For example, if 10^{-9}s is taken as a typical relaxation time, then $\delta v = 10^8$ Hz (100 MHz). This corresponds to a line width of nearly 6 mT. A still shorter relaxation time will increase the width of the resonance line further. Widths of ~ 10 MHz are not unusual in ESR spectra. This may be compared with the normal line width of NMR absorption for a solution which is typically of about 1 Hz. The mechanisms which dominate the relaxation processes in ESR differ considerably from those in NMR. In ESR much more efficient relaxation mechanisms are present giving shorter relaxation times and hence the ESR line widths are broader compared to those of the NMR. The line width arising from relaxation times is known as the 'natural (or inherent) line width'. Experimentally T_1 values typically are about 10^{-6}s.

The line broadening is usually analyzed in terms of the contributions arising from the spin-lattice and spin-spin relaxation times. The inherent line width is given by

$$\delta B \approx \frac{\hbar}{g\beta_e(2T_1)}; \quad \hbar = \frac{h}{2\pi} \tag{5.4}$$

The observed line width is usually much greater than that expected from the T_1 value. It is, however, useful to retain the concept of relaxation time while discussing the line width. A new relaxation time T_2 based on the line width of the resonance line is defined. The spin-spin relaxation time T_2' encompasses all additional sources of broadening. Thus the observed line width is generally defined in terms of a relaxation time T_2 such that

$$\delta \mathrm{B} = \left(\frac{\hbar}{g\beta_e}\right)\left(\frac{1}{T_2}\right) \tag{5.5}$$

where $\frac{1}{T_2} = \left(\frac{1}{T_2'}\right) + \left(\frac{1}{2T_1}\right)$

Each relaxation mechanism contributes about $h/2\pi\tau$ to the energy spread of the ESR line. In liquid samples, if $T_2 \sim 10^{-8}$ s and $T_1 \sim 10^{-6}$s, the former leads to a spread of about 0.3 mT and the latter, about 0.003 mT.

Broadening due to spin lattice relaxation results from the interaction of the paramagnetic ion with the thermal vibrations of the lattice. The variation in spin-lattice relaxation time for different systems is quite large. For some compounds, it may be sufficiently long to allow the measurement of ESR spectra at room temperature, while for others it may not be possible to record the spectra at room temperature. Since relaxation times generally increase with decrease in temperature, as the thermal motions of the lattice are reduced, many complexes of transition metals require to be cooled to liquid nitrogen or liquid helium temperature before well-resolved spectra are obtained.

Spin-spin interaction results from the small magnetic fields that exist on neighbouring paramagnetic ions. As a result of these fields, the total field at the ions is slightly altered and the energy levels are shifted. A distribution of energies results which leads to broadening of the signals. Since this effect varies as $(1/r^3)(1 - 3\cos^2\theta)$, where r is the distance between the ions, and θ is the angle between the magnetic field and the z–axis, this type of broadening is highly dependent upon the direction of the field. The effect can be reduced by increasing the distance between the paramagnetic species by diluting the paramagnetic compound with an isomorphous diamagnetic material, for example, a small quantity of the copper(II) complex can be doped in the corresponding diamagnetic zinc(II) complex.

In solids, the spin-lattice relaxation is less important than the interaction between the spins. The random flipping of neighbouring electron spins in dilute paramagnetic crystals leads to the spins getting out of phase with each other and it results in the broadening of the resonance signal. In more concentrated samples, stronger broadening occurs. The major cause of relaxation in liquids as noted before is due to modulation of the spins by dipolar interactions. Random fluctuations occur due to interactions between the paramagnetic molecule and neighbouring nuclear or electron spin dipoles.

For stable free radicals, broadening due to spin-lattice relaxation time T_1 is relatively unimportant. In the spectra of free radicals, the observed line width is usually a simple function of the spin-spin relaxation time T_2'. However, a characteristic feature of most organic free radicals in the condensed phase and in undiluted form is the extreme narrowness of the ESR spectral lines. The half width at half-height of an organic free radical is typically 0.01 mT, it is very much less than that expected from magnetic dipolar interactions. This is said to be due to the presence of large exchange forces. For transition metal ions, the ESR line width is determined exclusively by their small values of T_1.

The broadening of spectral lines can occur either by homogeneous broadening or by inhomogeneous broadening.

INHOMOGENEOUS BROADENING

The unpaired electron in different free radicals is subjected to slightly different magnetic fields. At any time, therefore, only a small fraction of the paramagnetic species is in resonance with the external magnetic field as the magnetic field is swept through the resonance. The observed line is hence a superposition of a large number of individual components each slightly shifted from the others thus producing an unresolved broad band. The local magnetic field varies from one paramagnetic centre to another but remains constant over a period for a given paramagnetic centre. The local magnetic fields resulting from anisotropic g and hyperfine interactions also give rise to inhomogeneity. The line shape may be highly unsymmetrical.

Some of the other causes of inhomogeneous broadening for a given spin system include inhomogeneous external magnetic field, unresolved hyperfine structure, anisotropic interactions in randomly oriented systems in the solid state and dipolar interactions with other paramagnetic centres. In some of these cases, the local fields giving rise to inhomogeneous broadening can be averaged out by rapid dynamic effects such as through tumbling, exchange and collisions. It would result in homogeneously broadened spectral lines.

HOMOGENEOUS BROADENING

The variation in the local magnetic fields can be either dynamic or spatial (static). In the former case, each paramagnetic species experiences a local magnetic field that fluctuates with time, whereas in the latter case, the local magnetic field varies from one paramagnetic species to another but remains constant for a given paramagnetic species over a sufficiently short interval. The static as well as time averaged magnetic fields can be considered to be the same at each dipole but the instantaneous magnetic field is not. For example, if a manganese(II) salt is added to an aqueous solution of a free radical, the paramagnetic Mn^{2+} ion would produce a strong and fluctuating local field. It is an example of a homogeneous type of broadening. If this solution is frozen, the broadening would become inhomogeneous because the local field would then be static for each radical. Homogeneously broadened lines generally have Lorentzian shape while inhomogeneously broadened lines have Gaussian shape.

Any process that increases the rate of the transitions between two electron spin states will decrease the lifetime and hence increase the line width as discussed before. This process applies equally to all the members of the spin system. Lifetime broadening is, therefore, a homogeneous broadening.

OTHER FACTORS AFFECTING LINE WIDTH

Many effects cause the width of one band to differ from that of another in a given spectrum, for example, molecular conformation of ethylamine radical influences the line widths of different protons in the molecule differently. Rapid exchange between various configurations of an ion pair with an anion or a cation radical can also lead to greater line broadening of certain resonances than those of the others. Dynamic processes such as hindered rotation, electron transfer, electron spin exchange, etc. lead to homogeneous line broadening. This topic is discussed in the following chapter.

ANISOTROPY

ANISOTROPY IN *g*-FACTOR

The treatment of ESR spectra so far has involved the so-called isotropic spectra obtained from isotropic systems such as solutions. Most systems are anisotropic and thus the energy level spacings and the observed spectra depend on the orientation of the sample in the external magnetic field. Thus in practice, many spectra are recorded for anisotropic systems which include unstable species obtained by irradiation of crystalline materials or of substrates trapped in other host matrices, solids including frozen solutions and paramagnetic point defects in single crystals. The description of systems showing anisotropic behaviour usually requires that each spectroscopic property be described by six independent parameters. Thus, the g-factor which we have regarded as a simple scalar quantity must be considered as a second order symmetric tensor. Such a tensor can be diagonalised to give three principal g-factors g_{xx}, g_{yy} and g_{zz}. For systems with spherical or cubic symmetry they are equal. The solution spectra qualitatively appear as isotropic since for such systems the anisotropic effects are averaged to zero by the rapid tumbling of the molecules. However, for lower symmetries the three g-factors may all be different.

In exactly the same way, in many oriented systems there may be anisotropy in the hyperfine splittings. The effects of anisotropy on g-values and hyperfine splittings in ESR spectra are exactly analogous to the effects of chemical shifts and dipolar coupling on the NMR spectra of solids discussed later. If a single crystal is studied by ESR, it is found that the spectra obtained depend on the orientation of the sample. By making measurements for various orientations, it is possible in principle to determine all the components of the g and hyperfine splitting tensors. This can yield valuable information about electron distribution.

The application of an external magnetic field generates an internal magnetic field in the sample. For convenience, B_{res}, is defined as the external magnetic field at resonance.

$$h\nu = g_{\text{eff}}\,\beta_e\,B_{\text{res}} \tag{5.6}$$

Local magnetic fields are accounted for by allowing the g-factor to vary as discussed earlier. The chief source of the local magnetic fields which cause the g-factor to deviate from the free electron value is due to the orbital angular momentum. In most free radicals (or in a solid) because of low symmetry which causes the complete removal of orbital degeneracy, the magnitude of spin-orbit coupling is very small. In a free radical, the electron is usually delocalized and it behaves very much the same way as an electron in free space having zero orbital angular momentum. Consequently, the g-values are nearly equal to the free electron (g_e) value of 2.00232. Small deviations, less than $\pm$ 0.05, often observed for most free radicals are accounted for by mixing in of low-lying excited electronic states with the ground electronic state. This mixing is in most cases not 'orientation independent' (isotropic) but is 'orientation-dependent' (anisotropic). That is, the induced local fields depend on the orientation of the molecule with respect to the external magnetic field. This is to be expected since most atomic orbitals have directional character and the interactions are hence directional. The anisotropy in g-values gives rise to asymmetry in the spectral line. The values of the ESR spectral parameters are thus strongly dependent on the orientation of the molecule in an external magnetic field.

There are two types of magnetic interactions involving the orbital angular momentum. Firstly, the interaction of the orbital angular momentum with the electron spin referred to as the spin-orbit coupling. It represents the interaction between the magnetic dipoles arising from the orbital motion of the electron and the spin of the electron. This interaction is approximately given by ξL.S, where L is the operator of the orbital angular momentum and ξ is the spin-orbit coupling constant which can be experimentally determined from atomic spectra. The magnetic moment generated by the orbital motion of the electrons interacts with the applied magnetic field in the same manner as does the magnetic moment produced by the electron spin and the energy of this interaction is given by $\mu_o L.B$.

The larger the magnitude of the orbital angular momentum L, the greater is the shift from the free electron g-value. The direction of the shift is strongly influenced by the sign of the spin-orbit coupling constant. If the spin-orbit coupling constant is positive, g is generally lower than the free electron g-value. Even when the ground state of the molecule has no orbital angular momentum associated with it, mixing of the excited electronic states with the ground state can occur in the presence of the applied magnetic field leading to some orbital angular momentum. The observed g-value differs from the free electron value by an amount which can be related approximately to the ratio of spin-orbit coupling constant to the separation between the electronic energy levels.

As noted before, anisotropy in g arises from the interaction of the spin angular momentum with the orbital angular momentum. The spin angular momentum is oriented with the field but the orbital angular momentum is associated with the electrons moving in molecular orbitals. Consider a situation where there is an orbital contribution to the magnetic moment of an electron in a circular molecular orbital that can precess about the z-axis of the molecule. Two different orientations of this orbital relative to the applied magnetic field are shown schematically in Fig. 5.2. The magnetic moment vector μ_L due to orbital angular momentum and the spin magnetic moment vector μ_S point in the same direction in Fig. 5.2 (a). On the other hand, in Fig. 5.2 (b), a different orientation of the molecule is shown. Here μ_L and μ_S do not point in the same direction and hence the net magnetic moment indicated by the arrow is in a different direction.

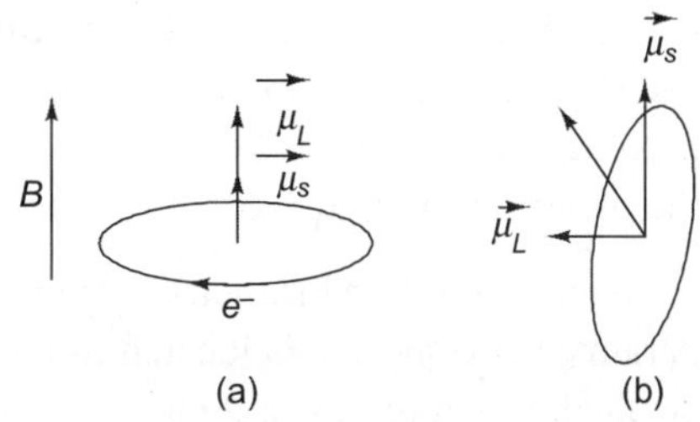

Fig. 5.2 Coupling of the magnetic moments arising from the spin and orbital angular momentum for two different orientations of the molecular orbital relative to the applied magnetic field B.

The result of spin-orbit coupling can thus be seen as a new effective spin vector quantized along the direction of the applied magnetic field or as a new magnetic field vector along which the electron spin is effectively quantized. This means that the interaction between the spin and the applied magnetic field depends on the orientation of both of them relative to a set of coordinates fixed in the molecule and not simply on their orientation relative to each other. The energy of interaction between the electron spin and the applied magnetic field is now given by

$$E = \beta \boldsymbol{S} \, g \, \boldsymbol{B} \tag{5.7}$$

where $\boldsymbol{S}$ and $\boldsymbol{B}$ are vectors and g is a tensor of rank 2 since it depends on both these vectors. The g-tensor gives us an effective spin, that is, $g.S = S_{eff}$. For different orientations, the g-tensor varies to incorporate the orbital effects. In Cartesian coordinates, the g-tensor may be represented by a third order square matrix as shown below,

$$g = \begin{pmatrix} g_{xx} & g_{xy} & g_{xz} \\ g_{yx} & g_{yy} & g_{yz} \\ g_{zx} & g_{zy} & g_{zz} \end{pmatrix} \tag{5.8}$$

where, x, y and z define the crystal axes. The g tensor is almost always symmetric, that is $g_{xy} = g_{yx}$, $g_{xz} = g_{zx}$, $g_{yz} = g_{zy}$ and thus it has only six independent components. The off-diagonal elements say g_{xz} gives the contribution to g along the x-axis of the crystal when the magnetic field is applied along the z-axis. Thus, each g component gives one scalar component of S with one component of B in an arbitrary set of coordinates fixed in the molecule. It is possible to transform g matrix to a diagonal form in the principal axes system X, Y, Z. If the molecule has rotation axes of symmetry, then these axes can be taken to coincide with X, Y or Z. For molecules of lower symmetry, the principal axes may be in any direction but should necessarily be orthogonal to each other. The three principal g-values are g_X, g_Y and g_Z (or g_{xx}, g_{yy} and g_{zz} respectively) and they may be obtained by making measurements on a crystal oriented successively in three mutually perpendicular directions with respect to the external magnetic field. Since the g-factor arises from the mixing of the spin-orbit interaction with the Z-component of the spin in an external magnetic field, it is expected that a different g-value will be obtained depending on whether the external field is along either of the molecular axes x, y or z.

In an octahedron or a tetrahedron, the x, y and z-axes are equivalent. Thus, if a paramagnetic species is located in a perfectly cubic crystal as, for example, at an octahedral or a tetrahedral site, the g-value is isotropic, that is, it is independent of the orientation of the crystal. This condition is also nearly obeyed for free radicals ($g \approx 2.0$). In a crystal of lower symmetry, the g-value depends on the orientation of the crystal in the applied magnetic field. As noted earlier, paramagnetic species of low molecular weight (< 1000) in solutions of low viscosity solvents generally exhibit a single isotropic g-factor. This results from an averaging process due to molecular tumbling giving $<g_{av}> = (1/3)(g_{xx} + g_{yy} + g_{zz})$. Many systems of higher molecular weight such as biopolymers in solution or liquid suspensions, may exhibit line broadening resulting from g-factor anisotropy.

For molecules of lower symmetry, three different principal g-components, namely g_{xx}, g_{yy} and g_{zz} could be obtained. The g-tensor is said to be anisotropic if $g_{xx} \neq g_{yy} \neq g_{zz}$. If the molecule has a two-fold or a higher-fold rotation axis of symmetry, then by convention, the symmetry axis is taken as the principal axis (z-axis). If the molecule has at least a single two - or higher-fold axis of rotation, then the X and Y axes would be equivalent but different from the z-axis. The molecule is then said to have an axial symmetry giving $g_{xx} = g_{yy} \neq g_{zz}$. It is usual in such cases to represent g_{zz} as $g_{\|}$ and $g_{xx} = g_{yy}$ as $g_{\perp}$. For example, many six coordinated copper(II) complexes have tetragonal symmetry. When the magnetic field and molecular axis are parallel, the signal appears at $g_{\|}$ and when they are perpendicular it appears at $g_{\perp}$. If the sample is a single crystal in which all the molecules are oriented similarly with their symmetry axes at an angle θ to the magnetic field B as shown in Fig. 5.3, the observed g-value is given by

$$g^2 = g_{\|}^2 \cos^2\theta + g_{\perp}^2 \sin^2\theta \tag{5.9}$$

If the symmetry is still lower, three principal g-values are needed and the observed g-value for an arbitrary orientation is given by

$$g^2 = g_X^2 \cos^2\theta_X + g_Y^2 \cos^2\theta_Y + g_Z^2 \cos^2\theta_Z \tag{5.10}$$

where the angles are those between the field and the molecular axes. The values of g and similarly a depend on the direction between the magnetic field and the principal (z) axis of the system.

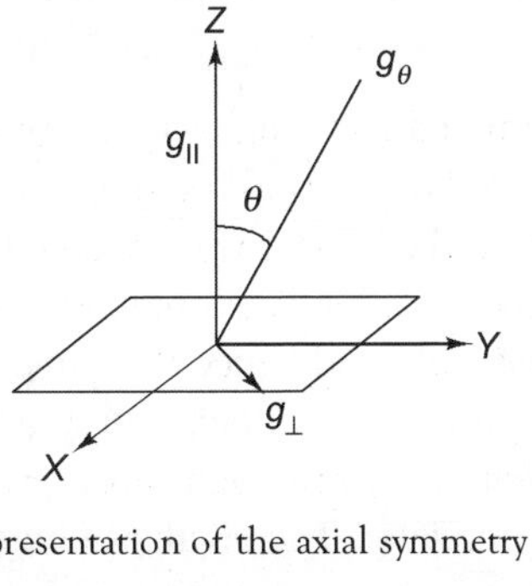

Fig. 5.3 Representation of the axial symmetry of g-factors.

In solution, the molecules continuously tumble usually at rates considerably greater than the frequency of the spectrometer. Any anisotropy is then averaged to zero during the time required to excite the system. If the solution is frozen or the sample is a powdered solid, all possible orientations still occur randomly but the molecular positions are fixed. Each molecule with a particular orientation has its own g-value and the observed spectrum is the sum over all the molecules. On a purely statistical basis, there will be relatively less number of molecules with principal axes parallel to the field and far more molecules with their axes perpendicular to the field. This factor combined with the angle-dependence of intensity gives a distribution of intensities as shown in Fig. 5.4. In the derivative spectrum it is possible to recognize the two principal g-values, though the resolution depends critically on the difference between $g_{\parallel}$ and $g_{\perp}$. For many rhombic systems ($g_X \neq g_Y \neq g_Z$) the difference between g_X and g_Y is not much and the spectrum appears to have an asymmetry in the $g_{\perp}$ line. If the three principal g-values are appreciably different, three major lines will appear in the spectrum.

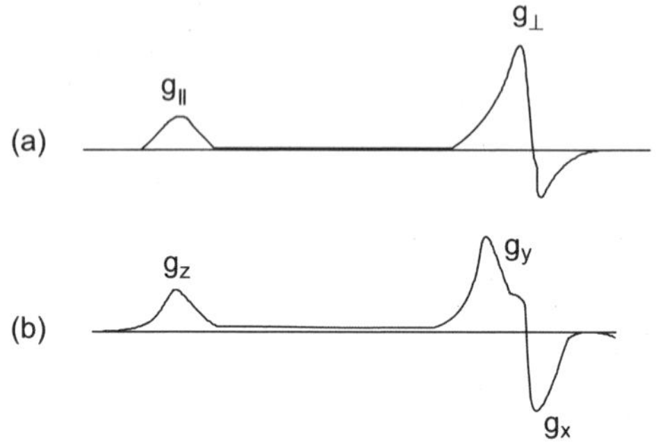

Fig. 5.4 Typical shapes of the first derivative ESR spectra for powders or frozen solutions of anisotropic systems (*a*) with $g_{\parallel} > g_{\perp}$ for axial systems and (*b*) rhombic systems with $g_X \neq g_Y \neq g_Z$.

The ESR spectra taken as a function of orientation of the crystal yield maximum amount of information. The analysis of such a spectrum requires a detailed knowledge of the nature of the anisotropic interactions. Free radicals oriented in solids and complexes of transition metal ions in single crystals constitute an important class of systems under this category. The anisotropy in g and a can be effectively used to obtain information on the ground electronic state of transition metal complexes.

Consider NO_2 radical trapped in a single crystal of KNO_3. For the crystal with the magnetic field parallel to the two-fold rotation axis (z-axis) of NO_2, a g-value of 2.006 was obtained. When the crystal was mounted with x or y-axis parallel to the magnetic field, a g-value of 1.996 was obtained. The molecule rapidly rotates about the z-axis in the solid state and hence the same result is obtained when the applied magnetic field is parallel to either x or y-axes. The differences in the g-values for different orientations are even more pronounced for complexes of transition metal ions discussed later. The a values obtained for the single crystal of NO_2 in KNO_3 are also anisotropic. When the two-fold rotation axis of thc molcculc was parallel to the applied magnetic field, the observed nitrogen hyperfine coupling constant was 176 MHz and for the orientation in which the two-fold axis was perpendicular to the magnetic field a value of 139 MHz was obtained.

The spectrum of NO_2 recorded by irradiating $NaNO_2$ with γ-rays at 77 K in an argon matrix is shown in Fig. 5.5 as a simple example. The broad and asymmetric ESR line shape observed in the spectrum is largely due to *g*-value anisotropy; the resonance position is dependent on the orientation of the molecule in the external magnetic field as discussed above and the spectrum is a superposition of resonances from different orientations. The spectrum has been analysed and the values $g_{\parallel}$ 1.992 and $g_{\perp}$ 2.003 and $A_{\parallel}$ 146 MHz and $A_{\perp}$ 21 MHz have been reported. The procedure for calculating the hyperfine coupling constants and *g*-factors from the spectrum is rather complicated and is not described here.

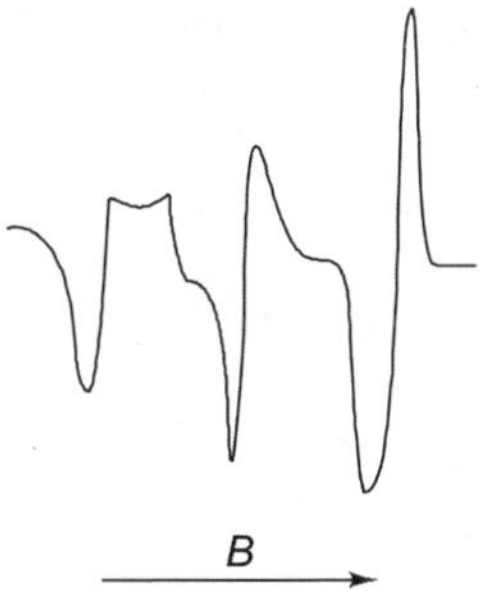

Fig. 5.5 The ESR spectrum of NO_2.

ANISOTROPY IN *A*-VALUES

In rigid systems, the interaction between the electron and nuclear dipoles gives rise to anisotropy in the electron-nuclear hyperfine interaction. The classical expression for the interaction between two dipoles apply. The energy of interaction of an electron magnetic moment with a nuclear magnetic moment is given by

$$E_{\text{dipolar}} = -g\,\beta_e\,g_N\,\beta_N\left[\frac{S.I}{r^3} - \frac{3(S.r)(I.r)}{r^5}\right] \tag{5.11}$$

For free radicals in solution only the isotropic interactions will be observed as discussed above. Alternatively, if the solution of the free radical is frozen, then the spectrum observed may be a superposition of the spectra of all possible orientations (as in a powdered sample) as discussed above. Unless the hyperfine splitting is large relative to the inherent line width, which is not often the case, much of the structure is lost in a broad and poorly resolved line.

Exactly the same considerations as discussed above for the *g*-values apply to the hyperfine splitting tensor *A*, which could be described by the three principal components A_{xx}, A_{yy} and A_{zz}. Some degree of anisotropy in the hyperfine splitting is usual in all ESR spectra. This is a consequence of the fact that the spin-spin interactions though essentially magnetic are transmitted through atomic orbitals and anisotropy present in all atomic orbitals excepting simple *s*-orbitals causes *A*-value anisotropy, since the interactions may be stronger along some molecular axes than along the others. In free radicals, the orbitals occupied by the unpaired electron usually have π or σ-character and the electron is often delocalized, resulting in very little directional character and consequently free radicals have almost isotropic *A*-values. Due to rapid molecular motions in solution, the observed hyperfine splitting is an average of the three principal components $a = (1/3)(A_{xx} + A_{yy} + A_{zz})$. The principle coordinate system for the *A*-tensor need not be the same as that for the *g*-tensor.

DYNAMIC PROCESSES 6

INTRODUCTION

In the discussion of the ESR spectra, we had implicitly assumed that the free radicals are completely independent and non-interacting. Such would be the case only at infinite dilution. We shall now discuss some of the effects of the paramagnetic centres interacting magnetically and chemically with each other and with the environment. Dynamic processes that alter the magnetic environment of the unpaired electron can give rise to line broadening and influence the spectral appearance. If the changes occur sufficiently slowly one observes ESR lines assignable to each of the distinct species (e.g. conformers). As the rate of fluctuations increases, the ESR lines broaden and finally coalesce into a single line (or set of lines), the position of which is the weighted average of the original line positions. Analysis of the spectra can provide us important information concerning the time-dependent phenomenon occurring in solution.

ESR spectroscopy can be used to study the dynamic processes. Time-dependent phenomena such as chemical or electron exchange processes or molecular motions such as internal rotation giving rise to two or more spatial configurations can affect the spectra. Electron exchange between a free radical and a diamagnetic molecule is another process which can cause changes in the spectra. ESR spectroscopy can be utilized to study such rate processes since the line widths and hyperfine splittings applicable to free radicals in solution are in the range 0.1 to 100 MHz, the time scale of the processes which can contribute to the line widths may be estimated to be 10^{-4} to 10^{-10} s. Significant line broadening occurs for dynamic processes with first order rate constants of 5×10^7 s^{-1}. The dynamic processes mentioned above contribute to the complexity of the spectra. However, an analysis of the spectra can yield valuable structural and kinetic information. These aspects are briefly outlined here qualitatively.

GENERAL MODEL FOR INTERCONVERSION

To begin with, a general model is discussed. Consider a free radical which can exist in either of the two distinct forms say A and B and each has a distinct ESR spectrum. For simplicity, equal concentration of each form is assumed and it is further assumed that each form gives rise to a single line in the spectrum one at the magnetic field B_a and the other at a higher field B_b (Fig. 6.1). The separation between the lines is $\delta B_0 = B_b - B_a$ and it generally depends on B. It means that the two species usually have different g-factors. If the rate of interconversion between the two forms A and B is slow, compared to the parameter $\gamma_e \delta B_0$ then a simple two line spectrum is expected as shown in Fig. 6.1(a). Here the lifetime τ_a and τ_b are long compared to $|\gamma_e\ \delta B_0|^{-1}$. With increasing rate of interconversion, broadening of the two lines is observed due to the reduced lifetime (τ_a or τ_b) of a given form. For free radicals in solution, if the half width at half height is taken typically as 0.01 mT, then from the Heisenberg's uncertainty principle, the mean lifetime τ would be nearly 5×10^{-7} s for line broadening to be significant (see Chapter 1). This determines the minimum rate required to observe the broadening effect. In the limit of slow interconversion, the relationship between the line width δB and the mean lifetime τ is given by

$$\delta B = \delta B_0 + \left(\frac{1}{2\tau\gamma_e} \right) \tag{6.1}$$

where γ_e is the magnetogyric ratio of the electron and δB_0 is the line width in the absence of interconversion and 2τ is the mean lifetime of the species A or B. It is the usual practice to define τ in terms of the mean lifetimes τ_A and τ_B as

$$\tau = \frac{\tau_A \tau_B}{(\tau_A + \tau_B)} \tag{6.2}$$

When $\tau_A = \tau_B$, the mean lifetime of the species A or B is 2τ.

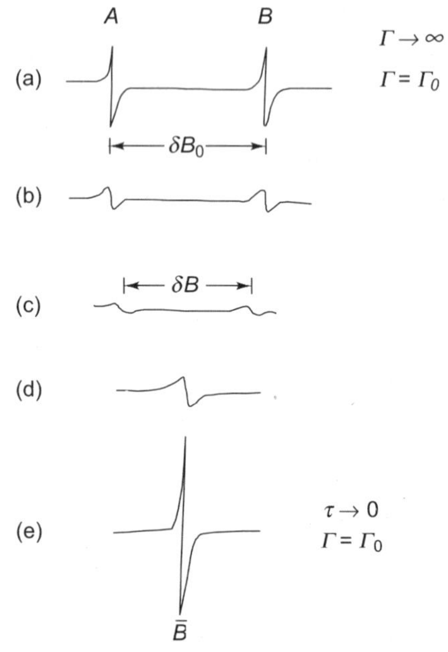

Fig. 6.1 Computed first derivative ESR spectra showing the effect of increasing rate of interconversion from (a) to (e) between the two forms A and B of a free radical : (a) slow rate, (b) moderately slow rate, (c) faster rate showing spectral line shifts, (d) coalescence and (e) fast rate limit.

As the rate of interconversion increases, the lines continue to broaden and begin to shift to their mid-point as shown in Figs. 6.1(a)–(e). When the rate of interconversion is sufficiently high, the two lines coalesce to a single line corresponding to the time averaged resonant field position.

For the situation shown in Fig. 6.1(c), the following equation relating the separation of the peaks to the lifetime could be employed.

$$(\delta B_0{}^2 - \delta B^2)^{1/2} = \frac{\sqrt{2}}{\tau \nu} \tag{6.3}$$

Here δB_0 and δB represent the separation between the spectral lines in the absence of interconversion and when interconversion is taking place, respectively. The coalescence phenomenon is a consequence of the uncertainty principle. When it is expressed as $\Delta \nu \tau = 1/2\pi$, where $\Delta \nu$ is the separation between the two lines in frequency units, then τ is the smallest time during which the forms A and B could be distinguished. For very fast interconversion, the half width at half height of this line (Γ) is proportional to τ and to the mean square of δB_0. The appropriate relation is

$$\Gamma = \Gamma_0 + \gamma_e \frac{<(\delta B_0)>^2}{4} \tag{6.4}$$

As the rate of interconversion increases to the fast interconversion limit, the line narrows to the limiting width δB_0, as shown in Fig. 6.1(e). The second order rate constant can be evaluated from the relation,

$$k = \frac{1}{2\tau[R]} \tag{6.5}$$

where [R] denotes the molar concentration of the free radical R.

If exchange occurs between equivalent ions, the lines broaden at the base and become narrower at the centre. When exchange involves dissimilar ions, the individual resonance lines merge together to produce a single broad line. The above general model is characteristic of many chemical processes and specific examples of such processes will now be briefly discussed.

ELECTRON SPIN EXCHANGE

Electron spin exchange represents a bimolecular reaction in which two free radicals exchange their unpaired spins. Very rapid electron exchange reactions can be studied by ESR spectroscopy by following the signal intensity of the paramagnetic reactant or the product as a function of time. Electron spin exchange processes are common in free radicals. In solution this is usually a bimolecular reaction in which the radicals exchange electrons during collisions. As an

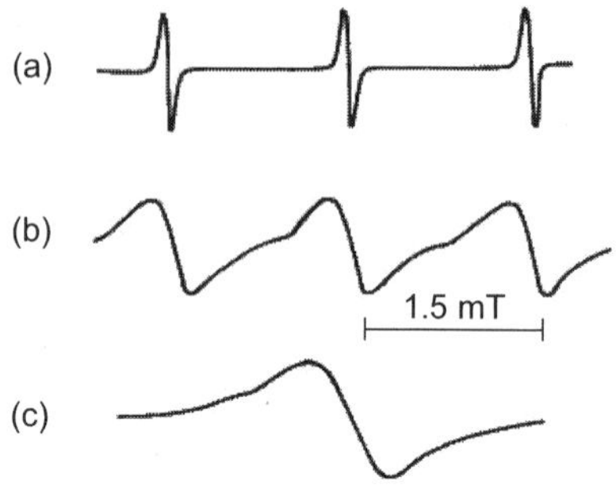

Fig. 6.2 ESR spectra of di-*t*-butyl nitroxide radical in ethanol at room temperature (25 °C): (*a*) 10^{-4} M, (b) 10^{-2} M and (c) 10^{-1} M solutions.

example, the ESR spectra of di-*t*-butyl nitroxide radical ($((CH_3)_3C)_2NO$ at different concentrations are shown in Fig. 6.2.

At very low concentration of the radical, the hyperfine triplet due to ^{14}N is observed [Fig. 6.2 (a)]. As the concentration of the free radical in solution is increased, the rate of the bimolecular process increases, the signals broaden and coalesce gradually to a single line which becomes narrower at still higher concentrations in the very fast exchange limit. The spectrum represented in Fig. 6.2 (c) is often referred to as the exchange narrowed spectrum, since the electron spins are exchanged so rapidly that the time averaged hyperfine field is close to zero. A similar exchange narrowed spectrum is observed for most free radicals in the solid state. The exchange narrowed spectrum of DPPH was discussed earlier. The second order rate constant for electron spin exchange in di-*t*-butyl nitroxide radical in N, N-dimethyl formamide was found to be 7×10^9 L $mole^{-1}s^{-1}$, which indicates that spin exchange occurs with a high probability. Electron spin exchange is a dynamical process which in liquids produces greater broadening than dipole-dipole interaction.

ELECTRON TRANSFER

One of the first systems to be investigated for electron transfer between a radical and a diamagnetic species involved the electron exchange reaction that takes place between the paramagnetic naphthalene anion radical and the diamagnetic naphthalene molecule itself.

$$\text{Naph(1)}^- + \text{Naph(2)} \rightarrow \text{Naph(1)} + \text{Naph(2)}^-$$

As naphthalene is added to a solution of naphthalene anion radical, broadening of the hyperfine structure of the ESR spectrum of the paramagnetic species occurs. With rapid exchange, the electron is subjected to a distribution of different internal fields by the virtue of being on several different molecules, leading to broadening of the spectral line. If the rate of exchange is slow compared to the hyperfine splitting, then the hyperfine structure in the ESR spectrum of napthalene anion radical is unaffected. If the rate of exchange is very fast, then the ESR spectrum shows a single line. Examination of the ESR spectra has yielded the rate constants for this electron transfer reaction in different solvents. The electron transfer reaction between a radical and a diamagnetic species is very similar to the electron spin exchange in its effect on the spectrum.

Electron exchange results in uncertainty in the field or frequency required for the transition to take place. The additional width $\Delta\nu$ of the resonance line is given by the equation

$$\Delta\nu = \left(\frac{1}{2\pi}\right)\left(\frac{1}{\tau}\right) \tag{6.6}$$

where τ is the mean jump time of the electron. For a process first order in naphthalene concentration, the following equation has been obtained.

$$k[\text{Naph}] = \frac{1}{\tau} = 2\pi\Delta\nu \tag{6.7}$$

where [Naph] is the naphthalene concentration. Measurement of the line width as a function of the concentration of the neutral molecules enables us to obtain the second order electron transfer rate constants which is not dealt with here. The rate constant for electron exchange between naphthalene and potassium salt of naphthalene i.e. naphthalenate anion in tetrahydrofuran has been found to be 6×10^7 L mole^{-1}s^{-1}.

PROTON EXCHANGE

Proton exchange rates are sometimes large enough to display detectable effects. For example, in the spectrum of $^{\bullet}CH_2OH$ radical discussed earlier, at high pH, the OH doublet is clearly resolved (see Fig 3.17). With lowering of pH, the doublet lines are broadened which finally coalesce to a single line (see Fig. 3.7). The OH protons here rapidly exchange with H^+ ions in solution. The second order rate constant for proton exchange was found to be ~ 1.8×10^8 L mole^{-1}s^{-1}.

FLUXIONAL MOLECULES

The internal motions of the unpaired electron can have an effect on the ESR spectrum. As an example, the rotational isomerisation of terephthaldehyde anion is discussed. Terephthaldehyde anion which can be obtained by electrolytic reduction can exist as a *cis* or a *trans* isomer as shown in Fig. 6.3. At ambient temperature, the ESR spectra of both the isomers can be observed.

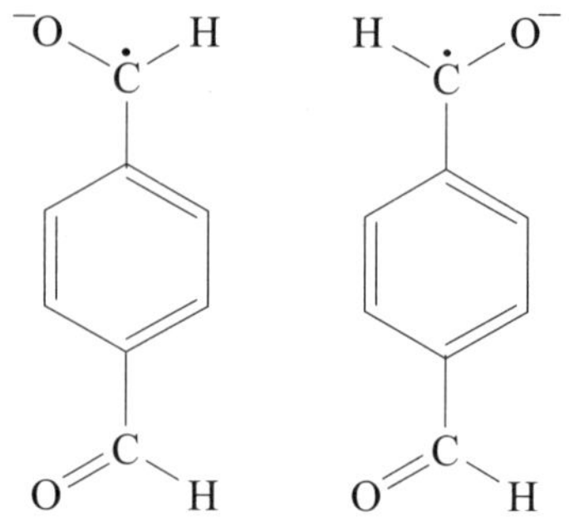

Fig. 6.3 Rotational isomers of terephthaldehyde anion.

The proton hyperfine splittings have been determined for both the isomers of terephthaldehyde anion. In each isomer, there are three pairs of equivalent protons, namely, two aldehyde protons and two protons *cis* and *trans* to an aldehyde oxygen atom. The lines are in the expected ratios of 1 : 2 : 1. The hyperfine splittings are $a_1 = 0.208$, $a_2 = 0.070$, $a_3 = 0.359$ mT for one isomer and $a_1' = 0.116$, $a_2' = 0.154$ and $a_3' = 0.301$ mT for the other. The a_3 and a_3' are assigned to the aldehydic protons and the remaining ones to the ring protons, which show changes in the distribution of the total unpaired electron spin density in the aromatic π orbitals. When the rate of isomerization is small compared to the hyperfine splitting, that is, $\tau_i >> 1/\delta B$, each isomer will produce its own characteristic spectrum. If the rate of interconversion is much greater than the hyperfine splitting, that is $\tau_i << 1/\delta B$, the observed spin distribution will be an average of that for the two rotational isomers and a simple average spectrum will result. Thermodynamic parameters can be evaluated from the rate constants determined at different temperatures. It was found that the free energy of activation for the interconversion was ~ 0.85 kJ mole^{-1} and the lifetime of the rotamers of terephthaldehyde was estimated to be at least 10^{-5} s.

A simple solid state example is $Cu(H_2O)_6^{2+}$ diluted with zinc flurosilicate crystal discussed in chapter 8. These ions can occur in any one of the three equivalent tetragonallly distorted octahedral configurations caused by the Jahn-Teller effect.

THE TRIPLET STATE 7

GENERAL REMARKS

In the discussion of ESR up to this point, we had considered free radicals which have just only one unpaired electron. However, there are many systems which have several unpaired electrons, such as transition metal ions. Further, molecules having diamagnetic ground states may have excited triplet states which have lifetimes long enough for their ESR spectra to be measured.

Very few molecules are found in the triplet ground state. Dioxygen molecule O_2 is one familiar example. Many transition metal ions exist quite stably in triplet and even higher multiplet states. Most triplet state molecules are obtained usually by photo-excitation. For example, when naphthalene is excited by ultraviolet light, one of the excited singlet electronic state of naphthalene decays to the metastable triplet state. By cooling the crystal to low temperatures or by diluting naphthalene in a solid inert matrix, the triplet state can be stabilized. Similarly, ultraviolet irradiation of diazodiphenylmethane leads to the formation of diphenylmethylene, which has a triplet ground state and it can be stabilized in solid matrices at low temperatures. These triplet state molecules are usually fairly long-lived and generally have lifetimes of 10^{-3} to 10 s. According to the selection rules, transition from the triplet electronic state to the singlet ground state is spin forbidden and hence the triplet excited state has an appreciable lifetime. The decay of the triplet state, therefore, continues for a relatively longer time after excitation and such a phenomenon is referred to as phosphorescence.

A molecule which contains two unpaired electrons with a total electron spin $S = 1$ is said to be in a triplet state. If the two electrons occupy the same spatial orbital, a singlet state is expected due to Pauli principle. However, if each electron occupies a different orbital, then both singlet and triplet states can occur. There are four possible combinations for the two electrons each having $m_s = \pm 1/2$. The electrostatic or exchange interaction between the two electrons separates one of the four states characterized by a total spin $S = 0$, that is, the singlet ground state ($m_s = 0$) from the other three with $S = 1$, the triplet state. The triplet consists of three distinct sublevels with the same energy and these sublevels can be distinguished by the components of their spin angular momentum about a chosen axis, namely, M_S, which has three possible values, + 1, 0, – 1. For a system of two interacting electrons, which are on different spatial orbitals, four configurations could be written : $\alpha(1)\,\alpha(2)$; $\alpha(1)\,\beta(2)$; $\alpha(2)\,\beta(1)$ and $\beta(1)\,\beta(2)$, where 1 and 2 indicate the two electrons and α and β refer to the up and down spins, respectively.

These configurations for a molecule of finite size can conveniently be combined into states which are symmetric or antisymmetric with respect to exchange of electrons. The three symmetric states defining the triplet state $S = 1$ are

$$\alpha(1)\,\alpha(2) \qquad (M_S = 1)$$

$$(1/\sqrt{2})\,[\alpha(1)\,\beta(2) + \beta(1)\,\alpha(2)] \qquad (M_S = 0)$$

$$\beta(1)\,\beta(2) \qquad (M_S = -1)$$

The antisymmetric state defining the singlet state $S = 0$ is

$$(1/\sqrt{2})\,[\alpha(1)\,\beta(2) - \beta(1)\,\alpha(2)]$$

The electronic configurations of singlet and triplet states are shown in Fig. 7.1. The three configurations of the triplet state are degenerate in the absence of a magnetic field, as stated before.

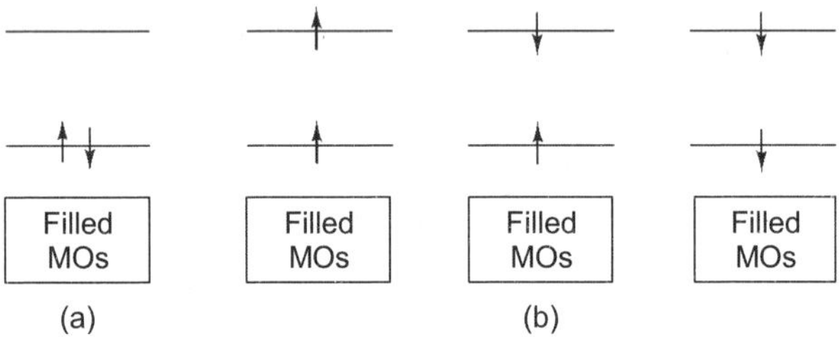

Fig. 7.1 (*a*) Singlet and (*b*) triplet levels corresponding to $M_S = 1, 0, -1$ for a system of two interacting electrons.

For a system of two interacting electrons, the singlet state lies lower in energy than the triplet state, separated by an energy difference of hJ where $J > 0$ is the exchange energy and h is the Planck's constant. The ESR spectroscopy of the triplet state molecules is described here. The ESR spectrum of naphthalene triplet which is a typical organic triplet is discussed as a representative system. Optical studies have indicated that the singlet-triplet splitting in napthalene is about 20,000 cm^{-1}. Some of the concepts employed here to interpret the spectra of the triplet state are common to the study of the ESR spectra of transition metal complexes described next.

SPIN TRANSITIONS IN TRIPLET STATE

In a magnetic field, there are three energy levels associated with a triplet, with $M_S = +1, 0, -1$ which may be thought of as arising from the electron spin orientations with (*i*) both the spins parallel to the field, (*ii*) both the spins antiparallel to the field and (*iii*) one spin parallel and the other spin antiparallel to the external magnetic field. The values of + 1, 0 and – 1 in units of $h/2\pi$ give the components of the total spin angular momentum (M_S) in the direction of the applied magnetic field B (see Fig. 2.3, chapter 2). If we neglect the dipolar interactions between the two electrons, in the presence of a magnetic field the degeneracy of $M_S = 1$, 0 and -1 is removed as shown in Fig. 7.2. The $M_S = 0$ level is not dependent on the external magnetic field. The energy of the other two states varies with the magnetic field as shown in Fig. 7.2. Only one signal is expected in the ESR spectrum under the selection rule $\Delta M_S = \pm 1$, since the energy difference between $M_S = -1$ and $M_S = 0$ sublevels is the same as that between the sublevels $M_S = 0$ and $M_S = +1$. The transition from $M_S = -1$ to $M_S = +1$ level is forbidden since it would be a $\Delta M_S = 2$ transition.

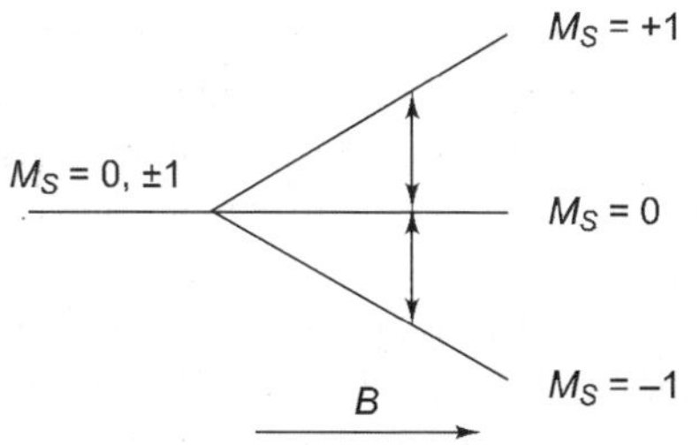

Fig. 7.2. Splitting of the triplet state in a magnetic field without zero-field splitting.

EFFECT OF DIPOLAR FIELD

For the triplet state molecules, the magnetic dipole dipole interaction between the spins (that is, the dipolar effect) is usually of the same order of magnitude as the interaction of the unpaired electrons with the applied magnetic field. Thus the energy levels are modified by the stronger dipolar field. For a long time, attempts to detect the triplet states by the ESR method were unsuccessful, since no account was taken of the dipolar interaction between the two electrons. The electron magnetic moment is about 10^3 times larger than the nuclear magnetic moment and the local field of the electron is enormous and dominates the feature of the spectrum.

The anisotropic broadening arising from the large dipolar effect of the two electron spins was the cause of the difficulty in obtaining the ESR spectra of the triplet state molecules. The dipolar interaction is so strong that the molecular rotation provides a very efficient relaxation mechanism such that the lines are broadened beyond detection. Therefore, a rigid matrix is required for the observation of the ESR spectra of triplet state molecules. Line broadening effects of the anisotropy can be eliminated if the system is prepared such that all the triplet state molecules are oriented in the same way in an external magnetic field, so that the angle between the molecular axes and the external magnetic field is the same for all the molecules.

The technique of reducing the effect of anisotropy is therefore to crystallize the compound to be excited to a triplet state in a large excess of another substance. To find a suitable host crystal for a particular triplet state molecule is not an easy task since most organic crystals do not accommodate substitute molecules in the same regular way as the host molecules are oriented in the lattice. For the study of the triplet state of naphthalene, single crystals of durene (1,2,4,5-tetramethyl benzene) containing naphthalene which enters the durene lattice substitutionally are utilized. Naphthalene and durene have similar geometry and dilution of naphthalene with durene greatly increases the lifetime of the naphthalene triplet state.

ZERO-FIELD SPLITTING

In addition to electron exchange which splits the states into singlet-triplet, there exists the anisotropic magnetic dipole-dipole interaction. For a system with two (or more) unpaired electrons, the three-fold degeneracy of the triplet spin state may thus be removed by the dipole-dipole interaction itself even in the absence of an external magnetic field. In a crystal or a frozen sample, dipole-dipole interactions are anisotropic and the levels with $M_S = \pm 1$ are shifted relative to that with $M_S = 0$ even in the absence of an external magnetic field. This effect is referred to as the zero-field splitting as shown in Fig. 7.3. The extent of zero-field splitting depends on the structure of the molecule, the extent of spin-orbit coupling, etc. If the number of unpaired electrons is even (i.e. the total spin S=1,2,...), then the spin degeneracy may be completely lifted due to zero-field splitting. When a steadily increasing magnetic field is applied to the sample, the $M_S = \pm 1$ levels diverge as shown in Fig. 7.3 (*a*). The important consequence of this is that the two allowed transitions $M_S = -1 \rightarrow 0$ and $M_S = 0 \rightarrow +1$ which are of equal energy in the absence of zero-field splitting are now of quite different energy, and two signals are observed. If the zero-field splitting is small compared to $g\mu B_o$, this results in a pair of signals near $g = 2.0$ but if the zero-splitting is large (Fig. 7.4) as is often the case with transition metal ions, the apparent g-values may be very different from 2 and it may be difficult to observe the expected transitions.

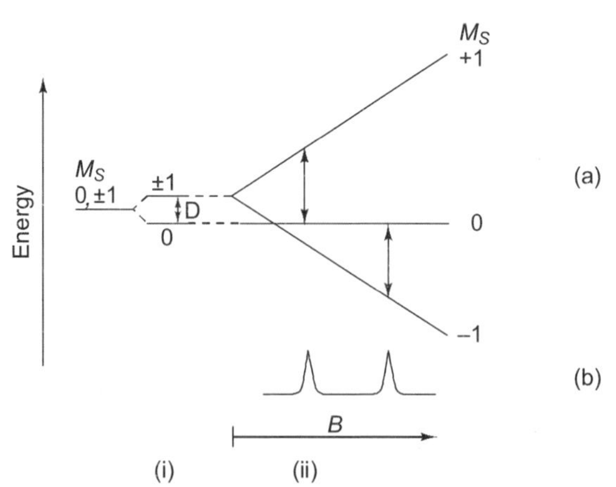

Fig. 7.3 (*a*) (*i*) The zero-field splitting (moderate) of a triplet state and (*ii*) the effect of a varying applied magnetic field on state energies. The allowed transitions ($\Delta M_S = \pm 1$) between the levels. (b) The resulting spectrum showing the fine structure.

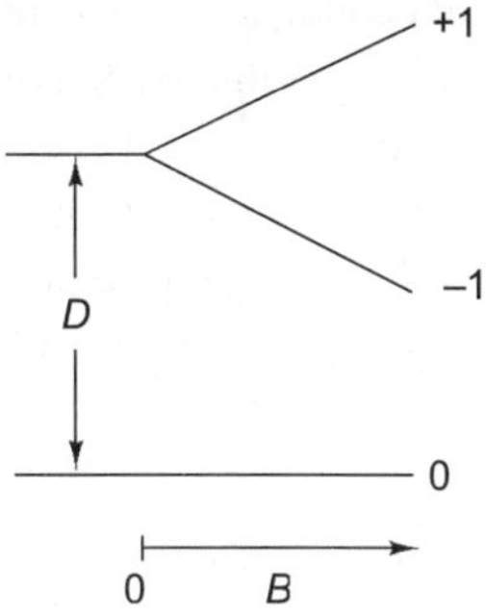

Fig. 7.4 Large zero-field splitting of the triplet state.

The net effect of the dipolar interactions is to raise both $M_S = +1$ and -1 states with respect to $M_S = 0$ state as shown in Fig. 7.3(a). Here, the dipolar splitting D is shown to be smaller than the splitting due to the applied magnetic field. Even for small magnetic fields, the energy levels of both $M_S = +1$ and $M_S = -1$ would be above $M_S = 0$ as shown in Fig. 7.3. The magnitude of the dipolar shift is independent of the applied magnetic field. The spin-orbit coupling may also disturb the energy level pattern but $M_S = 0$ state will be unaffected. The spin-orbit coupling may be in the same sense as the dipolar interaction or opposed to it. Further if there is a strong internal electric field within the crystal, it will result in further perturbation of the $M_S = +1$ and -1 energy levels but not $M_S = 0$ level. The net result of these three perturbations is usually considered under the zero-field effect, since they produce a shift of the $M_S = \pm 1$ states with respect $M_S = 0$ state even in the absence of an external magnetic field. The magnitude of zero-field splitting is usually large. For example, the observed separation between the $M_S = 0$ and $M_S = \pm 1$ sublevels for the naphthalene triplet is about 3,000 MHz. It may be compared with the separation between the levels $M_s = +1$ and $M_s = -1$ of nearly 8,000 MHz for an external magnetic field of 0.3 T. There will now be two transitions ($\Delta M_S = \pm 1$) for the triplet state molecule at two different applied magnetic fields for a given microwave radiation. The spectrum of naphthalene triplet will consist of two lines as shown in Fig. 7.3(b). The term fine splitting is used when a spectral line is split because of zero-field splitting. The appearance of more than one line in the spectrum when $S > 1/2$ is known as fine structure. In general, if there are n parallel spins, there will be n equally spaced fine structure lines in the ESR spectrum. The separation between the fine structure lines is usually large, some 0.1–0.2 T or more and the line width is 0.1 to 0.01 mT.

The zero-field splitting strongly depends on the orientation of the molecule in the magnetic field. The triplet state anisotropy can be studied by forming dilute solutions of the triplet state molecules in a suitable host single crystal and by varying the orientation of the single crystal in the applied magnetic field. The dipolar interaction between the two unpaired spins of the triplet molecule depends on the average separation between them. Two terms D and E are employed, D as a measure of the dipolar interaction and E as a measure of the extent of deviation of the molecule from three-fold or higher-fold rotational axis of symmetry. At zero-field, the two energy levels $M_S = +1$ and -1 are separated by the quantity $2E$ and the average of these two energies differs from the energy of the $M_S = 0$ level by the quantity D. It is important to note that the values of D and E depend on the axis chosen as Z-axis. The ESR line positions depend only on the relative signs of D and E and their absolute signs are often not known. The values of D and E parameters are usually given in cm^{-1}, D/hc and E/hc.

SPECTRA OF NAPHTHALENE TRIPLET

For the application of an external magnetic field to the triplet state molecules in single crystals, the choice of X, Y, Z-axes depends on the symmetry of the molecule. For axially symmetric molecules such as naphthalene, the Z-direction is taken as perpendicular to the molecular plane. The Y-axis lies along the $C_9 - C_{10}$ bond and the X-axis lies in the plane of the ring and perpendicular to the Y-axis as shown in Fig 7.5. By suitable rotation of the crystals with respect to the applied

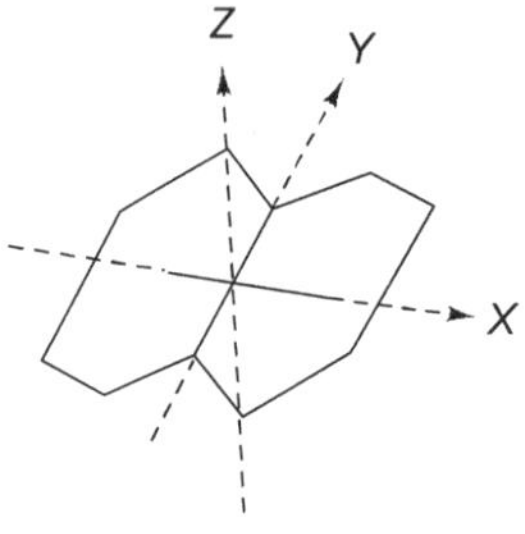

Fig. 7.5 Definition of the coordinate axes for naphthalene.

magnetic field, the same resonance signal could in fact be shifted over a range of nearly 0.2 T. It means that this signal in a polycrystalline sample would have a width of over 0.2 T. Thus, in a frozen glassy solution, where the molecules are randomly oriented, the resonance line is spread over a range of 0.2 T giving rise to an extremely broad and low intensity band rendering the detection of the signal difficult.

For naphthalene in its lowest triplet state, the spin energies as a function of the magnetic field parallel to Z-axis and the transitions observed are shown in Fig. 7.6 assuming that $D > 0$. Similar plots have been made when the magnetic field is parallel to X and Y axes. The spectrum of naphthalene triplet molecules in a single crystal of durene when the applied magnetic field is along the Z-axis shows two signals at 223.0 and 437.8 mT due to the transition between adjacent energy levels ($\Delta M_S = \pm 1$) as shown in Fig. 7.6. The transition energies are given by $h\nu = g_Z \beta_e B \pm D$. Two fine structure lines occur at a distance of $2D/g\beta_e$ in an applied magnetic field. For naphthalene triplet, if we neglect E since it is small ($0.014\ cm^{-1}$), it is noted from Fig. 7.6, that the difference between the two transitions is 214.8 mT and therefore D is $0.10\ cm^{-1}$. By studying the temperature dependence of the spectrum, the sign of D was found to be positive.

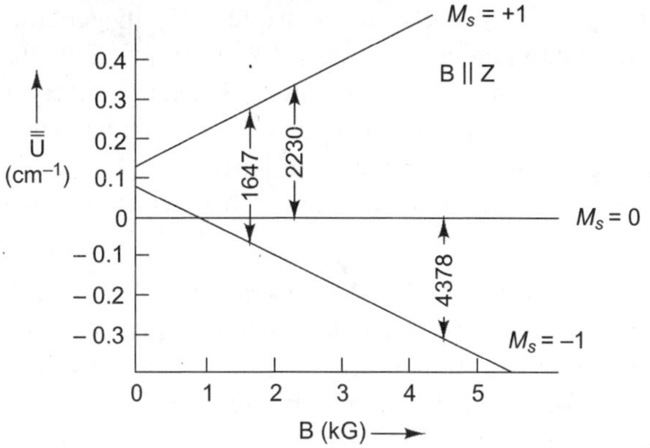

Fig. 7. 6 Spin energies for the naphthalene triplet state as a function of the applied magnetic field parallel to the Z-axis and ν = 9.272 GHz.

For each of the three axes of the molecule *X*, *Y* and *Z*, to which the applied magnetic field may be parallel, two transitions between the adjacent levels can occur. Therefore, under favorable conditions, one can obtain totally six lines representing the transitions between adjacent energy levels. In other words, there are three small sub-populations parallel to each of the three cartesian axes of the triplet state molecules which can be detected. Each of these three sub-populations gives rise to a pair of signals representing the two adjacent energy level transitions. These three pairs occur on the opposite sides of the centre of the spectrum. The ESR line positions for different orientations of naphthalene triplet molecules with respect to the external magnetic field are summarized in Fig. 7.7.

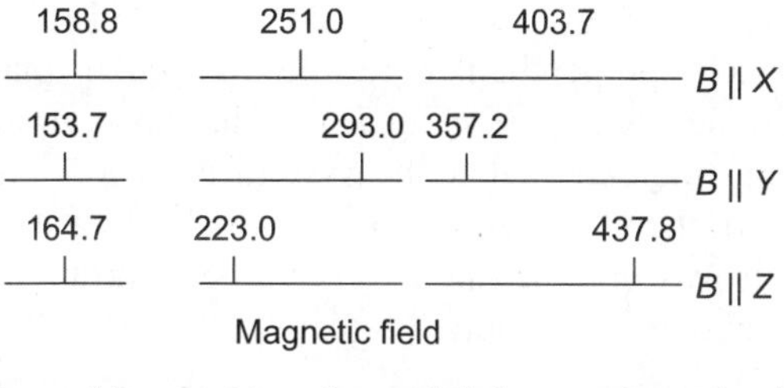

Fig. 7.7 The ESR spectral line positions for naphthalene triplet molecules for ν = 9.272 GHz for different orientations of the applied magnetic field. The lowest field transition in each case arises from $\Delta M_S = \pm 2$ type of transition. The values are in mT.

The transition between adjacent energy levels is more anisotropic and therefore more difficult to observe. It can be detected in some cases since the signals are inherently more intense than those of the transitions between non-adjacent energy levels. Although these transitions can be observed in randomly oriented samples, the intensity of the signals is weaker than that for the same number of triplet state molecules oriented in a single crystal. We measure the resonance arising from the triplet state molecules having the molecular axis nearly parallel to the applied magnetic field *B*. The intensity of the resonance due to other molecules of the triplet population is too weak to be detected.

The ESR spectrum characteristic of the triplet state is obtained only if the energy difference between the singlet and triplet states is larger than any magnetic interaction. The energies of the three sublevels of the triplet are, in general, functions of D, E, g and B. For naphthalene triplet g (isotropic) = 2.0030, $D = 0.1012$ cm^{-1} and $E = -0.0141$ cm^{-1} have been obtained. The magnitudes of D and E are related to strength of the interaction between the two spins. They become negligible when the two electrons are localized in parts of the molecule that are very far apart. The spectra of triplet states cannot be observed in solution unless the two spins are far apart.

Another consequence of zero-field splitting is that the selection rule $\Delta M_S = \pm 1$ may break down and more transitions may be observed. At low fields, quantum number M_S is no longer strictly applicable. In addition to $\Delta M_S = \pm 1$ transitions (Fig. 7.3), the resonance due to $\Delta M_S = \pm 2$ transition (between $M_S = -1$ and $M_S = +1$) has also been observed at low fields when the zero-field splitting is small compared to the microwave frequency for significantly high power levels of microwave radiation. These transitions occur at nearly one-half the magnetic field required to cause the $\Delta M_S = \pm 1$ transitions and are hence generally known as the half-field resonances. They are weakly allowed transitions. The half-field resonances are markedly less anisotropic than the $\Delta M_S = \pm 1$ transitions and are spread over a narrow range for different orientations of the molecule so that signals of considerable intensity can be detected. It is possible, therefore, to study the triplet states in glassy solutions at low temperatures. It is of practical importance since one is no longer restricted by the single crystal technique. The triplet states of many molecules have been studied in this manner. If the zero-field splitting parameters are sufficiently large compared to the microwave photon energy $h\nu$, $\Delta M_S = \pm 2$ transitions cannot occur. For naphthalene triplet, with the applied magnetic field perpendicular to the molecular plane, the half-field resonance is found at 164.7 mT (Fig.7.6). The observation of a half-field transition is a strong indication that the system is in a triplet state. The ESR spectrum of 1, 3, 5–triphenyl benzene dianion which has a triplet ground state is shown in Fig. 7.8. It was obtained in a rigid glassy solution and the spectrum is consistent with the above discussion.

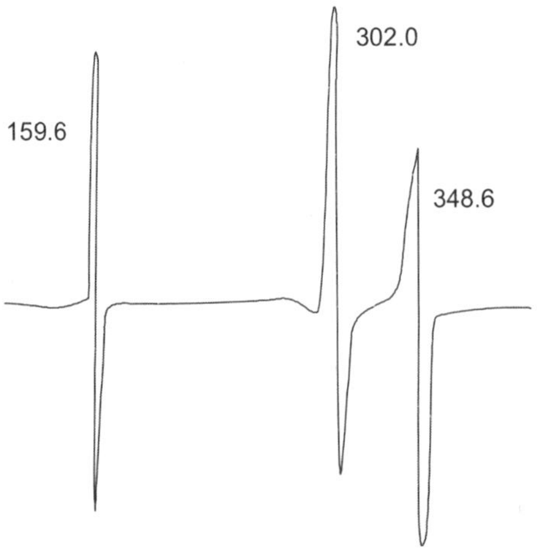

Fig. 7.8 The ESR spectrum of a rigid frozen solution of 1,3,5-triphenyl benzene dianion. The resonance at 159.6 mT is due to $\Delta M_S = \pm 2$ transition and the other lines due to $\Delta M_S = \pm 1$ transitions are also in mT.

TRANSITION METAL COMPLEXES 8

INTRODUCTION

Very few inorganic free radicals are stable under ordinary circumstances. A representative group includes O_2, NO, NO_2, etc. These radicals are exactly similar to the organic free radicals with regard to having the hyperfine structure and the interpretation of their ESR spectra. There is another class of inorganic free radicals, which are numerous. These are the complexes of transition metal ions, which have incomplete 3*d*, 4*d* or 5*d* orbitals, and the unpaired electrons reside in their *d* orbitals. They present some new features in the ESR spectra which are either not present or not important for organic free radicals.

The complexity of the subject prevents a detailed discussion of the ESR spectra of metal complexes. Instead only a qualitative picture is briefly presented here. The metal complexes are of interest because of their diverse roles in biochemistry and solid state chemistry. One of the most classic demonstrations of electron delocalization from the metal ion to the molecules coordinated to it (referred to as ligands) was provided by Griffiths and Owen from the ESR spectra of the chloro complexes of iridium, $(NH_4)_2IrCl_6$ and Na_2IrCl_6. The hyperfine splitting caused by the interaction of the unpaired electrons with the surrounding Cl^- ions was clearly seen in the spectra. These hyperfine splittings showed directly for the first time that not only the ligand orbitals participate in bonding but also that the *pi* bonding of the metal with the ligands is important. They were able to calculate the extent of delocalization of the unpaired electrons onto the chloride ions from the observed splittings. For the transition metal complexes, the calculated *g*-values, spin-orbit coupling constant and the separation between the terms (for example, $^3F-{}^3P$ or $^4F-{}^4P$) using the ESR spectral results were found to be lower than those for the corresponding free metal ion. These findings are often interpreted in terms of the covalent interaction of the metal ion with the ligands.

For the transition metal complexes, the nature of the metal, the arrangement of the ligands around the metal ion and the nature of the metal-ligand bonding can be investigated by studying the hyperfine structure and the *g*-values. The most complete results about the ordering of the energy levels and bonding in transition metal complexes are obtained, in general, from the ESR study of dilute single crystals. However, the requirement of single crystals places many implicit restrictions on the systems which can be studied. Consequently, much efforts have been devoted to analyzing the spectra of polycrystalline or frozen glass samples. Detailed information on the distribution of the unpaired electron can also be obtained from a study of dilute polycrystalline samples and from glasses obtained by freezing suitable solutions. ESR spectral studies may be used to supplement crystallographic studies; to determine the number of magnetically inequivalent ions in the unit cell and the symmetry of these ions at the crystal sites. ESR spectroscopy has provided very reliable evidence for the dynamic Jahn-Teller distortion in transition metal complexes.

Extensive studies have been carried out on the ESR spectra of transition metal complexes particularly of the first transition series. The spectra of the complexes of the second and third transition series and also of lanthanides and transuranic elements have also been investigated. Transition metal complexes of biochemical importance such as metalloproteins have been investigated. The ESR spectroscopy is a very useful technique for determining the structure of these classes of compounds and information on the oxidation state, the coordination number of the metal ion and the nature of metal-ligand bond have been derived.

FEATURES OF THE SPECTRA OF TRANSITION METAL COMPLEXES

The ESR spectra of transition metal complexes are markedly different from the spectra of small organic and inorganic free radicals in solution. In transition metal complexes, the energy levels available for the unpaired electrons may be split by the intrinsic field of the surrounding ligands. The type and extent of splitting depends on the nature of the ligand and their arrangement around the metal ion. In transition metal complexes at least five additional factors will have to be considered.

(*i*) One of the major differences between the ESR spectra of metal complexes and the solution spectra of organic free radicals is that the electron spin is strongly quantized along the molecular axis of the metal complex. Thus, the geometry of the metal complex introduces anisotropy into the ESR spectrum. When the crystal axis of the complex makes different directions with the external magnetic field, the ESR resonances are observed at different positions. This anisotropy can be quite large. The signals in the spectrum of an organic free radical are usually much sharper than those of a randomly oriented metal chelate spectrum.

(*ii*) Complications arise due to the degeneracy of the *d*-orbitals and the presence of more than one unpaired electron in the *d*-orbitals. The electronic configuration, zero-field splitting and the applied magnetic field determine whether or not all or any of the transitions can be observed or what their *g*-values would be. The ESR spectra of the complexes containing more than one unpaired electron can be difficult to observe or interpret. In particular, transition metal complexes with even electron configuration are very difficult to study. When zero-field splitting is very large for systems in which the number of electrons is even, no resonance may be observed, within the observable range and those with odd number of electrons may show only a single resonance corresponding to the $M_s = -1/2 \rightarrow 1/2$ transition.

Transition metal complexes contain several unpaired electrons and degenerate orbitals. These manifests themselves in an orbital contribution to the magnetic moments which leads to anisotropy in *g*-factors and secondly in zero-field effects.

(*iii*) Combined effects of coulombic repulsion between the *d*-electrons and spin-orbit coupling give effective *g*-factors which deviate widely from 2.0023 (an extreme range could be 0 to 9) and which vary with the orientation of the complex in the magnetic field. For example, in ferrihaemoglobin, the *g*-factor for the iron resonance vary from a value of 6 when the field is in the plane of the porphyrin ring, to a value of 2 when the applied magnetic field is normal to the plane of the porphyrin. Spin-orbit coupling effects also give rise to large zero-field splitting of 10 cm^{-1} or more by the mixing of the ground state with the excited electronic states. Rapid electron relaxation and *g*-value anisotropy are also manifestations of the effects of spin-orbit coupling. The spin-spin coupling is much larger for heavier atoms than for lighter ones. Partly for this reason, ESR spectra of the second and third row transition metal complexes are often more difficult to observe and interpret than those of the first row transition metal complexes.

(*iv*) We noted earlier that for the naphthalene triplet, the zero-field effects arise from dipolar interactions between the two electron spins. In transition metal ions, however, this term is employed to describe any effect that lifts the spin degeneracy, which includes dipolar interactions and spin-orbit coupling. A low symmetry crystal field often gives rise to a large zero-field splitting. Although tumbling of molecules averages zero-field splitting to zero in solution, the spin-spin interactions cause rapid relaxation and the spectra may be too broad to be observed. It is often necessary to isolate spin centres from each other by freezing the solution to glassy state. Another consequence of the arrangement of the ligands around a metal ion in a lower symmetry is that the *g*-value becomes anisotropic. Therefore *g* is, in general, a tensor quantity for the transition metal complexes, reflecting the anisotropy of the ligand field.

(*v*) The variation in the spin-lattice relaxation time for different transition metal complexes is rather large. For some complexes, the relaxation time may be sufficiently long to allow the measurement of spectra at ambient temperature and for other complexes this may not be possible. Since these relaxation times usually increase with decrease in temperature, some transition metal complexes may require cooling down to liquid nitrogen or liquid helium temperature before spectra are measured.

The ESR spectra of randomly oriented powdered samples provide a less detailed picture of their structure and electron distribution than studies in single crystals or in solution. Nevertheless, there are many systems, where studies as single crystals or as solutions are not feasible so that polycrystalline or amorphous materials must be utilized. Method of obtaining the g and a-values from the spectra of polycrystalline materials may be complicated.

It may be recalled that dilution is necessary by incorporation of the paramagnetic ion in an isomorphous diamagnetic crystal by say 1 : 100 (to 1000) to reduce the line broadening caused by neighbouring dipoles and to obtain fairly resolved spectra. The spin-spin interaction is reduced by increasing the distance between the paramagnetic ions by diluting them in an isomorphous diamagnetic host. For example, a copper(II) complex could be studied as a single crystal or as a powder by diluting it with the corresponding zinc complex or in a frozen solution. The electron spin of the given complex is isolated from another of the paramagnetic molecule by dilution of the solid and thereby the spin lifetime is increased. Chemical exchange processes also affect the line widths considerably. This effect can also be reduced by dilution. Dipolar broadening is still an important factor for pure solids but well-resolved spectra with hyperfine structure are readily observed in solution at room temperature. Host lattices such as MgO, TiO_2, Al_2O_3, etc. have been utilized to introduce the paramagnetic ions. This technique allows the observation of the spectra of paramagnetic ions in symmetry environments of one's choice and it provides a more systematic approach to the study of the effects of the crystal fields. Octahedral environments are provided by MgO, CaO, MnO, Al_2O_3, etc. while TiO_2 provides a tetragonal field. Host lattices such as ZnO, ZnS, CdS, etc. provide tetrahedral environments.

The hyperfine splitting in transition metal complexes arise not only from the nuclear spin of the central metal ion but also from the ligand nuclei. The resulting additional splitting of the lines in the ESR spectrum due to the ligand nuclei is called 'super hyperfine splitting'. The strength of coupling depends on the extent of delocalization of the unpaired electron onto these atoms, and decreases rapidly as the number of bonds increase. The hyperfine splitting due to the metal nucleus depends on the natural abundance and also on the magnetic moment of the metal isotope. Considerably large magnetic moment or magnetogyric ratio is necessary for the observation of the hyperfine structure due to the metal nucleus. When the unpaired electron interacts almost exclusively with a single metal atom, hyperfine splitting of 10 mT or more may be observed.

Crystal field and ligand field theories, which serve as a basis for the interpretation of bonding, magnetic and electronic spectral properties of the transition metal complexes, also serve as a basis for the interpretation of the ESR spectra of transition metal complexes. For a detailed description of the bonding and structure of the transition metal complexes from crystal field and ligand field theories, the books cited in the references may be consulted. A brief sketch of the electronic structure of the transition metal complexes and their ESR spectra is given here.

ENERGY LEVELS IN A METAL ION

Russell–Saunders Coupling

We shall first discuss the electronic states of free atoms and ions. We shall then see how the symmetry of the chemical environment around an ion determines its energy levels. The effect of ligand environment is treated as a small perturbation on the states of the free ion in the crystal field and ligand field theories.

Due to electrostatic interactions in many electron atoms, an electronic configuration can lead to states having different energies. We shall now determine the number of states of different energies that may arise for a given electronic configuration. The most convenient classification of the energy states is by the Russell-Saunders scheme. In this scheme, the individual orbital angular momenta l_i of the electrons are strongly coupled together to form the total orbital angular momentum L. The individual spin angular momenta s_i are similarly coupled together to form the total spin angular momentum S. Weak coupling occurs between L and S to give the resultant total angular momentum J (= $L + S$) which can have $2J + 1$ values varying from $L + S$, $L + S - 1$, ... to L – S. The $(2J + 1)$- fold degenaracy of the total angular momentum can be removed in an external magnetic field and the component of J about a reference direction is denoted as M_J.

Corresponding to the orbital designations s, p, d, f, ..., etc. for quantum number $l = 0, 1, 2, 3, \ldots.$, which determine the angular momentum of electrons in individual orbitals, the total orbital angular momentum $L = 0, 1, 2, \ldots$ are specified by the capital letters as $S, P, D, F\ldots$ (J is omitted in the sequence), respectively.

The term symbol denoting an atomic energy state is written as $^{2S+1}L_J$. The spin multiplicity is shown as a superscript on the left while the J value is indicated as a subscript on the right of the term symbol. For example, for the d^1 electron, $L = 2$ and $S = 1/2$, and hence $J = 5/2$ and $3/2$ giving the terms $^2D_{5/2}$ and $^2D_{3/2}$. The energy difference between these two states is as a result of spin-orbit coupling.

By analogy with the nomenclature for one electron configuration, the projections of L and S are called M_L and M_S. The sum of m_l values for all electrons of the set is M_L. Similarly, M_S is the sum of all m_s values. For a given value of L, M_L can have $(2L + 1)$ values ranging from $L, L - 1, \ldots$ to $-L + 1$, – L. For a given S, there are $(2S + 1)$ values of M_S differing by integers from $+S$ to $-S$, which denotes the spin degeneracy or spin multiplicity as it is called.

We shall now derive the term symbols for an atom containing two unpaired electrons which are equivalent as in the ground state of carbon $1s^2 2s^2 2p^2$. Since the two $2p$ electrons have the same n and l quantum numbers, the terms that are consistent with the Pauli's principle will have to be selected by utilizing the m_l and m_s values. Therefore, we write all the allowed combinations of m_l and m_s values for the two p electrons and determine M_L and M_S values for each of these. A term corresponding to a given value of L and S has $(2L + 1)(2S + 1)$ combinations of M_L and M_S values associated with it. Each allowed combination of M_L and M_S is called a microstate. The number of microstates for e electrons in n orbital sites (6 for p orbitals, 10 for d, etc.) is given by

$$\frac{n!}{e!h!}$$

where h is the number of holes $(n - e)$. Holes are vacancies in the subshell. For the p^2 configuration, the number of microstates is 15. A term corresponds to a set of microstates. Thus by writing all possible microstates (that is, all the arrangement of electrons) we can find the term. In Table 8.1 the possible 15 microstates which are consistent with the Pauli's principle are listed. We shall write a microstate as $(m_{l_1}^{\pm}, m_{l_2}^{\pm})$ as shown in Table 8.1 where the superscript + or – refer to m_s values + 1/2 or – 1/2, respectively. For the p^2 configuration, M_L can take 2, 1, 0, –1, –2 and M_S can take + 1, 0, – 1 values.

Table 8.1 Microstates for p^2 configuration

	M_S		
M_L	1	0	– 1
2		$(1^+, 1^-)$	
1	$(1^+, 0^+)$	$(1^+, 0^-)(1^-, 0^+)$	$(1^-, 0^-)$
0	$(1^+, -1^+)$	$(1^+, -1^-)(1^-, -1^+)(0^+, 0^-)$	$(1^-, -1^-)$
– 1	$(-1^+, 0^+)$	$(-1^+, 0^-)(-1^-, 0^+)$	$(-1^-, 0^-)$
– 2		$(-1^+, -1^-)$	

We begin with $M_S = 0$, $M_L = 2$ and associate with them $L = 2$, $M_L = 2, 1, 0, -1, -2$ and $S = 0$, $M_S = 0$. The microstates $(1^+, 1^-)$, $(1^+, 0^-)$, $(1^+, -1^-)$, $(-1^+, 0^-)$ and $(-1^+, -1^-)$ corresponding to the values of $M_L = 2, 1, 0, -1, -2$ arise from ^1D state. For $L = 0$, and $S = 0$, corresponding to ^1S term, we have $M_L = 0$, $M_S = 0$ and it corresponds to the microstate $(0^+, 0^-)$. The remaining nine microstates account for the ^3P term. Thus the configuration $2p^2$ gives rise to three terms ^3P, ^1D and ^1S. The ^3P term is further split by differing J values. The relative energies of the terms arising from the $1s^22s^22p^2$ configuration of carbon as determined experimentally are shown in Fig. 8.1. The separation between the adjacent terms of the free ion is of the order of ~10,000 cm^{-1} and is much larger than the spin-orbit coupling. The complexity of determining the terms increases with the number of electrons and with higher L values.

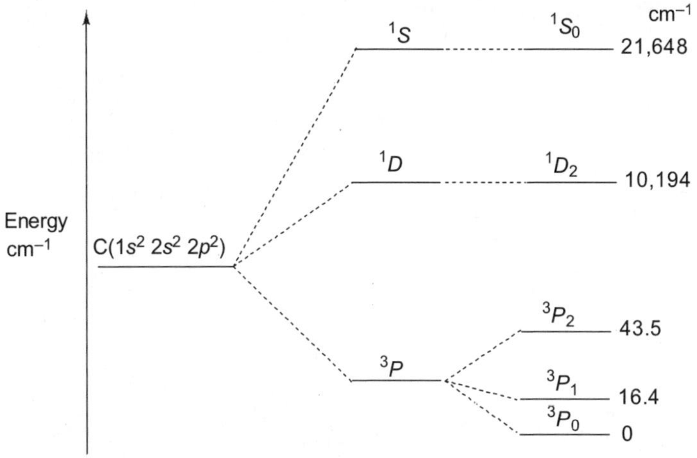

Fig. 8.1 Terms arising from the ground state configuration, $1s^2\ 2s^2\ 2p^2$ of carbon. The ^3P term is further split as a result of spin-orbit coupling.

Hund's Rules

The Hund's rules are used to identify the lowest energy term arising from the ground state electronic configuration. These rules can be given in three steps.

(*i*) The term with highest spin multiplicity has the lowest energy. The interelectron repulsion for the states with maximum number of parallel spins is the least. For atoms with two unpaired electrons triplet state generally lies lower in energy than the singlet state arising from the same electron configuration.

(*ii*) Among the states of same spin multiplicity, the term with highest L lies lowest in energy. Thus, the ^1D state lies lower in energy than the ^1S state.

(*iii*) For the given S and L values where the subshell is less than half-full, state with minimum J lies lower in energy and if the subshell is more than half-full, state with maximum J value lies lower in energy. This rule predicts the ground state of carbon as $^3P_0$.

Owing to the well-known equivalence of electrons and holes (that is, vacancies in the subshell), the electronic configurations p^n and p^{6-n} (similarly, d^n and d^{10-n}) will produce identical terms. For example, $C(1s^22s^22p^2)$ and $O(1s^22s^22p^4)$ both give rise to the same terms ^3P, ^1D and ^1S. However, the ground state term symbol for oxygen is $^3P_2$ whereas for carbon it is $^3P_0$.

The Russell-Saunders terms for a $3d^2$ configuration can be derived following the above procedure for the $2p^2$ configuration. The $3d^2$ configuration gives rise to ^1S, ^1D, ^1G, ^3P and ^3F terms (Fig. 8.2). Using Hund's rules, the lowest energy ground state is identified as $^3F_2$. The sequence of energy for the other terms are obtained from atomic spectroscopy.

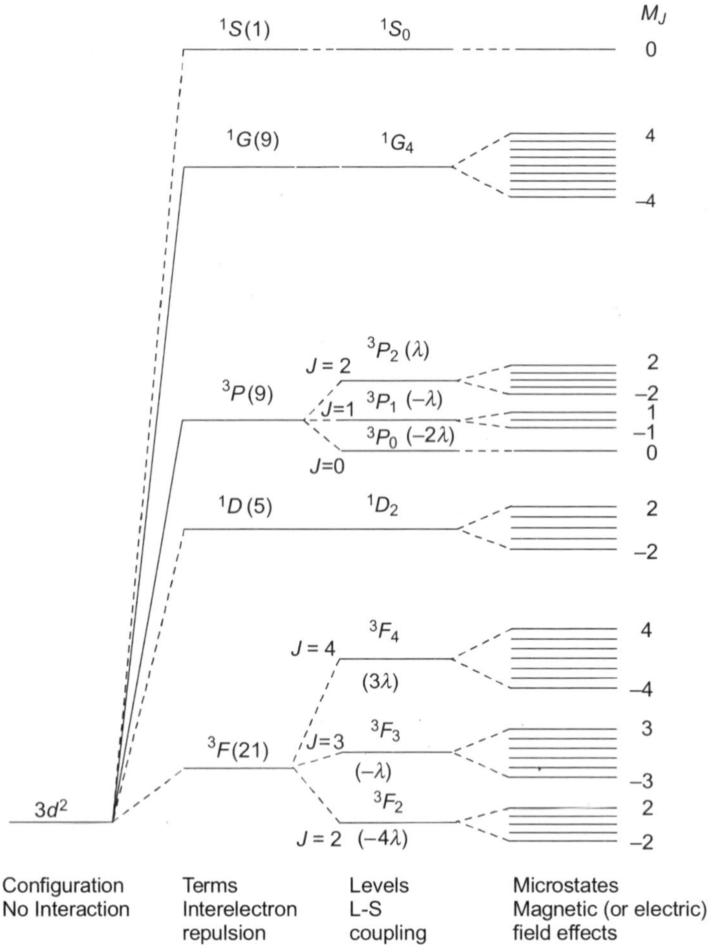

Fig. 8.2 Schematic representation of the terms arising from the ground state $3d^2$ configuration of a metal ion. λ denotes spin-orbit coupling constant. The number in the parenthesis indicates the total degeneracy of each level. Altogether there are 45 different levels.

The Landé interval rule provides a basis to obtain the separation between any two J levels of a Russell-Saunders term. According to this rule, the energy difference between any two J levels of a Russell-Saunders term is proportional to the larger J value of the pair. As an example, the splitting of the ground state ^4F term of Cr^{3+} (d^3) ion is shown in Table 8.2. Taking the J = 3/2 sublevel to be at 0 cm^{-1}, the sublevels corresponding to J = 5/2, 7/2 and 9/2 occur at 225, 560 and 960 cm^{-1}, respectively. It is seen that the Landé rule holds well, the numbers in the last column of the Table 8.2 being nearly equal.

Table 8.2 ^{4}F State in Cr(III) d^3

Level	*Energy, cm^{-1}*	*Interval ΔE, cm^{-1}*	$\lambda = \Delta E/J$
$^4F_{3/2}$	0		
$^4F_{5/2}$	225	225	90
$^4F_{7/2}$	560	335	96
$^4F_{9/2}$	960	400	89

In many cases, we wish to determine the ground state term symbol without constructing all the possible microstates. This is achieved by filling up the orbitals starting from the highest m_l value and adding one electron to each orbital before any orbital is doubly occupied. It is assumed that electrons preferentially occupy the orbitals with the highest m_l value. For the d^4 (Mn^{3+}) ion, it gives

m_l	2	1	0	− 1	− 2
m_s	+1/2	+1/2	+1/2	+1/2	

$(M_S)_{max} = 2$ $\qquad\qquad$ $(M_L)_{max} = 2$

We know that if $M_S = 2$, then $S = 2$ and if $M_L = 2$, then $L = 2$. The term symbol when $L = 2$ and $S = 2$ is ^5D.

Spin-Orbit Coupling

The interaction of the electron spin with its orbital motion is termed as spin-orbit coupling. The spin-orbit interaction energy is given by

$$E_{so} = \xi\, \mathbf{l.S} \tag{8.1}$$

where ξ is a constant depending on n, l and Z^4 where Z denotes the nuclear charge. The spin-orbit splitting decreases with n and l and it is proportional to Z^4. For some heavier elements, the spin-orbit interaction is so large that *RS* coupling is not valid and a better description of the states is to use *jj* coupling. While it is almost negligible for 1*s* electron of hydrogen atom (~ 1 cm^{-1}), the spin-orbit coupling is larger in heavier atoms (e.g. 580 cm^{-1} for 3*p* electrons of chlorine).

The magnitude of the spin-orbit interaction is usually described by two parameters λ and ξ. The parameter ξ is employed to describe the spin-orbit interaction of a single electron. The parameter λ is used to describe the energy of the spin-orbit interaction for the Russell-Saunders term. The values of ξ and λ are related by

$$\lambda = \pm\, \xi \,/\, 2S \tag{8.2}$$

where S is the total electron spin. The parameter ξ is a positive quantity. If the shell is less than half-full, the sign of λ is positive. If the shell is more than half-full, λ is negative. For half-filled shells, $L = 0$, the spin-orbit energy is zero.

EFFECT OF CRYSTAL FIELD ON *d* ORBITALS

In a solid or a liquid, a transition metal ion is usually surrounded by the ligands, which are either negatively charged such as Cl^-, Br^-, CN^-, etc. or negatively polarized such as OH_2, NH_3, etc. The ligands donate a pair of electrons each to the metal to form metal-ligand bonding. The most common arrangement of ligands around a metal ion is six ligands arranged at the corners of an octahedron or four ligands arranged at the corners of a tetrahedron. We shall confine in the following discussion to octahedral coordination.

The crystal field theory is conceptually quite simple and it assumes the interaction between the metal ion and the ligands as purely electrostatic in nature. In an octahedral complex, the six ligands may be visualized to coordinate along the Cartesian coordinate axes x, y and z (Fig. 8.3). The spatial orientation of the five d orbitals is shown in Fig. 8.4. In a free gaseous transition metal ion, the five d orbitals are degenerate. The metal ion at the centre of the complex will experience a non-spherical field due to the ligands. The electrostatic repulsion between the electrons of the metal ion and the electrons of the donor ligands is greater for the $d_{x^2-y^2}$ and d_{z^2} orbitals than for those in the d_{xy}, d_{yz} and d_{xz} orbitals.

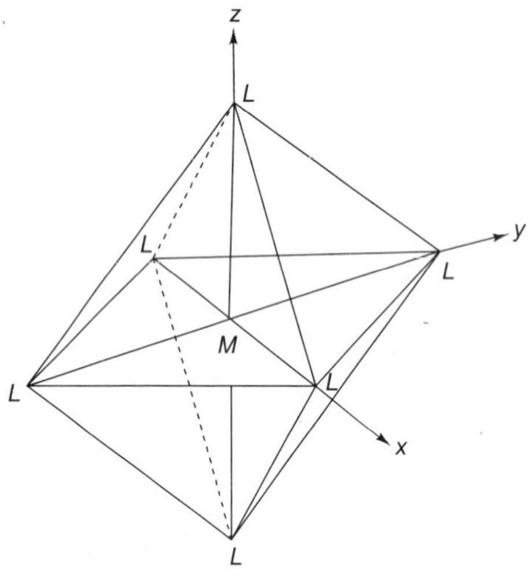

Fig. 8.3 The arrangement of six ligands around the central metal atom in an octahedral ML_6 complex.

As a result, the two orbitals $d_{x^2-y^2}$ and d_{z^2} known as the e_g orbitals will be raised in energy. The other three orbitals d_{xy}, d_{yz} and d_{xz} known as the t_{2g} orbitals will be lowered in energy relative to the energy of the d orbitals in a spherical field. The difference in energy between the two sets of orbitals is termed as the crystal field or ligand field splitting and it is denoted by the symbol 10 Dq or Δ as shown in Fig. 8.5. Thus, when an octahedral complex is formed, the energy of the e_g orbitals is raised by 6 Dq relative to their energy in a spherical field, while the corresponding decrease in energy for each of the t_{2g} orbitals is 4 Dq. The value of 10 Dq could be obtained experimentally from electronic spectral measurements. The mean energy remains unchanged relative to the energy of the d orbitals in a spherically symmetric environment.

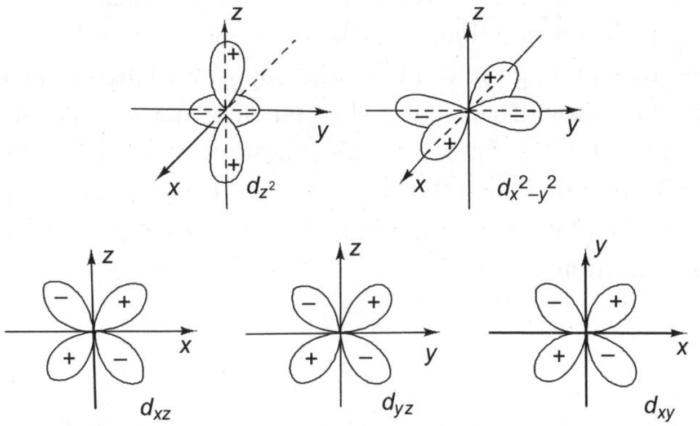

Fig. 8.4 The 3d atomic orbitals. The sign of the wave function changes from one lobe to the next in a given orbital.

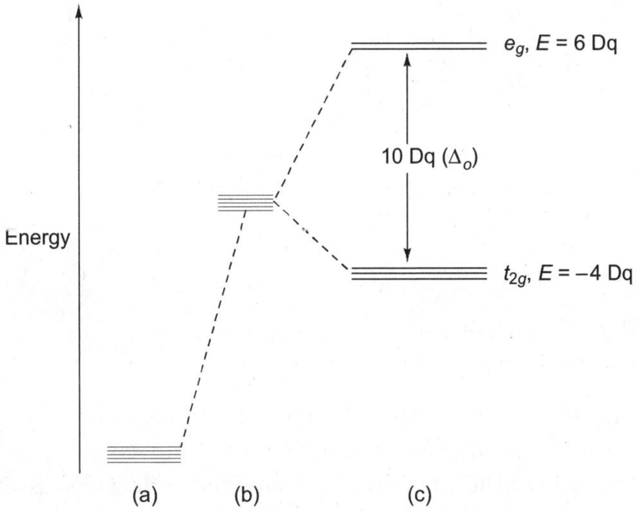

Fig. 8.5 Crystal field splitting in an octahedral complex: (a) free metal ion, (b) spherical field and (c) splitting of d orbitals into two groups in an octahedral field.

The capital letters T, E and A (or B) are group theoretical symbols used to designate triply, doubly and non-degenerate energy states, respectively. The small letters t, e and a (or b) are used similarly to represent the triply, doubly and non-degenerate orbitals, respectively. Subscripts such as 1 and 2 are added to indicate the symmetric and antisymmetric property of the orbitals with respect to plane of symmetry and similarly g and u to denote the symmetric and antisymmetric property with respect to centre of inversion appropriate to the point group symmetry of the complex. Books on group theory should be consulted for details.

The splitting of the d-orbitals would be altered by changing the coordination geometry of the complex and it is also sensitive to the nature of the ligand as well as the metal. A distortion of the octahedral geometry to lower symmetry may cause further splitting of the *d*-orbital energy levels as shown in Fig. 8.6. For an octahedral structure, there are three four-fold rotation axes (C_4) coincident with the *X*, *Y* and *Z*-axes shown in Fig. 8.3. There are four three-fold rotation axes (C_3) which are the lines joining the centre of the opposite triangular faces through the centre of the octahedron. There are eight triangular faces giving four C_3 rotation axes. Tetragonal distortion represents elongation (or contraction) along one of the three four-fold rotation axes (C_4 axes) of an octahedron. Distortion of the octahedral structure along one of the four three-fold rotation axes (C_3) is referred to as trigonal distortion.

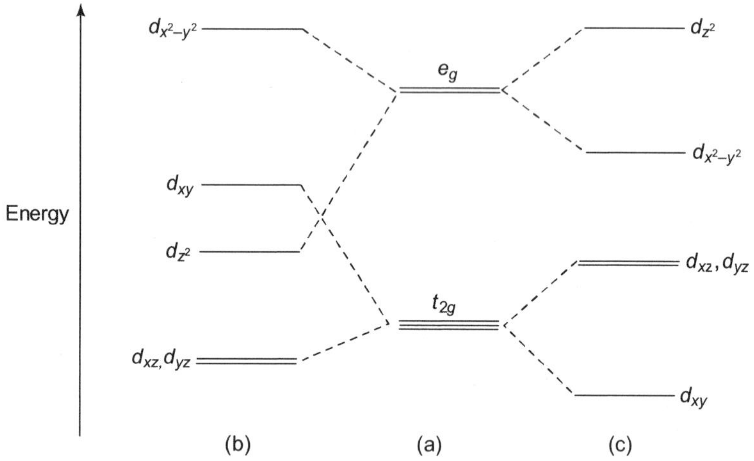

Fig. 8.6 Crystal field splitting in an (a) octahedral geometry with the ligands on the Z-axis (b) elongated and (c) compressed.

If the two ligands in an octahedral complex along the *Z*-axis are moved either away from or towards the metal ion, the resulting complex is said to be tetragonally distorted. The presence of Jahn-Teller effect discussed later may favour such a distortion. If the tetragonal distortion progresses sufficiently such that the ligands along the *Z*-axis are removed to infinity, a square planar complex results [Fig. 8.6 (b)]. Two of the most common arrangements for four-coordinated complexes are the tetrahedral and square planar geometries. For a tetrahedral complex, the 10 Dq value is 4/9 times that of the octahedral complex. Thus, only a weak field situation occurs for tetrahedral complexes (*vide infra*).

The arrangement of electrons in the *d* orbitals depends on whether the 10 Dq value required to promote an electron to the higher energy e_g orbitals is higher or lower than the energy involved in spin-pairing. For the d^1 configuration, the crystal field stabilization energy (CFSE) is – 4 Dq. For d^2 and d^3, the CFSE is –8 and –12 Dq, respectively. For these configurations, the electrons remain unpaired and enter the t_{2g} orbitals according to the Hund's rule. For the d^4 configuration, when the splitting of the orbitals is small compared to the energy required to cause the pairing of electrons, the fourth electron goes to one of the e_g orbitals. This situation is referred to as the weak field (or high spin) case. The net CFSE is – 6 Dq if the electronic configuration is $t_{2g}^3 e_g^1$. If the splitting of the *d* orbitals is larger compared to the spin pairing energy, it is then more favorable for the electrons to pair up in the t_{2g} level giving t_{2g}^4 configuration. This situation is known as the strong field (or low spin) case. The CFSE for the d^4 configuration in the strong field case is –16 Dq.

The magnitude of 10 Dq in octahedral complexes varies from 6,000 to 45,000 cm^{-1}. A change in the oxidation state of the metal ion from divalent to trivalent would result in an increased value of Dq by nearly 50%. The magnitude of 10 Dq is also dependent on the number and arrangement of the ligands around the metal ion as well as on the nature of the ligands. On going from 3*d* to 4*d* transition series or from 4*d* to 5*d* series, the value of Dq increases by nearly 50%.

The crystal field splitting pattern for d^1 and d^6 configurations is the reverse of that for d^9 and d^4 ions, respectively. There is an inverse relationship between d^n and d^{10-n} ions. For the same configuration $3d^n$, the splitting pattern between the octahedral and tetrahedral crystal fields is exactly the reverse. In the following section, the ESR spectra of the 3*d* transition series in octahedral or distorted octahedral environments are considered and no further reference is made to tetrahedral crystal fields, since the studies on the ESR spectra of complexes of tetrahedral structure are meargc.

A plot of the variation of energy levels with increasing crystal field strength (Dq), which is known as the Orgel diagram, is shown for d^1 and d^2 configurations in Fig. 8.7. The ^2D ground state of the $3d^1$ configuration is split in an octahedral field into two states $^2T_{2g}$ and ^2E_g as shown in Fig. 8.7. The ^2E_g state lies higher in energy by more than 10,000 cm^{-1} and thus does not influence the properties of the $^2T_{2g}$ state to a first approximation. The ground state for the gaseous V^{3+} (d^2 ion) is ^3F. The octahedral crystal field removes the seven-fold degeneracy of this state

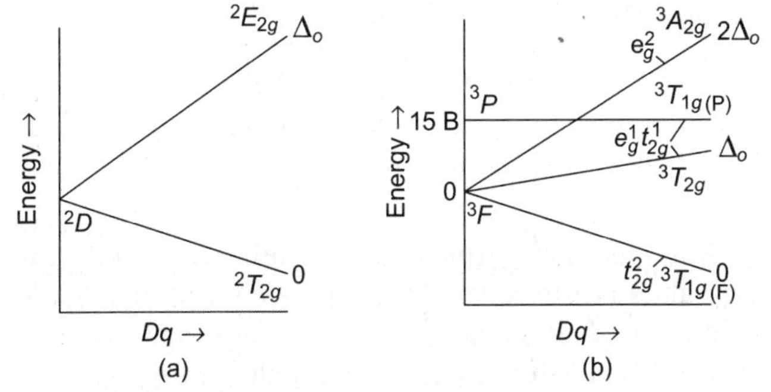

Fig. 8.7 Orgel diagram for (a) d^1 and (b) d^2 configurations in an octahedral field.

into three-fold degenerate states ${}^3T_{1g}(F)$ and ${}^3T_{2g}$ and one non-degenerate state, ${}^3A_{2g}$, as shown in the Orgel diagram (Fig. 8.7). The ${}^3T_{1g}(F)$ becomes the ground state. The ${}^3P$ state transforms as ${}^3T_{1g}$ state in the octahedral field. The two T_{1g} states coming from ${}^3P$ and ${}^3F$ terms will repel each other and the magnitude of their interaction is inversely proportional to the difference in energy between the ${}^3F$ and ${}^3P$ levels.

EFFECT OF CRYSTAL FIELD ON *g*-VALUES

The *g*-value of an unpaired electron in an isolated gaseous atom or an ion within the framework of the Russell-Saunders coupling is given by the relation,

$$g = 1 + \frac{J(J+1) + S(S+1) - L(L+1)}{2J(J+1)} \tag{8.3}$$

For a *s* electron ($L = 0$, $S = 1/2$, $J = 1/2$), the *g*-value is obtained as 2.0 from equation (8.3). The actual value of 2.00232 for the free electron is obtained after relativistic correction. The ESR spectra of halogen atoms in the gaseous state which have ${}^2P_{3/2}$ as the ground state have been obtained. The experimental value of $g = 4/3$ is the same as that derived from the equation (8.3). This formula works well for cases in which there is very little or no influence of the surrounding atoms on the orbital moment of the unpaired electrons i.e. for free gaseous atoms and for the compounds of the lanthanide elements.

For the 3*d* series, the behaviour of the unpaired electrons is considerably influenced by the ligands with the result that the orbital contribution may be severely reduced or even completely quenched. Any electronic configuration which is orbitally degenerate and contain the unpaired electrons will show an orbital contribution and have a *g*-value different from 2.0. Examples of such configurations are tetrahedral d^3 and octahedral d^7 (high spin) or d^5 (low spin). It usually happens that the excited states have an orbital momentum and to the extent that they get 'mixed into' the ground state the magnetic moment and *g*-value depart from the spin only value. The extent of mixing is inversely related to the energy of the excited state and the general formula is

$$g = 2.0023\ (1 - f\,\lambda/\Delta E) \tag{8.4}$$

where λ is the spin-orbit coupling constant, f is a numerical factor and ΔE is the energy difference between the ground state and the excited state which mixes with the ground state.

Ions such as Mn^{2+} and Fe^{3+}, which have an orbitally non-degenerate ground state (${}^6S$), give *g*-values nearly equal to the free electron value. In a strong ligand field, the splitting of the *d*-orbital energies will be large and $\Delta E >> \lambda$ and $g \sim g_e$ (free electron value). This is termed as the decoupling of L and S.

The nature of the ESR absorption is determined by the ground state electronic configuration of the metal ion in its crystalline environment in the solid phase or by the local environment in solution. The crystal field and spin-orbit coupling determine the magnitude of the splitting of the ground state in an applied magnetic field and also the extent of interaction between the electron spin and the lattice and hence it determines the spin-lattice relaxation time and the width of the ESR line. Thus the measured g-factors, hyperfine splittings and to some extent the line width provide information on the ground state of the paramagnetic species in its crystalline environment. For ions of the first transition series, $3d^1$ to $3d^9$, the magnitude of the spin-orbit coupling increases along the series from 150 to 850 cm^{-1} and the value of crystal field splitting 10 Dq is about 10,000 to 15,000 cm^{-1} for divalent metal ions in an octahedral field. In transition metal complexes, lower symmetry fields and spin-orbit coupling remove the degeneracy of the energy levels so that in most cases, only the lowest ligand field level need be considered. However, the higher energy levels may influence the ground state considerably.

JAHN-TELLER AND KRAMERS THEOREMS

It is appropriate to state here the two theorems concerning the orbital and spin degeneracies. Jahn-Teller theorem states that for a non-linear molecule such as an octahedral complex, in a degenerate electronic state distortion occurs to lower its symmetry and thus remove the degeneracy. Configurations that might lead to observable Jahn-Teller distortions are d^1, d^2, high spin d^6, d^7 and low spin d^4 and d^5.

Jahn-Teller theorem does not predict which type of distortion would occur. The ligands along the Z-axis may be compressed or elongated. The splitting of the d orbitals due to Jahn-Teller distortion is similar to that shown in Fig. 8.6 for a weak tetragonal distortion of an octahedral field. The t_{2g} orbitals are less influenced by the ligands since they are not involved in the dominant σ bonding and the t_{2g} level should show somewhat smaller Jahn-Teller effect. Since the splitting of the e_g level is somewhat greater than the splitting of the t_{2g} level, we should expect weak field d^4, strong field d^7 and d^9 to show the greatest effect. Copper(II) is an excellent example of the d^9 species showing Jahn-Teller distortion. Both Cr(II) and Mn(III) (d^4 ions) show Jahn-Teller distortion.

The other theorem concerns the spin degeneracy. When a system contains an odd number of unpaired electrons, the spin degeneracy of every level, in the absence of an external magnetic field, remains at least doubly degenerate and this is known as the Kramers degeneracy theorem. For an even number of unpaired electrons, the spin degeneracy may be lifted entirely by the crystal field and this phenomenon is known as zero-field splitting which was discussed earlier. Therefore, a metal ion with odd number of d electrons always has as its lowest level at least a doublet known as Kramers doublet. The degeneracy can then be removed by a magnetic field and an ESR transition should be observable. In a transition metal complex, the electron has orbital angular momentum and the angular momentum tends to be bolstered by weakly interacting with the spin. There is thus a competition between the quenching effect of the crystal field due to the ligands (coordinated to the metal) and the sustaining effect of the spin-orbit coupling.

SPECTRA OF FIRST TRANSITION SERIES : A SURVEY

This section contains selected examples of the ESR spectral studies of the complexes of the first transition series to highlight the application of the principles outlined above. Some of the most favorable and widely investigated valence states of the $3d$ group are V^{4+} ($3d^1$), Cr^{3+} ($3d^3$), Mn^{2+} ($3d^5$), Fe^{3+} ($3d^5$), Co^{2+} ($3d^7$) and Cu^{2+} ($3d^9$). We shall begin with transition metal complexes of $3d^1$ and $3d^9$. For a paramagnetic metal ion with $S = 1/2$ such as Ti^{3+}, Cu^{2+}, etc., the situation is somewhat similar to that of the free radicals.

The orbital containing the unpaired electron may have some s character. To this extent it is able to enter the nucleus directly since s-electrons have finite probability of being found origin to nucleus. For transition metal complexes this implies a relatively low symmetry for the system since the unpaired electrons are formally d electrons and mixing of d and s orbitals can only occur when they have the same symmetry classification for Fermi contact interaction.

(a) $3d^1$ and $3d^9$ Ions

A simple example of a $3d^1$ ion in an octahedral field is Ti^{3+} in CaO. In an octahedral field, Ti^{3+} ion has a lower energy triplet state, which splits into two Kramers doublets by tetragonal distortion as shown in Fig. 8.8. When the distortion from the octahedral symmetry is small, the separation between the ground state Kramers doublets is small (~ 10^2 cm^{-1}). They are hence extensively mixed by

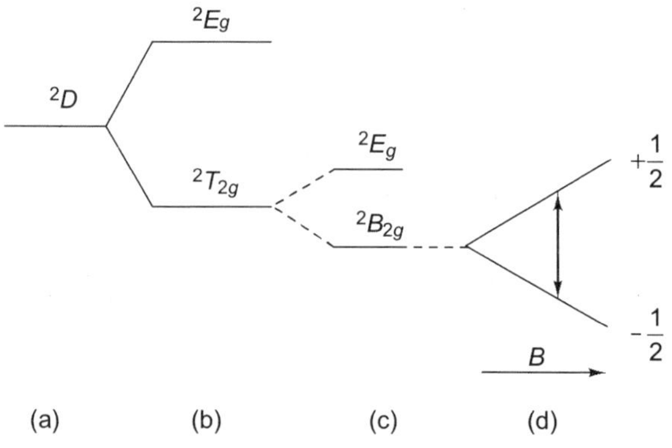

Fig. 8.8 Splitting of the ground state of a $3d^1$ ion : (a) free ion, (b) splitting in an octahedral field, (c) tetragonal distortion and (d) effect of magnetic field.

spin-orbit coupling leading to rapid spin-lattice relaxation. The spectra of Ti^{3+} complexes can only be obtained at very low temperatures. Thus the inability to obtain the ESR spectra of octahedral complexes of $3d^1$ configuration is easily predicted by the crystal field theory. Tris(acetylacetonato)titanium(III) is an example of a $3d^1$ ion in an octahedral field with large trigonal distortion and its spectrum in tris(acetylacetonato)aluminum(III) as the diamagnetic host has been measured at 77 K. As a result of large trigonal distortion, considerable splitting of the $^2T_{2g}$ state occurs and the electron spin lifetime is increased. For this complex, $g_{\parallel} = 2.000$ and $g_{\perp} = 1.921$ have been reported.

The origin of the anisotropy effect lies in the mixing of the excited states with the ground state. The states which can be mixed depend on the relative orientation of the spin and magnetic field axes. For a d^1 system, in an axially symmetric field, such as VO^{2+}, since λ is positive both $g_{\parallel}$ and $g_{\perp}$ values will be less than the free electron value, and $g_{\parallel}$ will be smaller than $g_{\perp}$. Similarly for Cu(II), the d^9 configuration can be treated as a hole, the hole occupying $d_{x^2-y^2}$. Since λ is negative, $g_{\parallel}$ will be greater than $g_{\perp}$.

When distortion from cubic symmetry is large, ESR spectra can be obtained readily for d^1 system. The most widely studied example is VO^{2+} which forms six coordinate complexes with tetragonal symmetry. The *d*-orbital splitting for distorted octahedral oxovanadium(IV) complexes is shown in Fig. 8.9. Since crystal field splitting is large, spin-orbit coupling will be weak. The spin-lattice relaxation time T_1 is, therefore, long and the ESR spectra of VO^{2+} complexes can be measured at room temperature and they show rather narrow lines. There will now be two (or three) principal *g*-values, the magnitude of which depends on the indirect orbital contributions from the spin-orbit coupled mixing in of the excited states. For VO^{2+} systems both *g*-values are less than the free electron value. Typical ranges are $g_{\parallel}$ 1.93 – 1.97 and $g_{\perp}$ 1.96 – 2.0. In solution only the average *g*-value is obtained which is usually reported as g_{iso}. Typical A values range from 8 to 20 mT.

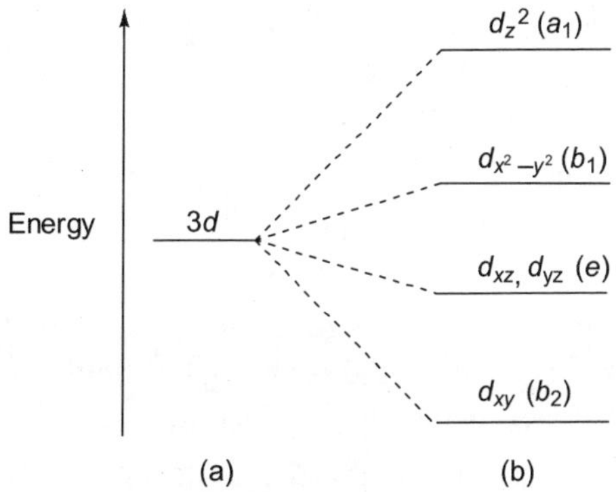

Fig. 8.9 Splitting of *d* orbitals in six coordinated oxovanadium(IV) complexes; (a) gaseous ion and (b) tetragonal crystalline field.

An additional feature of V^{4+} complexes is the prominent octet hyperfine splitting caused by the nuclear spin of $^{51}V(I = 7/2)$. A typical spectrum of an oxovanadium(IV) complex, namely, bis(acetylacetoneethylenediamine) oxovanadium(IV) in tetrahydrofuran at room temperature is shown in Fig. 8.10.

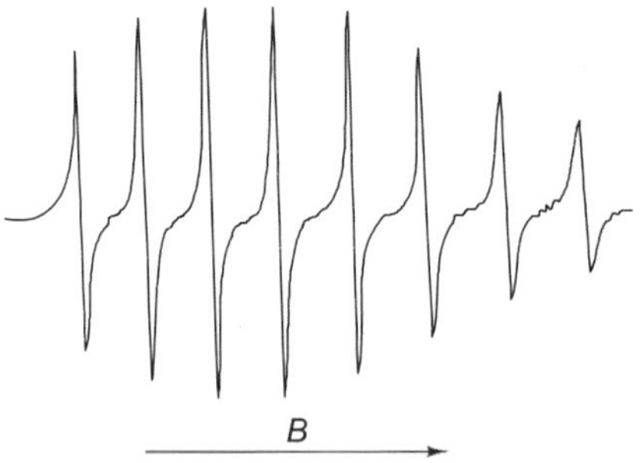

Fig. 8.10 ESR spectrum of bis(acetylacetoneethylenediamine)oxovanadium(IV) in tetrahydrofuran at room temperature.

An interesting complex of the d^1 ion in a symmetry corresponding to Archimedean antiprism is $Mo(CN)_8^{3-}$. The spectrum of this complex at ambient temperature and in aqueous solution shows a sextet due to the hyperfine interaction with $^{95,\,97}Mo$ (I = 5/2). The natural abundance of the two isotopes ^{95}Mo and ^{97}Mo are relatively low, about 25% each. It may be difficult to detect the hyperfine interaction with Mo nucleus, so that only a single line arising from $3d^1$ electron may be observed in the spectrum of the complex. Enrichment of ^{13}C (I = 1/2) gives eight equivalent ^{13}C nuclei so that a hyperfine structure with 9 lines may be seen.

The Cu^{2+} ion is the most important example of the $3d^9$ ion which has been very widely studied. The majority of the complexes have distorted octahedral structure (due to Jahn-Teller distortion) in which two bonds are longer than the remaining four in the plane. The unpaired electron lies in the $d_{x^2-y^2}$ orbital and all the other orbitals are doubly occupied. The d^9 configuration is conveniently treated by the hole concept with a single positive hole in $d_{x^2-y^2}$. This has the effect of inverting the orbital energies to give the pattern as shown in Fig. 8.11. The splitting of Cu^{2+} energy levels shown in Fig. 8.11 follows that of Ti^{3+} ion and the order of the levels is, however, inverted compared to that of Ti^{3+} (Fig. 8.8.).

The ^2D term of the free ion d^9 is split in an octahedral field into $^2T_{2g}$ and ^2E_g states. The ground state ^2E_g which is doubly degenerate is split by a large Jahn-Teller effect making the measurement of the ESR spectrum at room temperature possible. The single spin is associated with some orbital angular momentum ($L \neq 0$) but in the usual ligand fields, the energy level separations are large, the electron is mainly localized in a single d orbital, and hence S and L cannot interact strongly. In tetragonal complexes, the ground state is $d_{x^2-y^2}$ and sharp ESR lines are found. Typical range of g-values are $g_{\parallel}$ 2.1 – 2.35 and $g_{\perp}$ 2.02–2.07.

Hyperfine coupling is also expected to be anisotropic. There are two isotopes of copper with high abundance, both of which have the same nuclear spin (I = 3/2). Thus a four-line splitting is expected for both isotopes. Data quoted usually refer to ^{63}Cu (69.1%) and ^{65}Cu (30.9%) signals are generally not seen. Increasing covalency of the metal-ligand bonds leads to a decrease in *g*-value and an increase in *A* value.

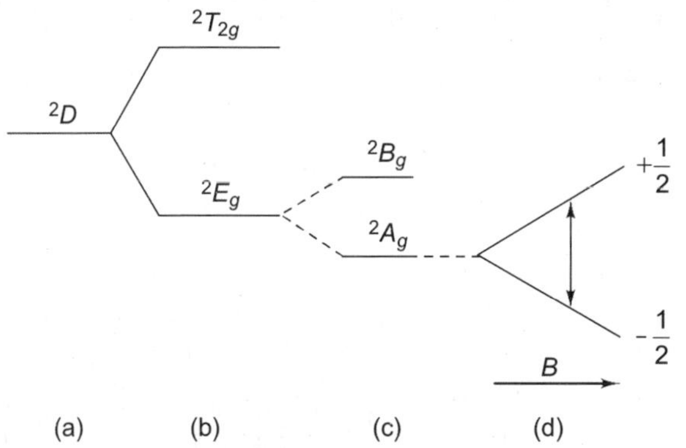

Fig. 8.11 Energy level diagram for hexacoordinated Cu(II) ions. (a) Free ion, splitting in (b) an octahedral field, (c) tetragonal field and (d) of ^2A_g in a magnetic field.

The ESR spectrum of bis(salicylaldiminato)copper(II) is interesting. The spectrum shown in Fig. 8.12 exhibits four main groups of lines resulting from the hyperfine coupling of ^{63}Cu (I = 3/2) with the unpaired electron. The hyperfine structure in each of the four groups consists of eleven lines in the intensity ratio 1 : 2 : 3 : 4 : 5 : 6 : 5 : 4 : 3 : 2 : 1. The two nitrogens split the resonance signal into five peaks of relative intensity 1 : 2 : 3 : 2 : 1. The splitting by the two equivalent protons will give rise to three lines for each of the five lines with an intensity ratio of 1 : 2 : 1. The total number of lines expected from the splitting by the two equivalent nitrogens and two hydrogens is 15 since $(2n_N I_N + 1)(2n_H I_H + 1) = 5 \times 3 = 15$. Due to considerable overlapping of some of the lines, only eleven lines are found for each group in the spectrum shown in Fig. 8.12.

The spacings between the corresponding lines in the four groups of lines which originates from Cu nuclear spin I = 3/2 are equal. The value of the magnetic field corresponding to the centre of the spectrum gives the *g*-value. For the metal complex in solution *a*(Cu) = 9.04 mT and *a*(N) = 1.43 mT have been found. The deuteration of N–H″ group did not affect the spectrum showing that no hyperfine splitting arises from N–H″ hydrogen. Similarly when the H′ hydrogens were replaced by the methyl groups, the ESR spectrum showed four main groups, each group consisting of five lines arising from the hyperfine interaction with nitrogens only. The hyperfine interaction due to N–H″ proton and that by the methyl group protons is apparently either non-existent or too small to be detected.

The ESR spectrum as discussed above provides evidence for the delocalization of the unpaired electron onto the ligand in the complex. It is interpreted as due to covalency in the metal-ligand bonding, since only by mixing of the ligand and metal ion orbitals, the ligand contribution to the molecular orbital of the complex which contains the unpaired electron is obtained. From g and a values, the nature of the bonding of the ligand to the metal could be derived. The large hyperfine interaction with nitrogen atoms is a convincing evidence of considerable delocalization of the unpaired electron on to the ligands. From the results, considerable covalent σ and π bonding of the ligands to the metal has been derived.

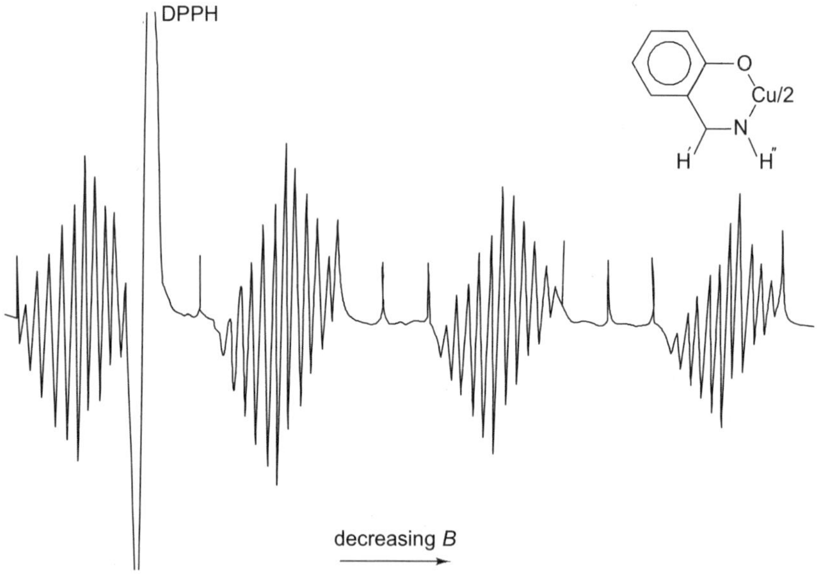

Fig. 8.12 First derivative ESR spectrum of bis(salicylaldiminato)copper(II) with isotopically pure ^{63}Cu.

Octahedral complexes of copper(II) have received special attention because of the expected large Jahn-Teller distortion. The best evidence for the presence of Jahn-Teller effect in copper(II) complexes has come from X-ray structural studies. If Jahn-Teller distortion is larger, one then expects to observe an ESR spectrum with $g_{\parallel}$ and $g_{\perp}$ components. If there is rapid interconversion amongst the distorted structures, the system will exhibit an isotropic g-factor. The two effects are distinguished as static and dynamic Jahn-Teller distortions.

First reliable evidence for dynamic Jahn-Teller distortion in transition metal complexes has been obtained from ESR spectroscopy. For example, when the ESR spectrum of $Cu(H_2O)_6SiF_6$ diluted with the corresponding diamagnetic zinc compound was measured at 90 K, the spectrum was found to consist of one band with poorly resolved hyperfine splitting and a nearly isotropic g-value. Although $Cu(H_2O)_6SiF_6$ has trigonal symmetry, the orbital degeneracy is not destroyed. Jahn-Teller distortion will therefore occur. However, there are three tetragonal distortions that will destroy the orbital degeneracy. These are the elongations or compressions along the three mutually perpendicular C_4 axes connecting the trans ligands. As a result, three ESR transitions, one for each species are expected. Since only one transition was found, it was concluded that rapid interconversion occurs between the three distorted structures. At lower temperatures, below 50 K, due to slow rate of interconversion among the structures, the spectrum becomes anisotropic and gives three sets of lines corresponding to the three different tetragonally distorted structures. Complexes with tetragonal elongation are very common than those with tetragonal compression. The single crystal ESR spectra of $[Cu(H_2O)_6]SiF_6$ diluted with the corresponding diamagnetic zinc complex have been studied. At 90 K, the values $g_{\parallel}$ 2.221 and $g_{\perp}$ 2.230 and at 20 K, g_z = 2.46, g_y = 2.10 and g_x = 2.10 were obtained. Similar observations have been made for $Cu(BrO_3)_2.6H_2O$ whose ESR spectra have been studied by diluting the sample with the corresponding zinc(II) salt. The ESR spectrum is isotropic with a *g*-value of 2.22 at 90 K. The dynamic Jahn-Teller effect is replaced by a static Jahn-Teller effect in this compound below ~ 10 K.

In dimeric copper(II) complexes, the identification of Cu-Cu dipolar interaction is of interest. In copper(II) dinuclear complexes, the value of zero-field splitting parameter *D* is generally small. However, the spectral appearance is modified. Additional transitions due to $\Delta M_s = \pm 2$ may be found. In the *X*-band spectra, the $\Delta M_s = \pm 1$ transitions are associated with fields of nearly 0.3 T, while $\Delta M_s = \pm 2$ transitions give absorptions at half-field values of ca. 0.15 T. The presence of 'half-field' band is a useful criterion for dinuclear or polynuclear complex formation (see chapter 4). Due to zero-field splitting, the spectrum may spread over a larger magnetic field up to 0.6 T.

The ESR spectrum of a polycrystalline sample of typical dimeric copper(II) complex is shown in Fig. 8.13. The intense bands at 0.26 and 0.38 T are assigned

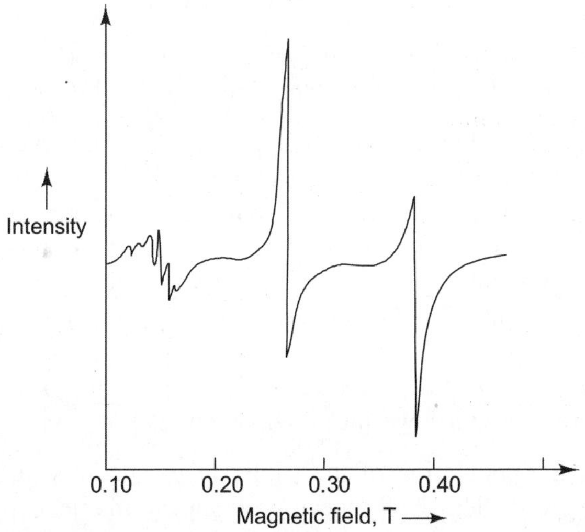

Fig. 8.13 The ESR spectrum of a polycrystalline dimeric copper(II) complex.

to the two g components. The seven-line hyperfine splitting due to the two copper atoms ($I = \frac{3}{2}$) are observed for the half-field peak at ~ 0.15 T, corresponding to $\Delta M_S = +2$ transition. Another high field peak is not observed since it is out of the spectrometer range (see chapter 4).

A novel copper(II) complex which has attracted much attention is copper(II) acetate which exists in the dimeric form bridging through the acetate groups. The ESR spectra of a polycrystalline sample of copper acetate dihydrate is shown in Fig. 8.14. The resolution is significantly increased by measuring the spectrum at low temperatures. The 'half-field' band occurs at ca. 0.15 T. The copper hyperfine coupling is usually seen more clearly in the spectra of the dinuclear copper(II) complexes diluted in a diamagnetic host such as the dinulear zinc(II) complex.

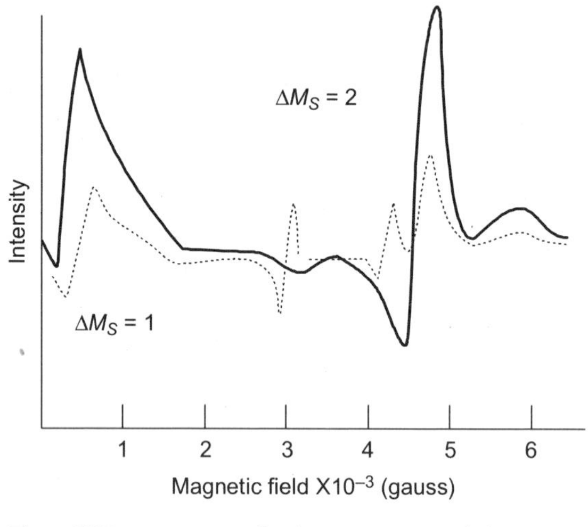

Fig. 8.14 The ESR spectra of dimeric copper(II) acetate dihydrate $[Cu(O_2CCH_3)_2(OH_2)]_2$ at room temperature (——) and at 77 K (........).

Consider a polycrystalline dinuclear complex containing two copper atoms per molecule with a total spin $S = 1$. The spin-spin interaction is measured in terms of the exchange integral J and the zero-field splitting. The energy level diagram for the dinuclear complex is shown in Fig. 8.15. The exchange interaction between the two $S = 1/2$ spin states gives a lower energy singlet ($S = 0$) and an upper energy triplet ($S = 1$) state. The lower level is diamagnetic but ESR transitions are observed for the upper triplet state. For the dimeric copper(II) acetate, the separation between $S = 0$ and $S = 1$ levels is $2J = 520$ cm^{-1}. The intensity of the ESR signals decreases with lowering of temperature.

Below 20 K, the population of the triplet level becomes negligible and no spectrum is observed. The ESR spectra of some dimeric copper complexes recorded at higher temperatures exhibit a seven line hyperfine splitting in the intensity ratio of 1 : 2 : 3 : 4 : 3 : 2 : 1 in agreement with the coupling between two identical metal nuclei both with $I = 3/2$. For dimeric copper(II) acetate, zero-field splitting parameters D and E of 0.35 and 0.007 cm^{-1}, respectively have been obtained. The values, g_x=2.053, g_y =2.093 and g_z =2.344 and A_x, A_y < 0.001cm^{-1} have been reported. In the powder spectrum, $g_{\parallel}$ and $g_{\perp}$ peaks arise from the two ΔM_S = ± 1 transitions averaged over all the orientations. When the exchange

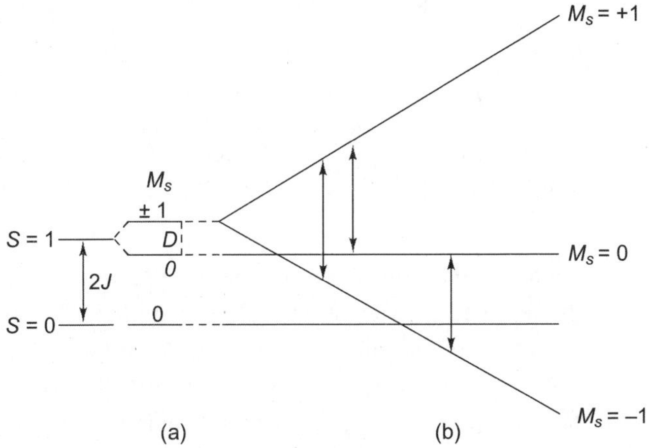

Fig. 8.15 The splitting of the spin energy levels of a dimeric copper(II) complex : (a) zero-field splitting and (b) splitting in an applied magnetic field. Transitions in the triplet state obeying ΔM_S = +1 and +2 are shown.

parameter J is smaller than the microwave radiation frequency of the ESR transitions, it is possible to observe the transitions between the electronic states S = 0 and S = 1 of a dimeric copper(II) complex.

Two general shapes observed for the ESR spectra of copper(II) complexes are shown in Fig. 8.16. The two axial spectra (*b*) and (*c*) may be utilized as a criteria for distinguishing the ground state, between $d_{x^2-y^2}$ and d_{z^2} respectively and also the *g*-values ($g_{\parallel} >> g_{\perp} > 2.0$ and $g_{\perp} >> g_{\parallel} \approx 2.0$). This simple interpretation of the ESR spectra is applicable if all the tetragonal axes of the local copper(II) ion environments are aligned parallel to the unit cell.

The anisotropy in the ESR parameters can be measured by studying the spectra of single crystals. However, information about the anisotropy can also be obtained sometimes from a study of the powders and frozen glasses since the spectra obtained are not those of a motionally averaged system. Consider a system with a three- or higher-fold axis of symmetry. The *g*-values can now be represented by $g_{\parallel}$ and $g_{\perp}$. The *g*-value for any orientation is given by

$$g^2 = g_{\parallel}^2 \cos^2\theta + g_{\perp}^2 \sin^2\theta \quad (8.5)$$

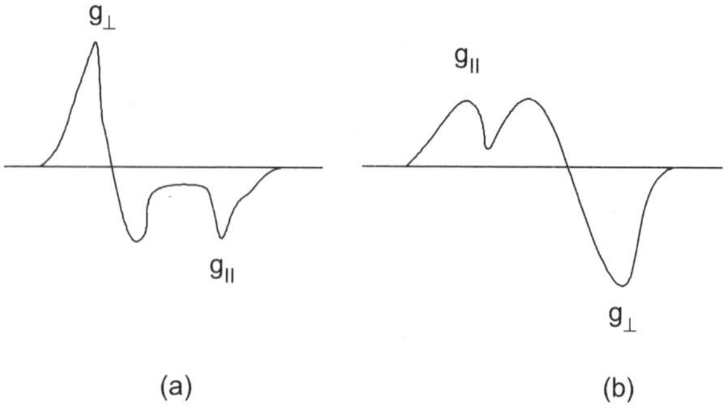

Fig. 8.16 The different types of expected first derivative ESR spectra of octahedral copper(II) complexes for the powdered sample: (a) axially elongated and (b) axially compressed.

where θ denotes the angle the principal axis, that is, the $g_{\parallel}$ axis makes with the external magnetic field. Since all orientations of the crystallites in a powdered solid are equally probable, absorption will occur at all fields between that associated with $g_{\parallel}$ and $g_{\perp}$, where $\parallel$ refers to the Z-axis parallel to the principal axis of rotation and $\perp$ refers to the X and Y axes perpendicular to the principal axis of rotation. In a bulk sample containing different orientations of the crystallites, far more molecules that have $g_{\perp}$ axis aligned with the applied magnetic field are possible than the molecules of the system that have $g_{\parallel}$ axis aligned with the magnetic field. If the probabilities of the various orientations and the transitions corresponding to each of these are taken into consideration, the derivative spectra as shown in Fig. 8.16 are expected. The two spectra (a) and (b) may be used as a criterion for distinguishing between $d_{x^2-y^2}$ and d_{z^2} ground states. This, of course, is an idealized situation and often the overlapping features generated by $g_{\perp}$ and $g_{\parallel}$ render it difficult to obtain their values. When the system is orthorhombic, and $g_x > g_y > g_z$, the powder spectrum obtained for nuclear spin I = 0 is expected to be similar to that shown in Fig. 8.17. The spectra for other systems are quite complex and rather difficult to assign and only in relatively simple cases, the values of g and A can reliably be obtained. Computer programs are available to simulate the ESR spectra of the powder samples for simple Cu(II) systems.

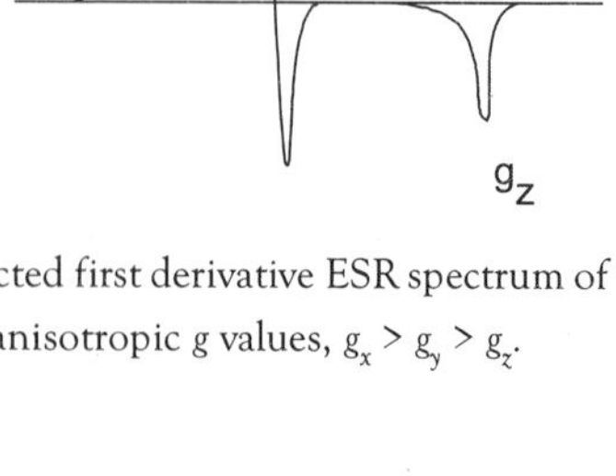

Fig. 8.17 The expected first derivative ESR spectrum of a copper(II) complex showing anisotropic g values, $g_x > g_y > g_z$.

(b) $3d^2$ and $3d^8$ Ions

The ESR spectral studies on octahedral complexes of d^2 ions are very few because of extensive spin-orbit coupling in the $^3T_{1g}$ ground state. For an ion which has two unpaired electrons ($S = 1$), the two transitions $M_S = 0 \rightarrow +1$, and $M_S = -1 \rightarrow 0$ ($\Delta M_S = \pm 1$) are degenerate and only one signal is expected. Zero-field splitting removes this degeneracy in M_S as shown in Fig. 8.18. Consequently, two signals are expected in the spectrum when zero-field splitting occurs (see Chapter 7).

A representative of $3d^2$ ions is V^{3+}. The seven-fold orbital degeneracy of the ^3F state of the gaseous $3d^2$ ion is split in an octahedral field and the $^3T_{1g}$ state lies lowest as shown in Fig. 8.18. The lowering of symmetry lifts the degeneracy of the T_{1g} state to leave an orbital singlet as the lowest state. There is extensive zero-field splitting due to spin-orbit coupling with the upper components of the T_{1g} state. In certain cases, the magnitude of the zero-field splitting may exceed the energies of the usual ESR transitions. When this happens, it may be difficult to

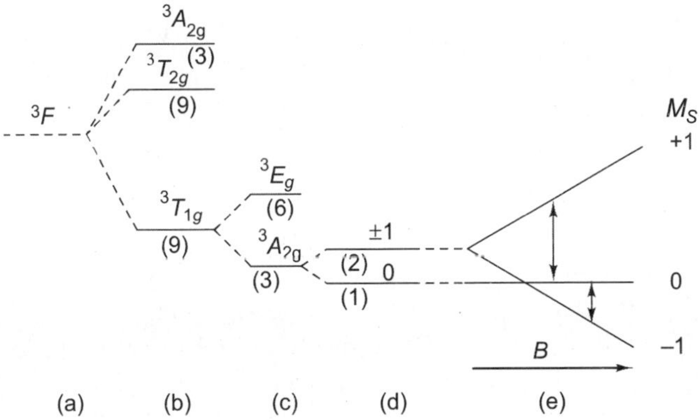

Fig. 8.18 Ground state term ^3F of a $3d^2$ ion: (a) free ion, (b) splitting in an octahedral field, (c) effect of trigonal distortion, (d) zero-field splitting and (e) effect of applied magnetic field. The value in the parenthesis denotes the total degeneracy of the level.

observe the ESR signals for $\Delta M_S = \pm 1$ transitions but it is possible to observe the weak $\Delta M_S = \pm 2$ transition. It is split into eight lines by the nuclear spin of ^{51}V (I = 7/2). For example, the spectrum of V^{3+} in Al_2O_3 has been studied at 4 K. The half-field resonance ($\Delta M_S = \pm 2$) splits into eight components. The g-values, $g_{\parallel} = 1.92$ and $g_{\perp} = 1.63$ and $D = +7.85$ and $A = 102$ cm^{-1} have been reported.

The Ni^{2+} ion is the most important example of the $3d^8$ ions. The lowest orbital level in an octahedral field for $3d^2$ ion is $^3A_{2g}$ (Fig. 8.19). Spin-orbit coupling mixes the excited state with the ground state which causes zero-field splitting making it difficult to detect the ESR spectra. A low symmetry crystal field also gives rise to a large zero-field splitting. It results in two transitions ($\Delta M_S = \pm 1$) as for the $3d^2$ ion. For example, Ni^{2+} in MgO shows two broad ESR lines

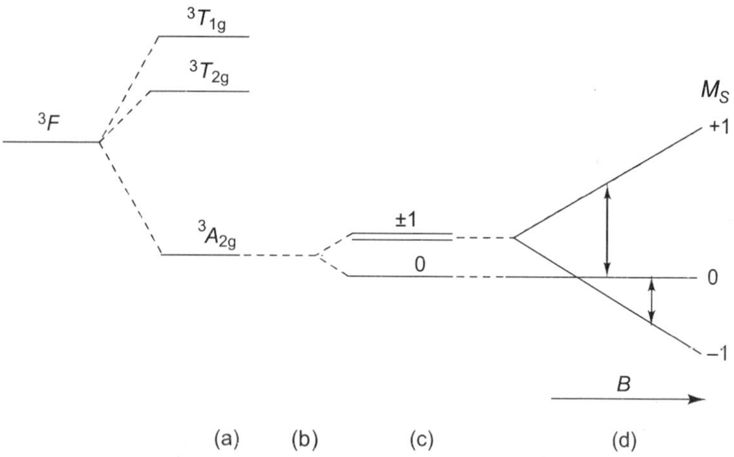

Fig. 8.19 Energy level scheme for the triplet state ^3F of a d^8 ion: (a) octahedral field, (b) tetragonal field, (c) zero-field splitting and (d) effect of magnetic field.

corresponding to $\Delta M_S = \pm 1$ transitions and a sharp band corresponding to $\Delta M_S = \pm 2$ transition. In some other systems too, sharp peaks due to $\Delta M_S = \pm 2$ have been observed in the ESR spectra. However, since zero-field splitting is highly anisotropic, it provides an efficient relaxation mechanism. Accordingly the spectra of nickel(II) complexes are difficult to obtain at room temperature and liquid nitrogen or liquid helium temperature must usually be employed. It has been found that for most $3d^8$ ions, the g-value is nearly isotropic but the values of D and E depend strongly on the nature of the lower symmetry crystal fields. The ESR spectra of hydrated nickel(II) sulphate and tris(acetylacetonato)nickel(II) complexes have been studied.

(c) $3d^3$ and $3d^7$ Ions

The d^3 system has been extensively studied and the most important metal ion of the $3d^3$ configuration is Cr^{3+}. In an octahedral geometry which is strongly favoured for d^3, the ground state is an orbital singlet $^4A_{2g}$ and hence has no spin-orbit coupling. The spin degeneracy is resolved by spin-orbit coupling through mixing via the $^4T_{2g}$ state or lower crystal fields or both into two Kramers doublets. Figure 8.20 illustrates the situation for the $^4A_{2g}$ ground state of a Cr^{3+} ion. The two Kramers doublets split again in an external magnetic field and three $\Delta M_S = \pm 1$ transitions are possible. Thus when zero-field splitting is small, three transitions can sometimes be detected as shown in Fig. 8.20 and the zero-field splitting parameter D can be evaluated from the spectrum. When zero-field splitting is large, compared to the microwave frequency, only one resonance may be observed.

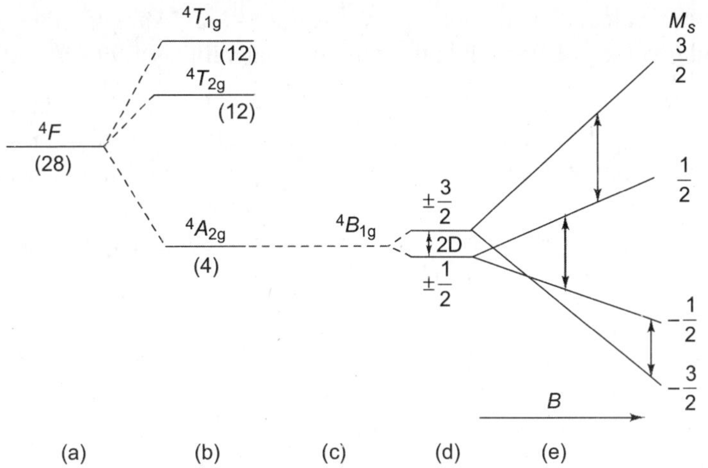

Fig. 8.20 Splitting of the ground state ^4F term of a d^3 ion: (a) free ion, (b) splitting in an octahedral field, (c) effect of tetragonal field on $^4A_{2g}$, (d) zero-field splitting and (e) effect of applied magnetic field. The value in the parenthesis denotes the total degeneracy of the level.

For the $3d^3$ ion, the ESR spectra are easy to observe and the g factors are close to 2.00 and nearly isotropic. There is no orbital angular momentum and the excited states are far removed from the ground state and hence there is very little spin-orbit interaction. The ESR spectra of the complexes of the $3d^3$ ion are observed at ambient temperature because of longer spin-lattice relaxation time and the observed line widths are small, which makes accurate determination of g, A, D and E possible. The ESR spectra of many crystals containing Cr^{3+} and V^{2+} in an octahedral symmetry have been investigated. For example, a study of Cr^{3+} in CaO has been made and very nearly isotropic g ($g_{\|}$ 1.970, $g_{\perp}$ 1.975) and A values have been reported. The D value was found to be small (0.136 cm^{-1}).

In octahedral complexes of d^3 ions, the metal electrons occupy the t_{2g} orbitals, and hence the ligand hyperfine splittings are usually small because the t_{2g} orbitals are involved only in *pi* bonding. The g-value for this system can be calculated from crystal field theory using the relation,

$$g = g_e - \frac{8\lambda}{\Delta E\left({}^4T_{2g} - {}^4A_{2g}\right)} \tag{8.6}$$

For $V(H_2O)_6{}^{2+}$, substituting the energy difference ΔE = 11,800 cm^{-1} and the spin-orbit coupling constant λ = 56 cm^{-1} gives the value of g as 1.964 in good agreement with the experimental value of 1.972. Since covalency becomes more significant as the charge on the central metal ion increases, the crystal field approximation becomes less applicable and somewhat larger deviation occurs between the calculated and observed g-values. For example, for Mn^{4+}, the calculated g-value is 1.955 while the experimental value is 1.994.

The commonest example of a high-spin d^7 ion is Co^{2+}. It has three unpaired electrons and therefore behaves like the d^3 case. The ESR spectra of the complexes of Co^{2+} have been studied in a wide range of crystalline environments. ^{59}Cobalt (natural abundance 100%) has $I = 7/2$ and octet hyperfine splitting expected due to spin of the metal nucleus is usually seen. The ground state for an octahedral high spin d^7 complex is $^4T_{1g}(F)$ (Fig. 8.21).

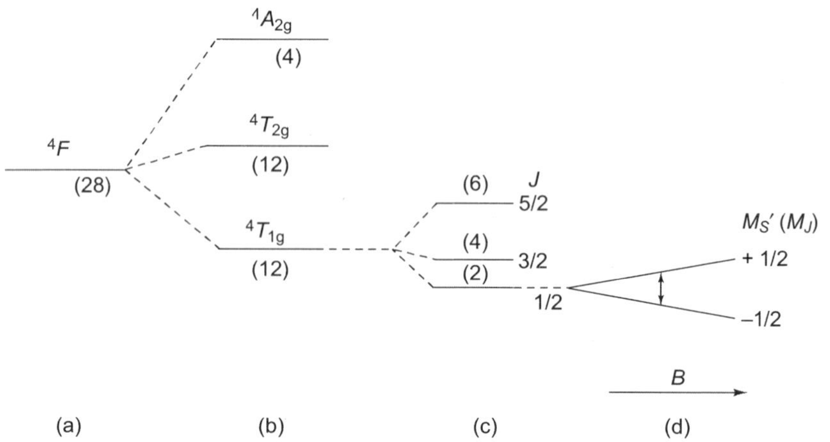

Fig. 8.21. Schematic energy level diagram of the ground state ^4F of a d^7 ion: (a) free ion, (b) under octahedral field, (c) zero-field splitting and (d) effect of external magnetic field. The value in the parenthesis denotes the total degeneracy of the level.

When the ground state is an orbitally degenerate T state in a cubic field, the lower symmetry fields and spin-orbit coupling leads to the removal of the orbital as well as spin degeneracy. As a result, three Kramers doublets will occur. The lowest doublet is separated from the next higher lying doublet by more than 200 cm^{-1} (kT). The system, therefore, behaves like a simple system having an effective spin $S' = 1/2$. The spin-lattice relaxation time is very short at room temperature due to strong spin-orbit coupling and the consequent necessity of working at very low temperatures (below 20 K) ensures that only the lowest energy level is significantly populated and an effective spin $S' = 1/2$ is appropriate. In distorted octahedral fields, both large spin-orbit coupling and low symmetry fields will result in marked anisotropy in g and A values. For example, for Co^{2+} ion in TiO_2, $g_x = 2.030$, $g_y = 3.725$ and $g_z = 5.860$ have been reported. There is pronounced orbital contribution for complexes of octahedral geometry. For axially symmetric systems, the two g-values usually give an average close to 4.3 but the individual values of $g_{\parallel}$ and $g_{\perp}$ may lie in the range 1–8.

The low spin d^7 state ($S = 1/2$, $t_{2g}^{\,6}\, e_g^{\,1}$) is found in most octahedral complexes of cobalt (II). The same considerations as in the $3d^9$ case apply. The ground state is

^2E_g. The orbital degeneracy in an octahedral field is not removed by spin-orbit coupling in the ^2E_g state and there are no nearby doublet states. However, Jahn-Teller distortion will lift the degeneracy. Working at low temperature is usually necessary for the measurement of the ESR spectra. The *g*-factors are slightly above 2.0 and show marked anisotropy in *g* and *A* values. For example, Ni^{3+} in Al_2O_3 shows an isotropic *g*-factor (2.146) above 50 K. At liquid helium temperature, the spectrum becomes strongly anisotropic presumably because each of the statically distorted configurations contribute individually. An example of low spin octahedral complex of $3d^7$ with tetragonal distortion is $Fe(CN)_5NOH^-$. Its spectrum has been measured in the diamagnetic host $Na_2Fe(CN)_5NO.2H_2O$ and $g_\parallel$ 2.0069 and $g_\perp$ 2.0374 have been reported.

(d) $3d^4$ and $3d^6$ Ions

Typical metal ions with $3d^4$ electronic configuration are Cr^{2+} and Mn^{3+}. The ground state of the free ion is ^5D, which goes over to ^5E_g in a weak octahedral field. Tetragonal distortion results in further splitting into an orbital singlet and a large Jahn-Teller distortion is expected in the E_g state (Fig. 8.22). Zero-field splitting of ± 2 and ± 1 levels results in four ESR transitions when the zero-field splitting is small as shown in Fig. 8.22 and when the splitting is large, however, none of these transitions is observed. The zero-field splitting in d^4 complexes is often large because of Jahn-Teller distortion. These factors usually make it very difficult to observe the spectrum. Thus there are very few ESR spectral studies on metal complexes of high spin d^4 electron configuration. From the ESR spectra of $CrSO_4.5H_2O$, the parameters, $g_\parallel$ 1.95, $g_\perp$ 1.99, D 2.24 cm^{-1} and E 0.10 cm^{-1} have been obtained.

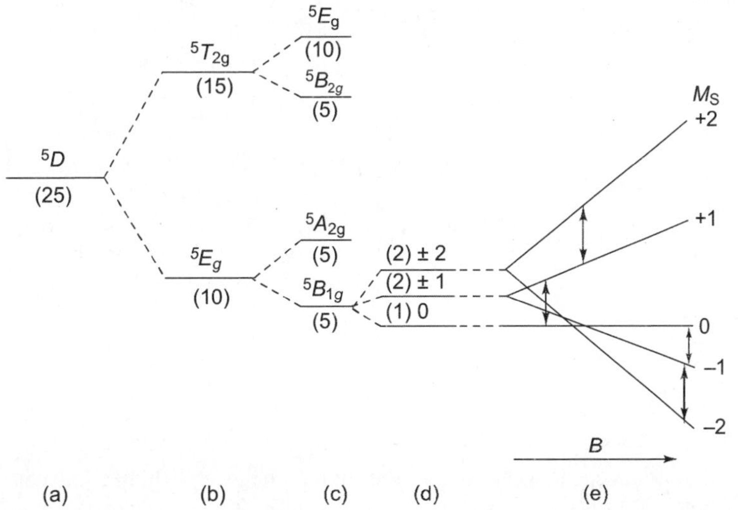

Fig. 8.22 Schematic energy level diagram of a $3d^4$ ion in a weak crystalline field: (a) gaseous ion, (b) in octahedral field, (c) in tetragonal field, (d) with zero-field splitting and (e) effect of magnetic field. The value in the parenthesis indicates the total degeneracy of the level.

Important metal ions with $3d^6$ electrons are Fe^{2+} and Co^{3+}. This *d* electron configuration has not been studied extensively. For octahedral geometry, the high spin state has $S = 2$ and the low spin state has $S = 0$. Co^{3+} usually occurs in the low spin state which is not of interest since it corresponds to a diamagnetic ground state.

The lowest energy level of a high spin $3d^6$ ion in an octahedral field is an orbital triplet ($^5T_{2g}$). Spin-orbit coupling is very large in the ground state and there are nearby excited states that can mix with the ground state. The splitting of the energy levels is shown in Fig. 8.23. In a distorted octahedral field, zero-field splittings are large and hence it may be difficult to observe the ESR spectrum. The spin-orbit coupling within the $^5T_{2g}$ state leads to short spin-lattice relaxation time and the ESR spectra can be observed only at lower temperatures (below 20 K). Small departure from octahedral symmetry will lead to large zero-field splitting and resonance absorption is possible when the separation of the singlet levels is not greater than the microwave quanta. With the ground state $J = 1$, two transitions would be observed if zero-field effects are small. The occurrence of $\Delta M_S = \pm 2$ transitions, which are forbidden for regular octahedral symmetry is an evidence for distortion from octahedral symmetry.

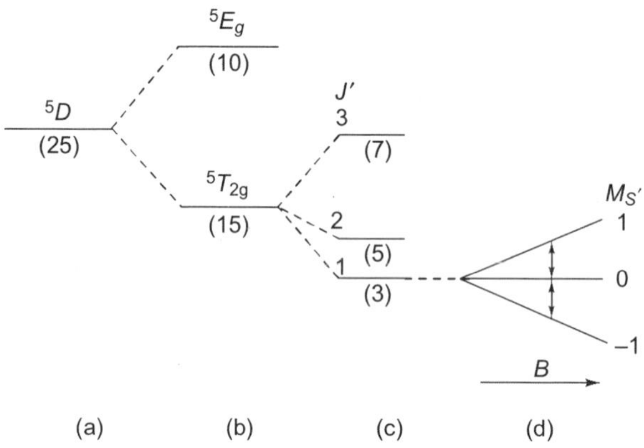

Fig. 8.23 Splitting of the ground state ^5D of $3d^6$ configuration in a weak crystalline field: (a) free ion, (b) under octahedral field, (c) zero-field splitting and (d) effect of applied magnetic field. The value in the parenthesis denotes the total degeneracy of the level.

(e) $3d^5$ Ion

This *d* electron configuration has been very extensively investigated and important metal ions having $3d^5$ configuration are Mn^{2+} and Fe^{3+}. If the crystal splitting is not large, the ions exist in the high spin state with all the five *d* orbitals half-filled. The d^5 system is unique in that the ground state is an orbital singlet (^6S) for the free ion, which is six-fold degenerate in spin and no other sextet states exist. For Mn^{2+} and Fe^{3+} ions in CaO, the *g*-value of 2.009 and 2.0052.

respectively have been obtained. Observed g-values are very close to 2.00 because of the very small amount of spin-orbit coupling. The ground state of octahedral Mn^{2+} is $^6A_{1g}$ which is split by zero-field effects. Spin-orbit coupling mixes the excited state $^4T_2$ which are split by the crystal field into the ground state (by second order effects) and it gives rise to a relatively small zero-field splitting in Mn^{2+} complexes, about 0.5 cm^{-1} in certain Mn(II) porphyrins. Dipolar interaction of the electron spins is small in comparison to the mixing of higher state with the ground state in the complex. Because of the absence of orbital angular momentum, Mn^{2+} has a long relaxation time and hence relatively narrow ESR lines are observed. The ESR spectra of Mn^{2+} complexes can usually be observed in solution at room temperature. Nuclear hyperfine structure from $^{55}Mn(I = 5/2)$ is readily observed in single crystals and for solutions of Mn^{2+} salts.

The energy level diagram and the spectrum expected for octahedral Mn(II) complexes in a weak tetragonal field are shown in Fig. 8.24. The zero-field splitting leads to three Kramers doublets ($M_S = \pm 5/2, \pm 3/2, \pm 1/2$). Each of these states is split into two singlets by the external magnetic field producing six levels. A five-line spectrum due to the transitions $-5/2 \rightarrow -3/2, -3/2 \rightarrow -1/2, -1/2 \rightarrow 1/2, 1/2 \rightarrow 3/2, 3/2 \rightarrow 5/2$ is expected. Hyperfine splitting due to the Mn nucleus (I = 5/2) splits each line into six hyperfine components giving a total of thirty

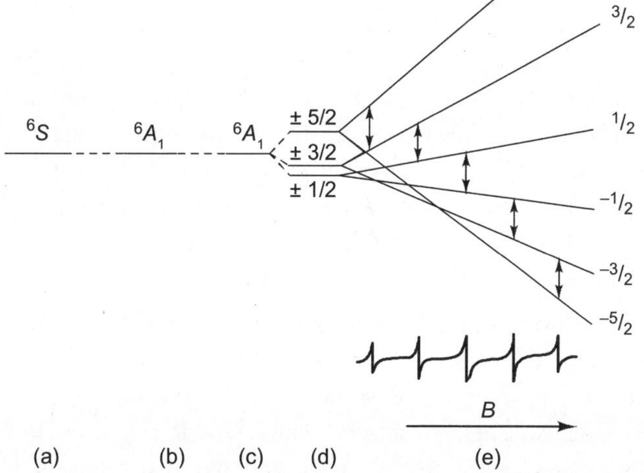

Fig. 8.24 Ground state energy levels and allowed electron spin transitions for a $3d^5$ ion in a weak crystalline field: (a) free ion, (b) under octahedral field, (c) tetragonal field, (d) zero-field splitting and (e) with applied magnetic field parallel to the tetragonal axis.

peaks. In powder or in solution, only the signal corresponding to the transition $M_S = 1/2 \rightarrow M_S = -1/2$ is expected, since this has a much smaller angular dependence. Hence, in the absence of hyperfine splitting, only a single line is seen, g is isotropic and has a value close to 2.00. Hyperfine coupling is usually seen for Mn^{2+} (I = 5/2) giving a six-line spectrum. The value of A usually lies in the range 5-10 mT and is also isotropic. If the symmetry is lowered, zero-field splitting may become much larger than the spectrometer frequency with the result that effectively only $M_s = 1/2 \rightarrow -1/2$ transition lies within the range normally scanned. For axial symmetry, the system is now anisotropic and while $g_{\parallel}$ remains closc to 2.0, $g_{\perp}$ becomes close to 6.0. Powders and frozen solutions therefore show two major resonances with these g-values.

The ESR spectrum of Mn^{2+} in MgO, for example, consists of five sets of sextets at room temperature due to hyperfine splitting arising from ^{55}Mn (natural abundance 100%). The hyperfine splitting is greater than the zero field splitting. Ligand ^{19}F hyperfine structure has also been observed for the complex MnF_6^{4-} present in MnF_2 and K_2MnF_4. The spectrum of MnF_2 in a diamagnetic host, ZnF_2 has been measured at room temperature and the g-value was found to be 2.002. The values of A 96 cm^{-1}, D ~ 1.86×10^{-2} cm^{-1} and E ~ $0.41 - 10^{-2}$ cm^{-1} have been reported.

With Fe^{3+}, both low spin and high spin forms occur. With stronger ligands such as CN^-, 10 Dq will be larger leading to a low spin form with $S = 1/2$. With weaker ligands, 10 Dq is smaller and a high spin form is produced. The observed g-values for the high spin octahedral Fe^{3+} complexes are close to 2.0. The ESR spectrum is easily observed at room temperature.

A low spin d^5 ion has the same ordering of the energy levels as a $3d^1$ ion in an octahedral field and the ground state is $^2T_{2g}$. Spin-orbit coupling splits the term into three closely spaced Kramers doublets. However, due to the presence of large spin-orbit interaction, the ESR spectrum may be observed only at liquid helium temperature. The situation is similar to that of d^1, except that we now have a positive hole. The energy level diagram under octahedral and tetragonal crystal fields, the effect of spin-orbit interaction and magnetic fields are shown in Fig. 8.25. For Fe^{3+} in $K_3Co(CN)_6$, from the ESR spectrum, $g_{\parallel} = 0.915$ and $g_{\perp} = 2.20$ have been obtained.

Large deviation from octahedral symmetry, as for example, in 2,2′-bipyridyl and 1,10-phenanthrolene complexes of iron(III) (low spin) causes an orbitally singlet state to lie lowest in energy well-removed from the orbitally non-degenerate excited states (Fig. 8.25). Low spin derivatives of ferric hemoglobin complexes are examples of such systems where tetragonal distortion is large. Longer relaxation times result and the ESR spectra can be obtained at higher temperatures. The spectra of bipyridyl and phenanthrolene complexes of iron(III) have been studied.

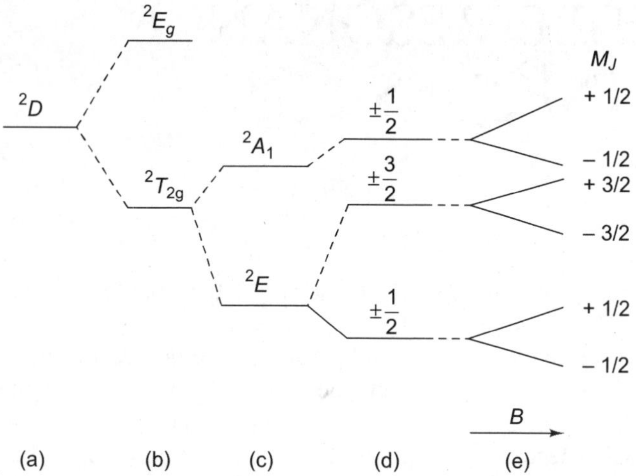

Fig. 8.25 Splitting of the ground state ^2D of a low-spin $3d^5$ ion: (a) free ion, (b) under octahedral field, (c) with tetragonal distortion, (d) with inclusion of zero-field splitting and (e) effect of applied magnetic field.

The complex ion $IrCl_6^{2-}$ referred earlier in the introduction belongs to the low spin $5d^5$ system in an octahedral field. $IrCl_6^{2-}$, t_{2g}^5 has one unpaired electron. In the iridium complexes, $(NH_4)_2IrCl_6$ and $Na_2IrCl_6.6H_2O$ diluted with the corresponding platinum salt, the ESR spectra exhibit hyperfine interactions. The electron spin of iridium in the complex is $S' = 1/2$ and the ground state is $^2T_{2g}$ (See Fig. 8.25). The nuclear magnetic moment of ^{193}Ir (abundance 62%) is very small ($\mu = 0.17$ in units of β_N) compared with that of ^{35}Cl ($\mu = 0.82$ in units of β_N) (abundance 75.4%) ($I = 3/2$). Therefore, hyperfine splitting of the electron spin with Ir nucleus $I = 3/2$ is not observed. The hyperfine interaction with six chloride ions should lead to the splitting of the electron spin resonance into 19 lines. In practice, however, the spectrum is complicated. The results indicated that the electron spin is located on six chlorine atoms to the extent of nearly 30% of its time, about 5% on each chlorine atom. This observation provided one of the earliest demonstrations of the delocalization of the metal electron onto the ligands.

DOUBLE RESONANCE TECHNIQUES 9

INTRODUCTION

An important group of special methods are the double irradiation methods, which are used to simplify the spectra, enhance the intensity of specific resonances and provide structural information. The double resonance techniques encompass all techniques in which the sample is irradiated at more than one frequency. The double resonance technique in ESR spectroscopy is exactly similar to that in NMR spectroscopy discussed later and the observation of the spectrum is made at one frequency while irradiating at the same time with another frequency. There are many multiple resonance techniques that have been applied to ESR spectroscopy but we shall discuss briefly two of them here. There are two broad possibilities in ESR spectroscopy, since the second irradiation may either be at nuclear spin or at electron spin resonance frequencies. The two double resonance techniques in ESR spectroscopy are the electron-nuclear double resonance (ENDOR) and electron-electron double resonance (ELDOR). The ENDOR technique is technically simpler and is used more often than the ELDOR.

ELECTRON-NUCLEAR DOUBLE RESONANCE

ENDOR stands for electron-nuclear double resonance but the term is generally applied to one rather special type of experiment which is used to study the hyperfine structure in the spectra of free radicals and transition metal complexes. The ENDOR method provides great simplification of the complex hyperfine structure in the ESR spectrum and it was introduced by G. Feher in 1956 to improve the resolution of an ESR spectrum. The ENDOR method is useful when *(i)* the hyperfine lines are not resolved in the ESR spectrum or the spectrum is complicated with too many lines, *(ii)* the hyperfine lines are resolved but more precise values of the hyperfine coupling constants are needed and *(iii)* the identity of the interacting nucleus is to be established by measurement of the nuclear *g*-factor. ENDOR has been used sometimes to study the ligand hyperfine splittings in transition metal complexes. The ENDOR spectrum of a transition metal complex in which several nuclei are coupled to the electrons is less complex than its ESR spectrum. One of the earliest applications of ENDOR to the measurement of ligand hyperfine splittings was for the complex FeF_6^{3-}.

A simple description of the ENDOR experiment for a system of a single unpaired electron interacting with a nucleus of spin 1/2 is given here. The unpaired electron is assumed to have isotropic hyperfine interaction with the nucleus. It is further assumed that the electron and nuclear spins relax through completely independent processes so that the saturation of the electron spin resonance does not alter the nuclear spin populations. For the ENDOR to be possible, the ESR line being observed should be inhomogeneously broadened.

The first order energy level diagram for a system which has a single electron spin and a nucleus with spin 1/2 in the presence of an applied magnetic field is shown in Fig. 9.1 (see also Fig. 3.3, Chapter 3). Here the electron spin is shown first, keeping in mind that for electron down spin is stabilized in the field. Assume that from the resonant field positions B_k and B_l of the two hyperfine lines, we have found the hyperfine splitting constant (Fig.3.3). For the system under consideration, the electron spin energy is split by the applied magnetic field into two widely separated states, say by some 9,500 MHz in a magnetic field of 0.34 T. In the same external magnetic field, the nuclear spin (if assumed to be hydrogen) is split by 13 MHz only. It is to be noted that the magnitude of electron and nuclear spin energy level spacings thus differ by a factor of 730.

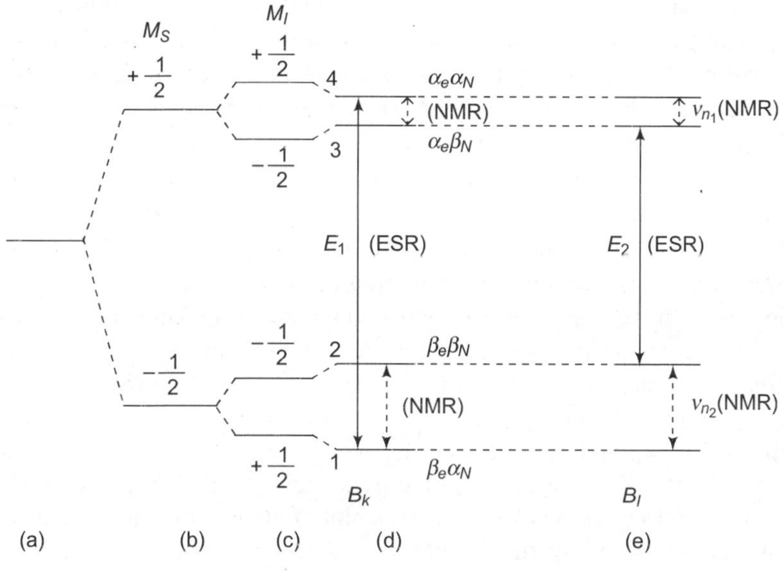

Fig. 9.1. Energy level diagram for a system of one unpaired electron $S=1/2$ and a nucleus with $I=1/2$ in a constant external magnetic field. The ESR transitions are shown by full line arrows and the NMR transitions with dotted line arrows at frequencies ν_{n_1} and ν_{n_2} that correspond to the selection rules $\Delta M_S = 0$, $\Delta M_I = \pm 1$. (a) Free electron, (b) in the presence of an applied magnetic field, (c) hyperfine interaction with nuclear spin $I = 1/2$ included, (d) microwave saturation of the ESR transition, E_1 and (e) microwave saturation of the ESR transition, E_2. The terms α_e β_N, etc. refer to the spin of the electron and the nucleus, respectively.

There are two allowed electron spin transitions denoted as E_1 and E_2 and two transitions involving nuclear spins denoted as ν_{n_1} and ν_{n_2}. In an ENDOR experiment, the ESR spectrum is observed while the nuclear spin transitions are irradiated simultaneously. The situation is similar to that in an all-nuclear INDOR experiment (Chapter 17) in that the populations of the various energy levels are perturbed. Enhancement of the intensity of the ESR transition is observed resulting from the redistribution of population of the various states.

The transitions in the ESR region, as shown in Figs. 9.1 (d) and (e) which change the electron spin give rise to a doublet due to coupling with the nuclear spin I=1/2. Similarly, the NMR transitions involving the nuclear spin may also be predicted to show two resonances. However, it is difficult to observe the NMR spectrum of the free radical owing to the very small concentration of the radical present in ESR experiments and the very efficient relaxation by the unpaired electron leading to extreme broadening of the signal. However, the effect of a strong radiofrequency radiation which causes the NMR transitions can be observed in the ESR spectrum. The ENDOR experiment is carried out as follows.

The sample is placed in a special microwave cavity cell. At low microwave power, the magnetic field is scanned to find the two ESR transitions. The instrumental parameters are optimised to maximize the ESR signal. The magnetic field corresponding to the two ESR transitions E_1 and E_2 (1→4 and 2→3) are B_k and B_l, respectively as stated before. The sample is irradiated with an intense microwave radiation at a frequency suitable for causing electron spin transition say E_1 (1→4). The microwave power level for the transitions say that at B_k corresponding to the 1→4 transition, $\beta_e\alpha_N \rightarrow \alpha_e\alpha_N$, is greatly increased by several folds so as to partially saturate the 1→4 transition. A high power radiofrequency radiation from a *rf* generator is now applied to the sample, through a range appropriate to the nucleus being investigated while still looking at the ESR signal under conditions of microwave saturation. The sample now experiences a radiofrequency magnetic field. At two frequencies of the radiofrequency radiation at which the photon energy ($h\nu$) matches with the energy of the nuclear spin transitions, say ν_{n_1} and ν_{n_2}, the recorder will trace out a spectrum such as that shown in Fig. 9.2. The transitions at the frequencies ν_{n_1} and ν_{n_2} correspond to the selection rule $\Delta M_I = \pm 1$ and $\Delta M_S = 0$. This plot of the variation in ESR absorption intensity as the radiofrequency radiation is swept through is referred to as the ENDOR spectrum (Fig. 9.2).

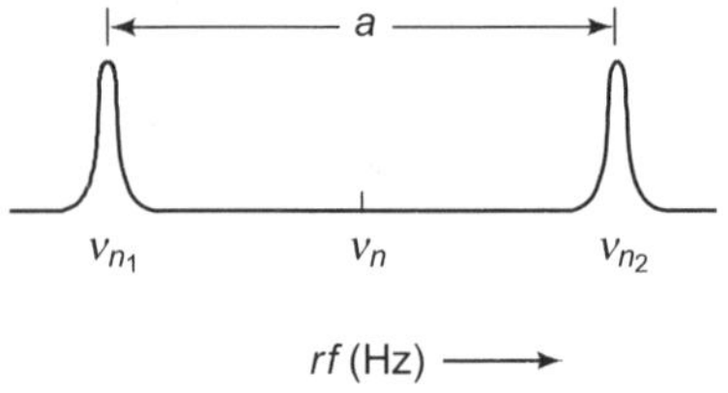

Fig. 9.2. ENDOR spectrum showing the change in the ESR signal intensity for a system with S=1/2 and I=1/2. The signals are separated by the hyperfine coupling constant a. ν_{n_1} and ν_{n_2} are the NMR frequencies of the nucleus.

In ENDOR experiments, the absorption of the *rf* power at the usual NMR frequencies is not observed. Rather, we observe an increase in the intensity of the ESR transition due to spin redistribution of the populations of different states. Typically enhancement of either line occurs both at ν_{n_1} and ν_{n_2}. Since one observes the enhancement as the *rf* field is scanned, the recorder traces out 'ENDOR lines'. Thus, ENDOR could be considered as a special case of NMR with the unpaired electron serving as the detector.

It is to be noted that the difference between $\nu_{n_1} - \nu_{n_2}$ is numerically equal to the hyperfine coupling constant but now obtained with ESR signals. The ENDOR spectrum thus provides an alternate method for obtaining the hyperfine splittings. The NMR signals are much sharper than the ESR signals although they may considerably be broadened by the shorter relaxation time due to the unpaired electron. The ENDOR technique thus combines the sensitivity of the ESR technique with the resolution of the NMR spectrum. Nuclear hyperfine couplings are often easier to observe and the hyperfine coupling constants in inhomogeneously broadened spectra are more precisely obtained from the sharper spectral lines of the ENDOR. Further the mean of the frequencies ν_{n_1} and ν_{n_2} will be equal to $\nu_n = g_N \beta_N B_k / h$ which is the NMR frequency of the bare nucleus in the applied magnetic field B_k. The identity of the nucleus causing the hyperfine splitting is now easily established from the value of g_N, the nuclear *g*-factor.

If the experiment is repeated but with the magnetic field set at B_l, the ENDOR spectrum would again consist of two signals separated by the hyperfine coupling constant and symmetrically situated in first order about the NMR frequency ν_n of the nucleus in the magnetic field B_l. The relative intensities of the two peaks may not be the same in the two ENDOR spectra. We have noted that for nuclei with $I = 1/2$, each hyperfine coupling gives a pair of lines in the ENDOR spectrum at $\nu_n \pm a/2$. For nuclei with $I = 1$ (e.g. ^{14}N), a characteristic four-line pattern of frequencies occur. When the nucleus has spin $I \geq 1$, it becomes necessary to include the quadrupolar interaction term. It then becomes difficult to predict the intensity of the ENDOR lines. For radicals in which there are *n* sets of *m* equivalent protons, there will be 2*n* lines in the ENDOR irrespective of the number of protons in the set. The ENDOR spectrum thus does not reveal how many nuclei contribute to it. For example, the ^1H ENDOR spectrum of $Ti(C_8H_8)(C_5H_5)$ clearly showed two doublets as shown in Fig. 9.3 which provided the hyperfine coupling constants of the protons of the C_8H_8 and C_5H_5 rings directly.

The ENDOR spectrum is much better resolved; the smallest observed ENDOR linewidths correspond approximately to typical NMR linewidths. The main advantage of the ENDOR method is that one can measure very small hyperfine splittings under conditions where the ESR spectrum contains so many lines that may overlap. A series of ENDOR spectra, each with as few as two lines, can greatly simplify the task of spectral assignment and this is particularly important when small couplings have to be identified and measured so that spin densities at a number of different nuclei can be determined. Finally, all double resonance methods are used to study relaxation phenomena. An ESR spectrometer of high sensitivity is required for the study since the intensity of the lines in the ENDOR spectrum typically represents a change in the intensity of 1% of the normal ESR spectral intensity.

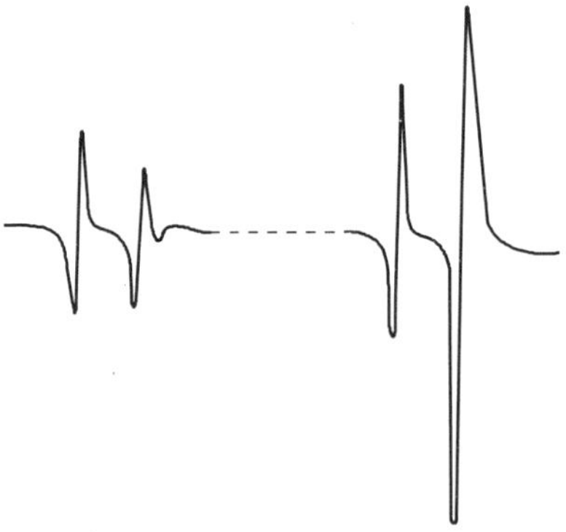

Fig. 9.3 ^{1}H ENDOR spectrum of $Ti(C_8H_8)(C_5H_5)$.

The higher sensitivity of ENDOR arises from the following:

(i) The energy of the ESR transition is greater than that of the NMR transition. Larger population differences occur for the more widely spaced levels.

(ii) The rate of absorption of energy is much higher at microwave frequencies than at radio frequencies.

(iii) The magnetic field of the electron which is typically of the order of 10^3 to 10^5 mT acts at the nucleus in addition to the applied magnetic field. This generates greater population differences than would be possible if the external magnetic field alone governed these differences.

ELECTRON-ELECTRON DOUBLE RESONANCE

The ELDOR technique is the observation of the reduction in intensity of one hyperfine transition when a second hyperfine transition is simultaneously irradiated to saturation. Simultaneous ESR transitions in an applied magnetic field for two different hyperfine transitions require the irradiation simultaneously at two different microwave frequencies. The intensity of the ESR signals will be a function of the difference between the two microwave frequencies. Accurate measurement of the coupling constants has been made by the ELDOR technique.

The ELDOR technique is somewhat different from the ENDOR. The simplest case of ELDOR is that of a single electron interacting with a single nucleus of spin 1/2 and the energy level diagram is shown in Fig. 9.4. The two ESR transitions shown therein have no level in common. However, they may be coupled by the following two mechanisms.

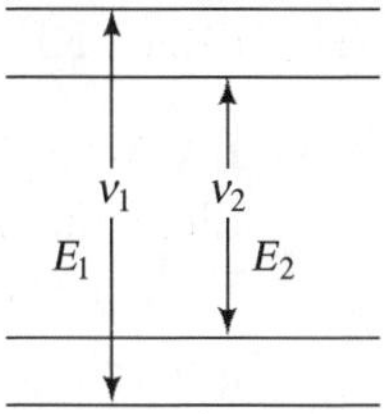

Fig. 9.4 ELDOR experiment. The intensity of the ESR transition observed at the frequency ν_2 using a low microwave power is monitored as a function of the high microwave power applied to the sample from a source operating at frequency ν_1.

(i) Rapid nuclear relaxation occurs due to dipolar coupling between the electron and the nuclei. The flipping of an electron spin under suitable conditions can cause simultaneously a flip of the coupled nuclear spin. This mechanism is important at low temperatures and at low concentrations.

(ii) At high temperatures, or at high concentrations, spin exchange or chemical exchange will tend to equalize the populations of all the spin levels.

In an ELDOR spectrum, the intensity changes in the monitored resonance is plotted against the frequency difference and a peak will be observed at the energy difference ($E_2 - E_1$) which is equal to the hyperfine splitting constant. For a more complicated system, the spectrum may exhibit several lines and each line is a direct measure of the hyperfine splitting constant. This is particularly useful for measuring very large splittings or when the spectrum is complicated by numerous hyperfine splittings. The ELDOR spectra also contain lines corresponding to the forbidden transitions, that is, $\beta_e\ \beta_N$ to $\alpha_e\ \alpha_N$ and $\beta_e\ \alpha_N$ to $\alpha_e\ \beta_N$ in Fig. 9.1 and these contain information about the mean nuclear resonance frequency as well as the hyperfine coupling constant. It is also useful for the accurate measurement of coupling constants and the assignment of hyperfine splittings.

ELDOR can be used to aid the assignment of overlapping resonance peaks in a manner similar to double resonance used in NMR (see Chapter 17). This technique is chiefly useful to study relaxation mechanisms. ELDOR was used to distinguished between the two nearly identical ^{14}N hyperfine splittings in DPPH utilizing the differences in the relaxation times of the various nuclei present.

GENERAL REFERENCES

1. J.A. Weil, J. R. Boulton and J. E. Wertz, Electron Paramagnetic Resonance: Elementary Theory and Practical Applications, Wiley-Interscience, New York, 1994.
2. R. S. Drago, Physical Methods in Chemistry, Saunders, Philadelphia, 1992.
3. I. N. Levine, Molecular Spectroscopy, Wiley, New York, 1975.
4. C. N. Banwell and E. M. McCash, Fundamentals of Molecular Spectroscopy, Fourth Edition, Tata McGraw-Hill, New Delhi, 1994.
5. M.M. Done and J.H. Freed (ed.), Multiple Electron Resonance Spectroscopy, Plenum, New York, 1979.
6. A. Carrington and A. D. McLauchlan, Introduction to Magnetic Resonance, Chapman and Hall, London, 1979.
7. A. Abragam and B. Bleaney, Electron Paramagnetic Resonance in Transition Metal Ions, Oxford University Press, London, 1970.
8. N.M. Atherton, Principles of Electron Spin Resonance, Ellis Harward/ Prentice Hall, Hemel Hempstead, 1993.
9. B. A. Goodman and J. B. Raynoor, Adv. Inorg. Chem. Radiochem. **13**, 135, 1970.
10. M.C.R. Symons, Chemical and Biological Applications of Electron Spin Resonance Spectroscopy, van Nostrand, New York, 1978.
11. E. Heilbronner and P.A. Straub, Huckel Molecular Orbitals, Springer-Verlag, New York, 1966.
12. D.N. Sathyanarayana, Electronic Absorption Spectroscopy and Related Techniques, Universities Press, Hyderabad, India, 2001.

EXERCISES

1. Calculate the magnetic field needed for an ESR instrument employing electromagnetic radiations of frequency (a) 35.0 GHz and (b) 10 mm.
2. Find the ratio of equilibrium populations of α ($M_s = 1/2$) and β ($M_s = -1/2$) spin states at 20 °C for an isolated electron in a magnetic field of 0.34 T.
3. Determine the relative population difference $\dfrac{\delta N}{N} = \dfrac{(N_\beta - N_\alpha)}{(N_\beta + N_\alpha)}$ for electron

   spins in an applied magnetic field of 0.34 T at 300 K (g = 2.0023).
4. Calculate the *g*-value of a radical showing a four-line ESR signal centred at 0.3292 T in a spectrometer operating at 9.223 MHz.
5. A free radical has a g-value of 2.20025. Calculate the field at which the resonance would occur in an ESR spectrometer operating at 9,300 MHz.

6. The ESR spectrum of hydrogen atom exhibits two lines at 0.3573 and 0.3066 T in a spectrometer operating at 9,302 MHz. Find the hyperfine coupling constant for the hydrogen atom. The signal for atomic hydrogen was observed at 9.50 GHz. Assuming the *g*-value to be 2.0023, calculate the magnetic field applied. Obtain the coupling constant.

7. Why is ESR spectral line of a sample observed at liquid nitrogen temperature nearly four times as intense as the same line at 300 K?

8. The hyperfine splitting in a free radical is given as 1420.4 MHz and g = 2.002. Express the splitting in mT.

9. A radical containing two equivalent protons shows a three-line spectrum in the intensity ratio 1:2:1. The spectral lines are found at 0.3302, 0.3325 and 0.3348 T. (a) Find the hyperfine coupling constant for the proton. (b) If the operating frequency of the ESR spectrometer was 9,319 MHz, determine the *g*-value of the radical. Find the coupling constants in MHz.

10. A radical containing two inequivalent protons with hyperfine splitting constants of 2.0 and 2.6 mT gives a spectrum centred at 0.3325 T. Find the fields at which hyperfine lines occur and their relative intensities.

11. A radical containing three equivalent protons shows a four-line spectrum with intensity 1:3:3:1. The lines occur at 331.4, 333.6, 335.8 and 338.0 mT. Find the hyperfine coupling constant for each proton and the *g*-value of the radical given that the spectrometer operates at 9.332 GHz.

12. Draw the energy level diagram for the interaction of an electron in the presence of a magnetic field with three equivalent nuclei of spin I = 1/2.

13. For the anion radical of the following molecules give the number of lines expected in a fully resolved ESR spectrum and their relative intensities: (a) toluene, (b) o-xylene, (c) 1-hydroxyethyl radical, (d) p-nitroaniline, (e) cyclooctatetraene radical anion and (f) triphenylmethyl radical.

14. Draw the ESR spectrum expected for (a) $^{\bullet}NH_2$ radical, given $a_N = 1.61$ mT and $a_H = 0.35$ mT and if $a_N = 2a_H$ what will be the intensity distribution and (b) $^{\bullet}NH_3$ radical $a_N = 1.25$ mT and $a_H = 0.45$ mT.

15. Predict the number of ESR lines expected and their intensity pattern in a completely resolved ESR spectrum of each of the following radicals:

    (a) $^{\bullet}CH_2D$, (b) $(CF_3)_2N^{\bullet}O$, (c) $^{\bullet}CF_2H$, (d) $CH_3{}^{\bullet}CHCH_3$, (e) $^{11}BH_3$, (f) $H_2\dot{C}(OCH_3)$ (g) p-nitrobenzoate dianion and (h) p-dinitro benzene radical anion.

16. A certain free radical shows ESR spectrum with the centre of the line at 0.3319 T and 0.3336 T respectively when the applied magnetic field is parallel and perpendicular to the molecular axis. Calculate the *g*-values of the free radical in the two directions given that the spectrometer operating frequency is 9.302 GHz.

17. A radical containing three inequivalent protons with hyperfine coupling constants 2.11, 2.87 and 2.89 mT shows a spectrum centred at 332.8 mT. Find the magnetic fields where the hyperfine lines occur and what are their relative intensities.

18. $(R_3Ge)_2$ on photolysis gives a radical which shows an ESR spectrum with nine lines. Identify the radical (for Ge I = 9/2).

19. The hyperfine coupling constant for the $^{\bullet}CH_3$ radical is 2.3 mT. Predict the hyperfine splitting in the ESR spectrum of $^{\bullet}CD_3$ radical. Find the overall width of the spectrum in each case. Sketch the ESR spectrum of $^{\bullet}CD_3$.

20. Predict the ESR spectrum of partially deuterated ethyl radical, $^{\bullet}CH_3CD_2$ using $a(CH_2) = 2.24$ mT and $a(CH_3) = 2.68$ mT.

21. The free radical formed by the photolysis of a solution of H_2O_2 and isopropyl alcohol at 110 K gives an ESR spectrum consisting of seven lines in the intensity ratio of 1:6:15:20:15:6:1 and a hyperfine splitting of 2.0 mT with a line width of 1.0 mT. Identify the radical.

22. The free radical generated by mixing an acidified solution of $TiCl_3$ with an acidified solution of ethanol and H_2O_2 gives an ESR spectrum which is shown in Fig.1. Identify the free radical.

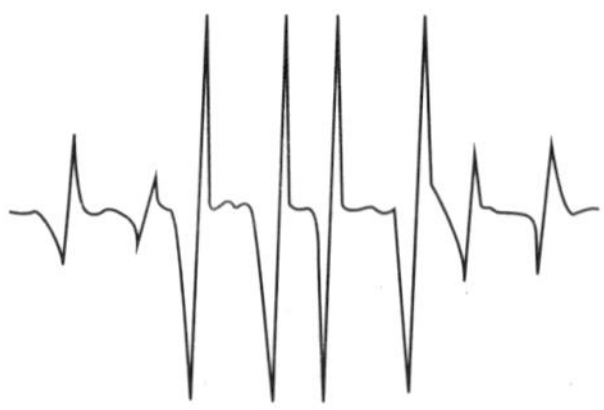

Fig. 1

23. From the spectrum given below (Fig. 2), find the values of *g* and *a*. What type of spin system would give rise to this type of hyperfine structure in the spectrum?

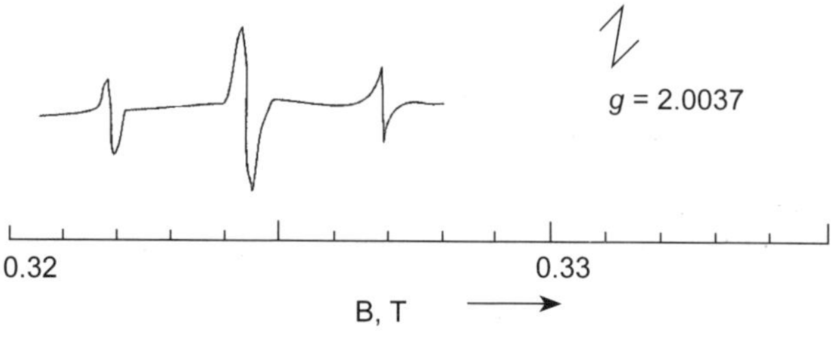

Fig. 2

24. The ESR spectra of (a) vanadyl acetylacetonate, $OV(acac)_2$ and (b) $OV(H_2O)_5^{2+}$ show two *g* values and two patterns of eight lines at 70 K. Explain.

25. Find the number of hyperfine lines expected in a high resolution ESR spectrum of bis(ethylene diamine)Cu^{2+}.

26. Irradiation of liquid ethane with highly energetic electrons gave an ESR spectrum with relative intensities of 1:2:3:1:6:1:3:2:1. Identify the free radical.

27. The ESR spectrum of NO_2 is a triplet at –120 °C. At –34 °C, the triplet is split into 15 lines. Is either of the formulations $NO_2.N_2O_2$ or $(NO_2)_x$ consistent with the splitting?

28. In the ESR spectrum of hypothetical radicals SCl_2 and SCl_3 (for S, I = 0 and for Cl, I = 3/2), how many hyperfine lines are expected? Explain the expected ESR transitions and their relative intensities.

29. The proton hyperfine coupling constants observed for different dinitrobenzene radical anions are given below in mT. Using the benzene value and the symmetry, obtain the probability of finding the unpaired electron in the $2p_\pi$ orbital of each of the carbon atoms.

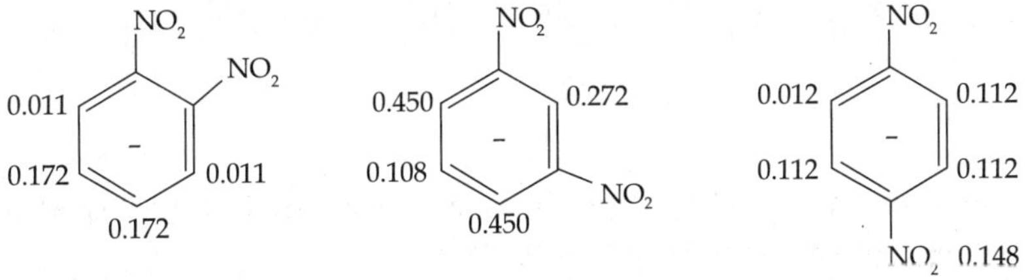

30. Predict the form of the ESR spectrum of anthracene anion using its spin densities given below.

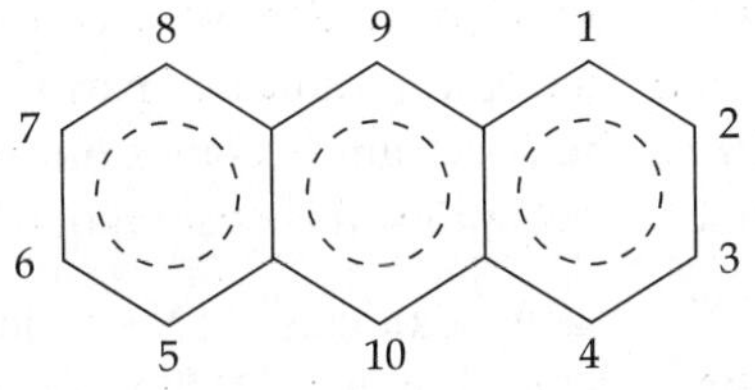

ρ_1 0.096
ρ_2 0.048
ρ_9 0.194

31. (a) The proton hyperfine coupling constants for naphthalene anion are 0.495 and 0.187 mT. Compare the observed ratio of the coupling constants with that obtained by the simple HMO theory and the McConnell equation. The lowest unoccupied HMO of naphthalene is given below.

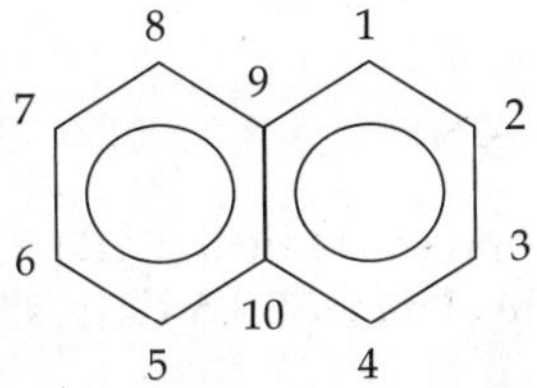

$$\psi_6 = 0.425\ (\phi_1 + \phi_4 - \phi_5 - \phi_8) + 0.263\ (-\phi_2 - \phi_3 + \phi_6 + \phi_7)$$

where ϕ *s* are the carbon $2p_\pi$ atomic orbitals.

32. (a) Among th diatomic molecules, Li_2, B_2, C_2, N_2, which is paramagnetic?

 (b) In naphthalene anion, how do you confirm that the larger coupling is to the α or β protons?

 (c) Does the ESR spectrum of a radical in solution depend on the sign of the proton coupling constants?

33. In an experiment, single crystals of succinic acid were irradiated with X-rays. The ESR spectrum obtained is shown in Fig. 3. Identify the free radicals formed and obtain approximately the hyperfine splitting constants.

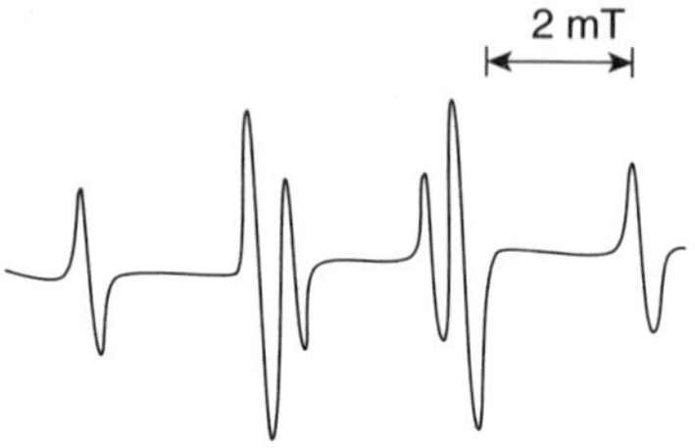

Fig. 3

34. Predict the number of hyperfine lines in the well-resolved ESR spectrum of the following transition metal complexes:

 (a) $Cu(NH_3)_4{}^{2+}$ (b) $Cu(H_2O)_6{}^{2+}$

 (c) $Ti(H_2O)_6{}^{3+}$ (d) $[Cr(H_2O)_4(NH_3)_2]^{2+}$ and (e) $[Ti(H_2O)_4(NH_3)_2]^{2+}$

35. The ESR spectrum of $Cu(H_2O)_6SiF_6$ diluted with the corresponding diamagnetic zinc salt showed nearly isotropic *g*-value and partly resolved hyperfine structure at 90 K. At 20 K, the spectrum becomes anisotropic and shows three sets of hyperfine lines. Account for this observation.

36. Tetracyanoethylene (TCNE) easily oxidizes SCN^- or iodide ion at room temperature. The spectrum produced by the ^{14}N hyperfine splitting consists of nine lines in the intensity ratio 1:4:10:16:19:16:10:4:1. Predict the spectrum of the TCNE anion.

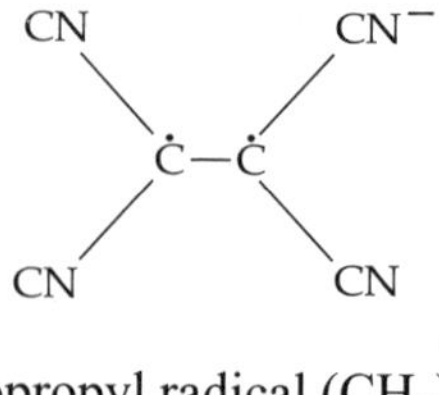

37. The ESR spectrum of isopropyl radical $(CH_3)_2{}^{\bullet}CH$ given below (Fig. 4) was obtained using a radiofrequency of 9.5 GHz. Calculate the hyperfine splitting constant.

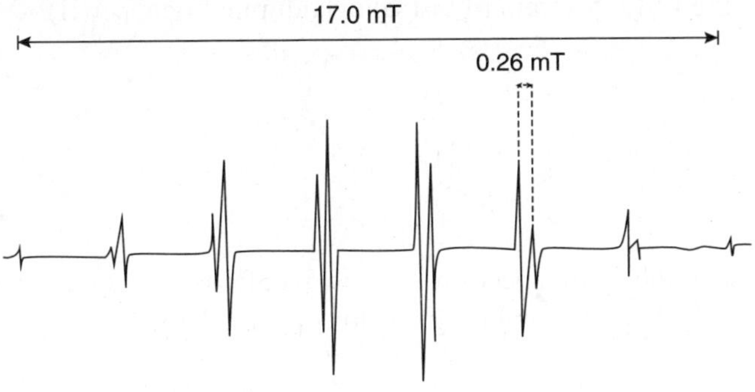

Fig. 4

38. (a) Draw the energy level diagram (energy versus applied magnetic field) for a system with $S = 1$. Show the possible transitions.
    a. with small zero-field splitting
    b. with zero-field splitting comparable to the ESR spectrometer frequency.
    c. with zero-field splitting much larger than the spectrometer frequency.

    (b) Find the number of ESR singnals expected for octahedral Ni(II) complexes.

39. What type of ESR spectra you would expect for
    a. an aqueous solution of Cu^{2+} and
    b. the same solution after it has been frozen?
40. Predict the number of ESR lines for (a) $Co(H_2O)_6^{2+}$ and (b) $Cr(H_2O)_6^{3+}$ Indicate how the zero-field splitting and Kramers degeneracy applies for these two complex ions.
41. Why ESR is a much more sensitive technique than NMR?
42. What are the factors that influence anisotropic interactions in transition metal complexes?
43. Account for the fact that (a) while CoF_6^{4-} gives a narrow ESR signal only at 20 K, CrF_6^{3-} gives sharp ESR lines even at room temperature and (b) the *g*-values of CoF_6^{4-} are anisotropic and deviate from 2.0 while CrF_6^{3-} gives *g*-values nearly isotropic and close to 2.0.
44. Predict the ESR spectrum of an aqueous solution of $Mo(CN)_8^{3-}$ ion at room temperature. ^{97}Mo which has I = 5/2 is less than 10% abundant. Predict the spectrum expected when carbons are enriched with ^{13}C.

45. Predict the ESR spectrum of bis(salicylaldiminato)copper(II).

(a)

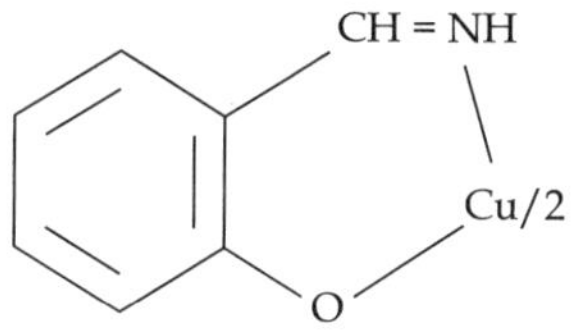

(b) Predict the ESR spectrum of bis(5–chlorosalicyldoxime) copper(II). Why the hyperfine splitting due to copper is large?

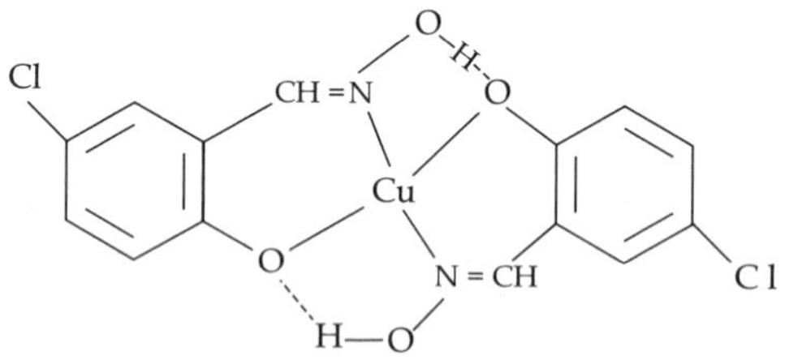

46. Carry out the Huckel molecular orbital calculations for 1,3-butadiene. Determine the π-electron densities. The ESR spectrum of butadiene anion show hyperfine splitting of $a_1 = a_4 = 0.762$ mT and $a_2 = a_3 = 0.279$ mT. Show that it is consistent with π electron charge density.

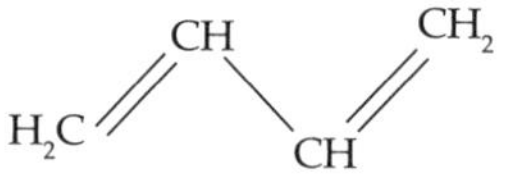

47. Sketch the expected ESR spectrum of vinyl radical $HC = CH_2$ using the observed values of hyperfine splitting $a_1 = 1.57$ mT, a_2 (*trans*) = 3.42 mT and a_3 (*cis*) = 6.85 mT. How many hyperfine lines are observed?

48. What are the advantages of using a Q band (35 GHz, 1.25 T) spectrometer?

49. Why $K_2[NO(SO_3)_2]$ does not show ESR spectrum in the solid state, while in solution it shows a triplet?

50. How does the *g*-value of a transition metal complex depend on its geometry?

51. Predict the ESR spectrum of $Ni(NH_3)_6^{2+}$. Comment on the value of the spin-orbit coupling constant in the complex.

52. Predict the ESR spectrum of Mn^{2+} ion in an octahedral field in solution.

53. Show how the ESR hyperfine lines arise and obtain the frequencies when an unpaired electron interacts with two non-equivalent nuclei, both with $I = 1/2$ assuming $a_1 > a_2$.

54. Predict the *g*-value for the Jahn-Teller distorted octahedral copper(II) complex.

55. Calculate the interaction energy of an electron with no orbital angular magentic moment in a magnetic field of 2.0 T and the corresponding frequency.

56. Which of the following species show the same number of hyperfine lines with the same intensity ratio?

(a) $\dot{C}D_3$ and $C_6H_6^+$

(b) $^{\bullet}NH_2$ and $CH_3\,\dot{C}H_2$

(c) H_2^+ and HD^+

(d) 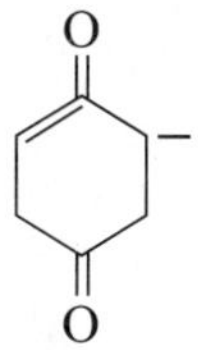 and CH_4^+

57. The doublet lines of ^1H occur at 326.86 and 377.66 mT. Estimate the line positions of the ^2H triplet given $g = 2.002$ for both species.

58. The ESR spectrum of a radical produced when H is abstracted from diethyl ether is given below.

Interpret the spectrum (Fig. 5).

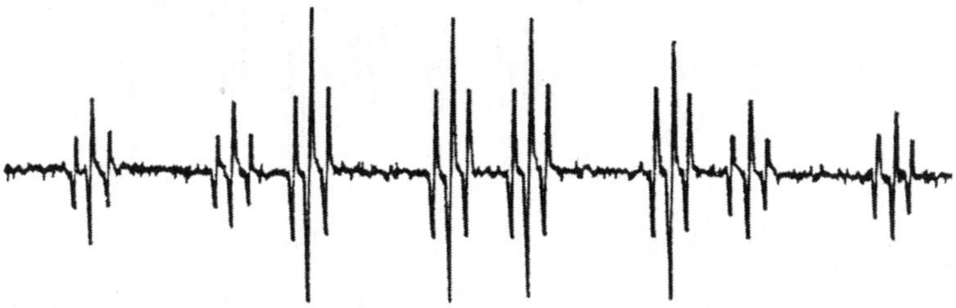

Fig. 5.

59. The ESR spectrum of condensed benzene 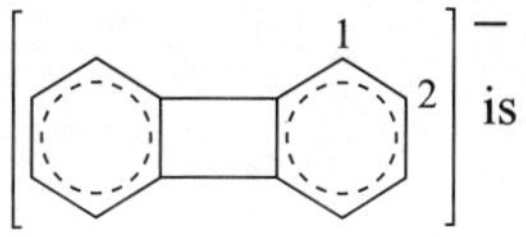 is given below (Fig. 6). Interpret the spectrum and find a values.

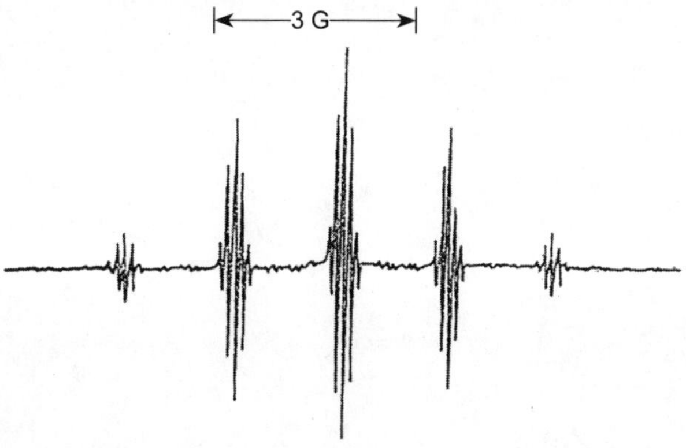

Fig. 6

60. The ESR spectrum of the potassium salt of the biphenyl anion,

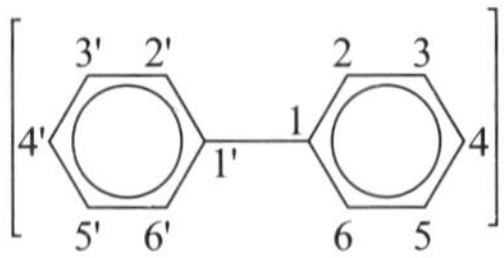

is given below (Fig. 7). Interpret the spectrum and obtain the *a* values.

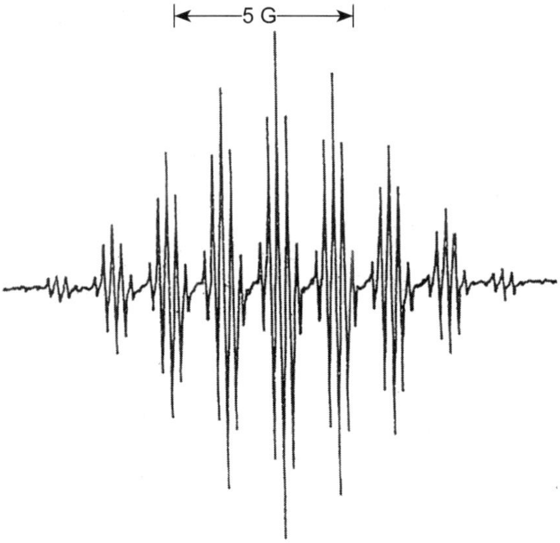

Fig. 7

Part Two

Nuclear Magnetic Resonance

GENERAL PRINCIPLES 10

NUCLEAR SPIN AND MAGNETIC MOMENT

Nuclear magnetic resonance (NMR) spectroscopy is probably the single most widely used and important spectroscopic technique. This may be traced to its wide applicability, its relative ease of use and the amount of chemical and structural information that can be obtained from its use. The technique is usually associated with proton and carbon nuclei but there are other elements to which it can be profitably applied.

The extensive chemical and biochemical applications of NMR lie in the elucidation of structure, identification of compounds, study of molecular conformation, dynamic processes and the interaction between molecules. In some cases, the NMR method alone may be sufficient for obtaining the structure of an unknown compound, while in other cases the NMR spectrum may complement the results obtained from other methods. NMR spectroscopy is a very powerful technique for studying the molecules at atomic level. A large number of metal nuclei which are of interest to inorganic chemists have become available for NMR experiments, due to the rapid developments in the experimental techniques. The most common applications of NMR spectroscopy are to liquid samples. Up to the end of the 1970s, NMR was used almost exclusively for structural analysis in chemistry. Since then, however NMR has become important in the investigation of biochemical processes *in vivo* and *in vitro*, as an imaging method (NMR tomography) in the observation of morphological changes in tissues and as a diagnostic aid in clinical practice.

NMR spectroscopy deals with the magnetic properties of the atomic nucleus. The existence of the nuclear magnetic moments and the illustration of their behaviour in a magnetic field were demonstrated by the Stern-Gerlach experiment with a stream of silver atoms and the molecular beam experiment of Rabi. The NMR was found in bulk materials in 1946 by Purcell, Torrey and Pound at the Harvard University. They reported NMR absorption in paraffin wax while Bloch, Hansen and Packard at the University of Stanford found NMR in water. Just six years later, they were jointly awarded the Nobel Prize in physics for their discovery. The importance of NMR was highlighted by the award of a Nobel Prize to Richard R. Ernest from the ETH, Zurich for his contributions to the development of the experimental techniques.

A nucleus consists of protons and neutrons both of which are called nucleons. The simplest nucleus is that of the hydrogen atom, i.e. the proton which has an intrinsic spin of ½. Neutron, which is a constituent of all nuclei except hydrogen nucleus also has an intrinsic spin of ½. The spin of nucleons of the same kind can pair up just as electrons do. Thus, each nuclear isotope possesses a varying spin value. If the spins of all the nuclear particles, i.e. protons and neutrons are paired, there will be no net spin and the nuclear spin quantum number I will be zero. However, it is difficult to predict exactly how many protons and neutrons will be paired except in simple cases such as ^{2}H. The observed nuclear spins have been rationalized using some empirical rules.

(i) Nuclei having even number of both neutrons and protons have zero spin (^{4}He, ^{12}C, ^{16}O, etc.).

(ii) Nuclei with odd number of both protons and neutrons have integral spins (^{2}H, ^{14}N, ^{10}B, etc.).

(iii) Nuclei with odd-even or even-odd combination of protons and neutrons have half-integral spins (^{1}H, ^{15}N, ^{19}F, ^{31}P, ^{11}B, ^{17}O, ^{27}Al, ^{35}Cl, etc.).

The nucleus of an element possesses an intrinsic (spin) angular momentum, I whose magnitude is given by $[I(I + 1)]^{1/2}\ h/2\pi$, where the nuclear spin quantum number or more commonly the nuclear spin I has the possible values of 0, 1/2, 1, 3/2, … . According to quantum theory, the angular momentum of a nucleus with spin I can have its components as $I_z\, h/2\pi$ along a particular reference direction say the Z-axis. The Z-component of the spin angular momentum vector I_z is quantized and it is designated as m_I, the magnetic spin quantum number, which has (2I + 1) values ranging from + I to – I in integral steps. Therefore, m_I can have the values:

$$m_I = I, I - 1, I - 2, \ldots\ldots\ldots\ldots - (I + 1), - I \quad \text{for integral spins,}$$

$$\text{and} \quad m_I = I, I - 1, I - 2, \ldots\ldots\ldots\ 1/2, - 1/2 \ldots\ldots - I \quad \text{for half-integral spins}$$

For a nucleus of spin I, the (2I + 1) components along a reference direction possess the same energy and they are said to be degenerate. If an external magnetic field is applied, the degeneracy is lifted and (2I + 1) different energy levels result. This space quantization is illustrated in Fig. 10.1. The splitting of the energy levels under the influence of a magnetic field is proportional to the field strength.

For a nucleus with I = 1/2, the two values of magnetic quantum number, m_I = + 1/2 and – 1/2, indicate the allowed orientation of the nuclear magnetic moment vector in an external magnetic field. The value of + 1/2 corresponds to alignment of the vector with the external magnetic field and – 1/2 opposed to it as shown in Fig. 10.1. Similarly for nuclei with I = 1, m_I has three values + 1, 0, – 1. The value of m_I = – 1 for instance represents a nuclear magnetic dipole opposed to the external magnetic field and is of higher energy, while the dipole for m_I = + 1 is in the same direction as the applied field and is therefore of lower energy. There is no net dipole along the field direction for m_I = 0 and is, therefore, unchanged in energy.

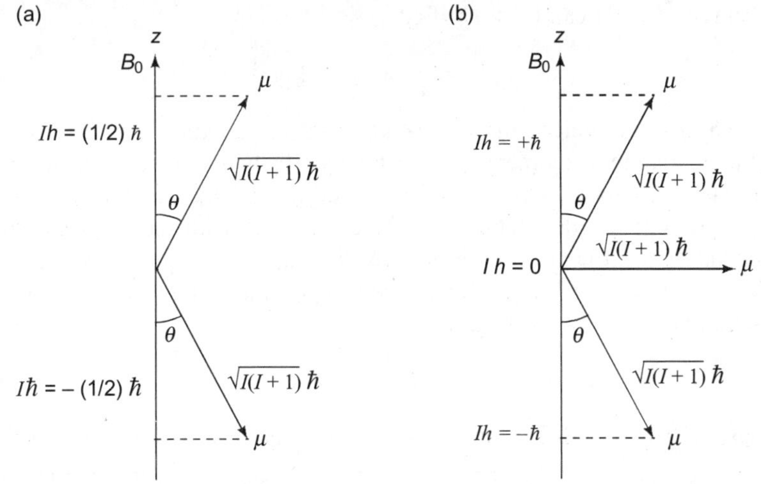

Fig. 10.1 Space quantization of nuclear spin angular momentum (a) I = 1/2 and (b) I = 1. The relationship between the magnetic field B_0, the nuclear magnetic moment μ and its component (μ_z) along B_0

A nucleus with a non-zero spin can be visualized as a charged spinning particle. A charged particle spinning about an axis constitutes a circular electric current, which in turn produces a magnetic dipole characterized by its magnetic moment, μ. The magnetic moment μ due to nuclear spin is directly proportional to the angular momentum vector, I of the spinning particle and is collinear with it (spin axis). It is convenient to define a ratio between them, which is called the magnetogyric ratio, γ.

$$\gamma = \frac{2\pi}{h}\frac{\mu}{I} = \frac{\mu}{I\hbar}\left(\hbar = \frac{h}{2\pi}\right) \tag{10.1}$$

γ has a characteristic value for each magnetically active nucleus, positive for parallel and negative for antiparallel vectors and it is measured in radian s^{-1} Tesla^{-1}. By analogy with the relation that describe the magnetic properties of the electrons, the magnitude of the magnetic dipole moment of the nucleus is given by

$$\mu_N = g_N \frac{eh}{4\pi m_p}\left[I(I+1)\right]^{1/2} \tag{10.2}$$

where g_N is a dimensionless constant called the nuclear g-factor, m_p is the mass of the proton and e is the protonic charge. The nuclear g-factor is characteristic of each nucleus and it is not possible to predict its value. Experimental values of g_N are available and it varies from – 2 to + 6 (see Table 10.1). The nuclear magnetic moments are much smaller than the magnetic moment of the electron. The nuclear dipole moment is expressed in terms of the nuclear Bohr magneton β_N where $\beta_N = eh/4\pi m_p = 5.051 \times 10^{-27}$ J T^{-1}.

The relation (10.2) can be simplified as

$$\mu_N = g_N \beta_N [I(I+1)]^{1/2} \quad (10.3)$$

This enables all nuclear magnetic moments to be expressed in terms of the basic unit of magnetic moment, β_N the nuclear Bohr magneton and the nuclear g-factor. The quantity $g_N I$ is commonly called as the nuclear magnetic moment. The magnetic dipole will interact to different extents with the external magnetic field depending on the magnitude of the dipole. The magnetic dipole moment defined by the equation (10.3) will have its components along the external magnetic field (z-axis) governed by m_I values

$$\mu_Z = g_N \beta_N m_I \quad (10.4)$$

Table 10.1 NMR frequencies and other properties for selected nuclei

Isotope	*I*	g_N	*Relative sensitivity at constant field*	γ $10^7 rad\ T^{-1}s^{-1}$	*Natural abundance,%*	*NMR frequency B = 2.3487 T*
^1H	1/2	5.585	1.000	26.752	99.985	100.0
^2H	1	0.857	0.010	4.107	0.015	15.35
^{10}B	3	0.600	0.020	2.875	18.83	10.75
^{11}B	3/2	1.792	0.165	8.584	81.17	32.08
^{13}C	1/2	1.405	0.016	6.727	1.108	25.14
^{14}N	1	0.404	0.001	1.933	99.63	7.22
^{15}N	1/2	–0.566	0.001	–2.710	0.37	10.13
^{17}O	5/2	–0.757	0.029	–3.627	0.037	13.56
^{19}F	1/2	5.257	0.834	25.177	100.0	94.08
^{29}Si	1/2	1.110	0.079	–5.320	4.70	19.90
^{31}P	1/2	2.263	0.066	10.840	100.0	40.48

RESONANCE FREQUENCIES

The energy of an isolated nuclear magnetic moment in an external magnetic field of strength B is the scalar product of the two vectors,

$$E = -\mu B \cos\theta \quad (10.5)$$

where θ is the angle between the nuclear magnetic dipole vector and the direction of the external magnetic field.

If μ_z is the component of μ in the direction of the magnetic field (along z-axis) of strength B, the equation (10.5) becomes

$$E = -\mu_z B \qquad (10.6)$$

Substituting for μ_z from equation (10.4), we get

$$E = -g_N \beta_N m_I B \qquad (10.7)$$

The difference in energy ΔE between the neighbouring levels where m_I differs by unity is

$$\Delta E = g_N \beta_N B \text{ Joules} \qquad (10.8)$$

We can cause transitions between these two levels by applying an electromagnetic radiation of appropriate frequency ν.

The resonance frequency ν is given by

$$\nu = \Delta E/h = |g_N \beta_N B/h| \text{ Hz} \qquad (10.9)$$

where β_N is in J T^{-1} and B is in Tesla. This is the basis of the NMR spectroscopic technique. A change in nuclear spin orientation between the two energy levels may be associated with the absorption or emission of radiation at the frequency given by the equation (10.9). It is easy to see that the nuclear spin energy levels are evenly spaced as shown in Fig. 10.2 and the spacing increases as the strength of the external magnetic field is increased.

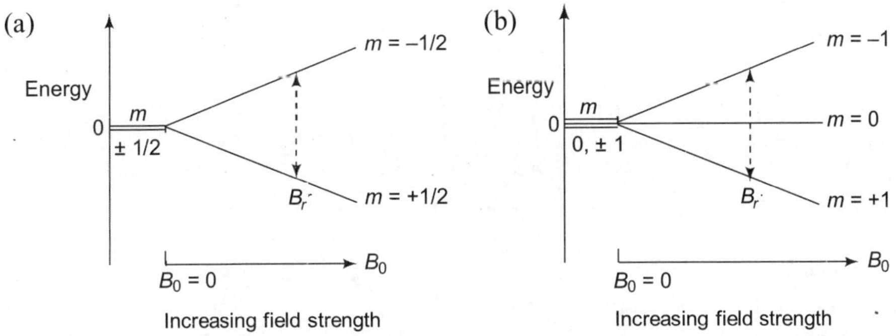

Fig. 10.2 Spin energy levels for (a) ^1H (I = 1/2) and (b) ^{14}N (I = 1) nuclei in a magnetic field B_0. The magnetic field strength at resonance is B_r.

In ESR spectroscopy, two different energy states arise from the alignment of the electron spin moments (arising from $m_s = \pm 1/2$ for a free electron), with the external magnetic field. We should note here the difference in the ground state for ESR and NMR. In ESR, the lowest energy β-spin state ($m_s = -1/2$) has its moment aligned with the external magnetic field whereas in NMR the lowest energy state corresponds to $m_I = +1/2$ (α_N), since the sign of the charge on the electron is opposite to that of the proton.

For a given magnetic field strength, the resonance frequency is chiefly determined by the nuclide since every nuclide (e.g., ^1H, ^2H, ^{13}C, etc.) has a characteristic magnetic moment. For example, for proton, g_N = 5.585 and using a magnetic field strength B of 1.5 T, we obtain

$$\nu = \frac{g_N \beta_N B}{h} = \frac{5.585 \times 5.051 \times 10^{-27} \times 1.5}{6.626 \times 10^{-34}} = 64 \text{ MHz}$$

Similarly for ^{31}P (I = 1/2), substituting the values of g_N, β_N and B, we obtain 25.67 MHz as the NMR frequency. A typical NMR frequency of 60 MHz corresponds roughly to ~ 10^{-2} J mol^{-1}.

A typical magnetic field strength used for ^1H NMR is 9.4 T (1 T = 10^4 Gauss). For hydrogen nucleus, equation (10.9) predicts a resonance frequency of (ν=) 4 × 10^8 Hz or 400 MHz when the external magnetic field is 9.4 T. Magnetic fields in the range 1.4 to 14.1 T are commonly employed giving proton resonance frequencies of 60 to 600 MHz. This falls in the radiofrequency region (~ 10^6 to 10^9 Hz) of the electromagnetic spectrum. Since proton is the most popular NMR nucleus, NMR spectrometers are usually classified by their proton frequencies rather than the strength of their magnetic fields. Table 10.1 summarises the NMR properties of several magnetic nuclei and of these nuclei, ^1H has the largest magnetogyric ratio (and the largest magnetic moment). The difference in resonance frequencies between nuclides (see Table 10.1) almost always exceeds the spectral ranges, so that it is rare to find overlapping spectra from different nuclides.

POPULATION OF ENERGY LEVELS

When placed in an external magnetic field, a collection of magnetic nuclei occupy (2I + 1) levels according to the Boltzmann distribution. The tendency of the nuclei to align with the magnetic field and thus occupy the lowest energy level is opposed by thermal motions, which tend to equalize the populations in the (2I + 1) levels Let us consider protons in a magnetic field strength of 1.4 T at room temperature. The ratio of the populations of the two levels designated by the subscripts α (lower) and β (upper), is

$$N_\beta/N_\alpha = \exp(-\Delta E/kT) \qquad (10.10)$$

where k is the Boltzmann constant, T is the temperature in degrees Kelvin and ΔE is the energy gap between the two levels. One finds that ΔE = 2.65 × 10^{-25}J, kT = 4.14 × 10^{-21}J and $\Delta E/kT$ = 6.4 × 10^{-5}. Thus, the energy needed to reorient the spins is extremely small compared to the thermal energy kT so that there is little tendency for the spins to be ordered in the lower energy level.

For very small values of $\Delta E/kT$, the approximation $e^{-x} = 1 - x$ may be used to show that the fractional excess population in the lower level is

$$N_\beta/N_\alpha \approx 1 - (\Delta E/kT) \qquad (10.11)$$

For $I = 1/2$, $\Delta E \approx 0.02$ J mol^{-1} in a field of 1.5 T at room temperature, $T = 300$ K,

$$N_\beta = N_\alpha \exp(-1.2 \times 10^{-5}) \tag{10.12}$$

There is thus a definite but very small excess of nuclei in the lower level, about 12 in a million in a magnetic field of 1.5 T. The net absorption of energy and hence the intensity of the spectroscopic transition is dependent on the difference in population between the two levels. Thus, in NMR spectroscopy, one detects only one nucleus in every 10^5 to 10^6 nuclei. Therefore, it is obvious that NMR signals must be very weak. The use of strong magnetic fields to maximize the transition energy ΔE is very desirable.

LARMOR PRECESSION

For treating the behaviour of a large population of spins, the classical approach is simpler. For example, the magnet moment vector μ will always be somewhat across the field and will therefore be under the influence of a torque tending to turn it in the field direction due to its interaction with the external magnetic field. This torque causes the μ vector to precess about the applied magnetic field B as shown in Fig 10.3 and the precession is known as Larmor precession. The angular velocity of Larmor precession is given by

$$\omega = \frac{\mu B}{I} \text{ radians } s^{-1}$$

$$\text{or} \quad \nu = \mu B / 2\pi I \text{ Hz} \tag{10.13}$$

Equation (10.13) can be derived from classical magnetodynamics.

This is generally written as

$$\nu = \gamma B / 2\pi \text{ Hz} \tag{10.14}$$

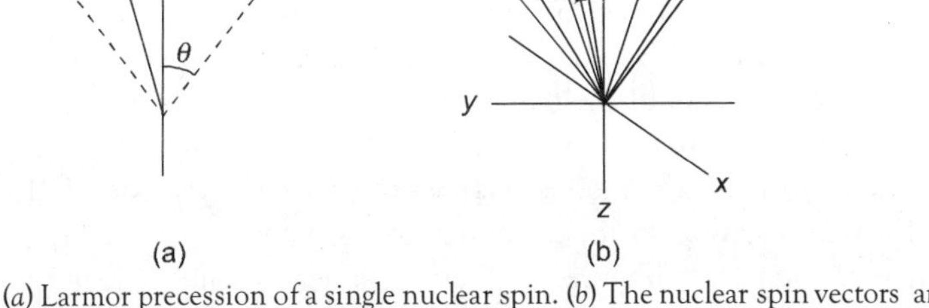

Fig. 10.3 (*a*) Larmor precession of a single nuclear spin. (*b*) The nuclear spin vectors arising from different atoms can be regarded as being spread out evenly over a conical surface, but can be drawn with the same origin. The corresponding cone along the –*z*-axis is not shown.

A proton exposed to an external magnetic field of 2.35 T will precess nearly 100 million times per second i.e. ν = 100 MHz. Thus, the Larmor frequency is directly proportional to the applied magnetic field and it also depends on γ (or μ) which varies depending on the nucleus.

One important feature of the Larmor equation (10.13) is that the angle θ does not occur in the equation. Therefore, the nuclear spin precesses at the Larmor frequency around the magnetic field, which depends only on its own characteristic properties. Substituting for μ and I from equation (10.3) in equation (10.13), we obtain

$$\nu = \frac{g_N \beta_N \left[I(I+I)\right]^{1/2} B}{\left[I(I+1)\right]^{1/2} (h/2\pi)\, 2\pi} = \frac{g_N \beta_N B}{h} \text{ Hz} \tag{10.15}$$

Comparison of the equation (10.15) with equation (10.9) reveals that the Larmor frequency is equal to the separation between the spin energy levels. If the electromagnetic radiation has the same frequency as that of the precessing nuclear spin, it can interact coherently with the nuclear spins and energy can be exchanged. For any other frequency of radiation, there will be no interaction. It is to be noted that it is the magnetic dipole vector of the electromagnetic radiation that interacts with the system to give resonance and not the electric dipole vector as in most other forms of spectroscopy.

The Larmor frequencies for selected isotopes in a magnetic field strength of 2.3487 T are given in Table 10.1 along with g_N and natural abundance of the isotopes. In NMR, only one type of nucleus can be studied with a particular radiofrequency. Besides the restriction imposed by I, the nuclei should also have high natural abundance and high relative sensitivity (higher magnetic moment or magnetogyric ratio). For these reasons, ^{1}H and ^{19}F are highly suitable for NMR, since their relative abundance is almost 100 percent and have high magnetic moment. ^{11}B has a natural abundance of 81%. However, since I = 3/2, it is not very suitable. The ubiquity of protons and the above factors explain the popularity of ^{1}H as NMR nucleus. The most abundant nuclei ^{12}C and ^{16}O have zero spin and hence do not show any resonance signal. This is very beneficial since it leads to relatively simple spectra.

The proton magnetic moment oriented in the direction of B is slightly more populated and hence there is a net macroscopic magnetization. If B is along the z- direction, the component of the macroscopic magnetization along the z-axis, M_z is non-zero. The component in the xy–plane, M_{xy} is zero because the individual nuclear magnetic moments randomly precess around B. If a second magnetic field B_1 rotating in the xy-plane at right angles to the main field B is applied, it can interact with the magnetic moment μ and affect the resultant macroscopic magnetization. The nuclear magnet will attempt to precess about this direction in addition to the direction of B_0. The vector B_1 will be relatively ineffective because it is smaller and the orientation of the nuclear magnet relative to it will be changing continuously as a result of precession about B. When the frequency of B_1 is the Larmor frequency, B_1 flips some nuclear magnetic moments from a lower energy orientation and forces them to precess in phase. This decreases M_z and creates a non-zero rotating component of magnetization in the xy-plane. When these conditions are met, the rotating magnetic field B_1 of the *rf* radiation and the precessing nuclear magnet are said to be in resonance and absorption of energy from the field B_1 by the nuclear magnets can occur. When the frequency of B_1 differs from the Larmor frequency, B_1 does not interact with the nuclear magnetic moment and has no effect on the macroscopic magnetization. The B_1 is obtained from radiofrequency radiation that is polarized in the xy–plane (also refer chapter 2).

The selection rule for NMR transition is $\Delta m = \pm 1$; the transition probability, as noted earlier, increases with γ and B. For nuclei with spin I, although there are (2I + 1) nuclear spin energy levels (Fig.10.2), they are equally spaced and the selection rule allows only transitions between adjacent levels and therefore a single NMR frequency results.

THE NMR SPECTRUM

The major components of a NMR spectrometer are a strong static magnetic field, a source of radiofrequency radiation to excite nuclei in the sample and a method for detecting the NMR signal. In a typical experiment, the sample solution in a glass tube of approx. 15 cm length and 5 or 10 mm in diameter is placed between the pole faces of a magnet. A strong homogeneous magnetic field is applied causing the nuclei to precess. Radiofrequency radiation is then imposed from a radiofrequency transmitter. The applied magnetic field is varied in a small range. When the frequency from the radiofrequency transmitter is equal to the Larmor frequency, the energy can be transferred to and from the source and the sample and the two are said to be in resonance. The net result of this resonance is that some nuclei are excited from the lower energy say $m_I = + 1/2$ state to the higher energy $m_I = - 1/2$ state and absorption of energy from the source at the frequency equal to the Larmor frequency occurs. The NMR spectrum is presented as a plot of the intensity of the absorption against the strength of the applied magnetic field B_0 (at constant frequency of the radiofrequency radiation).

The magnetic field experienced by the nucleus in a molecule differs very slightly from the external field such that the exact resonance frequency is characterized by the chemical environment of the nucleus. As an example a low resolution NMR spectrum of C_2H_5OH is shown in Fig. 10.4. The three kinds of hydrogen atoms have different resonance frequencies, and hence give rise to three separate peaks identifiable by their integrated areas, which are in the ratio of 1 : 2 : 3 showing the number of protons of each type. Typical line widths in the high resolution ^1H NMR spectra of small molecules in solution are around 1 Hz.

The NMR peaks of nuclei with spin quantum number greater than ½ are often broad, 10–100 Hz or more, due to nuclear quadrupole moment. The majority of NMR studies are, therefore, performed on spin ½ nuclides, the most popular ones being ^1H, ^{13}C, ^{19}F and ^{31}P.

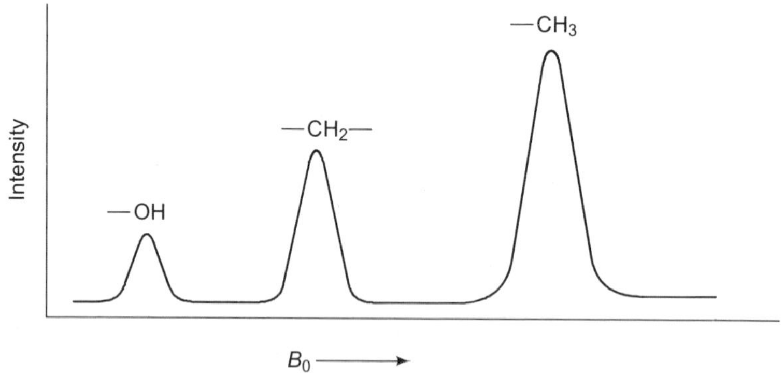

Fig. 10.4 Low resolution NMR spectrum of ethanol.

The relative NMR signal strength obtainable from solutions of different elements at equal concentrations and at a fixed applied magnetic field is proportional to γ^3 NI(I + 1), where γ is the magnetogyric ratio, N is the natural abundance and I is the nuclear spin quantum number.

There is a dramatic difference between the NMR spectra of solids and those of liquids and solutions. In solids, the rigid orientation of the nuclei with respect to the applied magnetic field B_0 as well as to the neighbouring nuclei leads to severe line broadening due to dipolar interactions. The spectra of solids have peaks several kHz wide which are generally not easily analysed. This topic is discussed later. In a liquid due to random translational and rotational motions of the molecules, the dipolar interactions between nuclei average to zero. In this situation, spectra with signals of line widths less 1 Hz result. This is referred to as high resolution NMR spectroscopy.

RELAXATION PROCESSES

The probabilities for stimulated absorption and emission of radiation are equal and they are much greater than the probability for spontaneous emission of radiation at radiofrequencies as discussed earlier. The probability of a transition occurring from one spin state to another is directly proportional to the population of the state from which the transition takes place. Since in the presence of a magnetic field, the lower state is more populated than the upper state at equilibrium, upward transitions predominate slightly over the downward transitions and a very small net absorption of energy from the radiation beam occurs. The probability of absorption of radiation is given by

$$p \propto B\,\rho(\nu)\,(n_{\text{lower}} - n_{\text{upper}})$$

where $\rho(\nu)$ is the radiation density and B is the Einstein's coefficient for stimulated absorption of radiation. During NMR absorption, the nuclei are raised from a lower level to an upper level. As a result, the very small population difference between the two states $n_{\text{lower}} - n_{\text{upper}}$ tends to decrease further. The intensity of the NMR absorption, which depends directly on this population difference, may progressively decrease and finally disappear causing no further absorption. The system is then said to be saturated. This problem, however, does not arise in most cases unless high radiofrequency power is used. This is because the equilibrium populations can be re-established if the system loses its absorbed energy; this does not occur spontaneously but only as a result of the interactions with the surrounding fluctuating magnetic fields of appropriate frequency. The mechanism by which the excess spin energy is shared either with the surroundings or with other nuclei is referred to as a relaxation process; the time taken for a fraction $1/e$ (0.37) of the nuclei to dissipate excess energy is known as the relaxation time. If the radiofrequency power is too high, the normal relaxation processes will not be able to compete with the excitation and thermal equilibrium will not be achieved. The Boltzmann population difference between the energy levels will decrease to zero and the intensity of the absorption signal will also therefore become zero. Saturation of NMR signal is less likely to occur when efficient relaxation processes operate to restore thermal equilibrium. Relaxation processes are thus essential to observe magnetic resonance. A detailed study of the relaxation processes can provide useful information on molecular structure and motion which is discussed later.

Two different relaxation processes as in ESR are known. In the first, the excess spin energy equilibrates with the surroundings (referred to as the lattice) and this process is called the spin-lattice relaxation and is characterized by the spin-lattice relaxation time T_1. Such a relaxation is the result of the lattice motions (for example, atomic vibrations in the solid or molecular tumblings in liquids and gases) having approximately the right frequency to interact coherently with the nuclear spins. The magnitude of T_1 for nuclei with $I = ½$ varies greatly from 10^{-2} to 10^{2} s for solids and 10^{-4} to 10 s for liquids, the overall shorter time for liquids is due to the greater freedom of molecular movement leading to greater fluctuations of the magnetic field in the vicinity of the nuclei. Highly viscous liquids and paramagnetic ions can cause T_1 to be less than 10^{-2} s. There is an additional relaxation mechanism for nuclei with $I > 1/2$ that involves the interaction of the electric quadrupole moments of the nucleus with the surrounding electric field gradients.

By imparting the spin energy to the lattice, the spin lattice relaxation process helps to maintain the Boltzmann distribution of molecules among the energy levels. As mentioned earlier, the spin system is coupled to the "lattice" which refers to other atoms in the molecule or surroundings including solvent molecules. While the spin system can hold only an extremely small amount of energy, the lattice has tremendous capacity to hold excess energy. We know that the "lattice" obeys

Boltzmann distribution law. The effective coupling of the lattice with the spin system makes the spin system also to obey the Boltzmann law although nuclear particles do not obey Boltzmann statistics.

Neighbouring nuclear magnetic moments can exchange spin orientations by interaction between the magnetic moments with no change in the number of nuclei in the excited state but the lifetime of the nuclei in the excited state would decrease. This relaxation referred to as the spin-spin relaxation is characterized by the time T_2. For solids, T_2 is usually very short, of the order of 10^{-4} s, while for liquids $T_2 \sim T_1$.

CHEMICAL SHIFT

11

SHIELDING CONSTANT

In a static magnetic field B_0, the energy difference between the two nuclear spin states + ½ and – ½ for a nucleus with I = ½ when the direction of B_o coincides with the z-axis, is given by

$$\Delta E = 2\mu_z B_0 \tag{11.1}$$

It provides the necessary condition for the observation of the spectral line and thus forms the basis of the NMR experiment.

We need an energy quantum

$$h\nu = 2\mu_z B_0 = \frac{\gamma h B_0}{2\pi} \tag{11.2a}$$

or, the radiation of frequency

$$\nu = \frac{\gamma B_0}{2\pi} \tag{11.2b}$$

The NMR frequency of a nucleus is thus determined chiefly by its magnetogyric ratio γ and the strength of the applied magnetic field B_0. Thus the protons resonate at 400 MHz in a field of 9.4 T, while carbon-13 nuclei resonate at 100.6 MHz in the same field. However, not all protons, similarly not all ^{13}C nuclei, have identical resonance frequencies. The resonance frequency depends slightly on the local electron distribution. It permits us to distinguish between, for example, the three types of hydrogen atoms in ethanol as noted in the previous chapter.

In a molecule, the magnetic field at the nucleus is due not only to the external magnetic field but also due to the magnetic fields of the electrons around the nucleus and on the other nuclei. In this chapter, the effects of electrons are considered and the effects of other nuclei are considered in the next chapter.

If any atom or molecule is placed in a magnetic field, an orbital motion of its electron is induced in such a way that these moving electrons produce an opposing secondary magnetic field (Lenz's law) which will act on all the nuclei present. This is the universal law of diamagnetism. The induced currents are proportional to the applied magnetic field B_0, so also the magnitude of the secondary field. Thus the local magnetic field at the position of a nucleus i is given by

$$B_i = B_0 (1 - \sigma_i) \tag{11.3}$$

where σ_i is a constant independent of B_0 but dependent on the chemical environment of the nucleus. The constant σ_i is known as the shielding constant or the screening constant. The nucleus is thus shielded from the external field by its surrounding electrons. The local field is now slightly smaller than the applied field. For protons in molecules σ is always positive. The value of σ ranges from 1×10^{-5} to 4×10^{-5} for protons to about 10^{-4} to 10^{-2} for heavier nuclei. As a result of nuclear shielding, the resonance condition, equation (11.2b) becomes

$$\nu = \gamma B_0 (1 - \sigma)/2\pi \tag{11.4}$$

i.e. the resonance frequency of a nucleus in an atom is slightly lower than that of a bare nucleus (without any of its electrons). Similar effects occur for nuclei in molecules except that the motion of the electrons in molecules is rather more complicated than in atoms and the induced fields may either reinforce or oppose the external magnetic field. Both the size and sign of the shielding constant of equation (11.3) are determined by the electronic structure of the molecule in the vicinity of the nucleus. The resonance frequency of the nucleus is therefore characteristic of its chemical environment. It is to be noted here that in ESR one varies the *g*-factor to describe the different resonance energies,

$$\Delta E = g\beta_e B_0 \tag{11.5}$$

whereas in NMR, the g_N is constant and shielding constants are introduced to describe the energy separation between any two adjacent levels.

$$\Delta E = g_N \beta_N B_0 (1 - \sigma) \tag{11.6}$$

CHEMICAL SHIFT

Due to the effect of shielding, the nuclear Zeeman levels become closer to each other as shown in Fig. 11.1. If an experiment is carried out by varying the magnetic

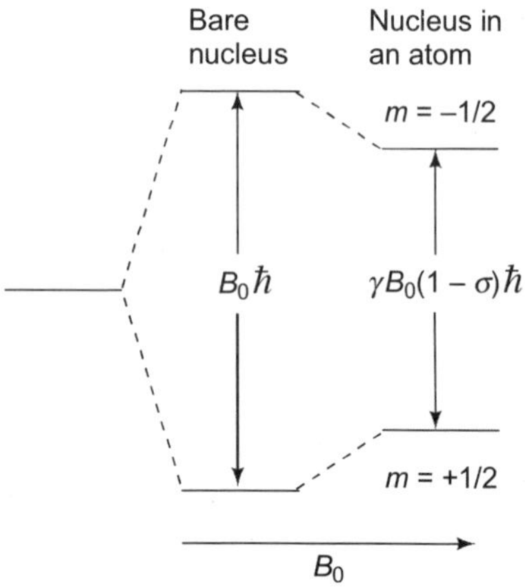

Fig. 11.1 Effect of shielding on the nuclear spin energy levels of a spin 1/2 nucleus.

field B_0 until the resonance is obtained at a fixed frequency as is done in practice, the applied field will have to be larger than it would have been the case if the nucleus were unscreened. For a given nucleus in various chemical environments, either in different molecules or in different chemical environments in the same molecule, there will be a corresponding set of different values of the shielding constant for the protons. As a result 'resonance' will occur in different parts of the spectrum for each type of chemically distinct protons. This shift of the resonance signal for different chemical environments due to variations in the shielding constant is referred to as the 'chemical shift'. Thus the change in transition frequency at a fixed B_0 or equivalently change in B_0 at a fixed spectrometer frequency due to shielding of the nucleus by the electrons of the molecule is called the chemical shift. A typical proton magnetic resonance spectrum of liquid ethanol under conditions of low resolution was shown in the previous chapter (Fig. 10.4). There are three types of protons in ethanol: they are the protons of the methyl, methylene and the hydroxyl groups. The spectrum clearly shows three types of protons. The three peaks have the relative intensities of 3:2:1 corresponding to three, two and one protons allowing immediate identification of the protons. Thus, with an applied magnetic field of say 2.35 T, the proton resonances of a molecule do not occur at $\nu = 100$ MHz but rather at $\nu + \Delta\nu$, where $\Delta\nu$ is, in general, less than 1200 Hz. Other magnetic nuclei are affected similarly and this phenomenon forms the basis of NMR spectroscopy

MEASUREMENT OF PEAK INTENSITY

The NMR signals have different intensities. The area under the NMR peak is generally proportional to the number of nuclei contributing to the peak as noted earlier. Thus the relative areas of the peaks give relatively the number of different types of nuclei present. Accurate measurement of the area under the curve greatly helps in the interpretation of the spectra and also provides a means of obtaining quantitative analysis. Most NMR spectrometers are equipped with electronic integrators, which record the relative areas of the spectral lines as step curve rising from left to right as it passes each peak as shown in Fig. 11.2. The extent of rise measures the integration and shows that the number of hydrogens in each signal is 1, 2 and 3 reading from left to right. The relative heights of the steps in the curve indicate the proton ratio in each group. Such integrators can be accurate to within 1 – 2 % provided the signal to noise ratio is high.

It is not always possible to determine the number of nuclei in a particular group based purely on intensities. This is most commonly encountered for ^{13}C spectra due to varying rates at which these nuclei relax to the ground state. Tertiary carbon atoms relax more slowly than the secondary and primary. This is one of the reasons why the integration of carbon-13 NMR signal is not feasible. Integration values may also be inaccurate for other nuclei particularly those that span large chemical shift ranges.

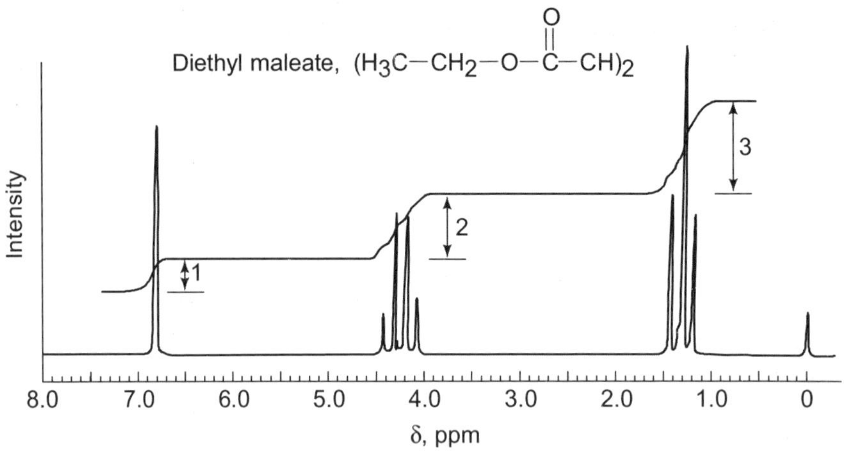

Fig. 11.2 ^{1}H NMR spectrum of diethyl maleate. Measurement of signal intensities showing the number of hydrogens of each type as 1 : 2 : 3 from left to right.

MEASUREMENT OF CHEMICAL SHIFTS

Actually shielding constants cannot be determined with high accuracy. To do so requires very accurate measurement of both the spectrometer frequency and the applied magnetic field. However, it is not feasible to make highly accurate measurement of B_0. Since absolute shifts are rarely required and difficult to determine, it is the usual practice to define the chemical shift in terms of the difference in resonance frequencies between the nucleus of interest (ν) and a reference nucleus (ν_{ref}) by means of a dimensionless parameter.

Chemical shift measurements are formally based on the reference position of a bare hydrogen nucleus (proton) as the primary standard: for this there is no shielding and hence $\sigma = 0$. Since this is an impracticable standard, it is necessary to choose some other substance as a secondary standard and measure the resonance position of other hydrogen nuclei from this.

The chemical shift may be measured in frequency units (Hz) from the chosen reference. In a C.W. instrument, the calibration is made on a chart paper by imposing a suitable audiofrequency, when a side band is observed at a distance equal to the audiofrequency from the reference compound tetramethylsilane (TMS) as shown in Fig. 11.3. For chemical shifts, however, the use of frequency units has the disadvantage that the reported chemical shift is dependent on the applied field or the radiofrequency radiation. It is highly desirable, therefore to have the values of chemical shifts in a unit, which is independent of the radiofrequency or the magnetic field strength. For recording the spectrum, the frequency of the *rf* radiation may be kept constant and the magnetic field strength varied. The resonance frequency of the sample and reference ν_s and ν_r, respectively will then occur at the values of the applied field given by equation (11.7)

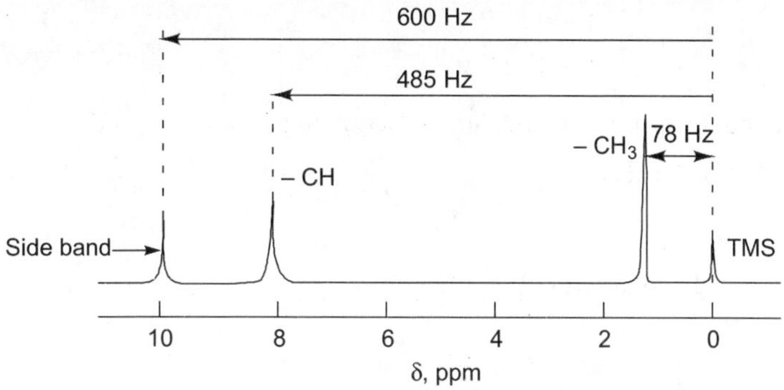

Fig. 11.3 Measurement of chemical shift with side band calibration by imposing an audiofrequency. Proton NMR spectrum of methyl formate with $Si(CH_3)_4$ as internal reference on a 60 MHz instrument with a 600 Hz side band.

$$\nu_s = (\gamma/2\pi)\, B_s\, (1 - \sigma_s) \tag{11.7a}$$

$$\nu_r = (\gamma/2\pi)\, B_r\, (1 - \sigma_r) \tag{11.7b}$$

where B_r and B_s are the magnetic fields at resonance for the reference compound and the sample respectively and σ_s and σ_r are the shielding constants for the sample and the reference, respectively. As the radiofrequency is kept constant, $\nu_s = \nu_r = \nu$, consequently,

$$\frac{1-\sigma_s}{1-\sigma_r} = \frac{B_r}{B_s} \tag{11.8}$$

Subtracting 1 from both sides and rearranging yields,

$$\frac{\sigma_r - \sigma_s}{1-\sigma_r} = \frac{B_r - B_s}{B_s} \tag{11.9}$$

Since σ_r is very much smaller than 1, the equation (11.9) becomes

$$\sigma_r - \sigma_s = \frac{B_r - B_s}{B_s} \tag{11.10}$$

The difference in the shielding constant between the reference and the sample defined in terms of the applied magnetic field in the equation (11.10) is the chemical shift. On the other hand, if the magnetic field B_0 is held constant and the frequency of the *rf* radiation is varied to scan the spectrum, the resonance frequencies of the sample and the reference are then given by

$$\nu_s = (\gamma/2\pi)\, B_0\, (1 - \sigma_s) \tag{11.11a}$$

$$\nu_r = (\gamma/2\pi)\, B_0\, (1 - \sigma_r) \tag{11.11b}$$

where ν_r and ν_s are the resonance frequencies for the reference and the sample respectively. From the equations (11.11a and b), we have,

$$\frac{1-\sigma_s}{1-\sigma_r} = \frac{\nu_s}{\nu_r}$$

Subtracting 1 from both sides and rearranging gives,

$$\frac{\sigma_r - \sigma_s}{1-\sigma_r} = \frac{\nu_s - \nu_r}{\nu_r} \tag{11.12}$$

Since $\sigma_r << 1$, equation (11.12) becomes

$$\sigma_r - \sigma_s = \frac{\nu_s - \nu_r}{\nu_r} \tag{11.13}$$

The radiofrequencies ν_r and ν_s are very large numbers, which are only very slightly different from the fixed frequency of the probe ν_o. The radiofrequencies commonly employed are 60 MHz and upwards. To express chemical shifts in a form independent of the value of the frequency of the *rf* radiation of the spectrometer, the chemical shift δ is defined as

$$\delta = \frac{\nu_s - \nu_r}{\nu_r} \times 10^6 \text{ ppm} \tag{11.14}$$

The frequency difference $\nu_s - \nu_r$ is divided by ν_r so that δ is a molecular property independent of the magnetic field used to measure it. The factor 10^6 is included to give numbers of convenient magnitude. The parameter δ which measures the position of the NMR resonance signal will now be independent of the applied magnetic field. It has no units and is expressed as fractions of the applied magnetic field in parts per million (ppm). For example, if the separation of the sample signal from the reference signal is 89.6 Hz at a fixed radiofrequency of 64 MHz, then the chemical shift is

$$\delta = \frac{89.6}{64 \times 10^6} \times 10^6 = 1.4 \text{ ppm}$$

The separation between the reference signal and that of the sample is directly related to the probe frequency (i.e. on the external magnetic field). For example at 200 MHz, it is 5 times greater than that at 40 MHz. The chemical shift δ is independent of the radiofrequency or the magnetic field employed. Thus, for the proton resonance of $CHCl_3$, δ value of 7.27 ppm is found regardless of whether the spectrum is measured at 100 MHz or 200 MHz. An increase in σ (greater shielding) leads to a decrease in δ; δ is thus a deshielding parameter.

Knowing the frequency of the radiofrequency radiation or the applied magnetic field, the value of the chemical shift in ppm can readily be converted into Hz or mT. The applied magnetic field in the above example is 1.5 T so that the corresponding measure of the signal separation in magnetic field units is $\sim$ 2.05 μT.

REFERENCE COMPOUNDS

Two types of references – internal and external – are used in NMR spectroscopy for calibrating chemical shifts. An internal reference is a compound that is soluble and exhibits preferably a single sharp NMR signal. The reference compound is dissolved in the sample solution under study. The magnetic field acts equally on the sample and the reference molecule, so that for obtaining the chemical shift, a determination of the difference in resonance frequency between the sample and the reference and a knowledge of the magnetic field or the radiofrequency is sufficient. Internal references are advantageous, accurate and are almost always used. However, a serious disadvantage of the internal reference is the possibility of the intermolecular interactions influencing the resonance frequency of the sample. By choosing relatively inert compounds, their effects can be made negligible.

The substance almost universally used for proton resonance as an internal reference in organic solvents is tetramethylsilane, $Si(CH_3)_4$, abbreviated as TMS. The reference signal is obtained by adding a small amount of TMS to the sample. This compound has several advantages.

(i) Its resonance is a sharp and intense singlet, since all the twelve protons are equivalent.

(ii) It is soluble in almost all the organic solvents.

(iii) The resonance position of both ^1H and ^{13}C of TMS is to the high field of almost all the other ^1H and ^{13}C resonances in other organic compounds and hence the TMS resonance, which appears at the low frequency end of the spectrum can easily be recognised. Most ^1H and ^{13}C chemical shifts are hence positive ($\delta > 0$).

(iv) It has a very low boiling point (27 °C) and hence it can readily be removed if required.

(v) It is chemically inert and inexpensive.

TMS is also used as a reference in carbon-13 NMR spectroscopy where it gives rise to a signal at high field. Some of the other reference compounds used for proton besides TMS are cyclohexane and benzene. For aqueous solutions, sodium salt of partially deuterated 3-trimethylsilylpropionic acid, $(CH_3)_3SiCD_2CD_2COONa$, the ^1H NMR signal of which in aqueous and methanol solutions occurs at exactly $\delta = 0.00$, is used as an internal reference. Acetonitrile and dioxane are also sometimes employed as internal standard for aqueous solutions. One should be careful when comparing the chemical shifts of compounds relative to different reference compounds and in different solvents because of the possible specific intermolecular interactions.

If the reference compound is not soluble or reacts with the sample or the solvent, an external reference is used. External reference is placed either in a sealed capillary tube inside the sample tube or in the space between the precision coaxial tubes. An external reference is advantageous in eliminating the possibility of intermolecular interactions or chemical reactions with the sample or the solvent. Further no difficulties arise concerning the solubility of the reference compound in the sample solution. However, there is a serious difficulty due to the difference in bulk magnetic susceptibility between the sample and the reference. With an external standard, when the compound is taken in a coaxial capillary, the field strength that exists in the sample solution differs from that in the capillary due to the different volume susceptibilities. The values of chemical shift must then be corrected. The difference in δ values caused by the susceptibility factor may be as much as 1 ppm. In NMR measurements, the field strength that exists within the NMR tube is influenced by the magnetic susceptibility of the solvent, and hence an internal standard is preferred. That is, the reference and the sample are present in the same solution so that both experience the same magnetic field strength.

For proton NMR spectra when TMS is used as an internal reference, the chemical shift of TMS signal is arbitrarily assigned a δ value of zero and the chemical shifts of other protons are measured with respect to this reference. In this, known as the δ-scale, two important factors are to be noted. (i) A sample less shielded than the reference gives a larger value of δ. (ii) A more shielded nucleus (that is, larger σ) will show its resonance at a higher field when the field is varied but at a lower frequency when the frequency is scanned. The NMR spectra are plotted on a chart paper with δ increasing from right to left, that is, the external magnetic field increasing from left to right. Thus more shielded nuclei (larger σ) with smaller δ appear towards the right hand side of the spectrum. Most ^{1}H chemical shifts of organic compounds lie between 0 to 10 ppm and to the low field of TMS.

As discussed above, a larger chemical shift δ (measured with respect to TMS) implies a nucleus that is less shielded and hence resonates at a lower magnetic field. Since it is convenient to think of increasing chemical shift in the direction of increasing shielding, an alternate scale known as the τ scale has also been introduced. Here TMS is assigned the chemical shift value of 10.00 ppm, with decreasing number corresponding to decreasing shielding. The δ and τ scales are related by the simple relation

$$\tau = 10.00 - \delta$$

For proton resonance, the small range of chemical shifts makes it necessary to correct for susceptibility if an external reference is used. If the sample solution is sufficiently dilute, conversions to internal TMS reference from various external references may easily be made by means of measured chemical shifts from TMS in different solvents.

When the compound is not sufficiently soluble in a non-polar solvent, polar solvents may be used. Care has to be taken since solute-solvent interaction may result in a spectrum for the solute, which is significantly different from that of the free solute in a non-polar solvent. For example, the spectra of aqueous solutions may exhibit features due to hydrogen bonding and/or exchange processes that may occur between the solute and water molecules. Aromatic compounds such as benzene when used as a solvent may form complexes with the solute, and the solute may be held in a specific orientation to the aromatic ring. There may be marked difference between the chemical shift of the protons when the sample is in an uncomplexed state and when in a solution of an aromatic solvent. When the chemical shift is dependent upon the concentration of the solute, the chemical shift of the protons in the free state may be determined by measuring the peak positions for various concentrations and extrapolating it to infinite dilution.

Figure 11.4 shows the 200 MHz ^{1}H spectrum of a mixture of solvents obtained using TMS as internal reference. Each of the compounds has a single group of identical protons and hence gives rise to a single chemical shift. Since the chemical shift of the reference compound namely, TMS is defined as 0 ppm, we can convert the chemical shift into frequency scale and vice-versa using the equation (11.14). For example, for the acetonitrile peak at $\delta = 2.0$ ppm in Fig.11.4.

$$\nu_{(\text{acetonitrile})} - \nu_{(\text{TMS})} = (\delta \text{ in ppm})\ (\nu_{(\text{TMS})}/10^6)$$
$$= (2.0 \times 200 \times 10^6) / 10^6 = 400 \text{ Hz}$$

$\nu_{(\text{TMS})}$ is the frequency of the spectrometer. On a 100 MHz spectrometer the chemical shift of acetonitrile would still be 2.00 ppm but the resonance frequency relative to TMS will be reduced in proportion to B_0 to 200 Hz.

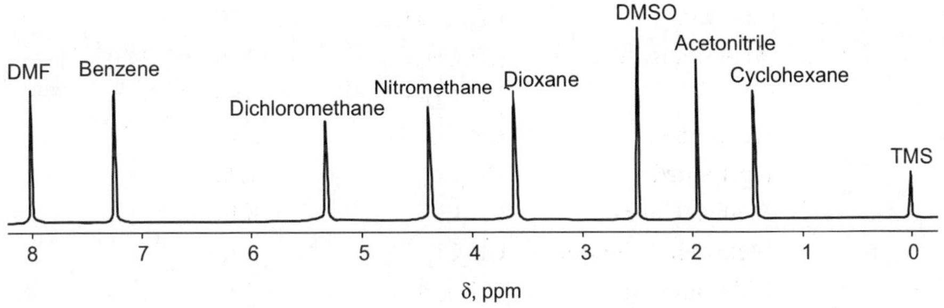

Fig. 11.4 ^{1}H NMR spectrum of a mixture of solvents with tetramethyl silane added as reference.

SIMPLE APPLICATIONS OF CHEMICAL SHIFT

Valuable information can readily be obtained from the NMR spectra by simply noting the number of peaks and their relative intensities as the following two simple examples illustrate. Phosphorous pentachloride PCl_5 shows a single ^{31}P resonance as expected from its trigonal bipyramidal structure found in the gas phase.

It is almost impossible for example to detect transition metal hydrides by conventional elemental analysis if organic ligands are also present. They may even escape detection in the infrared spectrum. In the NMR spectrum, they are readily detected by their characteristic chemical shifts which are found to the high field of TMS. The most common range is δ –10 to – 60 ppm. A resonance at such a high field is diagnostic of the presence of hydride ion coordinated directly to a metal ion. The shielding can be attributed to the electron density on the hydride ion from the non-polar nature of the M-H bond and from the shielding resulting from the electron density in the large and diffuse 4*s* and 4*p* orbitals of the metal. If the metal nucleus to which hydrogen is bonded has a spin, the hydride resonance will appear as a multiplet. For example, the ^1H NMR spectrum of $HRh(CN)_5^{3-}$ is a doublet (Rh, I = 1/2) and occurs at δ 15.6 ppm on the high field side of water. Similarly, the signals at δ –20.2 and –21.6 ppm in the ^1H spectrum of $[Rh(en)_2IH]^+$ and $[Rh(en)_2H_2]^+$ species, respectively are split due to coupling with Rh (I = 1/2) with a coupling constant of 27 Hz.

INTERPRETING CHEMICAL SHIFT

Table 11.1. Proton resonance of selected organic compounds

Compound		*δ, ppm*
Methane	CH_4	0.23
Ethane	$CH_3–CH_3$	0.88
Ethene	$CH_2 = CH_2$	5.29
Ethyne	$CH \equiv CH$	2.88
Cyclopropane	C_3H_6	0.22
Propene	$\underline{CH_3}\ C = CH_2$	1.71
Propyne	$\underline{CH_3}\ C \equiv CH$	1.80
Acetaldehyde	$\underline{CH_3}CHO$	2.20
	$CH_3C\underline{H}O$	9.80
Acetone	CH_3COCH_3	2.17
Acetonitrile	CH_3CN	1.98
Methyl chloride	CH_3Cl	3.10
Methylene chloride	CH_2Cl_2	5.30
Chloroform	$CHCl_3$	7.27
Methanol	CH_3OH	3.48
Acetic acid	$CH_3COO\underline{H}$	8.63
	$\underline{CH_3}COOH$	2.15
Ethyl acetate	$CH_3COOC\underline{H_2}CH_3$	4.12
	$CH_3COOCH_2\underline{CH_3}$	1.25
	$\underline{CH_3}COOCH_2CH_3$	2.03
Benzene	C_6H_6	7.27
Cyclohexane	C_6H_{12}	1.44
Dioxan	$C_4H_8O_2$	3.70
Nitromethane	CH_3NO_2	4.32
Dimethyl sulfoxide	$(CH_3)_2SO$	2.62

A highly symmetric compound can be recognised from its NMR spectrum by the smaller number of signals it exhibits as in infrared spectroscopy. Chemical shifts may be interpreted empirically using the data obtained from compounds of known structure. A list of chemical shifts of some common organic compounds is given in Table 11.1.

Figure 11.5 shows typical ^1H chemical shift range for various functional groups. When combined with empirical rules for predicting the substituent effects, such tables are useful for elucidating molecular structure from the observed chemical shifts. However, with NMR technique becoming increasingly sophisticated, more direct structural evidence is often obtained by spin-spin coupling, which will be discussed in the following chapter and nuclear Overhauser effects, which will be discussed later.

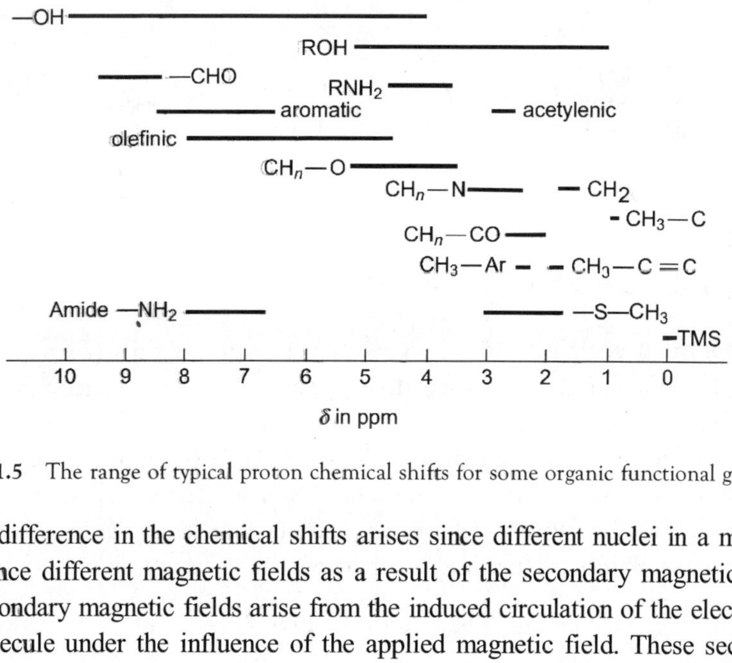

Fig. 11.5 The range of typical proton chemical shifts for some organic functional groups.

The difference in the chemical shifts arises since different nuclei in a molecule experience different magnetic fields as a result of the secondary magnetic fields. The secondary magnetic fields arise from the induced circulation of the electrons in the molecule under the influence of the applied magnetic field. These secondary magnetic fields may either oppose (diamagnetic) or reinforce (paramagnetic) the applied magnetic field at a particular nucleus in the molecule. In the latter case, the effective field experienced by the nucleus is higher than the applied field and the external magnetic field necessary to bring the nucleus into resonance will be lower than that if there were no secondary reinforcing magnetic field. The nucleus is then said to be deshielded and the resonance position shifts downfield. In the former case, the effective field experienced by the nucleus is less than the applied magnetic field and the external magnetic field necessary to bring the nucleus into resonance will be higher than that if there were no secondary opposing magnetic field. The nucleus is then termed as shielded and the resonance position moves upfield (Fig. 11.6). The terms diamagnetic and paramagnetic effects are used to denote respectively shielding and deshielding of a nucleus by the secondary magnetic fields. For historical reasons, the chemical shift positions are almost invariably referred to by the value of the applied field. The high frequency end of the spectrum on the left with high δ values is described as being downfield and the right side with low δ values is said to be upfield. It is, however, preferable not to use the terms upfield and downfield since these suggest that the magnetic field is varying which is not the case for modern spectrometers which work at a fixed magnetic field and varying frequency.

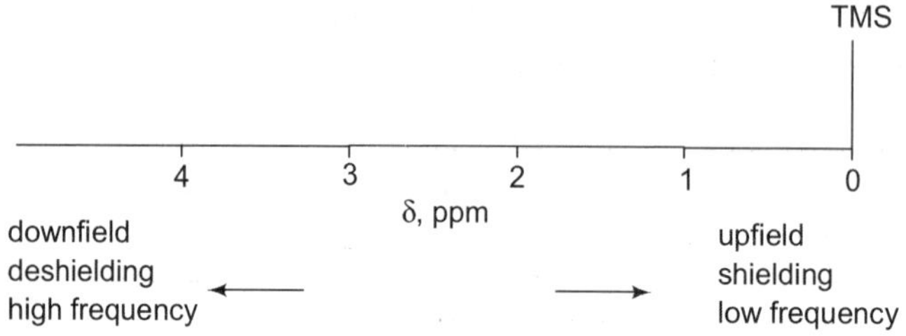

Fig. 11.6 Diagrammatic representation of the nomenclature of chemical shift variations.

ORIGIN OF SHIELDING CONSTANT

The factors that contribute to the value of the chemical shift (δ) will now be discussed. Chemical shifts are sensitive to subtle changes in electron density. Local effects arise from the electron distribution around the atom whose chemical shift is being measured while remote effects include contributions to the chemical shift from electron density on the other atoms. There are a number of semiempirical theories to account for these variations. Factors that may contribute to the chemical shift variations are listed below.

(a) Local diamagnetic shielding (from the atom in question).

(b) Effects of neighbouring atoms.

(c) Interatomic currents.

(d) Local paramagnetic shielding (from the atom under consideration).

(e) Contact interaction.

(f) Effect of external dipolar fields.

For proton chemical shifts, the first three terms (a) to (c) are important. The total range of chemical shifts for proton is much smaller than that for any other nucleus. This is clearly because the number of electrons in the vicinity of the proton is much smaller than that for any other nucleus. Further, the local paramagnetic contribution which appears to give a dominant contribution to the chemical shifts of other nuclei is negligible for protons, since there are no low lying *p* orbitals associated with the hydrogen atom. Contribution (f) arises chiefly from the neighbouring molecules and will be considered under solvent effects. Consideration of all the factors that determine the shielding constants allows a satisfactory interpretation of the experimental values of the chemical shifts.

(a) Local Diamagnetic Shielding

Lamb showed that the effect of an applied magnetic field on a free atom is to induce an electron current as shown in Fig. 11.7 that leads to a shielding factor σ_d given by

$$\sigma_d = \frac{4\pi e^2}{3mc^2} \int_0^\infty r\rho(r)\, dr \qquad (11.15)$$

where ρ (r) is the electron density as a function of the radial distance r from the nucleus and e, m and c are the usual constants. The magnitude of σ_d depends on the electron density close to the nucleus and it can be calculated from the Lamb's formula:

$$\sigma_d = \frac{e^2\ \mu_0}{12\ \pi\, m_e} \left\langle \frac{1}{r} \right\rangle \qquad (11.16)$$

where μ_0 is the vacuum permeability and r is the electron-nucleus distance. The shielding constant is inversely proportional to the atomic radius. Lamb's theory is inadequate for molecules, since it assumes that the electrons are free to move in any direction, whereas in a molecule, electronic motion is severely restricted. In a molecule, the perturbation of the spherical symmetry of the electron distribution due to the presence of other nuclei reduces the diamagnetic effect. This reduction can be treated as corresponding to a paramagnetic effect which strengthens the external field B_0. However, for protons local diamagnetic effect is of primary importance. The local diamagnetic shielding for the protons can be evaluated approximately from the Lamb's formula and it is in the range of a few ppm in reasonable agreement with the range of chemical shifts observed for protons. The Lamb's formula gives σ_d as 17.8 ppm for the proton in a free hydrogen atom. The greater the shielding, the smaller is the chemical shift value δ.

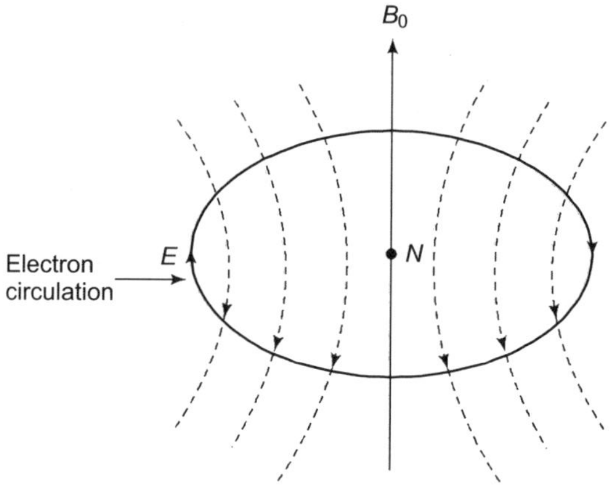

Fig. 11.7 Electron circulation around the nucleus N. Lines of flux due to secondary magnetic field arising from an applied magnetic field.

The equation (11.15) suggests an approximate correlation between the shielding constant and the electron density around hydrogen. For example, a more acidic proton such as –OH proton in phenol should be less shielded than the corresponding but less acidic proton in alcohol. This has indeed been found to be so; the chemical shift for the –OH proton of phenol is at 4 ppm lower field than that of alcohol. A strongly electronegative atom or group attached to (or near) a magnetic nucleus is expected to deshield the nucleus. Accordingly, the proton chemical shifts have been found to correlate with the electronegativity of the substituents. For example, the value of chemical shift increases with increase in the value of Pauling's electronegativity of the halogen atom in the series CH_3I, CH_3Br, CH_3Cl and CH_3F, the δ value being 2.0, 2.45, 2.84 and 4.12 ppm, respectively. Electronegative groups like fluorine in CH_3F withdraw electron density from the methyl group by inductive effect. Methane has a much smaller chemical shift (0.13 ppm) than methyl iodide. The chemical shift positions of ^1H and ^{13}C nuclei of the methyl group attached to various atoms/groups show that electropositive elements (Li, Si) shift the signals upfield and the electronegative elements (N, O, Cl) shift of the signals downfield because they donate and withdraw electrons, respectively. The shielding effect of the electronegative atoms decreases with distance. The effect of charged species on the values of chemical shift is very marked; protons adjacent to N^+ as in quaternary ammonium ions R_4N^+ are very strongly deshielded (high δ values) while carbanionic centres exert powerful shielding influences (low δ values).

The significance of other sources of proton shielding is demonstrated by the observation made on ethyl halides CH_3CH_2X. The expected change in the resonance frequency i.e. low field shift – with increasing electronegativity – is noted for $-CH_2$ protons, while the reverse trend was observed for the $-CH_3$ protons. In this case, the contribution of magnetic anisotropy of the C–C and C–X bonds is important and its effect on shielding constant dominates.

(b) Neighbouring Group Anisotropy

Diamagnetic currents arise from the movement of electrons within the atomic or molecular orbitals. The current so induced generates a small local field opposed to the applied magnetic field B. The magnitude of the diamagnetic current is determined solely by the ground state electronic wave function of the atom or the molecule and depends on the electron density close to the nucleus and provides the only contribution to an atom with spherically closed shell. The shielding due to the local diamagnetic screening effect is dependent strongly on the number of electrons: ~ 18 ppm for hydrogen, ~ 260 ppm for carbon, ~ 300 ppm for ^{19}F, ~ 960 ppm for phosphorus increasing to about 10,000 ppm for atoms in the fifth row of the periodic table. Thus the diamagnetic shielding of a proton by its 1s electron density is relatively small compared with the shielding of the nuclei of heavier atoms that have filled inner shells.

The presence of these field-induced magnetic moments in a molecule at atoms other than the one undergoing NMR transition can be felt at the nucleus being investigated. For example, consider the ^1H chemical shifts of acetylene, ethylene and ethane. Based on the hybridization of the carbon orbitals, one would expect $\delta (C_2H_2) > \delta (C_2H_4) > \delta (C_2H_6)$. As the *s* electron character increases in the order $sp^3 < sp^2 < sp$, the bonding electrons should be held more tightly to the carbon removing electron density from the hydrogen atoms and thus deshielding the protons. However, the order of the observed chemical shifts, C_2H_4 (5.9 ppm), C_2H_2 (2.85 ppm) and C_2H_6 (0.85 ppm) differ. It is traced to the shielding effects of the π electrons of the C ≡ C triple bond of acetylene and the C = C double bond of ethylene. The chemical shift (δ) of protons bonded to C = C in alkenes appears at δ 5.0–6.0 ppm, whereas alkyne protons appear at relatively low δ, 1.5 –3.5 ppm. Similarly, the chemical shift (δ) of aldehyde and aromatic protons is higher than that can be accounted for by the electronegativity effects alone. The explanation lies in the circulation of π electrons in the nearby groups of atoms under the influence of the applied magnetic field. The effect may be diamagnetic or paramagnetic which would shield or deshield the nucleus depending on the location of the nucleus relative to the neighbouring group. These effects are complex and direction-dependent (anisotropic). The effects of neighbouring group anisotropy arising from currents induced in the nearby groups of atoms are discussed here.

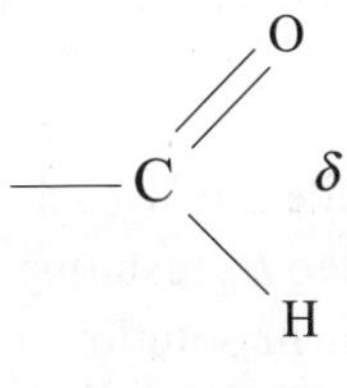

δ 9.5 – 10.0 ppm

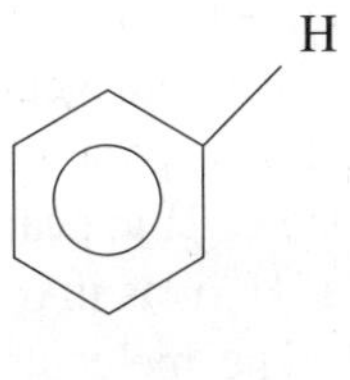

δ 7 – 8 ppm

To explain how a nucleus can be shielded or deshielded by the motion of the electrons in a nearby group, a highly simplified model is considered. It is assumed that the distribution of electrons in the neighbouring group has a cylindrical symmetry, for example, like the triple bond in acetylene. Consider a diatomic group H–Y, where Y is a group of atoms or a part of a molecule. In the presence of the applied magnetic field, the group H–Y as discussed above gives rise to an induced magnetic moment μ_Y that could be considered as a localized point dipole μ_Y at the centre of Y; the magnitude of μ_Y is proportional to the magnetic susceptibility χ_Y of the group Y. This induced moment gives rise to a magnetic field at the nucleus with a strength, that is, inversely proportional to the cube of the distance of the nucleus from the group of atoms. The field varies with the orientation of the molecule but it does not average to zero because the magnetic susceptibility also changes as the molecule presents different orientations to the applied field. Consequently, the shielding constant depends on three quantities: the difference in the magnetic susceptibilities parallel and perpendicular to the group $\chi_\parallel$ and $\chi_\perp$, respectively, the angle θ between the direction of μ_Y and H–Y bond axis and the distance r of the nucleus H from the group Y. Angle θ, χ and vector r from the nucleus to the nearby group are defined in Fig. 11.8.

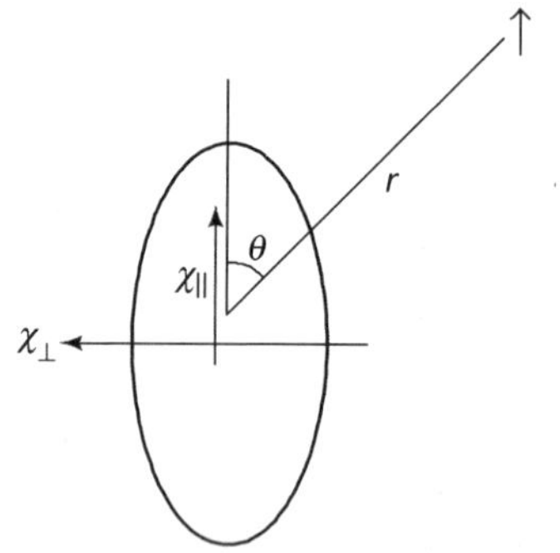

Fig. 11.8 Neighbouring group anisotropy: definition of θ and the vector r from the nucleus to the nearby group.

Using the relation

$$\Delta\chi = \chi_\parallel - \chi_\perp \tag{11.17}$$

the anisotropy of a group with an axis of symmetry can be defined as the difference in the susceptibilities parallel and perpendicular to the axis of symmetry. The magnetic moment μ_Y could be resolved into its cartesian components $\mu_Y(x)$, $\mu_Y(y)$ and $\mu_Y(z)$. In the geometric arrangements (a) and (b) (Fig. 11.9), the secondary field at H is parallel to B_0. The induced field reinforces the external field and the resonance of H appears at a lower field, i.e. H is deshielded. However, in the orientation of H–Y as in (c) the induced field at H is opposed to B_0 resulting in shielding of H. The molecules undergo rapid tumbling motion in solution and averaging takes place. The factor $1 - 3\cos^2\theta$ of equation (11.18) given below becomes zero if the components $\chi_Y(x)$, $\chi_Y(y)$ and $\chi_Y(z)$ of the susceptibility χ_Y have identical values, when the group Y is said to be magnetically isotropic. Only when the neighbouring group has a very high symmetry, e.g. tetrahedral, will the three magnetic moments become equal. When this is not the case, Y is said to possess magnetic anisotropy $\Delta\chi$. Depending on the orientation of Y, paramagnetic or diamagnetic shift of the resonance frequency of the nucleus H occurs.

The magnetic anisotropy contribution to the chemical shift can be calculated from the relation,

$$\Delta\sigma = \left(\frac{1}{3}\right)\frac{\Delta\chi\left(1-3\cos^2\theta\right)}{4\pi r^3} \tag{11.18}$$

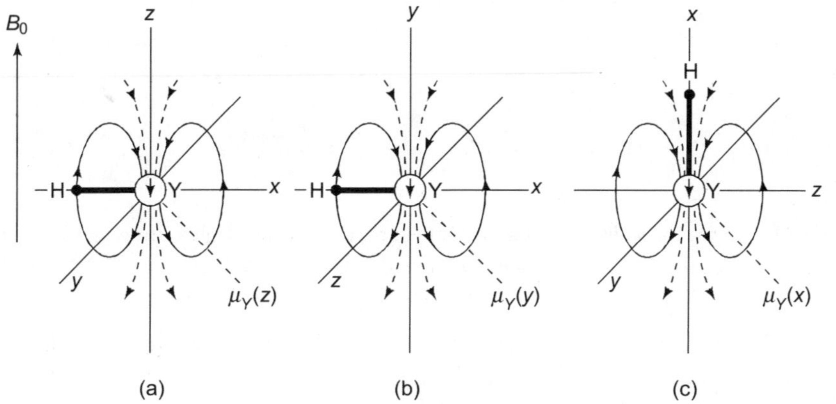

Fig. 11.9 Lines of flux generated by an induced magnetic moment at the centre of a cylindrically symmetric neighbouring group Y. The applied magnetic field in (*a*) and (*b*) is perpendicular to the H–Y axis and in (*c*) it is parallel to the H–Y axis.

if the magnitude and sign of the magnetic anisotropy $\Delta\chi$ of the group, equation (11.17), with axial symmetry is known. This expression (11.18) shows that the neighbouring group contribution may be positive or negative according to the relative magnitudes of the two magnetic susceptibilities and the relative orientation of the nucleus. The latter effect is easy to predict. The value of $1 - 3\cos^2\theta$ is positive if $54.7° < \theta < 125.3°$ and it is negative otherwise.

To find out whether the field arising from the induced magnetic moments shields or deshields a nearby nucleus, we need to simply average the dipolar field experienced by that nucleus over all orientations of the molecule as it tumbles in solution. The results of the equation (11.18) are generally represented graphically by shielding cones for various groups whose nodal plane $\Delta\sigma = 0$ is fixed at the magic angle of 54.7^0. Shielding is indicated by the + sign and protons in these cone-shaped zones are shielded (lower δ values). Deshielding is indicated by the – sign (see Fig. 11.10).

The magnitude of the neighbouring group effect depends on the magnetic anisotropy of the group itself and not on the nucleus being shielded or deshielded. Neighbouring group anisotropy is, therefore, relatively more important for protons with their small local diamagnetic shielding effects than for other heavier nuclei, for example, ^{13}C. It is important to note that chemical bonds are regions of high electron density and for proton chemical shifts almost all chemical bonds are magnetically anisotropic and circulation of electrons induced by the external magnetic field, therefore, gives important contributions to the proton shielding constants. The π bonds are especially effective in influencing the chemical shifts of nearby atoms. The C – C, C – Cl single bonds, C = C, N = O and C = O double bonds and C ≡ C and C ≡ N triple bonds etc. possess strong magnetic anisotropies and have a strong influence on ^1H chemical shifts.

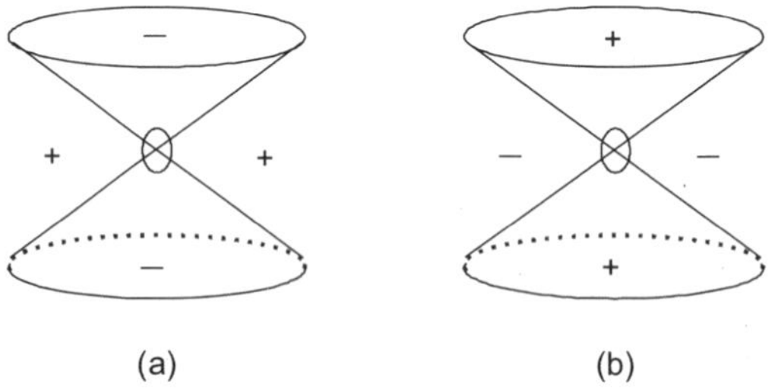

Fig. 11.10 Regions of shielding (+) and deshielding (–) due to neighbouring group magnetic anisotropy. (*a*) $\mu_{||} > \mu_{\perp}$ and (*b*) $\mu_{||} < \mu_{\perp}$. The axis of the cones coincides with the symmetry axis of the neighbouring group shown as an ellipsoid.

The observed chemical shift of acetylene (δ 2.85 ppm) is explained on the basis of neighbouring group anisotropy. In acetylene, dominant contribution to the shielding of nearby hydrogen nuclei comes from paramagnetic currents since the diamagnetic currents are almost isotropic. Acetylene is linear and an applied magnetic field induces paramagnetic currents when it is perpendicular to the axis of the molecule. When the bond axis and the direction of B_0 are perpendicular, the π electron circulation in acetylene is hindered. The induced paramagnetic current around the molecular axis is zero because of cylindrical symmetry of the electron distribution when the applied magnetic field is parallel to the symmetry axis. The pattern of shielding and deshielding resulting from the neighbouring group anisotropy is shown in Fig. 11.11. Protons lying on the axis of the group as in acetylene are shielded but a proton that lies perpendicular to the bond as a part of a larger molecule is deshielded. The situation is more complicated for C=C and C=O double bonds, since these groups do not have a cylindrical symmetry. In this non-linear group, the applied field can induce a paramagnetic current when it is parallel to the C=O or C = C axis. To a good approximation, the shielding effects of both the groups can be represented by the diagrams given as in Fig.11.11. When an alkene group is oriented such that the plane of the double bond is perpendicular to the direction of the applied magnetic field (Fig. 11.11), the induced circulation of π electrons generates a secondary magnetic field that is diamagnetic, which opposes the applied magnetic field B_0 around the carbon atoms. But in the region of alkene protons, the secondary field is paramagnetic, that is, it augments B_0. The alkene protons, therefore, resonate at lower fields (higher δ values) than expected. Any group held above or below the plane of the double bond will experience a shielding effect.

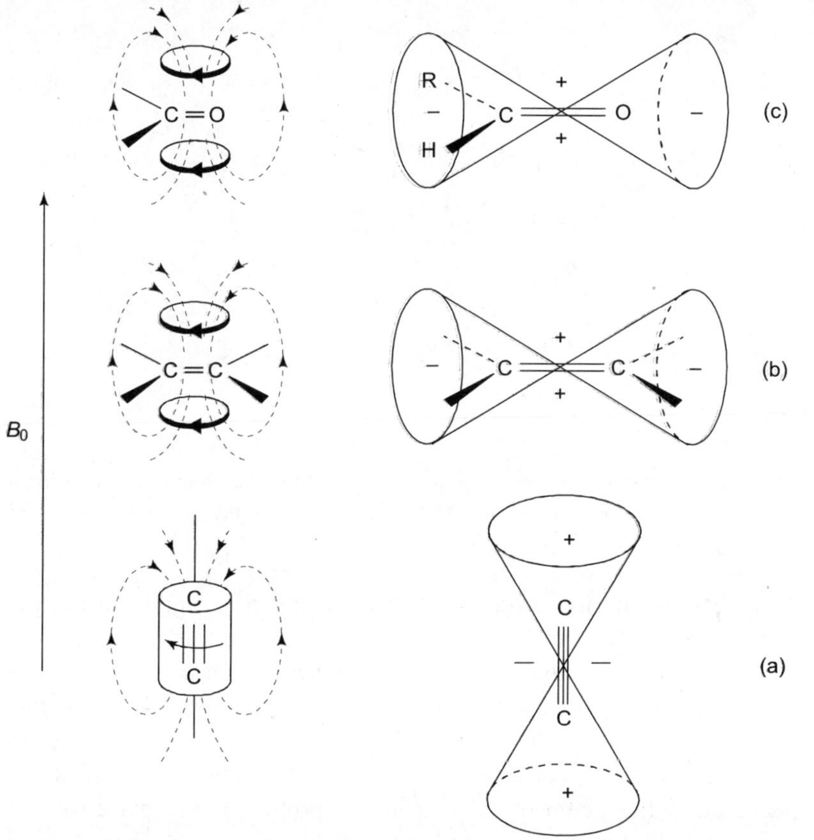

Fig. 11.11 Schematic representation of the neighbouring group anisotropic shielding and deshielding effects of (a) acetylenic C ≡ C, (b) ethylenic C = C and (c) aldehydic C = O groups.

For the carbonyl group a similar situation exists as in alkenes (Fig. 11.11). Accordingly aldehydic protons occur at high δ values and protons held above or below these cones will resonate at lower δ values. Aromatic and alkyl halides give a sharp signal around δ 10.0 ppm. Rather different values of magnetic anisotropies are often reported for various groups such as C=C, C=O, etc. For C=O group, for example, there is general agreement that a proton directly above the plane of the >C=O group is shielded but the magnitude of the anisotropies seems to be not settled.

(c) Ring Current

A special case of neighbouring group anisotropy occurs in aromatic compounds. In molecules such as benzene, the dominant contribution to the magnetic anisotropy comes from the extensive circulation of the π electrons in their delocalized molecular orbitals.

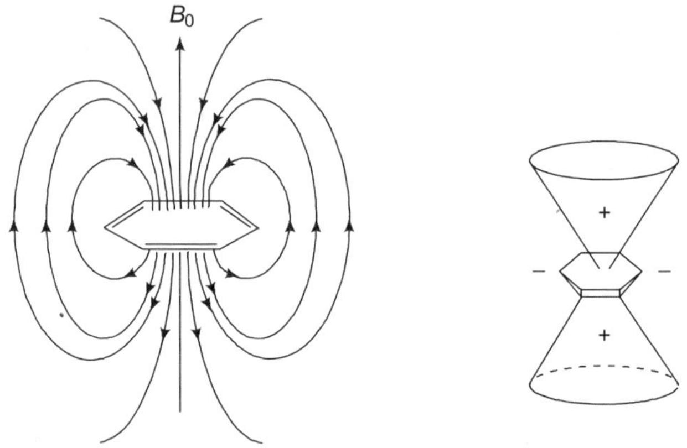

(a) (b)

Fig. 11.12 (*a*) Magnetic lines of flux arising from the induced ring current when the applied magnetic field is perpendicular to the plane of the aromatic ring, (*b*) shielding (+) and deshielding (–) regions associated with the benzene ring.

When the aromatic ring is oriented perpendicular to the applied magnetic field, as shown in Fig.11.12, the π electrons circulate relatively freely around the ring and will move in a direction such that the magnetic moment resulting from their motion opposes the applied magnetic field. It may be visualised as a point magnetic dipole situated at the centre of the ring. If we construct the lines of magnetic flux as shown in Fig. 11.12, it is seen that the sign of the ring current effect is highly dependent upon the geometry. An aromatic proton in the plane of the ring experiences a field that enhances the applied field; hence its resonance occurs at a lower field than might otherwise be expected. Thus the effect of the induced field is to deshield substantially the hydrogens attached to the aromatic ring which generally comes into resonance at δ 7.27 ppm in benzene. Accordingly, the chemical shift of benzene protons is about 1.7 ppm lower than that of ethylene and 1.4 ppm to the high frequency side of the olefinic protons in cyclohexa-1,3-diene, Fig. 11.13 (a), which is very similar to benzene from a consideration of bonding. On the other hand, a proton held over the aromatic ring would experience an upfield shift. For example, the protons of the four methylene groups in 1, 4-decamethylene benzene, Fig. 11.13 (b) $n = 10$, are held in a position directly above the ring. They experience a magnetic field arising from the aromatic ring current which is opposite in direction to that associated with the aromatic protons. These CH_2 protons are shifted to a high field $\delta \sim 0.3$ ppm relative to the normal CH_2 ($\delta \sim 1.3$ ppm) positions in paraffin chains by nearly one ppm. Similarly, the central methylene groups of [1, 8] paracyclophane (Fig. 11.13 (c)) are shielded by nearly 1 ppm more than normal methylene groups since they are held rigidly above the plane of the benzene rings. Another interesting example is dimethyl substituted pyrene (15, 16-dihydro-15, 16-dimethyl pyrene), Fig.11.13 (*d*). It has 14 π electrons and is hence aromatic according to the Huckel rule ($4n + 2$, $n = 3$) even though it does not contain a classical benzenoid ring system. The ring protons are deshielded as expected but the methyl groups, which are situated above and below the plane of the molecule are shielded by more than 5 ppm relative to ethane.

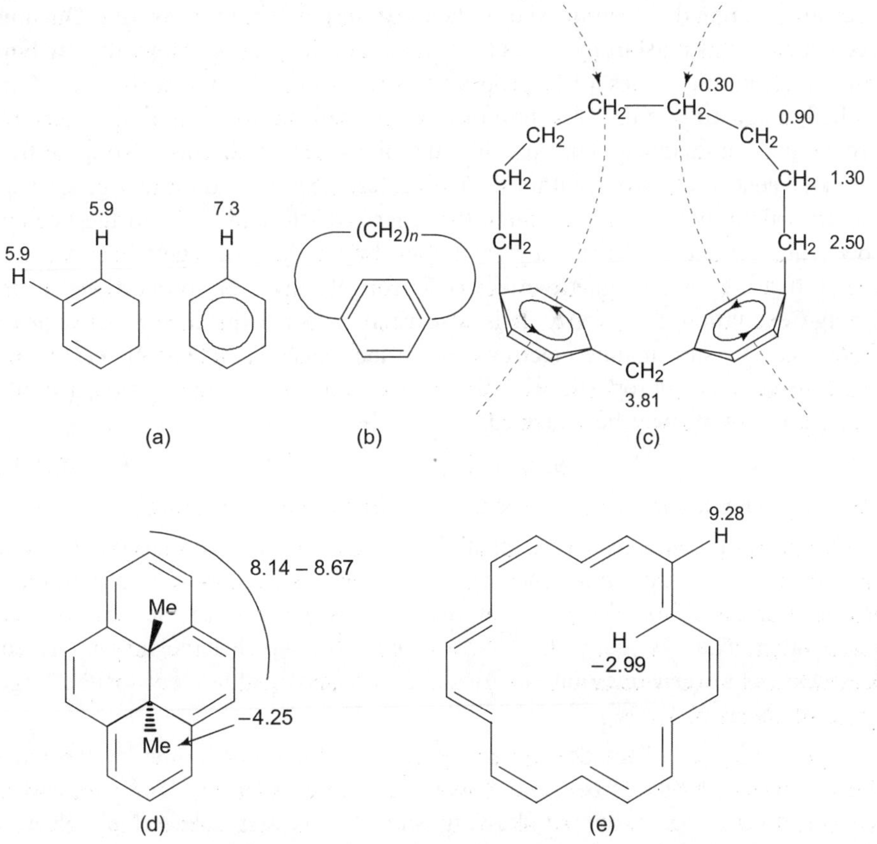

Fig. 11.13 Examples of ring current shifts of ^{1}H chemical shifts (in δ, ppm).

Annulene molecule [Fig. 11.13 (*e*)] is planar and aromatic with $4n + 2 = 18\ \pi$ electrons ($n = 4$). It shows two proton resonances, one from the 12 strongly deshielded exo protons at 9.28 ppm and another at – 2.99 ppm from the six endo protons. The latter set of hydrogen atoms lies within the current loop by the circulating π electrons and is shielded by the ring current effect. Negative δ values correspond to protons resonating to the low frequency side of TMS. While the outside hydrogens come into resonance at normal aromatic region, the inside hydrogens experience a weaker field than the applied field and come into resonance at a conspicuously high field.

The α-protons in naphthalene show signals at a lower field than the β protons since the contributions of the two rings are more important at the α position which is closer to both the rings. Comparatively little is known about ring currents around saturated or partly saturated rings.

(d) Local Paramagnetic Shielding

Local paramagnetic contributions σ_p arise from the anisotropy in the electron distribution around the atom whose chemical shift is being measured. Quantum mechanically this anisotropy arises by a mixing of the ground state with low lying excited electronic states of appropriate symmetry under the influence of the applied magnetic field. This provides a mechanism for anisotropic electron circulation. Paramagnetic currents, like the diamagnetic currents, also arise from the movement of electrons within the molecules. The induced current generates a magnetic field, which will reinforce the external field and deshield the nucleus. Low-lying excited states make a greater contribution than the higher energy states. The shift due to paramagnetic effect σ_p is roughly inversely proportional to the average excitation energy ΔE. It is also related to the distance r between the nucleus and its surrounding electrons. Since the magnetic field at the centre of a small current loop is proportional to the inverse cube of its radius like σ_d, a similar dependence for σ_p may be expected.

$$\sigma_p \cong (1/\Delta \mathrm{E})\cdot\langle 1/r^3\rangle \tag{11.19}$$

where $\langle\ \rangle$ indicates an average over the local electronic distribution.

The excitation energies to the empty higher energy hydrogen orbitals from the ground state are large and hence this effect makes negligible contribution to proton chemical shifts. The paramagnetic effect is the dominant feature for most nuclei other than ^{1}H (e.g. ^{13}C, ^{14}N, ^{19}F and ^{31}P) which have larger electron densities and lower excitation energies and accounts qualitatively for the larger range of chemical shifts.

A good example of the dependence of σ_p on r^{-3} is given by the ^{13}C chemical shifts of monosubstituted benzene derivatives (see chapter 15). The para carbon is deshielded by electron withdrawing substituents and shielded by electron releasing groups. Although this is the same trend as that observed for the ^{1}H chemical shifts in these compounds, the effect is paramagnetic not diamagnetic in origin. Electron donating groups delocalize their lone pairs into the ring and increase the electron density at the ortho and para carbons. The increased electron repulsion causes the orbitals around these atoms to expand (reducing r^{-3}) and hence the variation in the chemical shift, δ.

When the electron distribution has a cylindrical symmetry and the applied magnetic field is parallel to the symmetry axis, the magnetic field will not be able to mix the excited states with the ground state. Thus, the π electrons in acetylene, for example, do not give rise to paramagnetic deshielding when the magnetic field is along the molecular axis.

(e) Contact Interaction

These are the paramagnetic shifts produced by the unpaired electrons. Unpaired electrons give rise to large dipolar magnetic fields. The magnetogyric ratio of the electron is 660 times greater than that of the proton, which can result in substantial nuclear shielding and deshielding.

For transition metal complexes, which are paramagnetic, the proton chemical shifts usually span a range of over 200 ppm. These large chemical shifts arise from the interaction of the unpaired electron with the protons. The contact interaction is invoked to interpret the NMR spectra of paramagnetic transition metal complexes. These aspects are discussed later. If the metal complexes are diamagnetic, the chemical shifts for the protons span a range only slightly larger than those observed for organic molecules.

(f) Hydrogen Bonding

Intermolecular interactions are important in NMR spectroscopy since almost all the measurements are made in solution. For example, it has been observed that the chemical shifts of substances dissolved in aromatic solvents appear at higher fields than those dissolved in aliphatic solvents. This effect can be ascribed to the π electron ring current of the aromatic ring. Solvent effect becomes particularly important when the formation of weak complexes of solutes with solvent molecules takes place. While solvent effects on chemical shifts can be quite large, the effect of the solvent on the coupling constants discussed in the next chapter is usually very small. Mention may be made of the strong specific molecular interaction of hydrogen bonding. Proton chemical shifts are extremely sensitive to hydrogen bonding. Hydrogen bonding accounts for some of the largest observed deshielding effects in ^{1}H NMR. The formation of hydrogen bond causes the chemical shift of the bonded proton to move downfield by as much as 10 ppm. Intramolecular hydrogen bonds show chemical shifts at extremely low fields. An approximate linear relationship exists between the NMR shift and the strength of the hydrogen bond as measured by the enthalpy of its formation. Table 11.1 shows typical changes in the proton resonance on hydrogen bonding.

Table 11.2. Effect of hydrogen bonding on proton chemical shifts

Compound	*δ(bonded)*	*Shift, δ (bonded) –δ (free)*
C_2H_5OH ----- OC_2H_5 H	5.3	4.6
C_2H_5OH ----- $O{=}P(OC_2H_5)_3$	8.7	4.3
$C_6H_5NH_2$----- $O(CH_3)_2$	5.0	1.7
$C_6H_5NH_2$----- N(py)	5.3	2.0
C_6H_5SH ----- DMF	3.8	0.5
$H_3C—C{=}CH—C—CH_3$ | ‖ O—H-----O	15.5	—

A hydrogen atom involved in hydrogen bonding shares its electron with two electronegative elements. Consequently, it is itself deshielded and comes into resonance at a lower field. The X group may draw the hydrogen atom away from the electrons in the X—H ... Y bond reducing the electron density around it. Chemical shifts are also affected by the local electric fields arising from charged or polar groups. These can modify both paramagnetic and diamagnetic effects by polarising local electron distributions and the position of the monomer $\rightleftharpoons$ dimer, trimer ... equilibrium. Strong electric field of the *Y* atom in the positive charge usually deshields the nearby protons while negative charges often give rise to shielding. For example, the chemical shift of protons on the imidazole side chain of the amino acid histidine is higher by about 1 ppm when the ring is protonated.

Three typical examples of intramolecular hydrogen bonding are shown in Fig. 11.14. In all the three hydrogen bonded compounds protons are highly deshielded. Intermolecular hydrogen bonds, which are somewhat weaker, produce smaller shifts. For example, the hydroxyl proton resonance of ethanol moves by over 4 ppm. The chemical shift of OH in ethanol varies from 5.5 ppm to about 0.8 ppm, when the concentration is varied from 1 M to 0.001 M. The limiting chemical shift of –OH at infinite dilution (0.8 ppm) compares well with that of the monomeric ethanol in the gas phase (0.55 ppm). In the spectrum of phenol, the hydroxyl proton signal varies from δ 12.55 to 15.63 ppm for concentration varying from 1% to 100% w/v in CCl_4. The intermolecular hydrogen bonding is disrupted on dilution with CCl_4. Similar changes in the chemical shift of –OH are found with increasing temperature which shifts the equilibrium to the monomer side. Thus, at higher concentrations, OH and NH protons appear at higher δ than in dilute solutions; δ values are lowered with increasing dilution or temperature.

The NMR spectrum of intramolecularly hydrogen bonded systems is unaltered by varying concentration or temperature. Salicylates and enols of β-dicarbonyl compounds are examples of such systems. Salicylates show the OH resonance at very high δ (10 – 12 ppm) and enol OH appears even higher (δ 11–16 ppm). Carboxylic acids have stable dimeric structure due to stronger hydrogen bonding, which persists even in very dilute solutions: carboxylic OH appears between 10 and 15 δ.

It is easy to identify the hydrogen bonds. If the sample in $CDCl_3$ or CCl_4 solution is shaken with a drop of D_2O, the OH, NH and SH hydrogens exchange rapidly with deuterons and the signals of OH, NH or SH disappear from the spectrum.

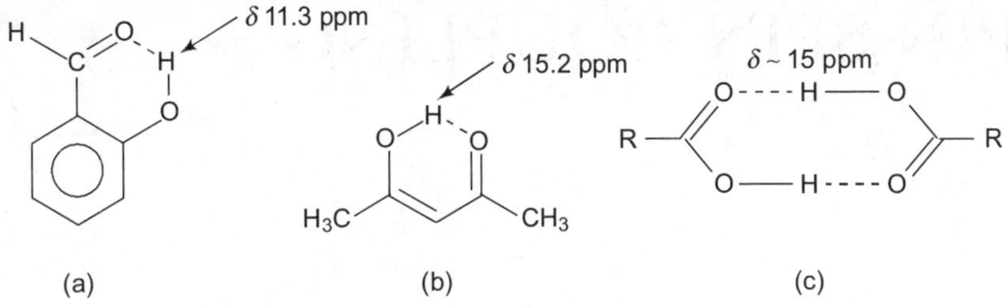

Fig. 11.14 The ^{1}H chemical shifts in (a) salicylaldehyde, (b) enol form of acetyl acetone and (c) carboxylic acid dimer.

Proton NMR spectroscopy is a sensitive method for studying hydrogen bonding, both qualitatively and quantitatively. Since hydrogen bonds are usually formed and broken very rapidly relative to the chemical shift difference between bonded and non-bonded forms, separate resonances are not observed for different species. Protons involved in hydrogen bonding say between acetone and chloroform exchange between 'free' and 'bonded' chloroform

$$(CH_3)_2CO + HCCl_3 \longleftrightarrow (CH_3)_2C{=}O \ldots HCCl_3$$

Since the rate of exchange is rapid on the NMR time scale, the resonance of $CHCl_3$ proton is found as a singlet. The position of the signal is dependent on the weighted average of the free and bonded $CHCl_3$. The variation of chemical shift with concentration and temperature can be studied to obtain the equilibrium constant and other thermodynamic properties for hydrogen bond formation. The reason for the larger downfield shift of the hydrogen bonded proton has not been fully understood.

When a fast exchange process of the type,

$$ROH_a + R'OH_b \longleftrightarrow ROH_b + R'OH_a$$

takes place, the observed ^{1}H resonance is a singlet (see Chapter 14). Its position is given by

$$\delta_{obs} = N_a\,\delta_a + N_b\,\delta_b \tag{11.20}$$

where N_a and N_b are the mole fraction of ROH and R′OH, respectively and δ_a and δ_b are the respective chemical shifts of H_a and H_b. For example, if a solution is made up of 2.0 moles of acetic acid and 4.0 moles of water, the mole fraction of H_2O is 0.8 and that of acetic acid is 0.2. The ^{1}H chemical shift of the solution should occur at 6.48 ppm, since the ^{1}H chemical shift of pure water and – COOH of acetic acid occurs δ 5.2 and 11.6 ppm, respectively.

SPIN-SPIN COUPLING 12

SCALAR COUPLING

The previous discussion would suggest that for different magnetic nuclei in a molecule a single spectral line would be observed for each entity in the NMR spectrum. However, the NMR spectrum usually consists of groups of lines. For example when the NMR spectrum of acetaldehyde (see Fig. 12.1) was examined under high resolution, it was found that the two lines one each expected for the CHO amd CH_3 groups splits into a number of closely spaced lines termed as multiplets. An important feature of the separations between the components of a multiplet which is measured in frequency units (Hz) is that they are independent of the applied magnetic field and not proportional to it as is the case for the chemical shifts. Ramsay and Purcell in 1952 showed that the splitting of the signals originate from an indirect coupling between the nuclear spins through the bonding electrons in the molecule. Multiplet patterns with characteristic relative intensities and spacings arise from the interaction between neighbouring nuclear magnetic moments through the chemical bond. This interaction splits the nuclear spin energy levels; several transitions then become possible in place of a single transition expected in the absence of this interaction. This type of interaction is known as the spin-spin coupling or scalar coupling. It gives us another extremely valuable information observed in most NMR spectra, namely, magnetic interactions between the nuclei. It provides information on the spatial positions of the magnetic nuclei in a molecule. The nuclear spin-spin interaction is not a direct dipole-dipole interaction and its magnitude is very small. Further the magnetic dipole-dipole interaction averages to zero in solution, while the spin-spin coupling is not.

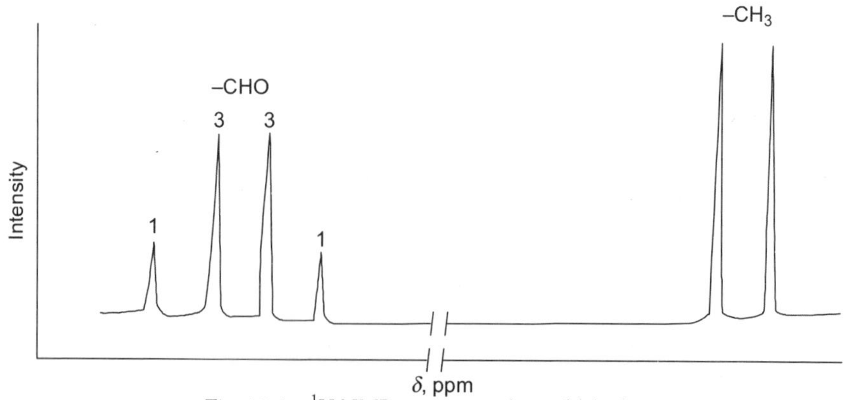

Fig. 12.1 ^1H NMR spectrum of acetaldehyde.

The information obtainable from the NMR spectrum includes the chemical shifts and spin-spin coupling constants. The centre of the multiplet pattern gives the chemical shift (δ) of the group. Chemical shifts and coupling constants are extremely sensitive to the molecular structure and form the basis for the application of NMR to the elucidation of structure. Systematic procedures have been developed for obtaining the chemical shifts and coupling constants from the NMR spectra.

Consider a system of two nuclei A and X each with $I = 1/2$, for example, the hydrogen fluoride in which the fluorine is ^{19}F or the formate ion ($HCOO^-$) in which the carbon is ^{13}C. In the former, both ^1H and ^{19}F NMR spectra in solution show a doublet separated by 530 Hz. In the latter ion both ^1H and ^{13}C NMR spectra contain two lines—a doublet as shown in Fig. 12.2. The splitting due to $^1H - {}^{13}C$ spin-spin interaction is 195 Hz and it is the same in both ^1H and ^{13}C the spectra. For doublets the value of chemical shift is the mid-point measured on the ppm scale and the coupling constant is the separation between the peaks measured in Hz as shown in Fig. 12.2.

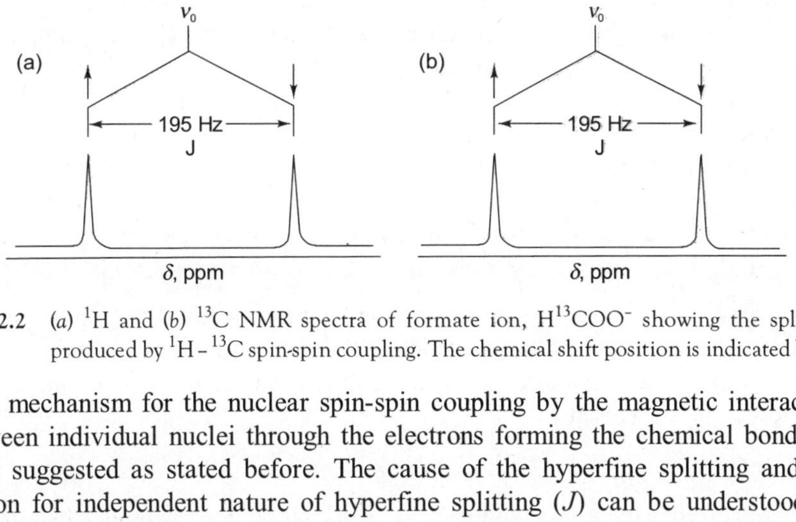

Fig. 12.2 (*a*) ^1H and (*b*) ^{13}C NMR spectra of formate ion, $H^{13}COO^-$ showing the splitting produced by $^1H - {}^{13}C$ spin-spin coupling. The chemical shift position is indicated by ν_0.

A mechanism for the nuclear spin-spin coupling by the magnetic interaction between individual nuclei through the electrons forming the chemical bond has been suggested as stated before. The cause of the hyperfine splitting and the reason for independent nature of hyperfine splitting (*J*) can be understood by considering a HF molecule. Figure 12.3 shows schematically the hydrogen fluoride molecule. Assume that hydrogen has its nuclear spin oriented parallel to the applied magnetic field B_0 (H↑). An electron near the hydrogen will be polarized to orient its spin parallel to that of hydrogen. The chemical bonding requires that the spins of the two electrons in the H—F bond be antiparallel. Now, if the second electron of the H—F bond is near the ^{19}F nucleus, it will tend to orient the spin of ^{19}F. Thus, the information about the spin of the hydrogen is carried over to the ^{19}F nuclear spin through the chemical bond. There is a slightly higher probability for the electron with its spin parallel to that of the nucleus to be in the vicinity of the nucleus. Consequently depending on the spin state of ^1H, the magnitude of the local field which determines the resonance frequency of ^{19}F is either increased or diminished and the NMR signal is split into a doublet. The same holds true for the

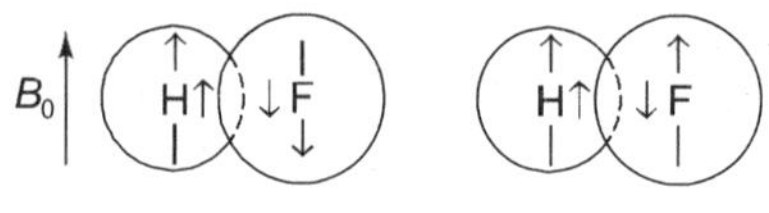

Fig. 12.3 Schematic representation of the coupling of nuclear spins through the bonding electrons in HF molecule.

^{1}H. The value of the coupling constant must therefore be the same for both nuclei. These magnetic interactions are so small that the parallel orientation of ^{1}H and ^{19}F nuclear spins is only very slightly higher in energy than that with antiparallel orientation of spins and occur with almost equal probability and the two spectral lines of the doublet will have almost the same intensity. The difference in their resonance frequencies J is proportional to the energy of interaction between ^{1}H and ^{19}F nuclei.

The magnitude of spin-spin coupling for a particular pair of nuclei depends on the following factors: (*i*) The nature of bonding, i.e. upon the number and bond order of the bonds intervening between the nuclei and upon the angle between the bonds. The coupling tends to decrease as the number of intervening bonds increase and is not usually observed over more than five or six bonds. (*ii*) The magnetic moments of the two nuclei and is directly proportional to the product $\gamma_A\gamma_X$ where γ_A and γ_X are the magnetogyric ratios of the two nuclei A and X respectively. (*iii*) The valence *s* electron density at the nucleus and therefore upon the *s*-character of the bonding orbitals. The interaction increases as the atomic number of either or both nuclei is increased in the same way as does the range of the chemical shift.

The energy E of the spin-spin interaction between two nuclei A and X say ^{1}H and ^{13}C (which need not be restricted to nuclei of spin 1/2 however) is proportional to the scalar product of the two nuclear magnetic moments μ_A and μ_X and can be expressed as

$$E = J_{AX}\boldsymbol{I}_A\boldsymbol{I}_X \tag{12.1}$$

where $\boldsymbol{I}_A$ and $\boldsymbol{I}_X$ are the so-called nuclear spin vectors of the two nuclei and J_{AX} is the scalar coupling constant between them. Equation (12.1) shows that the coupling constant is independent of the strength of the applied magnetic field. The coupling constants are therefore given in frequency units. If measurements are made at different B_0 fields, the splitting due to spin-spin coupling remains the same while the chemical shift varies.

THE ENERGY LEVELS OF COUPLED SYSTEMS

Figure 12.4(b) shows that the spin-spin interaction ($J > 0$) in formate ion stabilises the antiparallel arrangements of nuclear spins (H↑C↓ and H↓C↑) i.e. the energy is lowered and destabilises the parallel configurations (H↑C↑ and H↓C↓). Thus the two energy levels of the proton spin are each split into two, with energies determined by the relative orientations of ^{1}H and ^{13}C nuclear spins. Molecules containing a C↑ undergo ^{1}H NMR transitions (H↑C↑ → H↓C↑) at a frequency lower than the chemical shift because the transition is from a less favourable energy state to a favourable energy state (parallel to antiparallel spins). On the other hand, molecules containing C↓ resonate at a higher frequency (H↑C↓ → H↓C↓).

The ^{1}H resonance is split into a doublet because the magnetic moment of the ^{13}C nucleus is the source of a small local magnetic field whose direction is determined by the ^{13}C magnetic quantum number. When the ^{13}C is in its spin 1/2 state, $m = 1/2$ (denoted as ↑), its magnetic field at the position of the ^{1}H opposes the external field and shifts the ^{1}H resonance to a lower frequency as shown in Fig. 12.4. On the other hand, for ^{13}C in its spin – 1/2 state, $m = -1/2$ (denoted as ↓) the local field adds to the external field at the proton and hence the resonance moves to a higher frequency. Thus, a C↑ carbon shields the proton while a C↓ deshields it. The two components of the ^{1}H doublet correspond to two types of formate ions, those having C↑ and others having C↓. Since the difference in energy between the two orientations of the carbon nucleus is extremely small compared to the thermal energy kT, the two kinds of formate ions $H^{13}COO^{-}$ are equally possible and the two components of the ^{1}H doublet are equally intense. The same argument can be used to explain the splitting of the ^{13}C resonance by the ^{1}H. The nature of spin-spin coupling was illustrated using a heteronuclear system. Homonuclear coupling say between protons with different chemical shifts give rise to splittings in exactly the same way.

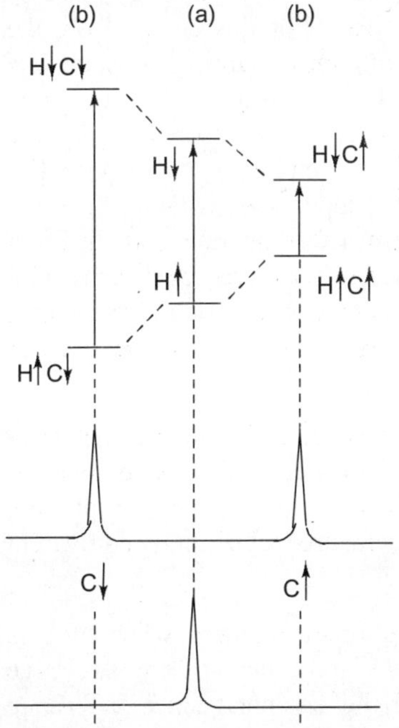

Fig. 12.4 The effect of ^{1}H – ^{13}C spin-spin coupling in formate ion, $H^{13}COO^{-}$ on the energy levels and spectrum of ^{1}H nucleus: (a) in the absence of spin-spin coupling and (b) in the presence of spin-spin coupling.

By convention, if the antiparallel arrangement of the nuclear spin is energetically favourable, then J_{AX} is taken as positive, $J_{AX} > 0$ (as in formate ion) and if the parallel spin configuration is lower in energy then $J_{AX} < 0$. The sign of the coupling constant does not affect the appearance of the spectrum.

Chemical bonding interactions are not as simple as just discussed and in some systems parallel orientations of nuclear magnetic moments may be favoured. Such systems are described by a negative spin-spin coupling constant. It is not possible to determine the signs of spin-spin coupling constants from the spectrum. Spin-spin coupling can occur where two nuclei are bonded together, as in ^{13}C—H or ^{19}F—H or where several bonds intervene as in H—^{12}C—^{12}C—H. The coupling constant is a sensitive parameter. The scalar coupling interaction, in general, is carried by the electrons through the chemical bonds and not through space. Some exceptions, particularly with non- hydrogen nuclei, have been suggested for through space coupling.

The spin-spin coupling is characteristic of the molecule. Observed J values for protons are less than 30 Hz. For other nuclei, J values may vary from 1 Hz to about 1000 Hz.

FIRST ORDER SPECTRA

Spectra originating from coupled nuclei, i.e. individual nuclei or groups of equivalent nuclei, may be treated by the first order analysis only when the following two conditions are fulfilled.

1. The chemical shift difference between the nuclei (say A and X) must be larger than the spin spin-coupling between them. For this comparison, the chemical shift difference $\delta_A - \delta_X$ and the spin-spin coupling constant J_{AX} must be expressed in the same unit say Hz. The separation in Hz is obtained by multiplying the chemical shift difference δ of the resonances by the spectrometer operating frequency and is written as $\nu_0\delta$ or $\Delta\nu$. The modification of the spectra from the first order appearance is then a function of the ratio $\nu_0\delta/J$ and is different for spectrometers operating at different frequencies. Since the chemical shift expressed in Hz increases linearly with the applied magnetic field, spectra obtained at higher magnetic field strengths are more likely to follow the first order analysis than those obtained at lower fields.

2. Coupling must involve groups of nuclei that are chemically and magnetically equivalent. Nuclei are said to be chemically equivalent if they have identical chemical shift, usually as a result of molecular symmetry, for example, the protons in CH_4 or fluorines in CF_3COOH. Nuclei are magnetically equivalent if they are equally coupled to the other nuclei.

Spectra can be satisfactorily analysed by the first order treatment when $\Delta\nu/J$ is greater than say 7 but still the spectra may show some deviations in relative intensities until $\Delta\nu/J$ is greater than 20. When the chemical shift difference between the nuclei say A and X i.e. $(\nu_A - \nu_X)$ is small, the ratio $(\nu_A - \nu_X)/J_{AX}$ will also be small, the spectrum will now be "second order". Second order spectra must be analysed by applying quantum mechanical treatment of spin-spin coupling. We shall only describe the nature of the spectrum in the second order analysis for selected systems later.

If the molecule has elements of symmetry such as a two-fold or a higher-fold rotation axis or a reflection plane, magnetic nuclei of the same species which exchange their positions under appropriate symmetry operations are symmetrically equivalent and they have the same chemical shift. The four protons of methane, for example, are chemically equivalent in this way. Likewise the two protons and two fluorines of 1, 1-difluoroethylene form chemically equivalent groups (Fig. 12.5).

Systems obeying the above rules are said to be weakly coupled and their spectra could be interpreted by the first order treatment. The spectrum of acetaldehyde (Fig. 12.1) follows first order analysis because the chemical shift difference between the methyl and aldehyde protons is large and each of the methyl protons is equally coupled to the aldehyde proton. In any one of the conformations, the three CH_3 protons appear to be chemically and magnetically inequivalent. However, rapid internal rotation of the CH_3 group about the $H_3C—C$ bond makes all the methyl protons equivalent. The spin-spin coupling between magnetically equivalent nuclei does not appear in the spectrum. In difluoromethane H_a and H_b are magnetically equivalent since they are by plane of symmetry equally coupled to F_a and F_b. However, in 1, 1-difluoroethylcnc, H_a and H_b are not magnetically equivalent since H_a and F_a are coupled by J_{cis} while H_b and F_a are coupled by J_{trans} and, in general, $J_{cis} \neq J_{trans}$ (Fig. 12.5).

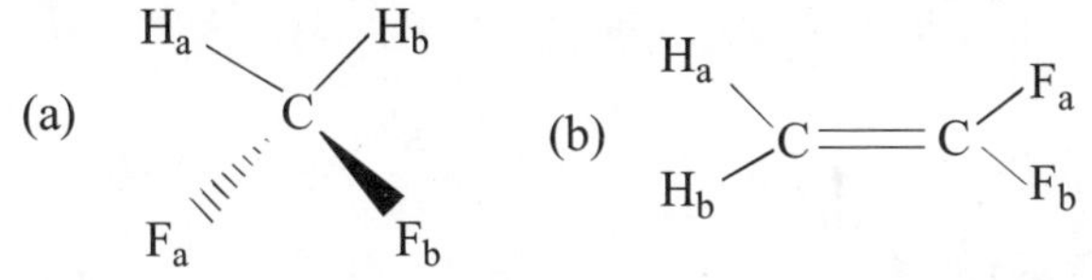

Fig. 12.5 CH_2F_2 and $CH_2=CF_2$; example of (a) magnetically and (b) chemically equivalent protons.

NOMENCLATURE FOR SPIN SYSTEMS

Analysis of the spin-spin splitting in NMR spectra is very similar to the analysis of the hyperfine splitting in the ESR spectra. The analysis of the spectrum is considerably simplified by restricting to nuclei with $I = 1/2$. A set of notations is used to describe the spin coupled systems. Each nucleus of spin 1/2 is denoted by some letters of the alphabet A, B, X, etc. The nuclei having small chemical shift difference relative to each other (strongly coupled) are given letters of the alphabet that are adjacent to each other (e.g. A, B, C). For nuclei having a large chemical shift difference, that is, weakly coupled groups of chemically and magnetically equivalent nuclei are represented by letters from different parts of the alphabet, for example, A, M and X. The number of nuclei in each group is denoted by a subscript, for example A_2X_3, AB_2, A_3B, etc. Systems in which there are both weak and strong coupling are represented by a combination of the above, for example, ABX. In a chemically equivalent group of nuclei, magnetic inequivalence is represented by using the same letter for each of the nuclei in the group to indicate chemical equivalence and primes to indicate magnetic inequivalence, for example, AA′XX′, AA′B, AA′BB′, and so on. The nuclei are generally labelled such that the sequence of letters in the alphabet matches with the sequence of the resonance frequencies in the order of increasing applied magnetic field strength. Nuclei with $I = 0$ and those with $I \geq 1$ that relax so rapidly that they behave as though they are not magnetic, for example, Cl, Br and I are not given any designation. Spin-spin coupling to nuclei with spin greater than 1/2 is usually not observed because of the short relaxation times for these nuclei which is discussed later. Generally, no significance is attached to the order in which the nuclei are listed. For example, AB_3 could be labelled as A_3B. Nuclei with different γ (magnetogyric ratio) are denoted by letters far apart in the alphabet. Even systems that might be classified as say ABC may sometimes be treated as say ABX for simplicity. The expected multiplet structure may be obscured if the splitting is smaller than the line width or when the molecule is undergoing a dynamic process, which may average out spin-spin interactions.

A few more examples are given below for a better understanding of the nomenclature.

Molecule	Designation
$CH_2=CCl_2$	A_2
CH_2F_2	A_2X_2
$CH_2=CF_2$	AA′XX′
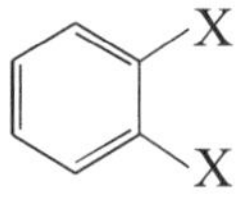	AA′BB′
$CH_2=CHCl$	ABX or ABC
$CH_3CH=CH_2$	A_3MXY
$CH_3CH_2NO_2$	A_3X_2
$^{13}CH_2F_2$	A_2M_2X

PATTERNS OF COUPLING

The NMR spectra of some simple systems are illustrated here. It is assumed that all pairs of spins are weakly coupled i.e. the difference in resonance frequencies of the two nuclei greatly exceeds their mutual spin-spin coupling, so that first order analysis is applicable.

(i) AX System

The interaction of nucleus A with a single spin 1/2 nucleus X leads to splitting of resonance A into two equally intense lines (a doublet) centred at the chemical shift of A with spacing equal to the AX coupling constant, J_{AX} as discussed above for the formate ion. The spectrum of X is also a doublet with the same splitting, as shown in Fig. 12.6. The energy level diagram along with the corresponding NMR spectra for a pair of nuclei with spin $I = 1/2$ with and without spin-spin coupling are shown in Fig.12.7. The allowed transitions are those in which single spin changes; therefore the transitions $A\uparrow X\uparrow \longrightarrow A\downarrow X\downarrow$ and $A\uparrow X\downarrow \longrightarrow A\downarrow X\uparrow$ are forbidden. Changing J_{AX} in Fig. 12.7 from positive to negative would only interchange the resonance of the $X\uparrow$ with that of the $X\downarrow$ and likewise in the A spectrum.

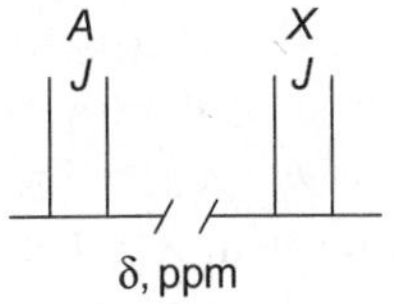

Fig. 12.6 Spin-spin coupling in an AX spectrum.

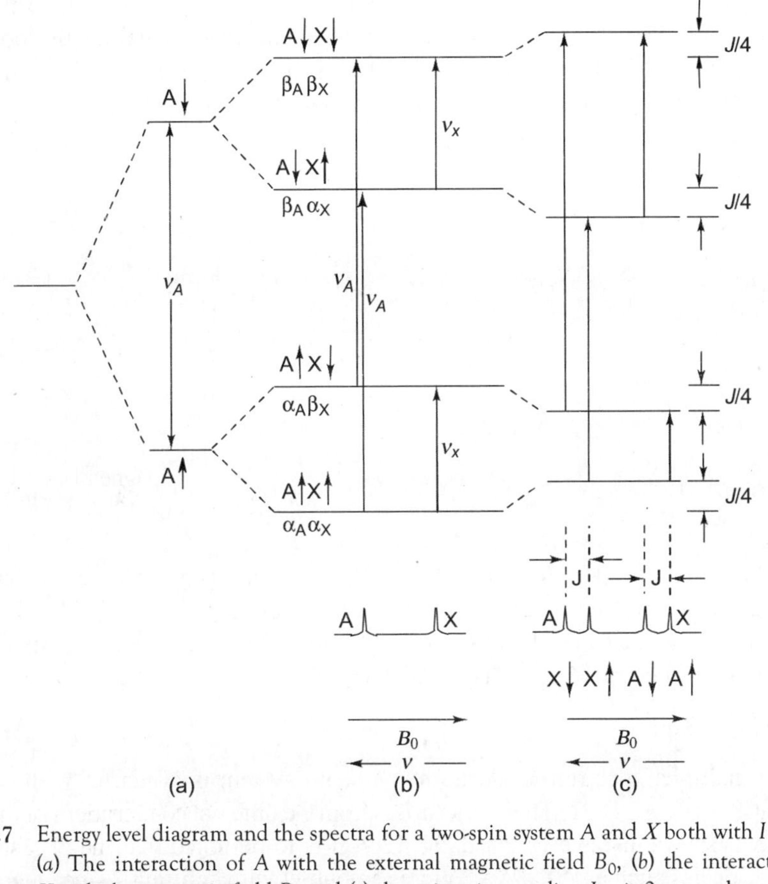

Fig. 12.7 Energy level diagram and the spectra for a two-spin system A and X both with $I = 1/2$. (*a*) The interaction of A with the external magnetic field B_0, (*b*) the interaction of X with the magnetic field B_0 and (c) the spin-spin coupling $J_{AX} > 0$. ν_A and ν_X are the resonance frequencies of the nuclei A and X, respectively.

Suppose the spin of X is α, then the spin of A will have Larmor frequency as a result of the combined effect of the external field, the shielding constant and the spin-spin coupling of A with X. The spin-spin coupling will cause one line in the spectrum of A being shifted by $-\frac{1}{2} J$ from the frequency it would have in the absence of coupling. If the spin of X is β, the spin of A will have a Larmor frequency shifted by $+ \frac{1}{2} J$. Therefore, instead of a single line from A we get two lines separated by J and centered on the chemical shift characteristic of A as shown in Fig. 12.6. The same splitting occurs in the X resonance; instead of a single line, the resonance is a doublet with splitting J which has the same value as for the splitting of A but centered on the chemical shift of X.

(ii) AX_2 System

This may be considered as an extension of the AX spin system. A simple way of describing the AX_2 spectrum is to regard that X_1 splits the single line due to A nucleus into a doublet, with spacing equal to the coupling constant J_{AX}, each line being split again into a further two-line spectrum by the interaction with X_2. The two central lines of the doublet of the doublets coincide giving a triplet centred at the chemical shift of A with spacing between the lines equal to the coupling constant and with relative intensities 1:2:1 as shown in Fig. 12.8. The central line of the triplet arises from two degenerate arrangements of the X spins ↑↓ and ↓↑.

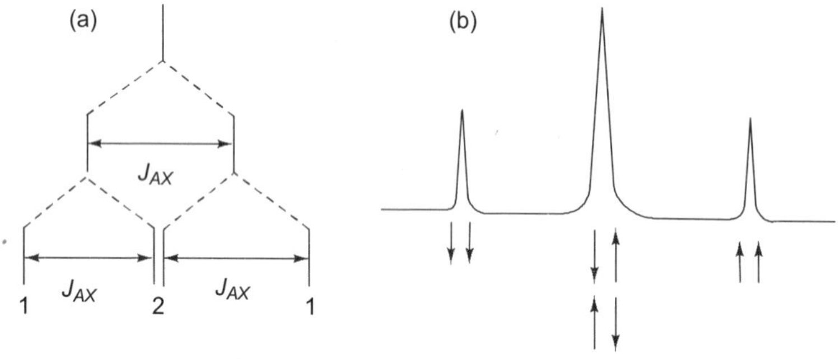

Fig. 12.8 (a) The hyperfine splitting diagram and (b) the NMR spectrum of A in an AX_2 spin system.

As expected, the ^1H NMR spectrum of CH_2F_2 exhibits a triplet with intensities 1:2:1 and with splitting equal to J_{HF} (the nuclear spin of ^{19}F is 1/2). The ^1H NMR spectrum of $CH_2{=}CF_2$ on the other hand has about ten lines. The analysis of such second order spectra will be discussed later.

(iii) AX_3 System

The multiplet spectrum of A in an AX_3 spin system is a quartet with relative intensities of 1 : 3 : 3 : 1. The quartet arises from the different combinations of the three X spins as shown in Table 12.1 and Fig. 12.9. It is to be noted that the X-resonance of AX_n species such as AX_2, AX_3, etc. is a doublet with splitting J since a group of equivalent nuclei resonates like a single nucleus. However, for the X-resonance of an AX_n species the intensity is n times as great as that of the AX species (Fig. 12.7).

Table 12.1. Spin-spin coupling in an AX_3 spin system

m_1	m_2	m_3	Σm_1
1/2	1/2	1/2	+3/2
1/2	1/2	–1/2	
1/2	–1/2	1/2	+1/2
–1/2	1/2	1/2	
1/2	–1/2	–1/2	
–1/2	–1/2	1/2	–1/2
–1/2	1/2	–1/2	
–1/2	–1/2	–1/2	–3/2

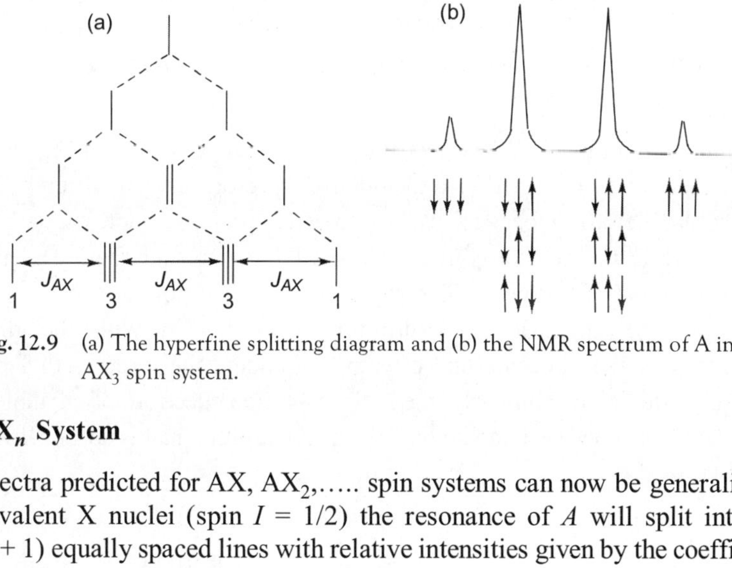

Fig. 12.9 (a) The hyperfine splitting diagram and (b) the NMR spectrum of A in an AX_3 spin system.

(iv) AX_n System

The spectra predicted for AX, AX_2,..... spin systems can now be generalised. For n equivalent X nuclei (spin I = 1/2) the resonance of A will split into n + 1 (= $2nI + 1$) equally spaced lines with relative intensities given by the coefficients in the binomial expansion,

$$1 : \frac{n}{1} : \frac{n(n-1)}{2 \cdot 1} : \frac{n(n-1)(n-2)}{3 \cdot 2 \cdot 1} \ldots$$

or equivalently by the $(n + 1)^{th}$ row of the Pascal's triangle given in Fig. 12.10. The rows give the relative intensities of the $(n + 1)$ lines in the A multiplet of an AX_n spin system (for X, I = 1/2) and the columns give the positions of the lines relative to the chemical shift positions in units of J_{AX}.

A						1					
AX					1		1				
AX_2				1		2		1			
AX_3			1		3		3		1		
AX_4		1		4		6		4		1	
AX_5	1		5		10		10		5		1

Fig. 12.10 Pascal's triangle.

When the first order analysis is applicable, the number of components in a multiplet, their spacing and their relative intensities can easily be determined. A nucleus or a group of nuclei coupled to a set of n nuclei with spin I will have its resonance split into $2nI + 1$ lines. The relative intensities of the $2nI + 1$ lines can be determined from the number of ways each spin state may be formed. The $2nI + 1$ lines are equally spaced with the frequency separation between the adjacent lines being equal to J, the coupling constant. For $I = 1/2$, there are $n + 1$ lines and the intensities of the $n + 1$ lines correspond to the coefficients in the expansion of the binomial theorem $(1 + x)^n$ as stated before. The first order procedure is widely used for the analysis of spectra.

The high resolution ^1H NMR spectrum of acetaldehyde given in Fig. 12.1 can now be understood. The methyl and aldehyde protons make up an A_3X spin system, the doublet arises because each of the CH_3 protons couples equally to the CH proton, while the quartet comes from the CH proton interacting identically with each of the CH_3 protons. The rapid internal rotation of the C–C bond averages out the chemical shift differences associated with the different conformations of the molecule and effectively renders the three methyl protons magnetically equivalent. Similarly, the high resolution spectrum of ethanol shown in Fig. 12.11 can now be understood. The ethyl protons make up an A_3X_2 spin system.

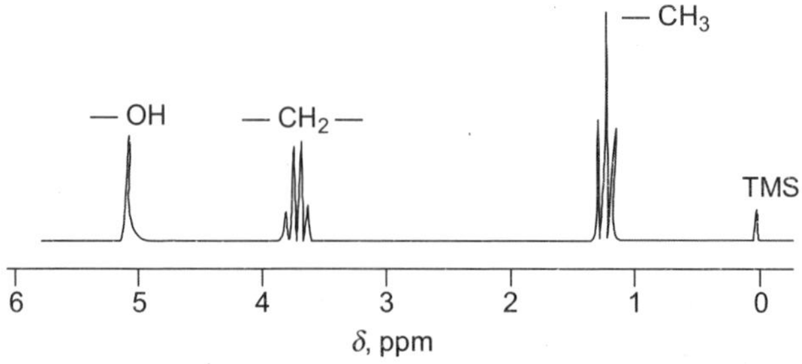

Fig. 12.11 High resolution NMR spectrum of ethanol.

In the case of ethanol, the coupling between OH and CH_2 can be observed if the sample is pure or if the spectrum is measured at low temperature when the intermolecular exchange is retarded or if the sample is dissolved in a highly polar solvent such as dimethyl sulfoxide when strong solvation reduces the exchange. Rapid proton exchange also occurs in carboxylic acids, amines, amides, etc. The exchange is acid-catalysed and base-catalysed.

(v) AMX System

The AMX spin system consists of three nuclei with three different chemical shifts and three distinct coupling constants, J_{AM}, J_{AX} and J_{MX}. Consider A, M and X to be protons. The notation AMX indicates that proton M has a chemical shift intermediate between A and X protons. The spectrum is first order like the AX spin system. A simple way to predict the multiplet structure of A is to begin the spectrum of A as a singlet at the chemical shift position. Then the AM coupling gives a doublet with splitting J_{AM}. When the AX coupling is introduced, each of the lines of the doublet then splits into a doublet with splitting J_{AX}. The stepwise procedure helps in arriving at the multiplet pattern as shown in Fig. 12.12. The order in which couplings are introduced is not important. The appearance of the doublet of doublets will depend on the values of the coupling constants.

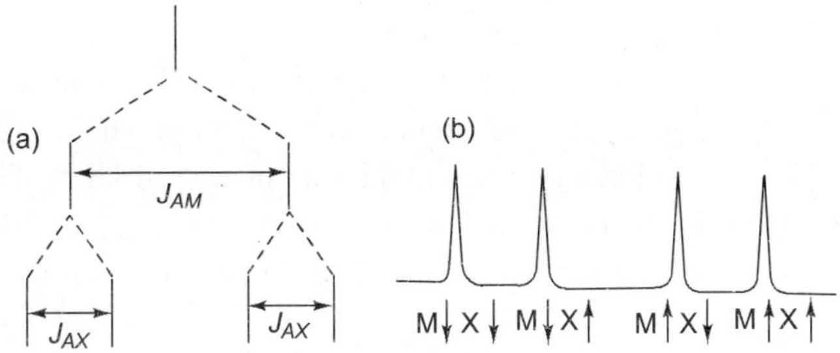

Fig. 12.12 (a) The hyperfine splitting and (b) the NMR spectrum of nucleus A in an AMX spin system, assuming $J_{AM} > J_{AX} > 0$.

When all the three protons are considered, the spectrum will consist of twelve lines of equal intensity when $J_{AX} \neq J_{MX} \neq J_{AM}$. The twelve spectral lines will be in three well-separated symmetrical 1:1:1:1 quartets as expected from the splitting pattern shown in Fig.12.13. Examples of AMX spin system include $F_2C = CFCl$, the three ring hydrogens of a monosubstituted furan e.g. furoic acid and many vinyl systems especially where the substitutent is an oxygen function as in vinyl acetate. The values of ν_A, ν_M and ν_X are given by the mid point of the quartets. The three values of the coupling constants can be directly obtained from the spectra.

If two of the coupling constants are equal for example, $J_{AM} = J_{AX}$, then the spectrum will consist of eleven lines and then the resonance due to H_A would appear as a 1 : 2 : 1 triplet instead of a quartet. If all three coupling constants are equal the spectrum will consist of nine lines and will appear as three sets of 1 : 2 : 1 triplets.

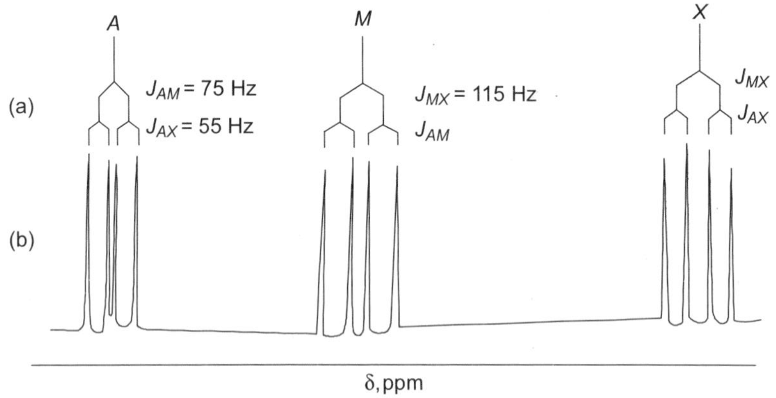

Fig. 12.13 (a) The splitting pattern for AMX spin system and (b) the appearance of a typical AMX spectrum.

The spin-spin coupling discussed above is valid for proton and other nuclei such as ^{31}P, ^{19}F etc. (see chapter 15), which have a natural abundance of 100%. However, the resonance signals due to coupling of other less abundant nuclei such as ^{13}C (1.1%) and ^{29}Si (4.7%) to protons can be observed in the ^{1}H NMR spectrum as the so-called satellite lines which is discussed later. If the natural abundance of NMR nuclei is below 1%, observation of satellite spectra becomes difficult. The ^{13}C satellites in ^{1}H NMR spectra is important for spectral analysis.

(vi) Spin System with $I \geq 1$

If a given nucleus A has a spin quantum number $I > 1/2$, its multiplet structure can be predicted exactly the same way as for a spin 1/2 nucleus. This is shown for the case of a spin $I = 1$ nucleus coupled to a spin 1/2 nucleus in Fig. 12.14. In the spectrum of HD, the H resonance is a triplet arising from $m = 1, 0, -1$ spin states of D and the resonance of D is a doublet. As another example, consider the NMR spectrum of NH_4^+ ion. The ^{14}N ($I = 1$) and ^{15}N ($I = 1/2$) NMR spectra of $^{14}NH_4^+$ and $^{15}NH_4^+$ both exhibit a quintet with relative peak intensities of 1 : 4 : 6 : 4 : 1. The NH coupling constant of the two isotopically substituted molecules are in the ratio of 0.713 : 1, which is the ratio of the magnetogyric ratios of the two nitrogen isotopes. However, the ^{14}N nuclei with $I = 1$ also possess electric quadrupole moment which can interact with local electric field gradient. This interaction leads to efficient relaxation of the quadrupolar nucleus in solution rendering the NMR lines so broad that the expected multiplet patterns may be partially or completely obscured. Quadrupolar relaxation may have the effect of decoupling A and X such that no splitting is observed in the spectrum (see Chapter 17). For example, ^{35}Cl and ^{37}Cl (both $I = 3/2$) rarely produce splitting in the NMR spectra of nearby nuclei. Although quadrupolar nuclei are generally less suitable for NMR studies than those with $I = ½$, some nuclei such as ^{2}H, ^{11}B, ^{14}N and ^{17}O are routinely studied. Coupling to more than one quadrupolar nucleus is rarely observed. The usual exception involves deuterium for which $I = 1$. Coupling to two deuterium nuclei gives rise to a five line spectrum with intensities 1 : 2 : 3 : 2 : 1.

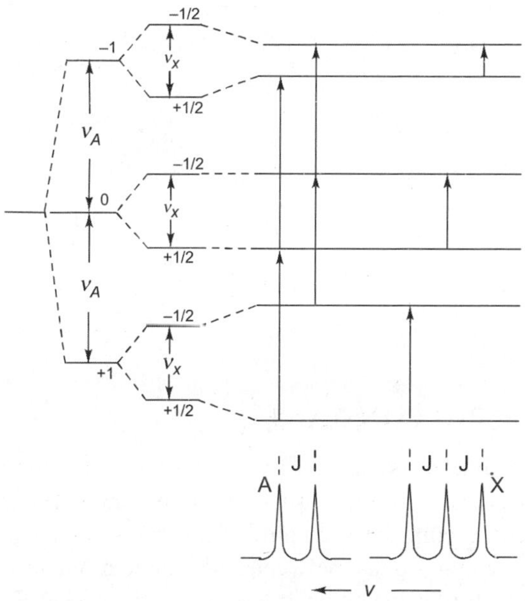

Fig. 12.14 Energy level diagram for a spin $I = 1$ nucleus (A) coupled to a spin 1/2 nucleus (X). The X spectrum is a triplet and the spectrum of A is a doublet.

The coupling to equivalent nuclei with $I > 1/2$ can easily be deduced by the 'stick diagram' discussed earlier. For example, the terminal protons of $^{11}B_2H_6$ (diborane) are predicted to show a quartet 1:1:1:1 due to coupling to the directly bonded ^{11}B ($I = 3/2$) while the bridge protons should exhibit seven lines with relative intensities of 1:2:3:4:3:2:1 arising from equal interaction with two symmetrically equivalent borons as shown in Fig. 12.15 (also see chapter 17).

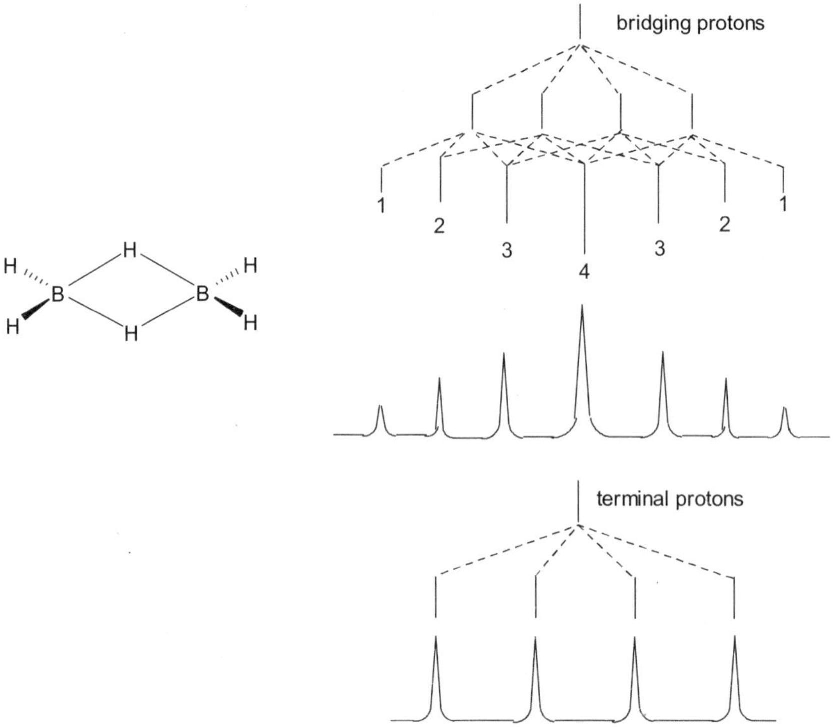

Fig. 12.15 The hyperfine splitting and the ^1H NMR spectra of the bridged and terminal protons of diborane, $^{11}B_2H_6$.

OBSERVED COUPLING CONSTANTS

The coupling of the nuclear spins is a through-bond interaction and the energy of the interaction decreases as the number of intervening bonds increases as mentioned earlier. In the commonly employed notations for coupling constants, the number of intervening bonds between the coupled nuclei is given as a superscript and the identity of the coupled nuclei as a subscript. For example, the spin-spin coupling in H–C–C–H is denoted by $^3J_{HH}$. The position of the coupled nuclei with respect to a double bond, namely, *cis* or *trans* is indicated as a subscript. The superscript or subscript or both may be omitted when there is no ambiguity. When the spectrum could be analysed by the first order procedure, obtaining the value of J from a multiplet is trivial. In more complex spectra extensive calculations may be needed to obtain the J values from the spectra. Knowledge of the magnitude of coupling constants for different class of molecules comes chiefly from experimental observations and empirical correlation.

The coupling constant is classified as geminal, vicinal or long range depending on whether the coupling occurs over two, three or more bonds (see Table 12.2). The coupling constants can be related to a number of physical parameters. The most important parameters influencing the coupling constant are the hybridization, dihedral angle and the electronegativity of the substituents. Coupling through more than three bonds which is referred to as long-range coupling is useful for stereochemical studies. Proton coupling constants through single bonds are usually attenuated rapidly so that, generally, ^4J < 0.5 Hz and is usually unobservable. Couplings are stereospecific so that certain geometric arrangements of nuclei may result in values of ^4J and even ^5J that are observable.

Table 12.2 Classification of spin-spin coupling

Type of coupling	*Classification*	*No. of bonds*	*Symbol*
	Geminal	2	^2J
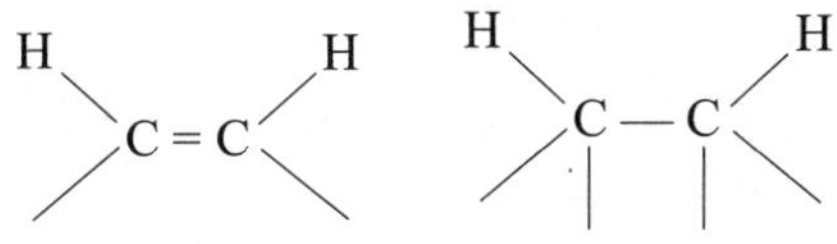	Vicinal	3	$^3J_{cis}$, ^3J
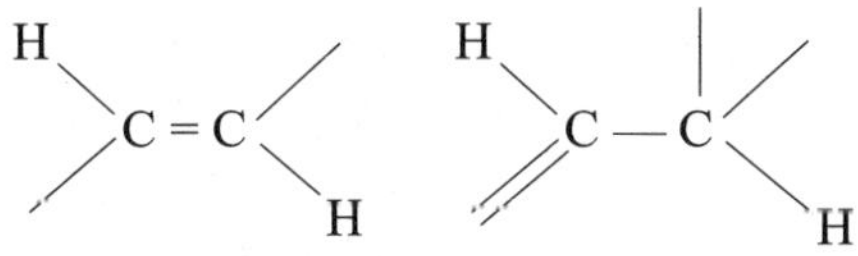	Vicinal	3	$^3J_{trans}$, ^3J
H—C—C≡C—H	Long range	4	^4J
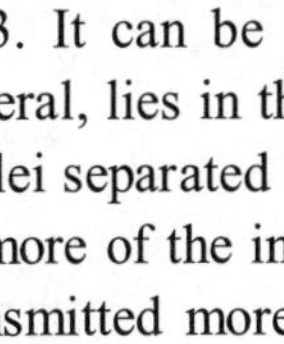	Ortho	3	$J_{1,2}$
	Meta	4	$J_{1,3}$
	Para	5	$J_{1,4}$

Some typical values of proton-proton coupling constants are given in Table 12.3. It can be seen that the magnitude of coupling constants for protons, in general, lies in the range 5–20 Hz. The spin-spin coupling constant J is small for nuclei separated by more than three bonds, although J value may be larger if one or more of the intervening bonds is a *pi* bond. Generally, spin-spin couplings are transmitted more effectively through multiple bonds rather than through single bonds. It is commonly observed that the *trans* coupling constants are considerably larger than *cis* coupling constants. For example, ^3J for ethane derivatives is usually less than 10 Hz while in ethylene derivatives $^3J_{cis}$ ~ 10 Hz and $^3J_{trans}$ ~ 17 Hz. Theory predicts that spin-spin coupling constants between protons separated by an even number of bonds are negative and those separated by an odd number of bonds are positive.

Table 12.3. Typical values of observed ^1H–^1H coupling constants (J_{HH})

Structure	*J value (Hz)*	*Structure*	*J value (Hz)*
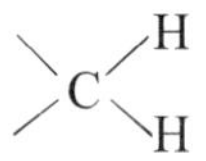	12–15	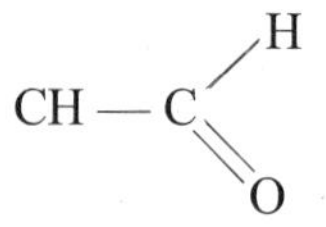	1–3
$H_3C—CH_2—$	6–8	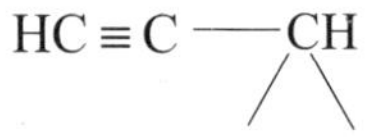	1–4
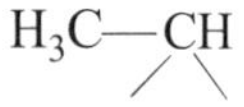	5–7		
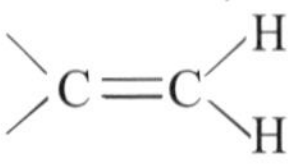	0.5–3	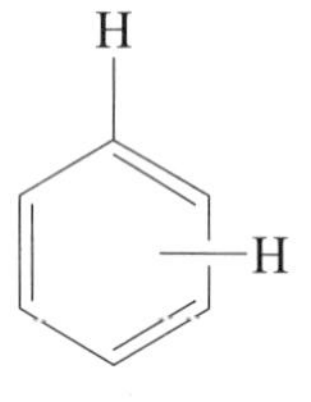	*o* 6.0–9.5 *m* 1.5–3 *p* 0–1
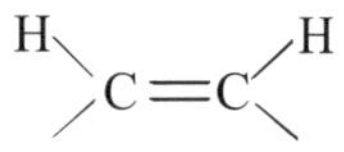	7–12	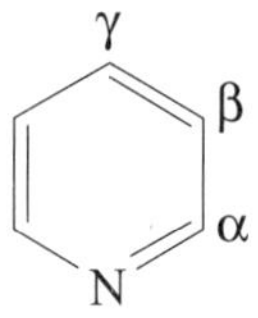	(α, β) 4.9–5.7 (α, γ) 1.6–2.6 (β, γ) 7.2–8.5
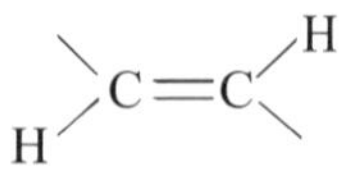	12–18	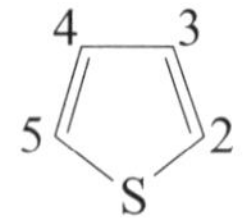	(2, 3) 4.7 (3, 4) 3.5 (2, 4) 1.5 (2, 5) 3.5
$—CH_2—CH_2—$	5–8		
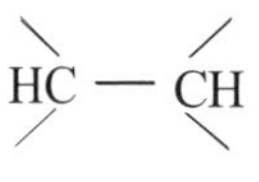	6–8	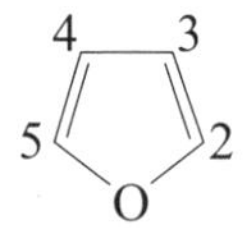	(2, 3) 1.8 (3, 4) 3.5 (2, 4) 0.1 (2, 5) 1.2
—CH=C=CH—	4–6		

The ^1H spectrum of methane is normally considered as a singlet since the 1.1% of $^{13}CH_4$ present is ignored and only $^{12}CH_4$ is considered. If we look carefully in the baseline, we would find a doublet with $J = 125$ Hz centred on either side of the intense singlet due to the main component. These two small lines are termed as satellites. Since 1.1% of carbon is ^{13}C, coupling to the protons will occur resulting in weak satellite peaks (0.55% of the total intensity) on either side of the main peak. Similarly, if the proton resonance signal of chloroform is recorded at high gain, a low intensity singlet is observed on either side of the signal at 104 Hz from the main signal (Fig. 12.16). These ^{13}C satellite signals of the $CHCl_3$ arise from the spin-spin coupling between ^{13}C and the proton that leads to doublet splitting in the proton NMR spectrum. The same splitting is found in the ^{13}C NMR spectrum of $CHCl_3$. The satellite structure is often seen in the proton NMR spectra of organic molecules. The ^{13}C satellites can be used to measure the coupling constants between magnetically equivalent nuclei. The analysis of ^{13}C satellite spectra also yields valuable information about spin-spin coupling in more complicated cases.

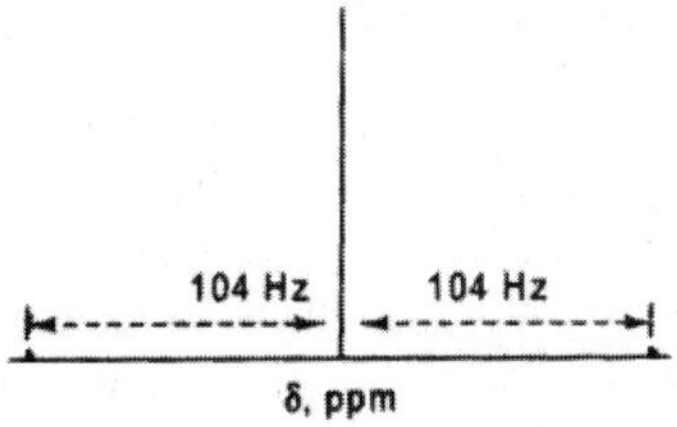

Fig. 12.16 ^1H NMR spectrum of $CHCl_3$ showing satellite signals due to $J(^{13}C-^{1}H)$208 Hz.

TWO-BOND COUPLING

Two-bond coupling $^2J_{HH}$ is also known as geminal coupling. It is found only in a methylene group in which the two hydrogens are not identical for some reason and their chemical shifts are different. Proton-proton coupling constant is usually the largest between geminal protons H–C–H. Many factors such as hybridization of the carbon and effect of substituents determine the magnitude and the sign of $^2J_{HH}$. Two-bond (geminal) proton-proton couplings vary over a wide range (–23 to 42 Hz) with large substituent effects, sp^2 hybridised CH_2 groups generally have smaller $^2J_{HH}$ than the methyl groups (Table 12.4).

Table 12.4 Two-bond $^1H—^1H$ coupling constants in Hz

	$H_2C{=}CHX$	$CH_3—X$
X = H	2.5	–12.4
X = Ph	1.3	–14.5
X = Cl	–1.3	–10.8
X = CN	1.0	–16.9

The electronegativity of the attached substituent alters the value of geminal coupling. In groups such as $–CH_2–X$, the geminal coupling varies from 9 to 12 Hz with increase in the electronegativity of X. These coupling constants cannot be measured directly since the two protons are chemically and magnetically equivalent and have identical δ values, but in the deuterium substituted –CHD–X the gem coupling between H and D (H–C–D) can be measured; J_{HH} can then be calculated from the relation J_{HH} (= $\gamma_H / \gamma_D \times J_{HD}$) = 6.53 J_{HD} since $J_{HH} \propto \gamma_H\, \gamma_H$ and

$J_{HD} \propto \gamma_H \gamma_D$. It is thus helpful to use partial deuterium substitution. The ^1H spectrum of –CHDX shows an extra 1 : 1 : 1 triplet splitting arising from $^2J_{DX}$. Thc gem coupling between H and D(H–C–D) can be measured.

The magnitude of J_{gem} also varies with the HCH bond angle, it is greatest in magnitude in the strain-free cyclohexanes and cyclopentanes (10–14 Hz). With increase in angular strain, J_{gem} decreases to 8-10 Hz in cyclobutanes and to 4-9 Hz in cyclopropanes.

THREE-BOND COUPLING

Three-bond coupling is also known as vicinal coupling. The most useful spin-spin coupling constants are those which involve nuclei separated by three bonds, for example, $^3J_{HH}$ (vicinal) in an H—C—C—H fragment. The usefulness of three-bond coupling constants lies chiefly in the conformational analysis. These coupling constants are found to vary with the dihedral angle ϕ between the two H—C—C planes (as defined in Fig. 12.17) according to the Karplus relation,

$$^3J = A + B\cos\phi + C\cos 2\phi \tag{12.2}$$

Although it is possible to calculate approximate values for A, B and C (including substituent and other effects) it is usual to treat them as coefficients to be determined empirically using conformationally rigid model compounds of known structure. Typical values are $A = 7$ Hz, $B = -1$ Hz and $C = 5$ Hz which give a ϕ-variation of the type shown in Fig. 12.17 known as the Karplus curve. The Karplus relation finds valuable application in studies of protein structure.

The *trans* proton-proton coupling constants across a C=C bond are up to a factor of two larger than the *cis* coupling constants ($^3J_{180} > {}^3J_0$). A clear distinction between *cis* and *trans* isomers can, therefore, be made in olefinic systems. In 1,2-disubstituted ethane, $J_{gauche} < J_{trans}$ has been found.

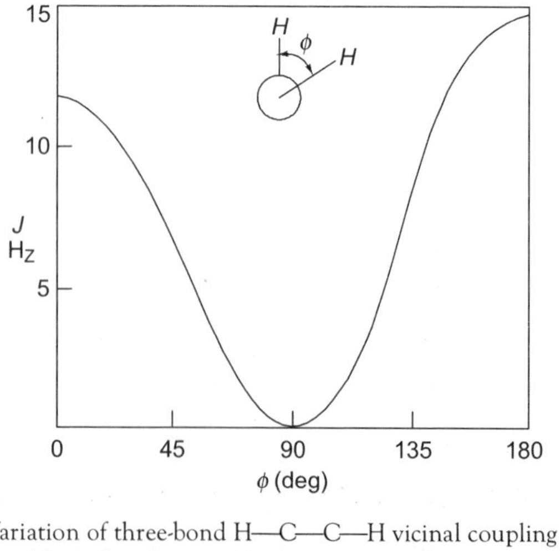

Fig. 12.17 Variation of three-bond H—C—C—H vicinal coupling constant with the dihedral angle ϕ as calculated using the Karplus equation.

The vicinal coupling constants depend, besides the dihedral angle ϕ between the C–H bonds, on the C – C bond length, the HCC valence angles, the presence of adjacent π bonds, the electronegativity of the substituent on the H–C–C–H moiety and the presence of adjacent π bonds. The more electronegative the substituent, the smaller the value of J_{vic}. In ethanes, the value of J_{vic} is ~ 8 Hz and in halogenoethanes, it is lowered to 6-7 Hz. The $^3J_{HH}$ values for the ring protons in cyclohexanes depend on whether axial or equatorial protons are involved.

According to Karplus rules, the largest vicinal couplings occur with protons in the *trans* coplanar positions (ϕ = 180°). Vicinal couplings for *cis* coplanar protons (ϕ = 0°) are almost as large. Very small couplings arise between protons at 90° to each other. As an example of the application of Karplus rules, in chair cyclohexane, diaxial protons have coupling constants of about 10 – 13 Hz consistent with their 180° orientation; diequatorial protons or axial-equatorial protons have coupling constants of 2–5 Hz corresponding to their 60° orientation. The vicinal coupling constants in some ethylene and ethane derivatives are given in Table 12.5.

Table 12.5 Three bond ^{1}H—^{1}H coupling constants (in Hz)

	$H_2C = CHX$		CH_3CH_2X
	cis	*trans*	
X=H	11.5	19.0	8.0
X=ph	10.7	17.5	7.6
X=Cl	7.5	14.8	7.2
X=CN	11.8	18.0	7.5

LONG-RANGE COUPLING

Proton-proton coupling constants as stated earlier are generally very small (< 1 Hz) when the nuclei are separated by more than three bonds. Some exceptions are shown in Fig. 12.18. Notably large $^4J_{HH}$ and $^5J_{HH}$ often occur when the coupling is transmitted along a zig-zag arrangement of bonds and/or through π-bonds.

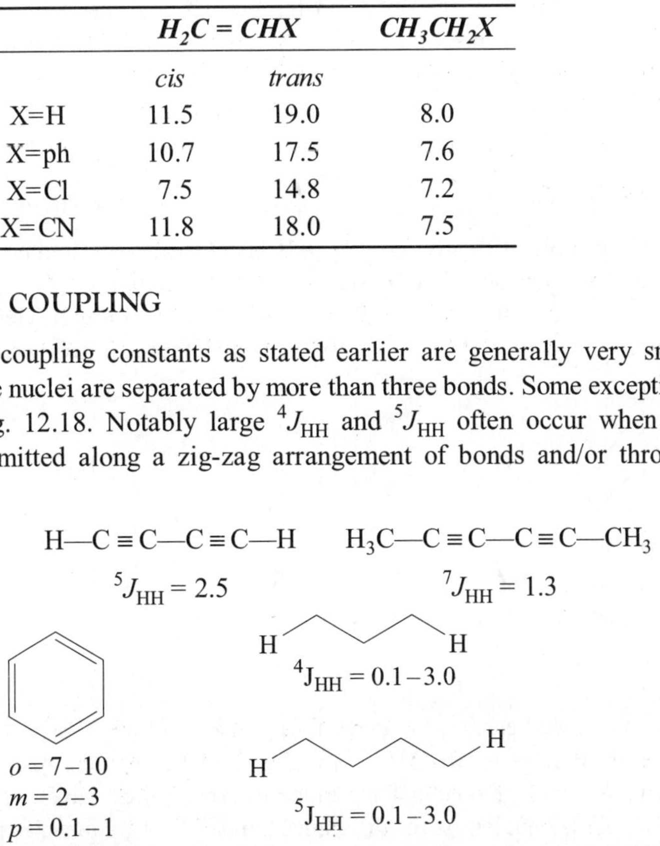

Fig. 12.18 Some examples of long range ^{1}H—^{1}H coupling constants (in Hz).

SECOND ORDER SPECTRA

First order rules can be applied only when the chemical shift difference $\nu_0\delta$ (Hz) between the individual groups of magnetically equivalent nuclei is large compared with the coupling constant between these groups. We then refer to the spectrum as "first order". If the chemical shift difference $\Delta\nu$ (or $\nu_0\delta$) is comparable to the coupling constant J or if the nuclei are chemically but not magnetically equivalent, then the system is said to be strongly coupled. The spectrum will now be "second order". In the second order spectra, the number, positions and intensities of resonance signals are changed and the resulting spectrum may appear incomprehensible at the first sight. More lines are observed in the spectrum than would be expected according to the first order rules. The intensity distribution in the lines of the two groups of signals is also greatly affected. Complicated spectra can often be simplified by measuring the spectra at higher magnetic field strengths, where the ratio $\nu_0\delta/J$ becomes sufficiently large. However, a number of spin systems as for example, the AA′ XX′ system will never transform into a first order spectrum even at the highest magnetic field. We shall briefly present the nature of the spectra obtained and describe how the chemical shift and coupling constants can be deduced from the observed frequencies and intensities of a second order spectrum. The presence of a simple splitting pattern is not always a guarantee that first order rules can be applied. Unlike first order spectra, all spectra of higher order depend on the relative signs of the coupling constants if spin systems of more than two nuclei are involved.

(i) AB System

Simple molecules giving only AB spectra are not numerous, although AB spectra are often observed as part of the more complex spectra. Spectrum of AB type is found in, for example, 1-bromo 1-chloroethylene (I), 2-bromo 5-chlorothiophene (II) and 1-amino-3, 6-dimethyl-2-nitrobenzene (III)

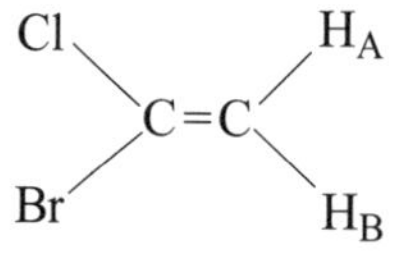

I

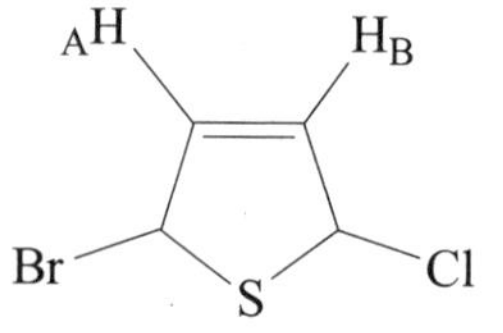

II

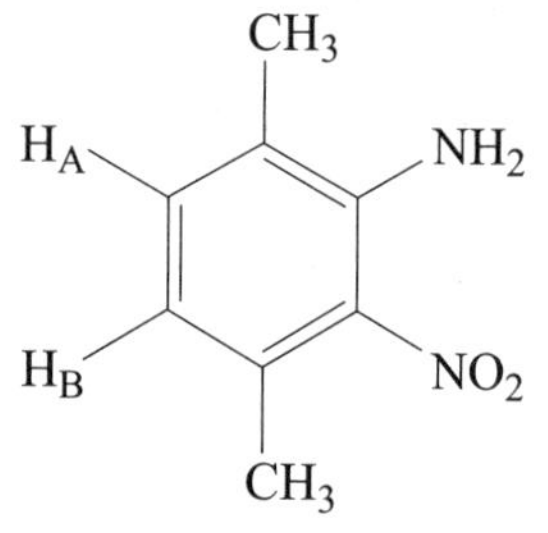

III

As mentioned earlier, the general appearance of an AB spectrum depends only on the absolute ratio of $|\Delta\nu / J|$. Figure 12.19 shows the results for an AB spin system. A series of spectra for a range of $\Delta\nu/J$ values have been calculated for a pair of spin 1/2 nuclei. As the difference in resonance frequencies is reduced keeping J constant, the inner component of each doublet steadily increases in intensity while

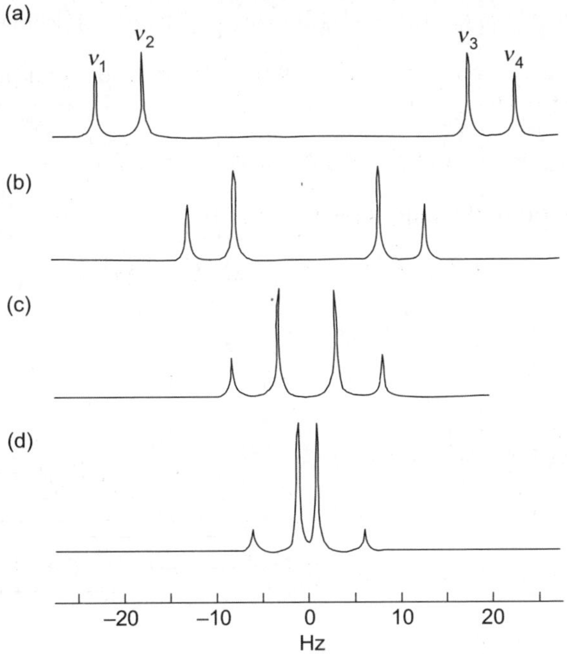

Fig. 12.19 The NMR spectra of a AB system as a function of $\nu_0\delta/J$. For J = 5.0 Hz and $\nu_0\delta/J$ equal to (a) 8.0, (b) 4.0, (c) 2.0 and (d) 1.0.

the outer components become weaker. The two doublets move together as their chemical shift difference decreases until at $\Delta\nu = 0$, when the inner lines coincide and the outer lines vanish, the spectrum degenerates to that of an A_2 system with frequency $(\nu_A + \nu_B)/2$. The spectrum thus changes from two simple doublets into a group of symmetrical four lines, the inner pair of the lines being stronger than the outer lines. At the intermediate stage, we may still describe the left hand pair qualitatively as predominantly A lines although the actual transition is a mixed one in which there is some change of spin of the *B* nucleus. It is not possible to determine which nucleus is more screened. By the analysis of the proton spectrum of 2-bromo-5-chlorothiophene, $|\nu_0\delta| = 4.7$ Hz and $J = 3.9$ Hz have been obtained.

The following relationships are useful for obtaining chemical shifts and coupling constant of *AB* quartets. A schematic diagram of the AB spectrum is given in Fig. 12.20 showing some of the relationships. The spacing between either the right hand pair or the left hand pair is equal to $J = \nu_1 - \nu_2 = \nu_3 - \nu_4$. The separation between the inner lines is $2C - J$. The spacing between the doublet centres is given by the equation

$$2C = [(\nu_0\delta)^2 + J^2]^{1/2} \quad (12.3)$$

$$\text{or} \quad 2C = \nu_2 - \nu_4 \quad (12.4)$$

The difference in the resonance frequencies of the two spins $\nu_0\delta$ in an *AB* spectrum is given by

$$\Delta\nu = \nu_0\delta = \sqrt{(2C - J)(2C + J)} = \sqrt{(\nu_2 - \nu_3)(\nu_1 - \nu_4)} \qquad (12.5)$$

If z is defined as the centre of the multiplet, that is, the mid-point between ν_1 and ν_4 or ν_2 and ν_3 then,

$$\nu_A = z - \tfrac{1}{2}\nu_0\delta \text{ and } \nu_B = z + \tfrac{1}{2}\nu_0\delta \qquad (12.6)$$

Further the ratio of the intensities is given by

$$I_2/I_1 = I_3/I_4 = (\nu_1 - \nu_4)/(\nu_2 - \nu_3) \qquad (12.7)$$

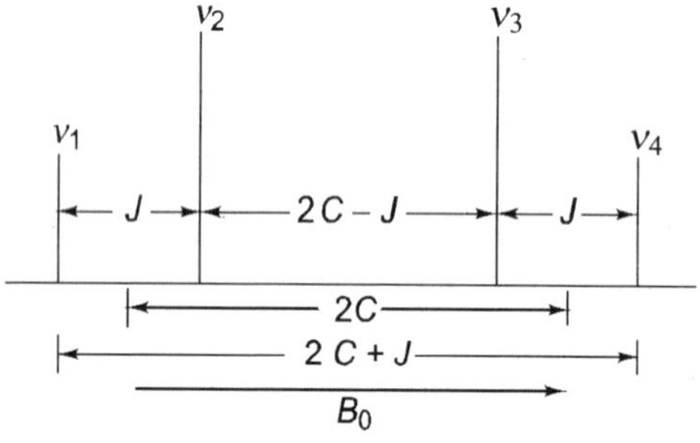

Fig. 12.20 Schematic AB spectrum. The absolute value of J is taken.

Figure 12.21 shows the energy levels and the spectrum for an AB system. The effect of spin-spin coupling is to mix the states $\alpha_A\beta_B$ and $\beta_A\alpha_B$ and hence modify their wave functions and energies. This results in a change in the transition probabilities for the four NMR lines, the inner lines becoming more allowed (i.e. more intense) and the outer pair of lines less allowed (i.e. weaker), the effect is more pronounced when the gap $\Delta\nu$ is smaller between $\alpha_A\beta_B$ and $\beta_A\alpha_B$ relative to J. In the limit $\Delta\nu = 0$, the outer lines are forbidden. The increase in the intensity of the lines nearest to the multiplet of the neighbouring group while that of the other lines decrease is called the 'roof effect'. The roof effect is useful in the assignment of coupling constants. The spectra in the two limiting cases may be referred to as AX (weak coupling) and AB (strong coupling) spectra.

For an AB system, the actual chemical shifts of the four lines in a 60 MHz spectrum were 6.13, 6.29, 7.58 and 7.74 ppm. In frequency units, these are 367.8, 377.4, 454.8 and 464.4 Hz, giving $\nu_0\delta/J \sim 9$.

Application of the above relations yield.

$$J = \nu_1 - \nu_2 = \nu_3 - \nu_4 = 9.6 \text{ Hz}$$

$$2C = 87 \text{ Hz}$$

z_0: Centre of the multiplet : (1/2) (367.8 + 464.4) = 416.1 Hz.

$$\nu_A = z_0 + 1/2\ \nu_0\delta = (416.1 + 43.5) = 459.6 \text{ Hz.}$$

$$\nu_B = z_0 - 1/2\ \nu_0\delta = (416.1 - 43.5) = 372.6 \text{ Hz.}$$

$$\delta_A = 7.66 \text{ ppm}, \ \delta_B = 6.21 \text{ ppm.}$$

$$\frac{I_2}{I_1} = \frac{I_3}{I_4} = \frac{96.6}{77.4} = 1.248$$

Intensity of I_1 and I_4 is given by $1 - \sin 2\theta$; I_2 and I_3 by $1 + \sin 2\theta$ and $J/2C = \sin 2\theta$

Hence, $I_1 = I_4 = 0.89$, $I_2 = I_3 = 1.11$

This spectrum is the same as the spectrum (a) of Fig. 12.19.

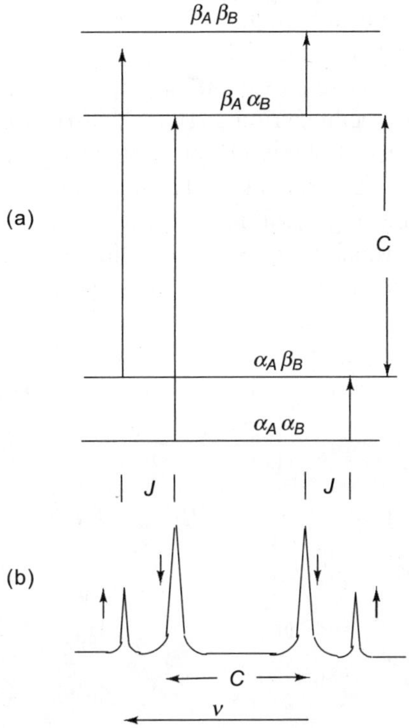

Fig. 12.21 (a) Energy level diagram and (b) spectra of a pair of strongly coupled spin 1/2 nuclei AB; compare with Fig. 12.7(c).

In practice *AB* systems that approach closely A_2 case is often met. The central lines may lie so close to one another that the spectral resolution is insufficient for their detection. Similarly, the intensity of the outer lines may be so small that the sensitivity of the spectrometer is insufficient for their detection. Such spectra are termed as deceptively simple spectra (see for example Fig. 12.19 (d)). Thus, the presence of simple splitting patterns is not always a guarantee that the spectrum is first order. For example, the ^{1}H NMR spectrum of furan shows two triplets at δ 6.37 and 7.42 ppm with coupling constant J = 3.2 Hz (see Fig. 12.22). The two triplets should not be interpreted as indicating that the two α protons are equally coupled to two β protons.

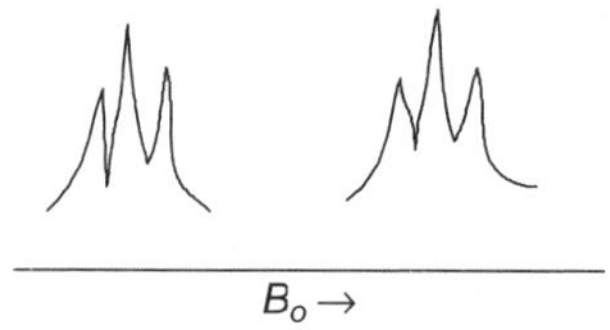

Fig. 12.22 ^{1}H NMR spectrum of furan.

(ii) AB$_2$ System

The effect of strong coupling is much more complicated when three or more spins as in the case of AB$_2$ system are involved. The multiplet structure would be difficult to recognise. The forbidden transitions may become allowed. The AB$_2$ spectra are found for compounds possessing a two-fold axis of symmetry. The most frequent examples are 1, 2, 3 or 1, 2, 6-trisubstituted benzene rings such as 2,6-dimethylpyridine (I) 1, 2, 3-trichlorobenzene (II) and pyrogallol (III). There are two different chemical shifts ν_A and ν_B and one coupling constant J_{AB}. The coupling constants of the ring protons with methyl or OH protons are negligibly small. In general, nine transitions are expected, of which four, ν_1 to ν_4, are in the A portion. In the B$_2$ portion, the lines ν_7 and ν_8 are well-separated whereas ν_5 and ν_6 are often not resolved or undetectable. In some cases, the lines ν_5 and ν_6 may coalesce giving a slightly broadened singlet. The line ν_9 arises from a forbidden transition and is therefore generally of very low intensity or undetectable and occurs beyond all the others on the B$_2$ portion of the spectrum.

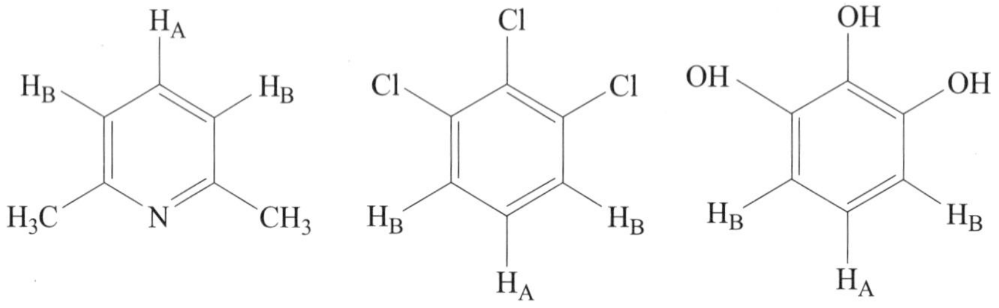

I II III

The precise appearance of the spectrum depends on the ratio of $\nu_0\delta/J$. The chemical shifts generally lie near the centre of gravity of the multiplets. The calculated AB$_2$ spectra for varying $\Delta\nu/J_{AB}$ from 10.0 to 0.5 are shown in Fig. 12.23.

The AB$_2$ spectrum is readily analysed when eight or nine lines can clearly be observed using the following relations.

$$\nu_A = \nu_3 \tag{12.8}$$

$$\nu_B = (\nu_5 + \nu_7)/2 \tag{12.9}$$

$$J_{AB} = [(\nu_1 - \nu_4) + (\nu_6 - \nu_8)]/3 \tag{12.10}$$

where ν_1, ν_2......, etc. are the positions of the spectral lines in Hz from TMS (see Fig. 12.23).

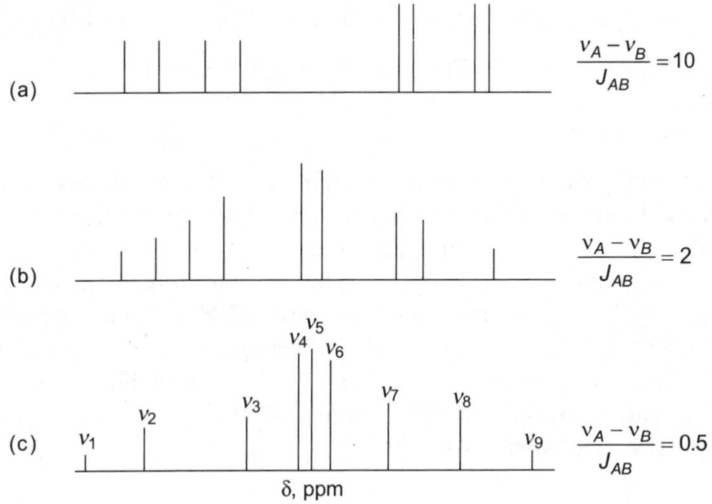

Fig. 12.23 Dependence of the AB_2 spectrum on the ratio of $\nu_o\delta/J$ of (a) 10.0, (b) 2.0 and (c) 0.5.

Since the coupling between the magnetically equivalent *B* nuclei does not influence the spectrum, the appearance of AB_2 spectrum is dependent only on the ratio $\nu_0\delta/J_{AB}$ and thus the spectral frequencies and their intensities in such a spectrum can be obtained on the basis of this ratio. The spectrum of pyrogallol is shown in Fig. 12.24 as a typical AB_2 spectrum.

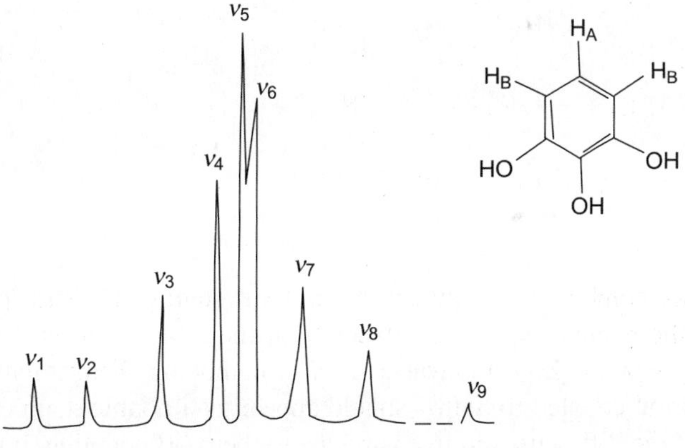

Fig. 12.24 ^{1}H NMR spectrum of pyrogallol.

As an example, consider an AB_2 spectrum which showed lines at $\nu_1 = 87.7$, $\nu_2 = 91.7$, $\nu_3 = 95.0$, $\nu_4 = 99.0$, $\nu_5 = 102.3$, $\nu_6 = 103.3$, $\nu_7 = 107.3$ and $\nu_8 = 110.0$

Hz. Let us determine the position of the lines v_A and v_B and the coupling constant J_{AB}.

The application of the above relations (12.8) – (12.10) yield,

$$v_A = v_3 = 95.0 \text{ Hz} \quad v_B = (102.3 + 107.3)/2 = 104.8 \text{ Hz}$$

$$J_{AB} = [(87.7 - 99.0) + (103.3 - 110.0)]/3 = 6.0 \text{ Hz.}$$

(iii) ABX System

Finally, the spin system, ABX is considered. Here, the X nucleus is weakly coupled to A and B and the A and B nuclei are strongly coupled. Such systems are characterized by three resonance frequencies, ν_A, ν_B and ν_X and three coupling constants J_{AB}, J_{AX} and J_{BX}. The ABX system is intermediate in complexity between AMX system (first order analysis) and ABC system (second order analysis). The ABX spectra occur frequently for example in trisubstituted aromatic compounds. Some examples of ABX system are 1, 2, 4-trichlorobenzene (I), 2-chloro 3-amino pyridine (II), 2-iodothiophene (III), 1, 2-dibromo 1-phenylethane (IV) and vinyl acetate (V).

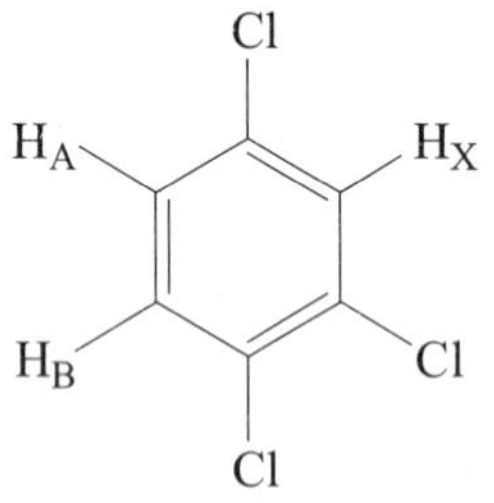

I

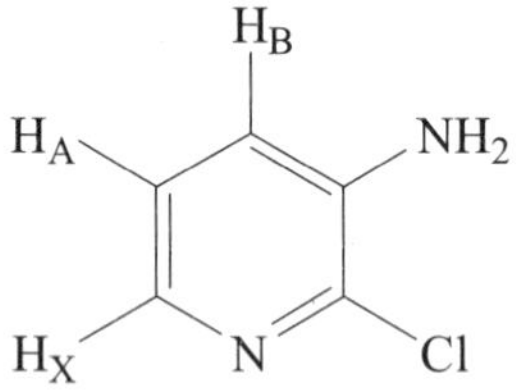

II

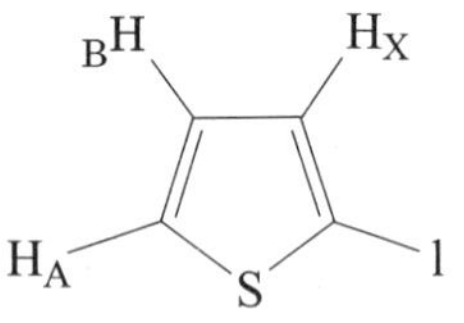

III

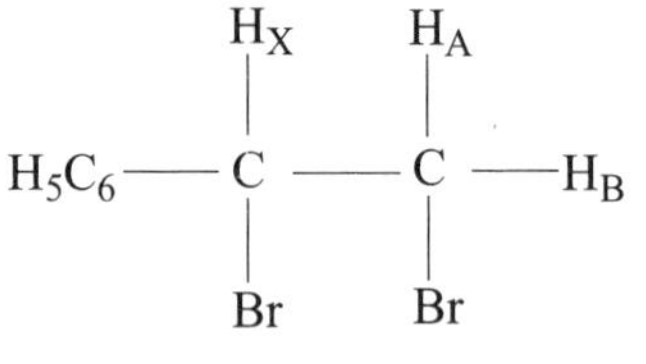

IV

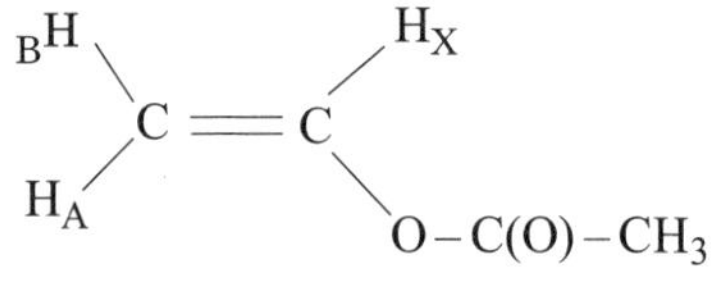

V

If we combine the treatment of an AB system with a simple first order splitting from the nucleus X, we can expect that the four AB lines will each be split into doublets by the X proton while the X proton being chemically well- shifted from A and B but coupled to both—should appear as a double doublet. The AB lines will inherit the differences in intensity due to their AB coupling, but the double doublet of the X proton should have four nearly equal intensity lines as shown in Fig. 12.25, which was computed for the parameters shown. Thus a first order approach to an ABX system predicts a twelve line spectrum with the X-region consisting of a symmetrical quartet and the AB region of two asymmetrical quartets. It should be noted, however, that this pattern of lines for an ABX spectrum is only one of the several possible patterns. There may be considerable overlapping of lines in the AB region with changes in the number of lines and their intensity distribution. There may also be changes in the position and in the number of lines in the X-region.

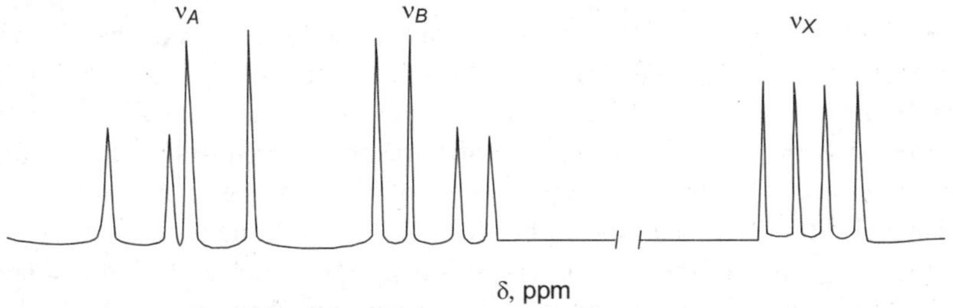

Fig. 12.25 Theoretical ABX spectrum calculated using the parameters : $\nu_A - \nu_B$ = 15.0 *Hz*, J_{AB} 5.0 Hz, J_{AX} 4.0 Hz and J_{BX} 2.0 Hz.

The spectrum of an ABX system often consists of twelve lines although it may contain upto fourteen lines or sometimes less than twelve lines if some of the lines overlap. The spectrum of an ABX system may be distinguished from that of an AMX system by the distortion of peak intensities in the ABX spectrum. Figure 12.26 shows the spectrum of 2-chloro 3-amino pyridine as typical of the ABX spectrum. The ABX spectrum may be analysed approximately as a first order AMX case. As $(\nu_A - \nu_B)$ decreases, there is greater departure from the first order spectra. In Fig. 12.26, the lines arising from the nucleus A and those from B can still be recognised by comparison with the spectrum in Fig. 12.25 but the transitions cannot strictly be called A or B transitions.

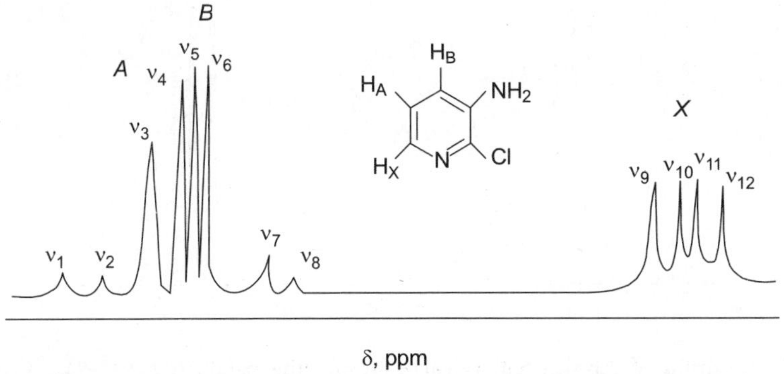

Fig. 12.26 The ^{1}H NMR spectrum of 2-chloro 3-amino pyridine as a representative of the spectra of ABX spin system.

The patterns of ABX spectra discussed above and more complicated versions of the same often occur in the NMR spectra of organic compounds. The ABX system represents the simplest spin system in which the relative signs of the coupling constants affect the appearance of the spectrum. Thus the relative signs of the coupling constants J_{AX} and J_{BX} can influence the appearance of an ABX spectrum. The appearance of the spectrum is insensitive to the sign of J_{AB}. In the AB and AB_2 cases, the spectrum is independent of the sign of the spin-spin coupling constant.

The analysis of an ABX spectrum discussed above was based on the assumption that all the eight lines of AB and either four or six lines of X are observed. Frequently some lines may coincide giving a spectrum which is not typical of ABX pattern. For example, in the spectrum given in Fig. 12.27, the AB portion shows only six lines. The appearance of an ABX spectrum is influenced by the chemical shift difference $\nu_A - \nu_B$ between the nuclei A and B. The AB portion as well as the X portion of the spectrum is sensitive to this parameter. For example, as ν_B approaches ν_A, the appearance of the spectrum changes drastically. When $\nu_B = \nu_A$, the AB region appears to be simply a doublet while the X region is a 1 : 2 : 1 triplet. It closely resembles the first order A_2X spectrum as shown in Fig. 12.28. Actually there are other weak satellites which are generally lost in the noise. The spectrum is an example of "deceptively simple spectra". Another example of a deceptively simple spectrum is the proton spectrum of 2, 5-dichloronitrobenzene (AA′X system). The AA′X spectrum is independent of the value of $J_{AA'}$. In some cases, it may lead to erroneous conclusions. Deceptively simple spectra are wide spread and not limited to ABX systems. The best way of resolving a deceptively simple spectrum is to obtain the spectrum at different applied magnetic fields or in different solvents.

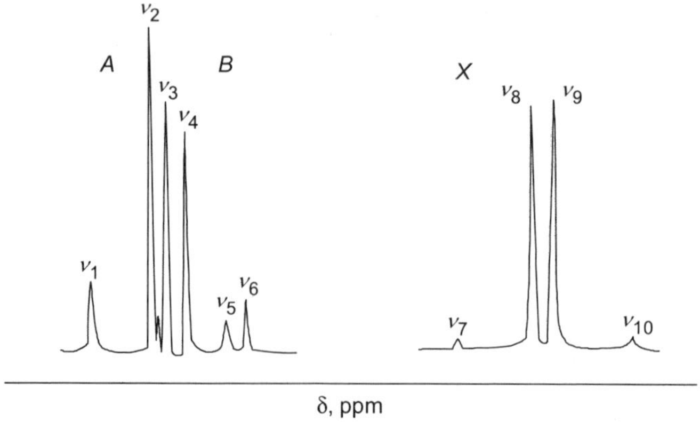

Fig.12.27 Spectrum of ABX system calculated using the parameters $\nu_A - \nu_B = 10.0$ Hz, $J_{AB} = 10.0$ Hz, $J_{AX} = 1.0$ Hz and $J_{BX} = -4.0$ Hz.

The methyl group of a long alkyl chain, ...CH_2 CH_2 CH_2 CH_3 should appear as a simple triplet in the spectrum. However, generally it occurs as a very blurred triplet. For example, the —CH_3 resonance of 1-hexanol is a broadened and

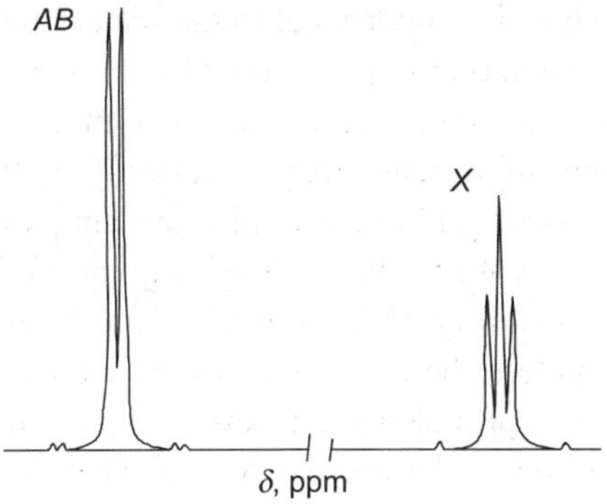

Fig. 12.28 Spectrum of ABX system calculated using the parameters $\nu_A - \nu_B$ = 10.0 Hz, J_{AB} = 10.0 Hz, J_{AX} = 1.0 Hz and J_{BX} = –4.0 Hz.

distorted triplet at δ 3.63 ppm. The broadening of the CH_3 resonance is attributed as due to its interactions with several non-neighbour CH_2 groups, even though there is no true coupling with these groups. True coupling occurs with the adjacent CH_2 group but this group serves to link spin-spin interactions with other CH_2 groups to the terminal CH_3 group. Such a higher-order effect is known as "virtual coupling". It leads to distortions in many first order spectra. Thus, in general, for virtual coupling to occur say in protons *P* they must be strongly coupled to their neighbouring protons *Q* which must in turn be strongly coupled to their neighbouring protons *R*; the coupling between *P* and *R* must also be zero.

ORIGIN OF SPIN-SPIN COUPLING

We had made a statement earlier about electron mediated spin-spin coupling. The chief source of spin-spin coupling in the molecules is thus an indirect interaction through the electrons involved in chemical bonding. There are three components of a coupling constant arising from the nucleus-electron interactions. Firstly, the magnetic moment of one nucleus interacts with the field produced by the orbital motion of the electrons, which in turn interacts with a second nuclear moment. Secondly, there is dipolar interaction involving the electron spin magnetic moments. The final contribution arises from the spins of electrons in orbitals (derived from *s* atomic orbitals) which have a non-zero probability of being found at the nucleus. This is known as the Fermi contact interaction and it is the most important term for proton-proton couplings but for other nuclei the situation is not so simple. All these terms depend on the magnetogyric ratios of the two nuclei involved.

(i) Contact Interaction

The electron has a strong dipolar magnetic interaction with the nearby nucleus. If it is purely anisotropic, it should average to zero for a molecule undergoing rapid tumbling motion in solution. This is so, except at electron nuclear separations comparable to the radius of the nucleus (10^{-14} m). At very small separations, the dipolar interaction of an electron and a nucleus is replaced by an 'isotropic' interaction known as the Fermi contact interaction, which is proportional to the scalar product of the two magnetic moments. Contact interaction is proportional to $-\gamma_e \gamma_n$ **I.S**, where **I** and **S** are the nuclear and electron spin angular momentum vectors, respectively. Since the electron has a negative magnetogyric ratio ($\gamma_e < 0$), a nucleus with $\gamma_n > 0$ is stabilised if the electron and nuclear spins are antiparallel and destabilised if they are parallel as shown in Fig. 12.29. The magnitude of the interaction is also proportional to the probability of finding the electron at the nucleus ($R = 0$) and therefore vanishes unless the electronic wave function has some *s* electron character. Since *p*, *d*, *f*, ... orbital wave functions have a node, there is no probability of finding an electron at the nucleus $R = 0$. That is, this isotropic interaction allows an electron to sense the orientation of the nearby nuclear spin. The contact interaction produces hyperfine splitting in the electron spin resonance (ESR) spectra of paramagnetic materials as discussed earlier (see Chapter 3). It also provides a mechanism for spin-spin coupling between pairs of nuclei.

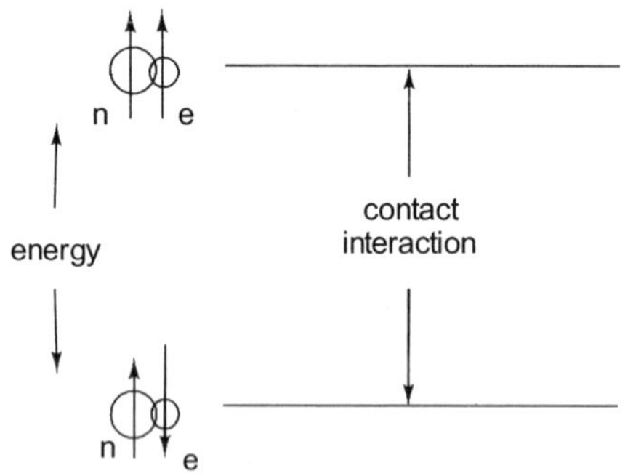

Fig. 12.29 Energy levels of an electron (*e*) coupling with a spin 1/2 nucleus (*n*) with $\gamma_n > 0$, by Fermi contact interaction.

The coupling of the nuclear spin to the electron spin by the Fermi contact interaction is the most important factor for proton spins but it is not necessarily the most important mechanism for other nuclei. These nuclei may also interact by a dipolar mechanism with the electron magnetic moments and with their orbital motion, and hence it is difficult to predict the sign of coupling constant.

To account for the magnitude of $^2J_{HH}$ as in CH_2, a pathway is needed that can transmit the spin alignments through the central C atom which may be ^{12}C with $I = 0$. In this case a H(1) nucleus with α spin polarizes the electrons in its bond and an electron with α spin is likely to be found closer to the C nucleus. The more favourable arrangement for two electrons on the same atom is with their spins parallel (Hund's rule), so the more favourable arrangment is for the electron of α spin of the neighbouring bond to be close to the C nucleus. Consequently, the electron with β spin of the bond is likely to be found nearer to the H(2) nucleus and hence that nucleus will have a lower energy if it is α. According to this mechanism, a lower Larmor frequency of H(2) will be obtained if its spin is parallel to that of H(1), i.e. $^2J_{HH}$ is negative.

The coupling in a CH_2 group may be pictured in terms of the interaction between two-centre two-electron bonds (Fig. 12.30). The spins of the electrons in one of the CH bonds are polarised via their contact interaction. The spin-spin coupling is rather more complex than the model just described. The signs of the coupling constants cannot, in general, be predicted. However, the appearance of the NMR spectra is usually independent of the sign of the coupling constants.

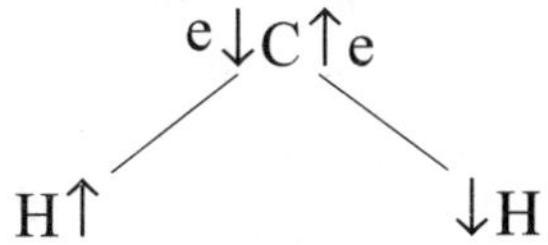

Fig. 12.30 A simple model of scalar coupling in a CH_2 group in a delocalised molecular orbital. The most stable configuration of nuclear and electron spins is shown.

(ii) Dipolar Interaction

Direct polar interactions between nuclei are important for understanding of the solid state NMR and relaxation processes which are discussed later (Chapter 18). In a system containing a number of magnetic nuclei ($I > 0$), direct dipole-dipole interactions between nuclear magnetic moments are expected. Due to dipole-dipole interaction, there will now be an additional field at a given nucleus. This field either opposes or adds to the applied magnetic field depending on how the other nuclei are situated with respect to the given nucleus. In the case of liquids, because of the rapid tumbling motion of the molecules, this interaction averages out to zero. However, dipole-dipole interactions are very important in solids (see Chapter 18).

AIDS IN THE ANALYSIS OF SPECTRA

When the spectrum becomes complex, there are several techniques, which can aid the analysis of the spectrum to obtain chemical shifts and spin-spin coupling constants.

(i) Varying Magnetic Field

Spin coupled spectrum becomes increasingly first order as the applied magnetic field B_0 is increased as noted above, since the chemical shift is directly proportional to the magnetic field strength while the spin-spin coupling constant is independent of it. There has been a continued trend towards higher magnetic fields in high resolution NMR. The commercial instruments have progressed from 60 to 950 MHz. With field strengths of 14.1 and 20.6 T, spectrometers with ^1H frequencies of 600 and 750 MHz, respectively are in use for synthetic and biopolymers. At higher field strength smaller chemical shift differences for groups of similar structure can be recognised. More information is thus available. A sensitivity enhancement is achieved by increasing B_0 since the Boltzmann distribution between the spin levels becomes more favourable (also see Chapter 13).

(ii) Isotopic Substitution

Analysis of a complex spectrum may be made from a comparison with the spectrum of an isotopically substituted molecule. The most important isotopic substitution is the replacement of hydrogen by deuterium. Isotopic substitution has only a very minor effect on the electronic structure of the molecule so that the chemical shift of the remaining nuclei will be practically unaltered. Deuterium is of use both for identifying proton signals, which disappear and for simplifying the spectra that remain. The spin coupling with deuterons (I = 1) is often eliminated by quadrupole relaxation. The coupling interaction is proportional to the product of magnetogyric ratios of the coupled nuclei.

$$J_{HX}/J_{DX} = \gamma_H/\gamma_D \sim 6.54 \tag{12.11}$$

Another advantage of isotopic substitution is the determination of coupling constants among the equivalent sets of nuclei. For example, coupling constants between nuclei in an equivalent set (H_2, CH_4, etc.) cannot be determined from the NMR spectrum because they give only one signal. By replacing one or more nuclei of the equivalent set by an isotope, the coupling constants within the equivalent set could be determined. For example, the proton spectrum of HD molecule gives a triplet with J = 43 Hz. The value of J in H_2 molecule is calculated as $6.54 \times 43 = 281$ Hz. Substitution by ^{15}N may sometimes be used to avoid broadening effects of the quadrupolar ^{14}N nucleus.

For simplifying the spectra spin decoupling methods are discussed later.

(iii) Computation of Spectra

Nowadays computers are used in the analysis of the complicated spectra arising from systems with a larger number of nuclei. The program LAOCOON (least squares adjustment of calculated on observed NMR transitions) developed by Castellano and Bothner-By is generally used and forms the basis of computer software which is commercially available for spectral analysis. The calculated spectrum is compared with the experimental spectrum.

EXPERIMENTAL ASPECTS: NMR

13

FT NMR SPECTROMETER

Earlier NMR spectrometers (1945-1970) used continuous wave (CW) methods using a weak fixed amplitude radiofrequency radiation. Continuous wave (CW) means that during the recording of an NMR spectrum the frequency of the radiation from a weak *rf* transmitter is varied continuously. However, spectra were obtained usually by keeping the frequency of the radiofrequency radiation fixed and gradually varying the magnetic field strength so as to bring the spins having different chemical shifts sequentially into resonance. The measurements on the CW instruments are thus in the frequency domain involving the measurement of the signal amplitude as a function of frequency. The time taken to sweep through each NMR peak should be long enough compared to its relaxation times T_1 and T_2 to avoid the appearance of distorted resonance signals arising from non-equilibrium spin states. One limitation of the above procedure is that at any given moment, only the nuclei say protons resonating at a particular chemical shift undergo excitation at the appropriate value of the magnetic field and it is necessary to sequentially excite the protons that have differing chemical shifts in a given molecule. A given set of protons will therefore be scanned for only a very small fraction of the total scan time, the remaining sweep time is actually wasted in recording the base line noise. Further, it is often necessary to distinguish splitting (arising form spin-spin coupling) that is only a fraction of a hertz in width. In order to distinguish spectral features separated by $\Delta\nu$ Hz , we need to spend a time $1/\Delta\nu$ s measuring each element of the spectrum of width $\Delta\nu$. If the spectral range is W, then the minimum time needed to record the spectrum would be $W/\Delta\nu$ s. Thus on a 500 MHz spectrometer, for a spectral width of 10 ppm (5,000 Hz) to obtain a resolution of 0.1 Hz would require 5,000/0.1 = 50,000 s (nearly 14 h) if the spectrum were to be recorded in the CW mode. This clearly imposes severe time limitation in CW spectrometers making it rather impractical at higher field strengths. The CW method is now completely replaced by the 'pulse techniques', which employ short intense pulses of the radiofrequency radiation of duration typically 10-50 μs. In the pulse Fourier transform (FT) NMR, the sample is subjected to a strong burst of *rf* pulse and here our *rf* field is polychromatic. Thus, the basis of FT NMR spectroscopy is polychromatic signal excitation.

A brief description of the NMR spectrometer as well as a pictorial model of magnetic resonance phenomena is given here. A schematic diagram of the Fourier transform NMR spectrometer is shown in Fig. 13.1. The major components of a NMR spectrometer are: (i) a magnet, (ii) a radiofrequency transmitter, (iii) NMR probe, (iv) a radiofrequency receiver and (v) a recording device (a computer). The method will be confined to NMR of liquid samples since for solids, the technique is somewhat different, and it is briefly dealt with later.

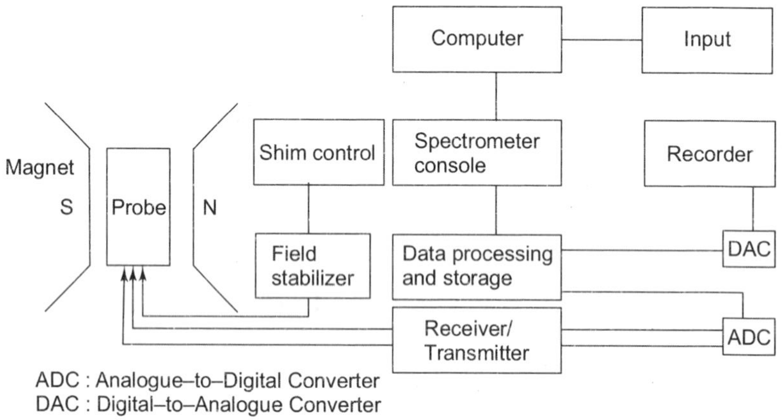

Fig. 13.1 Block diagram of a F T NMR spectrometer. The probe is parallel to the Z-axis of the magnet.

(i) Magnet

The basic requirements of a magnet are that it must produce a field that is strong, stable and highly homogeneous. The advantages of using a higher magnetic field are the following.

(a) Higher sensitivity : The intensity of the NMR signal increases with increased separation between the energy levels, which is determined by the applied magnetic field B_0. By increasing B_0 the Boltzmann distribution of population between the spin levels becomes more favourable. For biopolymers such as proteins or nucleic acids, NMR spectrometer with high magnetic fields is indispensible.

(b) The separation between the resonance signals i.e. the chemical shift differences (in Hz) increases linearly with B_0 while the splitting due to spin-spin coupling is independent of B_0 and hence higher magnetic fields reduce the overlap of NMR multiplets. That is, the spectral resolution increases with increasing spectrometer frequency, while the line width is governed by the relaxation times. An extremely high and uniform magnetic field is required for very high resolution.

(c) Higher magnetic fields give rise to first order spectra by minimising the coupling effects. For example, in contrast to a complicated ABC spectrum observed at 60 MHz for acrylonitrile, at 220 MHz, an AMX spectrum that can be analysed as a first order spectrum is obtained. The spectral resolution greatly improves and smaller chemical shift differences for groups of similar structure can be recognized.

(d) The use of higher field strengths is advantageous for the study of dynamic processes (see Chapter 14).

(e) The ratio of the height of a signal to the noise is called the signal to noise ratio. The simplest way to define sensitivity is by the signal to noise (*S/N*) ratio. The *S/N* ratio can be improved by signal averaging (*vide infra*). Resolution is generally determined by the width of the signal at half-height in Hz (see Chapter 14, Fig. 14.3).

The sensitivity of a nucleus towards NMR experiment depends on the magnitude of its magnetic moment μ which determines the energy difference between the nuclear spin states and hence the excess population in the lower energy state. At a field B_o, the signal strength is roughly proportional to

$$\frac{I+1}{I^2}\,\mu^3\,B_0^{\;2}$$

In practice, the dependence on $B_0^{\;3/2}$ has been found; it shows the importance of strong fields in improving the signal to noise ratio. If we compare a 60 MHz spectrometer with a 400 MHz instrument, which is equivalent to changing the magnetic field from 1.4 to 9.4 T, the signal to noise ratio increases by 1:800. These relations were the driving force behind the development of superconducting magnets, which are used to obtain much higher fields (above 5 T). Magnetic field strengths have now reached 22.32 T which corresponds to a ^{1}H resonance frequency of 950 MHz.

In recent years, superconducting magnets (solenoids) operating at liquid helium temperature are employed. It is intrinsically very stable and more economical than the electromagnets. Superconducting magnet requires regular supplies of liquid nitrogen and liquid helium. Since there is no resistance in the coils of the magnet, once the current is fed into it, it flows undiminished for years and no external source of power is needed. The coils are placed in liquid helium, its dewar is insulated from the atmosphere by another vacuum vessel which itself is placed within a liquid nitrogen dewar.

The natural abundance is a critical factor (see Table 10.1). In the earlier years, the NMR spectroscopy for example of ^{13}C was greatly hampered due to its low abundance.

The following are some of the other common methods for enhancement of sensitivity.

(i) Sample concentration: When the sample concentration in a given volume of the solvent increases, the number of NMR active nuclei also increases.

(ii) Temperature: Lowering of temperature increases the Boltzmann population difference between the ground and excited states.

(iii) *rf* power: The signal strength increases with increase in *rf* power. Care must be taken to avoid saturation.

(iv) Number of scans: The signal to noise ratio increases as the square root of the number of accumulated scans.

(v) Polarization transfer and NOE effects which are discussed later also contribute to the sensitivity enhancement.

(vi) Relaxation rate: Slow-relaxing nuclei cannot attain thermal equilibrium within the pulse interval which results in lowering of sensitivity. One way to deal with this problem is to introduce a sufficiently large delay time between the pulses.

The compound to be investigated is contained in a sample tube—a glass tube of nearly 15 cm length and 5 to 10 mm in outer diameter—between the pole faces. The pole faces of the magnet are usually 20–30 cm in diameter and the pole gap is 2–3 cm. The pole faces are machined to optical flatness. The sample tube fits the bore of the solenoid and spins about its vertical axis which is parallel to the *z*–axis. The probe is placed at the centre of the magnet. It houses the coils for transmitting the radiofrequency power to the sample and receiving the re-radiated energy. The transmitter and receiver are coupled through the sample nuclei.

In the FT experiment, pulse excitation always occurs at a constant magnetic field. Homogeneity of the magnetic field should be such that the signals have a natural line width. Homogeneity better than one part in 10^9 is needed to get line widths of about 1 Hz on high field say 500 MHz spectrometers. Any variation in the magnetic field experienced by the sample would lead to a spread of the resonance frequencies resulting in line broadening. Homogeneity is achieved by using a small size of the sample since it is easier to achieve homogeneity over a small area. However, large sample sizes are needed for nuclei with small magnetic moments. Stringent homogeneity is achieved by passing d.c. currents through small electrical coils referred to as "shim coils" which produce small magnetic fields. Shim coils are located on the pole faces of the magnet with geometries suitable for generating magnetic field gradients of various forms, which cancel the field gradients produced by the magnet itself. Considerable improvement in the effective homogeneity is achieved by spinning the sample tube about its own axis at rates up to 50 Hz. The sample tube is spun using a sample holder which acts as a turbine driven by compressed air. The frequency of spinning should be greater than the rate at which the magnetic field fluctuates. The sample behaves as though it experiences only an average field rather than a range of fields. The temperature must be carefully controlled.

The spectral data have to be accumulated and stored over long periods to obtain acceptable signal to noise ratio. During this time, there may be small drifts in the magnetic field due to slight electrical resistance in the solenoid, variations in the room temperature and other influences. A highly homogeneous field is of little value if there are significant fluctuations in the magnetic field or the radiofrequency. Because of the Larmor relationship, the field and the frequency need not be individually held extremely constant; only their ratio is important. The required stability is attained by employing a field/frequency control procedure using the ^{2}H NMR signal of a deuterated solvent, e.g. $CDCl_3$, C_6D_6, etc. and any variation in it is used to control the strength of a small supplementary magnetic field which compensates for the drift in the main magnetic field. The deuterium lock prevents changes in the static magnetic field (B_0) and the radiofrequency field (B_1) by maintaining a constant ratio between the two. It ensures long-term stability of the magnetic field.

(ii) Radiofrequency Transmitter

By means of a pulse transmitter, the radiofrequency radiation is generated for the nucleus of interest. A waveform generator produces a continuous voltage oscillating sinusoidally at the required frequency. This output is chopped into pulses by a switch, which is opened and closed by a pulse generator. The spectrometer computer controls the pulse generator. The pulses are amplified and sent to the NMR probe. In order to permit double resonance experiments of different types, a second *rf* source of variable frequency should be available. The duration of the pulse affects both the range of frequencies that can be excited and the resultant nuclear magnetization.

To record the spectra of nuclei other than ^{1}H using the instrument of magnetic field strength say 2.349 T, a different radiofrequency source appropriate to the nucleus being examined is required. For example, with a 2.349 T magnet, ^{1}H requires 100 MHz, but ^{19}F spectra can be recorded using a radiofrequency source at 94.07 MHz while ^{13}C spectrum requires 25.14 MHz source.

(iii) NMR Probe

The probe at the centre of the magnetic field is an electrical device with a sample holder and contains the transmitter and receiver coils and a spinner. The radiofrequency is applied to the sample through the transmitter coil. The probe contains the electronics designed to detect the tiny NMR signal. The central component of the probe is a wire that receives the *rf* pulse from the transmitter and dissipates it into the sample. It also receives the signal from the sample and transfers it to the receiver circuit. At resonance a voltage is induced in the receiver coil, which detects sample magnetization. The probe holds and spins the sample about its verical axis in order to average out field inhomogeneities. Another frequency has to be transmitted and received to operate the field/frequency lock.

(iv) Computer

A prominent feature of the FT NMR spectrometer is a digital computer. Practically all the functions of the spectrometer are under its control. It plays a central role in the experimental set-up. The computer controls both the radiofrequency transmitter and the receiver, stores and processes the incoming data and transfers the results to the recorder. The computer may also be used to manipulate, compare, add, subtract or simulate spectra or to control the more complicated pulse sequences. Nowadays, the recording device is always a computer.

RADIOFREQUENCY PULSES

In CW NMR experiments relatively weak *rf* fields (B_1) of the order of a few tenths of a microtesla (μT) are used for excitation. In pulsed NMR experiments stronger B_1 fields in the range of 0.01–0.4 T are used. To avoid saturation of the spin system, these strong fields are applied only for a few microseconds in the form of radiofrequency pulses. The duration of the pulse affects both the range of frequencies that can be excited and the direction of the resultant nuclear magnetization. A sequence of *rf* pulses is shown in Fig. 13.2.

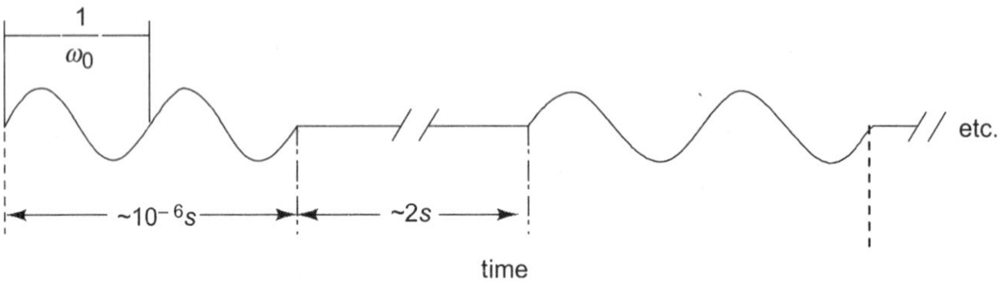

Fig. 13.2 A sequence of *rf* pulses.

A radiofrequency pulse is a burst of energy that is applied close to the resonance frequencies of interest by switching on the *rf* transmitter. A short intense radiofrequency pulse extending over the entire range of frequencies in which the nuclei of interest resonate is applied so that all nuclei falling within the region are excited simultaneously. As long as the pulse is on, a constant force is exerted on the sample magnetization causing it to precess about the *rf* vector. The net magnetization along the *Z*-axis (B_0) can be deflected in different directions by applying the pulse along one of the axes, that is along $+x$, $-x$, $+y$ or $-y$ axis. It induces (x, y) – magnetization. The extent by which the $+Z$ magnetization is turned is controlled by choosing pulse of suitable duration. Thus the term '90° (or $\pi/2$) pulse' actually refers to the duration of the pulse applied to turn the *Z*-magnetization vector by 90° as shown in Fig. 13.3. A subscript such as *x* or *y* is usually placed after the pulse angle as a subscript to indicate the direction in which the pulse is applied. A 90°_x pulse therefore refers to a pulse along the $+x$ axis to rotate the nuclear magnetisation by 90°. A 90° pulse equalizes the two populations and also converts the equilibrium magnetization completely into y magnetization, M_y. If it takes t μs to turn the equilibrium magnetization by 90°, it would need half that time i.e. $t/2$ μs to turn the magnetization by 45°. A 180° (or π) pulse, on the other hand, would require double that time, i.e. $2t$ μs to cause the Z-magnetization to become inverted so that it lies along the –Z axis. Thus different flip angles are achieved by the appropriate choices of the duration of the pulse. For a typical radiofrequency field strength, the duration of 90° pulse is of the order of 10 μs. The most commonly used pulses have 90° or 180° flip angles. A 180° pulse inverts the population difference Δn by interchanging the populations of the $m = + 1/2$ and $m = - 1/2$ states (Fig. 13.2). This will lead to a Boltzman excess of nuclei in the upper energy state and the spin system is then said to possess a 'negative spin temperature'. When the pulse is switched off, the relaxations that occur immediately after the excitation process are measured as exponentially decaying waves known as the free induction decay (FID) and converted to the usual NMR spectrum by Fourier transformation.

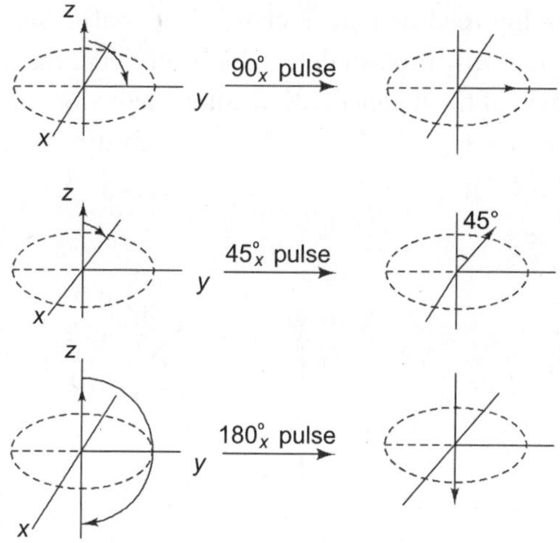

Fig. 13.3 (a) The position of the magnetic moment vector M in the rotating coordinate framework after the application of *rf* field B_1 in x-direction: (a) 90°_x pulse, (b) 45°_x pulse and (c) 180°_x pulse.

Although signals of maximum intensity result from a 90° pulse, small flip angles are advisable for data accumulation in order to reduce the recovery time for the *Z*-magnetization that is governed by the spin-lattice relaxation. The pulses used in FT NMR spectroscopy generally extend over the whole frequency range. These pulses are called hard pulses. In a number of cases, however, it is desirable to use soft pulses with a small frequency range, for instance, if solvent signals are to be saturated or if certain resonances are to be excited selectively.

The direction in which the nuclear magnetization is turned by a particular pulse is controlled by the direction in which the pulse is applied. It could be predicted by a simple rule. A pulse applied along a certain axis causes the magnetization to rotate about that axis in a plane defined by the other two axes. For example, a pulse along the *x*-axis will rotate the *Z*-magnetization around the *x*-axis in the *yz*-plane. Similarly, a pulse along the *y*-axis will rotate the bulk magnetization in the *xz*-plane. The final position of the magnetization will depend on the duration the *rf* pulse applied. Modern FT NMR spectrometers are equipped with pulse transmitters which can generate transverse magnetization in any direction of the (*x*, *y*)-plane in the laboratory frame.

The flip angle θ (in radians) by which the magnetization vector is deflected due to the pulse is given by

$$\theta = \gamma B_1 t_p \tag{13.1}$$

where B_1 is the magnetic field generated along the y-axis at right angles to the applied magnetic field and t_p is the duration of the pulse in seconds. The time t_p is usually chosen such that $\theta = 90°$. Since the spectrometers are normally set up to detect the components lying along the y-axis, only this component of the total magnetization will be recorded as signals. The pulse serves to convert the longitudinal (or z) magnetization to detectable transverse magnetization along the y-axis. The magnitude of the magnetization component along the y-axis is given by $M_0 \sin \theta$, where θ is the angle by which the magnetization is deflected away from the z-axis (see Fig. 13.4).

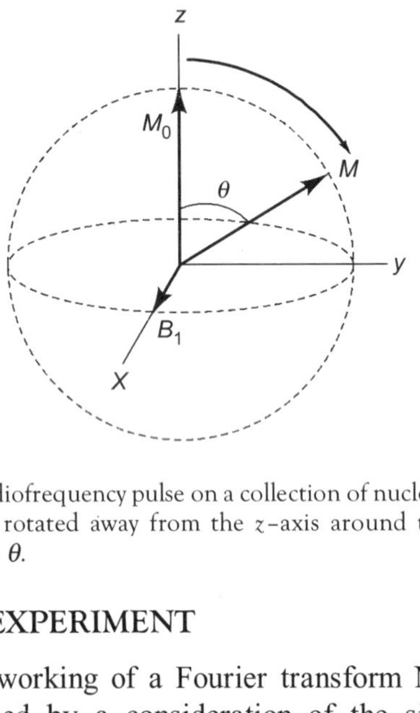

Fig. 13.4 The effect of a radiofrequency pulse on a collection of nuclear spins at equilibrium. The magnetization is rotated away from the z-axis around the direction of B_1 (y-axis) through an angle θ.

THEORY OF NMR EXPERIMENT

An insight into the working of a Fourier transform NMR spectrometer can conveniently be obtained by a consideration of the simple vector model of magnetic resonance.

LARMOR PRECESSION AND RELAXATION

Consider a sample consisting of a large number of identical non-interacting spin 1/2 nuclei. As noted earlier (Chapter 10), a nucleus of spin 1/2 in a strong magnetic field B_0 exists in one of the two orientations with equal and opposite projections on to the direction of the applied magnetic field (z-axis). There is a slight excess of nuclei in the $m = +1/2$ state which is lower in energy than that in the $m = -1/2$ state for a nucleus whose magnetogyric ratio is positive. The bulk magnetic moment vector of magnitude M_0 of a very large number of spins in the sample can be denoted by an upward arrow along the positive z-axis as shown in Fig. 13.5 with the length of the arrow proportional to the population difference. The nuclear spins in the sample, on an average, will be spread evenly around the cone describing precessional motion around their cone at a rate determined by the applied magnetic field. Depending on its magnetic quantum number ($m_I = \pm 1/2$) each spin contributes ($\pm$ 1/2) $\gamma(h/2\pi)$ to the total Z magnetization of the sample, where γ is the magnetogyric ratio. At thermal equilibrium, the total magnetization M_0 of the sample aligned along the positive + z direction is

$$M_0 = (1/2)\ \gamma\,(h/2\pi)\Delta n_{eq} \tag{13.2}$$

where the population difference is Δn_{eq}. In the perpendicular x and y directions however, the phases of the individual nuclear magnetic moments are random because there is no transverse magnetic field to align them and their vector sum vanishes (Fig.13.5).

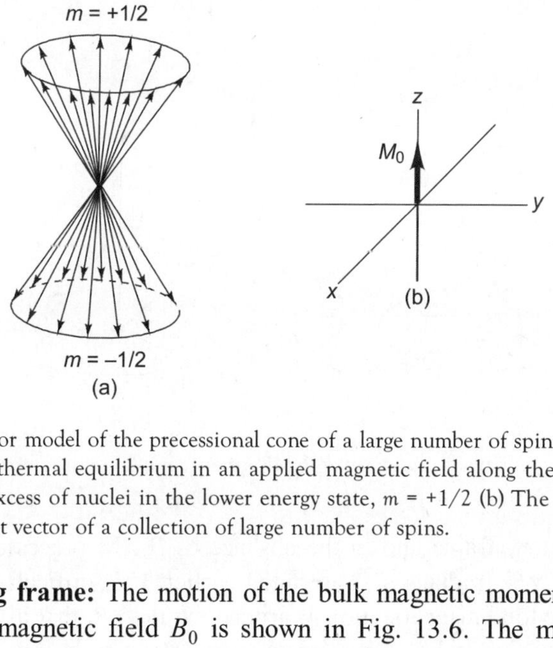

Fig. 13.5 (a) Vector model of the precessional cone of a large number of spinning nuclei of spin 1/2 at thermal equilibrium in an applied magnetic field along the z-axis. There is a slight excess of nuclei in the lower energy state, m = +1/2 (b) The net bulk magnetic moment vector of a collection of large number of spins.

The rotating frame: The motion of the bulk magnetic moment vector M_0 of the spins in a magnetic field B_0 is shown in Fig. 13.6. The magnetization M displaced from its position along the z-axis, experiences a force due to the applied magnetic field B_0 causing it to undergo Larmor precession around the z-axis at a frequency of γB_0 radians per second (or $\gamma B_0/2\pi$ Hz). The stronger the magnetic field and larger the magnetogyric ratio of the nucleus, the faster is the precession. In the NMR experiment, this precessional motion of the nuclei is measured. One way to do this is to displace the magnetization from its position along the z-axis by the application of a *rf* pulse. The magnetization M deflected away from the z-axis will have a certain component in the xy-plane; it is this component that is measured as the NMR signal.

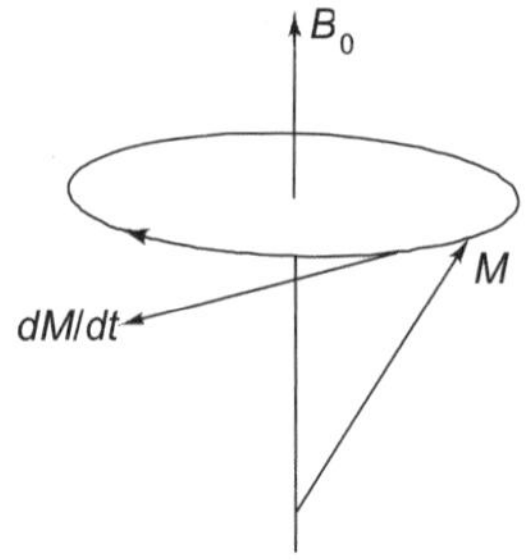

Fig. 13.6 The precessional motion of magnetization vector M in the applied magnetic field B_0.

In an NMR experiment, there is a strong static magnetic field B_0 along the z-axis and a relatively weak radiofrequency field $B_1(t)$ which rotates in the xy-plane at a frequency ω_{rf}. The total field B(t) is the vector sum of B_0 and $B_1(t)$. It is tilted slightly away from the z-axis and rotates around it with the frequency ω_{rf} as shown in Fig. 13.7 (a).

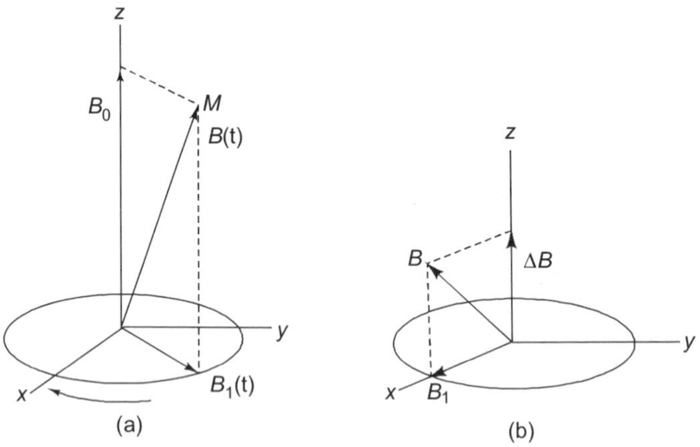

Fig. 13.7 The magnetic fields present in an NMR experiment: (a) in the laboratory frame and (b) in the rotating frame.

In the laboratory frame and in the absence of B_1, M precesses around B_0 at frequency $\omega_0 = \gamma B_0$ (radians s^{-1}) as stated earlier. It is difficult to visualize the precession of the total magnetization M around the field B_1 that itself is moving. In order to simplify the problem, we imagine ourselves to be rotating around the z-axis along with the radiofrequency field, i.e. at frequency ω_{rf}. In this rotating frame, the radiofrequency field appears stationary say along x-axis. The concept of rotating frame is convenient since NMR spectrometers detect offset frequencies ($\Omega = \gamma\Delta B$) rather than the actual resonance frequencies (ω_0). If ω_0 were identical to the radiofrequency ω_{rf}, then M like B_1 would appear stationary in the rotating frame. In general, the frequency with which M precess around the z-axis is reduced from ω_0 to

$$\Omega = \omega_0 - \omega_{rf} \tag{13.3}$$

Since $\omega = \gamma B$, an effective field ΔB existing along the z-axis in the rotating frame is

$$\Delta B = (\omega_0 - \omega_{rf}) / \gamma = B_0 - (\omega_{rf}/\gamma) \tag{13.4}$$

whose magnitude is determined by the difference between the precession frequency ω_0 and the radiofrequency ω_{rf}. In the rotating frame, M therefore precess around the resultant of B_1 and ΔB (Fig. 13.7*b*). Although B_0 is always much greater than B_1, the residual field ΔB may be comparable to B_1 if the transmitter frequency is close to resonance ($\omega_{rf} = \omega_0$). It is to be noted that ΔB becomes dependent on the chemical shift of the nuclei under consideration, when B_0 is replaced by the field actually experienced by the nuclei, $B_0 (1 - \sigma)$, where σ is the shielding constant. The precession frequency in the laboratory frame in the absence of $B_1(t)$ becomes $\gamma B_0 (1 - \sigma)$ which is the NMR frequency of the spins.

The most important feature of the pulsed NMR is the ability to excite simultaneously and uniformly nuclei having different chemical shifts (that is, different ΔB). For example, a typical range of ^{1}H resonance frequencies is 5 kHz in a 500 MHz spectrometer (= 10 ppm × 500); a 90° pulse of strength $\gamma B_1/2\pi >>$ 5 kHz will therefore rotate the magnetisation vectors of all the protons in the sample on to the y-axis, irrespective of their resonance frequencies. Starting from the equilibrium state with random phases, a pulse along the x-axis in the rotating frame causes the spins to come together to some exent producing a net y magnetization in the sample. This is known as coherence.

After a 90° pulse, when $B_1 = 0$, the only field remaining in the rotating frame is ΔB along the z-axis. M therefore precesses in the xy-plane at frequency (Fig. 13.8), equation (13.4)

$$\Omega = \gamma \Delta B = \gamma B_0 (1 - \sigma) - \omega_{rf} \tag{13.5}$$

where ω_{rf} is the radiofrequency field and the chemical shift has been introduced through the shielding constant σ. When several types of nuclei are excited, each gives rise to a magnetization that precesses with a frequency characteristic of its chemical shift. However, the spins will return to equilibrium once the pulse is switched off.

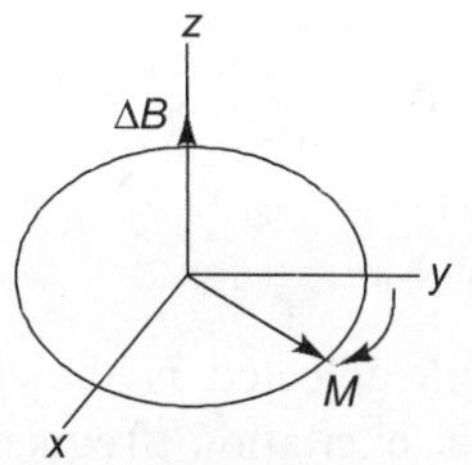

Fig. 13.8 Dependence of M following a 90° pulse. The magnetization precesses at frequency $\gamma\Delta B$ around the z-axis in the rotating frame.

The spin-lattice relaxation aids the recovery of the Z magnetization to its equilibrium value, while the spin-spin relaxation aids the decay of the xy magnetization to zero by randomization of the phases of the individual spins. The random local magnetic fields produce small time-dependent variations in the precessional frequencies of the individual spins which leads to loss of coherence in the sample as a whole.

Figure 13.9 (a) shows the variation of M during the interval following a 90° pulse. The longitudinal relaxation of M to M_0 follows a decreasing spiral.The variation of M with time is shown in Fig 13.9(b). The oscillating decay of transverse magnetization is detected by the NMR spectrometer through the voltage induced in the receiver coil on the x-axis which exponentially decays to zero with time constant T_2^*. This signal is known as the "free induction decay" (FID). The free induction decay curve is a plot of the decay of the x-y component of the magnetisation with time. It is an interferogram decaying with time (see Fig. 13.10). It is the sum of the individual oscillating voltages from the various nuclei in the sample each with characteristic chemical shifts and spin-spin coupling constants. It contains all the information required to obtain an NMR spectrum. The *rf* field B_1 is switched off during signal detection and hence the signal detected here is an emission signal. This yields a nearly linearly polarized *rf* field in the y-direction. In a FID, the amplitude decays asymptotically to zero while the precessional frequency remains unchanged. The amplitude represents the circular motion of the magnetization vector along the xy-plane which slowly relaxes back toward the z-axis. The Larmor frequency of the nucleus remains unaltered throughout the FID since it is inherent to the nucleus. Most of the relevant information is in the initial portion of the FID envelope. Fourier transformation of the FID yields the frequency domain, that is, the normal spectrum.

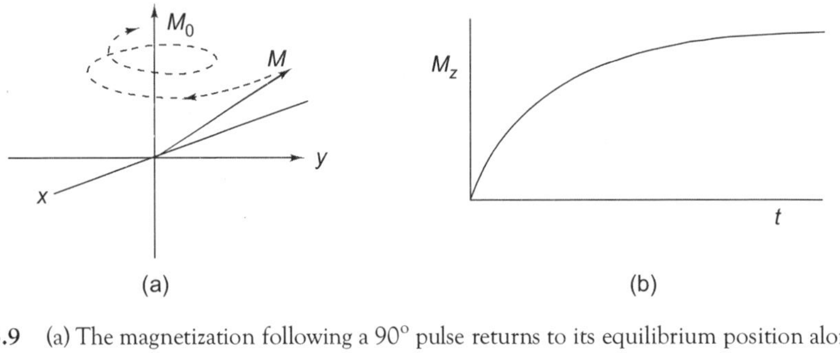

(a) (b)

Fig. 13.9 (a) The magnetization following a 90° pulse returns to its equilibrium position along the z-axis, longitudinal relaxation of M to M_0 follows a decreasing spiral and (b) variation of M_z with time.

The frequency of oscillation detected is the difference between the NMR resonance frequency and the excitation frequency, although the excitation frequency is only little different from the NMR frequency, the former can still excite the latter, since the excitation frequency is applied as such a short pulse (μs)

that it actually behaves as a spread of frequencies–as a consequence of the Heisenberg's uncertainty principle.

When the pulse is passed through the sample, the appropriate frequencies are absorbed by the sample causing the transitions. The pulse must therefore be short enough to cover the distribution of expected spectral frequencies. Accordingly $t_p << 4\pi\Delta\nu$, where $\Delta\nu$ is the range of frequencies utilized. For a typical compound, a pulse of 10^{-6} s is used and the free induction decay curve is converted into a series of data points and stored in the memory of the computer. The pulse is repeated again and the FID is measured and added to the earlier one in the computer. The process is repeated for a large number of pulses.

Interferograms of real NMR spectra are much more complicated since the spectra typically contain many NMR resonances over a larger range of frequency differences. Such interferograms are the 'free induction decay' (FIDs) curves as noted above. The Fourier transformation of the resulting FID curve gives the frequency spectrum (Fig.13.10).

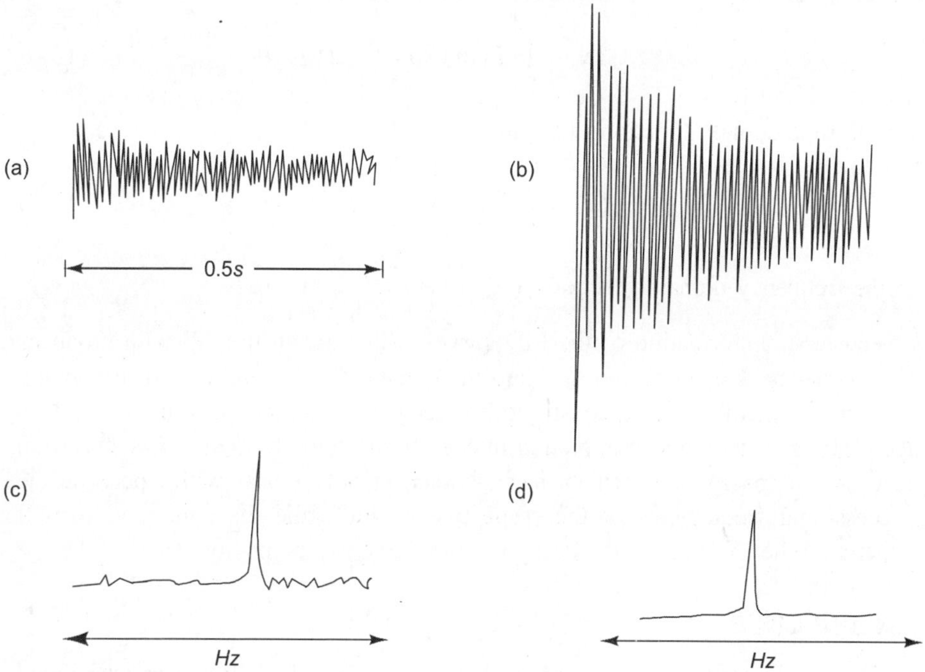

Fig. 13.10 The receiver signal of a single NMR peak : (a) in the time domain as free induction decay (FID), (b) sum of four FIDs shown in (a), (c) the Fourier transform of the single FID (a), and (d) the Fourier transform of (b) in the frequency domain. The improvement in the spectrum for four FIDs is two times.

THE NMR SPECTRUM

The detector is conventionally regarded as being set up to detect the signals along the y'-axis and hence the signal intensity will be maximum immediately after the 90° pulse. During the subsequent delay, the detector records the exponential decay of the signal with time in the form of a decaying beat pattern, the free induction decay (FID). The FID curve recorded on time scale contains all the information needed to characterize the NMR signal on the frequency scale. The FID curve is a representation in the time domain and the spectral trace is a function of time, $f(t)$. The other representation is on the frequency domain and the spectral trace is a function of frequency, $I(\nu)$. The data from NMR spectrometers are obtained in the time domain as FIDs, they are collected and stored in the computer as a function of time as mentioned above. The FID of an NMR spectrum is rather much complicated since it results from the superposition of the FIDs of the individual resonance lines including the noise. It is stored in the digital form in the computer memory and then transformed into the frequency domain spectrum. The conversion of the data from the time domain to the frequency domain is achieved by Fourier transformation, a standard mathematical process relating the time domain data $f(t)$ to the frequency domain data $I(\nu)$, Mathematically, the two representations of the receiver signal is expressed as,

$$f(t) = \int_{-\infty}^{\infty} I(\nu) \exp(-i2\pi\nu t)\, d\nu \tag{13.6}$$

for the time domain function $f(t)$ and

$$I(\nu) = \int_{-\infty}^{\infty} f(t) \exp(-i2\pi\nu t)\, dt \tag{13.7}$$

in the frequency domain function $f(\nu)$, where $i = \sqrt{-1}$.

Fourier transformation of the FID curves which are in time domain produces the frequency domain spectrum. The time domain signals and the corresponding frequency domain partners constitute Fourier pairs. Fourier transformation gives the NMR spectrum which is plotted in exactly the same form as a CW spectrum in which the positions at which the individual signals appear will depend on the precessional frequencies of the respective nuclei while the signal width will depend on the life time of the decaying transverse magnetization.

CALIBRATION

Spectra are recorded on calibrated chart paper. For ^1H NMR spectra, the signal from the standard is set on the zero line at the right hand side of the chart paper. The position of the resonance signal is measured relative to a reference compound or a standard. The usual standard for ^1H, ^{13}C and ^{29}Si spectra is tetramethyl silane abbreviated as TMS. If the reference reacts with the compound under study then an external standard may be used by placing a small sealed tube of the reference material inside the sample or by adding the material into the thin gap between the inner tube and the outer glass tube.

Calibration of a CW instrument is accomplished by modulating the reference signal by a known audiofrequency (e.g. 600 Hz) that is large compared to the width of the NMR lines. It results in the appearance of side bands at multiples of the audiofrequency. By altering the audiofrequency power applied, the intensity distribution in the side band could be varied. For most purposes, only the first side band is needed (see Chapter 11). The instrument calibration may also be checked by measuring the spectra of standard samples. In a FT NMR experiment, the spectral frequencies are directly available after data processing.

Most NMR spectrometers have built-in electronic integrators for measuring the relative areas under the peaks (integrated intensities). The integrated intensity of a band is a step function with the height of each step directly proportional to the area under the peak corresponding to that step (see Fig. 11.2).

ADVANTAGES OF FT NMR

Fourier transform NMR has several advantages. The principle advantage of FT NMR is its sensitivity. Excitation and detection of the whole spectrum in one step is much more efficient than sweeping slowly through the spectrum detecting one resonance at a time. In FT NMR spectroscopy all the nuclei in the spectral range of interest are excited simultaneously by a *rf* pulse. The FIDs can be obtained much more rapidly with certain relaxation time delay between the sucessive scans. For example, the continuous wave spectrum may take 100-1000 times longer time to obtain the spectrum than a single FID. The time saved can be used to improve the sensitivity by recording and adding together several hundreds or thousands of FIDs. The FIDs are recorded sequentially and they are loaded into the computer memory before Fourier transformation is carried out to obtain the spectrum. The intensity of NMR signals, which are identical in every FID builds up in proportion to the number of scans N recorded, while the electronic noise which varies randomly in each measurement adds up as $N^{1/2}$. The co-adding of FIDs N times therefore results in an improvement in the signal to noise ratio by $N/N^{1/2} = N^{1/2}$. Thus with 400 scans, an improvement in the signal to noise ratio by $\sqrt{400} = 20$ times is achieved. For a large number of transients, the improvement in sensitivity is very significant. Thus, greatly enhanced sensitivity in a very short measuring time is a major advantage of the pulse FT NMR. The time saving aspect of the pulse NMR is known as the 'multiplex or Fellgett advantage'. Improving the sensitivity by spectral accumulation of CW spectra requires long measurement time. The detection of many less sensitive and less isotopic abundant nuclei has become feasible with FT NMR. As a result of FT methods, obtaining NMR spectra of ^{13}C in natural isotopic abundance is now routine. Short-lived unstable molecules can be studied by pulse FT NMR.

Another major advantage of pulse NMR is that it opens up the possibility of a vast number of new experiments in which one can control the information contained in the FID by using pulse sequences designed to excite the spins in a specific manner. As examples, the determination of spin-lattice and spin-spin relaxation times and the two-dimensional NMR experiments which are discussed later may be cited.

SAMPLING PROCEDURE

The sample is taken in a thin walled 5 mm diameter pyrex glass tube to which approximately 0.5 ml of the sample solution is added. It is placed in the probe. Larger diameter tubes (10 mm) may be used for less sensitive nuclei so that more sample is present. The solution should fill the tube to a height of 2–3 cm. The sample size for routine ^{13}C and ^{31}P spectra is about 50–100 mg and for ^{1}H and ^{19}F NMR spectra, usually 5–10 mg or 10 μl of the sample dissolved in 0.5 ml of the solvent. An optimum concentration of 20–25 millimolar solution is recommended. Thus ^{1}H NMR spectra can be readily measured at much lower concentrations though higher concentrations are necessary for recording ^{13}C NMR spectra. Repetitive scanning and signal accumulation allow ^{1}H NMR spectra to be obtained with very small quantities. The reference compound is then added. The tube is spun at about 30 rps about its vertical axis to improve the effective homogeneity even further. Although the superconducting magnet is operated at the temperature of liquid helium (4 K), the sample itself is normally at room temperature.

Dilute solutions of the sample in inert solvents are utilized to obtain high resolution NMR spectra. Chemical shifts will depend on concentration, solvent and temperature, if there are solute-solvent interactions. For proton spectra, deuterated solvents are most often used partly to avoid the resonances from protons of the solvent from swamping the proton spectrum of the compound. However, more importantly, the resonance of the spin-active deuterium nucleus is utilized as a reference to ensure that the frequencies generated through out the experiment remains constant or locked. If protons are present in the solvent (say as impurities) then the solvent should not have resonances in the region of interest. Commercially available deuterated solvents are 98–99.8% isotopic purity. The deuterated solvents may have undeuterated species as impurities and the residual proton signals may be observed in the resulting spectra since the solvent is used in excess. For example, $CDCl_3$ and DMSO-d_6 may show signals at δ 7.26 and 2.5 ppm respectively due to residual protons. The more commonly used solvents are $CDCl_3$, CCl_4, CD_3CN, DMSO-d_6, C_6D_6 and D_2O. Frequently a mixture of solvents such as $CDCl_3$ and DMSO-d_6 may be used because it has solvating properties similar to but is less viscous than DMSO-d_6. Chloroform-d_1 is relatively inexpensive and is most commonly employed for ^{1}H NMR. A very useful method when trying to identify exchangeable hydrogen is to add a couple of drops of D_2O to the solution, shake thoroughly and record the spectrum. The solution should be free both of paramagnetic and insoluble impurities. Paramagnetic impurities at concentrations as low as 10^{-6} M can appreciably broaden the line.

VARIABLE TEMPERATURE NMR

Often it is desirable to record the NMR spectrum at different temperatures. The spectra of some compounds are temperature-dependent due to chemical exchange processes. Variable temperature studies help in obtaining information about such processes which is discussed next (see Chapter 14). The rotational motion and intermolecular forces are also affected by temperature. Changes in the temperature of the sample alter the Boltzman distribution of spins in the various energy levels. Lowering the temperature causes an increase in signal intensity and an increase in temperature results in a decrease in intensity. The solubility of a poorly soluble compound can be improved at higher temperatures. An unstable compound could be stabilized at lower temperatures.

Most NMR spectrometers are equipped with accessories to vary the temperature of the sample by passing preheated or precooled nitrogen past the spinning sample tube. The instrument may operate in the range of –190 to 200 °C. The solvent used must obviously remain liquid over the whole temperature range. Higher temperatures are achieved by passing the gas for example nitrogen or air through a heater and then around the sample tube. Low temperaures are produced by the evaporation of a controlled quantity of a coolant for example liquid nitrogen. Another method is to utilize Joule-Thomson effect, when a gas is passed directly from the storage cylinder through a small orifice within the probe. For high temperature measurements, dimethyl sulfoxide and nitrobenzene have been used as solvents. At lower temperatures, acetone-d_6 and carbon disulfide can be used down to about –100 °C.

DYNAMIC NMR SPECTROSCOPY

14

INTRODUCTION

A branch of NMR spectroscopy known as the dynamic NMR (DNMR) spectroscopy deals with the effects in a broad sense of chemical exchange processes on the NMR spectra and conversely with the changes in the environment of the magnetic nuclei that can be derived from the observation of the NMR spectra. These changes in the environment may be due to exchange between sites with different chemical shifts and/or different coupling constants. The spectra are affected profoundly by many time dependent processes. The most important kind of information obtained in this way concerns the rate constants for the processes studied and the time scale for the processes is such that first order or pseudo-first order rate constants in the range 10^{-1} to 10^{6} s^{-1} can be measured. Besides this very large span, the DNMR technique has several advantages compared with other kinetic methods. The NMR technique gives direct information about parts of the molecule affected by the exchange, which is not usually obtained by other methods. The kinetics of these processes are too fast to be determined by classical methods. The DNMR spectroscopy has given us insight into the dynamic, time dependent nature of many systems particularly those which are at equilibrium, or where only intramolecular motions are involved. They are different from the usual chemical reactions. Even degenerate systems in which the exchange leads to molecules indistinguishable from the original ones can be investigated. The changes observable in the spectra depend on the time scale of the experiment relative to the time needed for the molecule for rearrangement. Spectroscopic techniques such as infrared, Raman and electronic spectroscopy take 10^{-13} to 10^{-15} s which are too short to detect any intramolecular movements. For NMR spectroscopy interaction time is significantly larger, 10^{-1} to 10^{-9} s. The several examples given here demonstrate the variety of phenomena that can be studied by means of NMR spectroscopy. Thus we describe the experimental observations and their qualitative significance.

Variable temperature NMR spectroscopy has been used extensively to study many dynamic processes and interesting information about exchange processes have been obtained. When the activation energy for these rate processes are in the range of 25 – 100 kJ mol^{-1} (5 – 25 kcal mol^{-1}), the rate of the rearrangements can be brought into the range of 10–10^{6} s^{-1} at temperatures in the range of + 200 °C to – 150 °C. By a proper choice of the temperature many such rearrangements can be controlled so that they are slow enough at lower temperatures to allow the detection of individual molecules or environments within the molecule and rapid enough at higher temperatures for the signals from the different molecules or environments to be averaged out into a single line at the mean position. Thus by studying the NMR spectra over a suitable temperature range, the rearrangement process can be examined in detail.

In a vast majority of DNMR studies, the desired kinetic information has been obtained by the analysis of exchange broadened spectra and it is known as the "band shape analysis". The first application of this was a classical study of the hindered rotation of the dimethyl amino groups in N, N-dimethyl formamide (DMF) and N, N-dimethyl acetamide (DMA) in 1956 by Gutowsky and Holm.

SYMMETRICAL TWO-SITE EXCHANGE

Consider a two-site exchange A $\rightleftharpoons$ B with equal forward and backward rates. Figure 14.1 shows the spectra for a range of rate constants k. For very slow exchange, one observes two equally intense narrow peaks at the resonance frequencies of the two sites ν_A and ν_B. As k is increased, the two lines initially broaden and shift towards one another, broadening further until they merge into a single resonance. This occurs when k is similar in magnitude to the difference in the resonance frequencies $|\nu_A - \nu_B|$ of the two exchanging sites. Further increase in the rate of exchange leads to a sharp resonance at the mean frequency (1/2) $(\nu_A + \nu_B)$. Thus, if the exchange is sufficiently fast, the difference between the resonance frequencies of the two sites reduces to zero. These changes are observed by increasing the temperature.

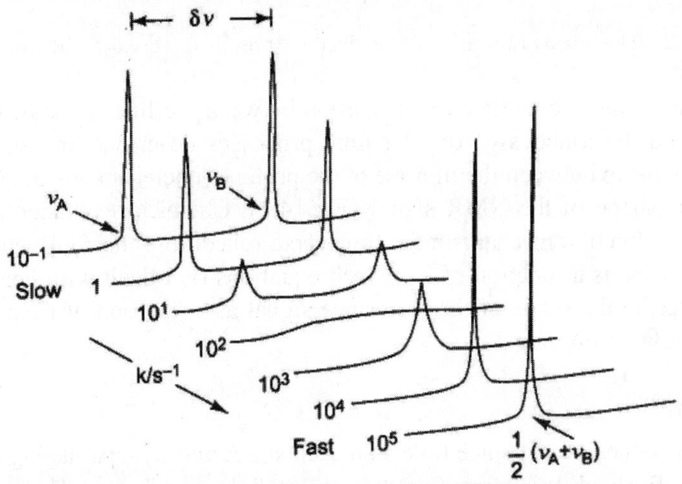

Fig. 14.1 NMR spectra for a pair of nuclei exchanging between two symmetrical sites with equal populations and for a range of the exchange rate constant k.

The temperature-dependent variation of the spectra described above is due to the small energy difference that exists between the magnetic sites of different

Larmor frequencies. To measure such small differences, the lifetime τ of the nuclei in each site must be sufficiently long. According to the uncertainty principle, the lower limit for the lifetime τ is given by

$$\tau\,\delta\nu \approx 1/2\pi \tag{14.1}$$

where $\delta\nu = \Delta E/h$ is the frequency difference. If the lifetime τ is very much larger than $2\pi\delta\nu$, two separate resonances are observed as shown in Fig. 14.2. If τ is much smaller than $2\pi\delta\nu$, then a sharp single resonance is found. If $\tau = 2\pi\,\delta\nu$, then the two resonances coalesce. It is important to note that in variable temperature NMR the coalescence of two peaks is frequency dependent. The greater the separation between the rotameric signals in question, the higher is the temperature required to achieve coalescence. Whether a process is slow, intermediate or fast is determined by the exchange rate constant k relative to the difference in resonance frequencies $\delta\nu$ of the exchanging nuclei. That is, the time scale of these changes is governed by $(\delta\nu)^{-1}$. Since $\delta\nu$ values are usually a few kHz, only relatively slow equilibria (seconds to microseconds) can be studied by NMR spectroscopy. Somewhat faster exchange processes can often be studied by ^{13}C NMR because of the much larger spread of the resonance frequencies (see Chapter 15).

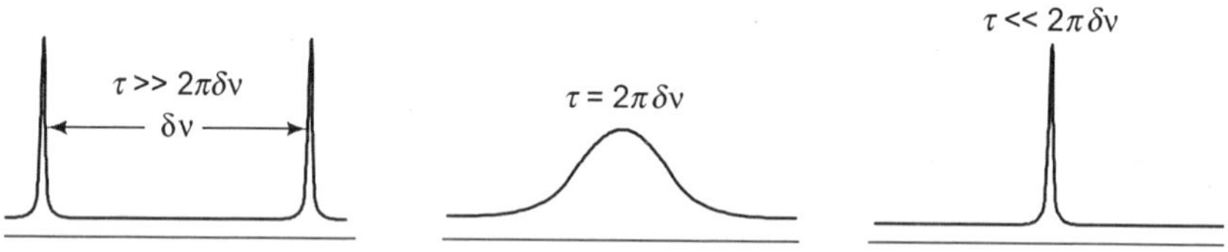

Fig. 14.2 The effect of the excited state lifetime τ on the observed NMR spectrum.

For establishing a quantitative correlation between the line shape of the NMR spectrum and the kinetics of the dynamic processes described above, we must find a relationship between the lifetime of the protons (nuclei) in positions A and B and the line shape of the NMR signal (Fig.14.1). Chemical exchange processes represent an effective mechanism for transverse relaxation, their influence on the NMR line shape is a function of T_2. Bloch equations (not dealt with here) can be used to describe the shape of the resonance signal as a function of frequency and transverse relaxation time T_2.

Slow Exchange

For large values of residence time τ on each site A and B, separate signals at ν_A and ν_B are observed. It is termed as the region of slow exchange. The resonances are exchange-broadened but are still found at frequencies ν_A and ν_B (see Fig. 14.1). In this limit, the increase in line width defined in Fig. 14.3 due to exchange is given by

$$\Delta\nu = k/\pi = 1/\pi\tau \tag{14.2}$$

i.e. the faster the exchange, the broader the line. This effect arises from lifetime broadening (refer Chapter 5). The exchanging nuclear spins do not necessarily have the same orientation so that extra relaxation is introduced leading to broadening of the peaks.

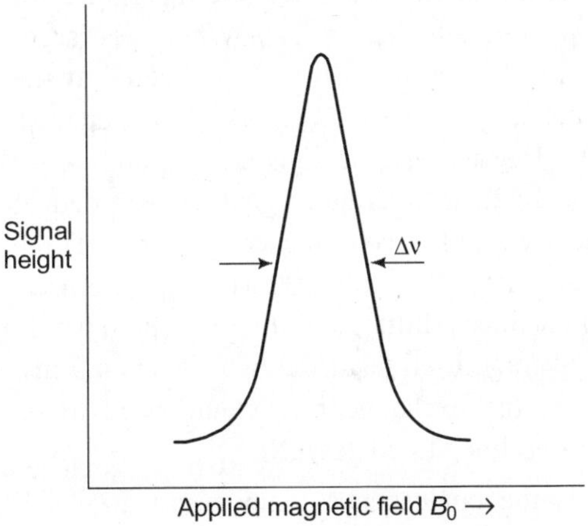

Fig. 14.3 The NMR absorption peak (Lorentzian curve). Definition of $\Delta\nu$, the width at half the maximum height.

Fast Exchange

If the lifetime is extremely short, the two resonances merge to form a single broadened resonance at the mean resonance frequency. It is known as fast exchange. In this limit, the extra line width ' 'ↄ chemical exchange is given by

$$\Delta\nu = \frac{\pi(\delta\nu)^2}{2k} = \frac{}{2} \quad (14.3)$$

where $\delta\nu = \nu_A - \nu_B$. The single line observed in the fast exchange becomes narrower and has a normal line width as the rate increases, due to the more effective averaging of the two environments A and B. The fast exchange causes the spins to experience an effective local field which is the mean of the two sites between which they hop. In the fast exchange, the resonance is found at the mean frequency, $(1/2)(\nu_A + \nu_B)$ since each spin spends nearly 50% of its time in each state.

Intermediate Exchange

At intermediate lifetimes, the shape of the lines would change in a more complicated manner and finally the two resonances coalesce. The condition for the two resonances just to merge into a single broad band when the valley between the peaks is smoothened out is

$$k = \frac{\pi\delta\nu}{\sqrt{2}} = 2.2\ \delta\nu \quad \text{or} \quad \tau = \frac{\sqrt{2}}{\pi\,\delta\nu} \quad (14.4)$$

The rate constants are obtained from the relation, $k = \dfrac{1}{\tau_C}$. At coalescence, where the multiplet structure is just lost, the coalescence residence time is τ_C. The coalescence of the signals under certain conditions is generally a strong indication of some averaging process such as an intramolecular rearrangement or a hindered rotation taking place.

Equation (14.4) gives the rate constant k at the coalescence temperature for the exchange between two equally populated sites ($p_A = p_B$). When k is larger than the right hand side of the equation (14.4), a single line is expected at the mean resonance frequency. On the other hand, when k is smaller, two separate resonances should appear (Fig. 14.1). That is, separate signals at ν_A and ν_B are observed for $k/\delta\nu > 2.22$. If $k/\delta\nu < 2.22$, an average signal is recorded at $(\nu_A + \nu_B)/2$. If the exchange is sufficiently rapid, a coalescence of (some of) the signals will be observed. Equation (14.4) shows that on a 100 MHz spectrometer the resonances of the two protons with chemical shifts separated by 1 ppm would merge when $k = 220$ s^{-1}, i.e. when the average lifetime at the two sites is 4.5 ms. On a 500 MHz spectrometer, however, the exchange rate would need to be 1100 s^{-1} (lifetime 910 μs) to cause the two lines to coalesce.

A particularly useful temperature is the coalescence temperature at which the two peaks just merge. At this temperature, the lifetime τ is equal to $\sqrt{2}/[\pi(\nu_A - \nu_B)]$ as given by the equation (14.4). Thus, the rate constant could be determined and hence the barrier or the activation energy for the process obtained. When the exchange rate is sufficiently high, the life times of the states will become less than the critical value and the signals cannot be separated. For exchange between more than two sites or for those involving unequal concentrations, the situation is more complicated but the principle remains the same.

For determining the rates of interconversion, consider as an example two resonances, which have a chemical shift difference of 3.00 ppm, corresponding to a frequency difference $\delta\nu$ of 192 Hz at the ^{13}C NMR spectrometer frequency of 64 MHz. At –38°C, the line broadening $\Delta\nu$ from slow exchange is 2.7 Hz giving $k = 2.7\pi = 8.5$ s^{-1} from equation (14.2). The two peaks coalesce at 9°C giving $k = \pi(192)/\sqrt{2} = 427$ s^{-1}. At 31°C, the two lines merge to give a single peak with an exchange broadening of 23.4 Hz. Using equation (14.3) in the fast exchange limit, we get $k = (1/2)\pi(192)^2/23.4 = 2475$ s^{-1}.

At an intermediate lifetime a broadened peak is obtained whose shape is determined by $\tau\nu_0\delta$ where δ is the chemical shift difference between the two resonances in the absence of exchange. The factors at the coalescence are more complex.

A complete line shape analysis over a large temperature range is always preferable for kinetic measurement. Line shape analysis by least squares fitting procedure can yield the activation energies for the process and also give information on the mechanisms involved. However, this requires a large effort. Consequently alternative to the use of complete line shape analysis, a number of approximate solutions are available. The best known of these is the relation (14.4) which gives the rate constant at the coalescence temperature T_c (K) for the exchange between two equally populated sites ($p_A = p_B$). This enables us to use the Eyring equation to obtain the energy barrier for the process at this temperature. On appropriate substitution, we get

$$\Delta G = RT_c\,[22.96 + \ln (T_c/\delta\nu)] \text{ kJ mol}^{-1}$$
$$= 8.3 \times 10^{-3}\, T_c\,[23 + 2.3 \log_{10} (T_c/\delta\nu)] \text{ kJ mol}^{-1} \quad (14.5)$$

where R is the gas constant. The value of the energy barrier thus obtained is temperature dependent $\delta\nu$ in Hz.

BARRIER TO INTERNAL ROTATION

Tertiary amides often exhibit two rotomeric forms, depending on the relative sizes of the R′ and R″ groups (Fig.14.4) The energy barrier about the amide C—N bond due to its partial double bond character as represented by the structure shown in Fig. 14.4 (b) to the resonance hybrid is of the order 40–100 kJ mol^{-1}. A 1 : 1 ratio of the two rotomers is expected when R′ = R″.

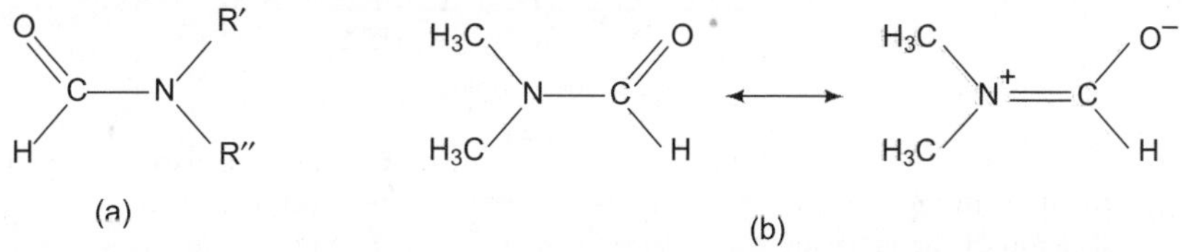

Fig. 14.4 (a) Structure of a tertiary amide and (b) resonance forms of N, N-dimethyl formamide.

The skeleton of N, N-dimethyl formamide (DMF), the simplest tertiary amide, is planar. The dimethyl group can undergo 180° rotations interconverting between the two forms of equal energy with identical forward and backward first order rate constants (Fig.14.5). N-Methyl resonance of N, N-dimethyl formamide is a doublet at 25 °C because the rotation about the C—N bond is restricted. The lone pair of electrons on nitrogen forms a partial double bond to the carbonyl carbon, and the two methyl groups lie in different chemical environments. At room temperature, the internal rotation is slow and the ^{1}H spectrum comprises of two equally intense sharp singlet methyl resonances *cis* and *trans* to the C=O group at δ 2.84 and 3.0 ppm. The origin of the chemical shift difference between the two methyl signals is the anisotropy of the carbonyl group. The resonance of the methyl group *trans* to the carbonyl group is downfield from the resonance of the *cis* methyl group which is more shielded. The resonance frequency of each group of methyl protons thus hops from ν_{cis} to ν_{trans} and vice versa with an average time jumps of $\tau = 1/k$. At higher temperature, rotation about the C—N bond takes place sufficiently rapidly to average out the two different environments of the methyl group and the two peaks coalesce at 118 °C. If the compound is heated, the rate of rotation and the line shapes change as shown in Fig. 14.6. From Arrhenius type temperature-rate plots, the values of energy of activation E_a and enthalpy ΔH have been obtained. The rate constants are also obtained from $k = 1/T_c$, where T_c is the coalescence temperature. The activation energy barrier to rotation in DMF was found to be about 86 kJ mol^{-1}.

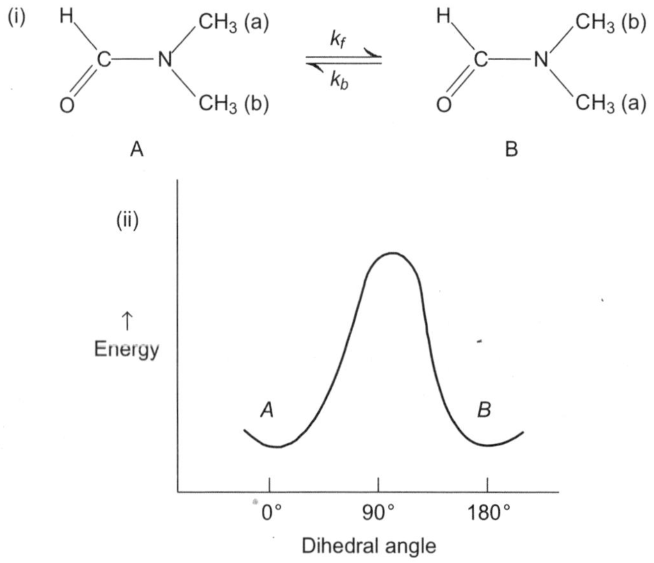

Fig. 14.5 (i) Symmetrical two-site exchange in N,N-dimethyl formamide. (ii) Barrier to internal rotation. In the transition state, the N – C == O plane is perpendicular to C_2N – C skeleton of the molecule.

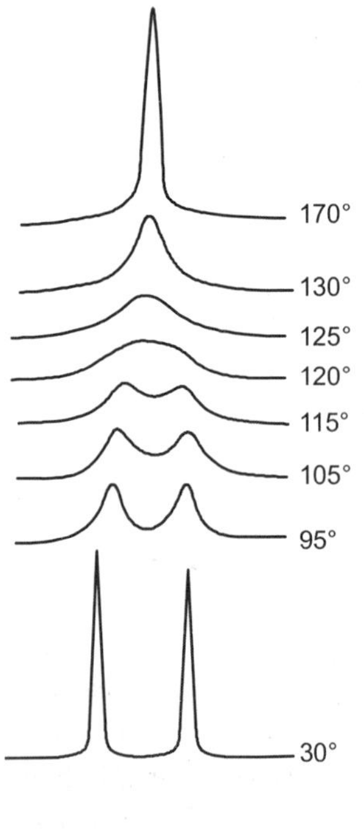

Fig. 14.6 Variable temperature ^1H NMR spectra of N,N-dimethyl formamide.

Another example of a symmetrical two-site exchange is N, N-dimethyl nitrosoamine (Fig. 14.7). The two conformations of the molecule interconvert by 180° rotations of the nitroso group around the N—N bond, which is a partial double bond. In the transition state, the NNO plane is perpendicular to the C_2NN plane of the molecule. The energy of activation for the hindered rotation has been found to be 96 kJ mol^{-1}. Variable temperature NMR spectra have been studied for thioamides, carbamates, nitriles, etc. and the activation energy for the hindered rotation has been determined.

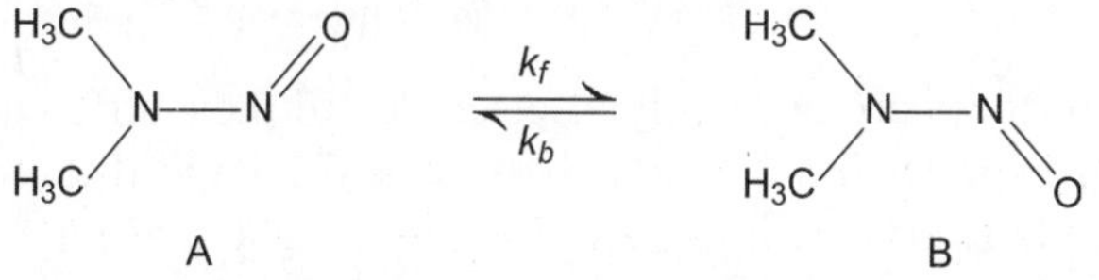

Fig. 14.7 Rotation over the N–N bond in N, N-dimethyl nitrosoamine.

UNSYMMETRICAL TWO-SITE EXCHANGE

Two-site exchange with equal populations is a special case of the more general situation in which two sites can be in any arbitrary concentrations. For example, if one of the methyl groups in dimethylnitrosoamine is replaced by a benzyl group (Fig. 14.8), the population of the two conformers differs and hence the forward and backward rates also differ. However, 180° flips of the nitroso group would still modify the chemical shifts of the methyl protons.

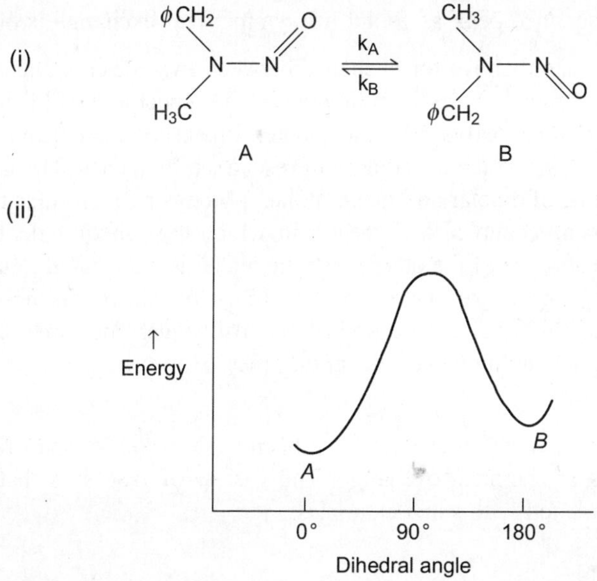

Fig. 14.8 (i) Benzylmethyl nitrosoamine as an example of unsymmetrical two-site exchange (ϕ denotes phenyl group). (ii) Barrier to internal rotation over the N – N bond.

If the fractional populations of the two sites are p_A and p_B (i.e. $p_A + p_B = 1$) and the two first order rate constants are k_A and k_B

$$A \underset{k_B}{\overset{k_A}{\rightleftharpoons}} B \tag{14.6}$$

Then at equilibrium $p_A k_A = p_B k_B$. The average lifetime of the two sites are $\tau_A = 1/k_A$ and $\tau_B = 1/k_B$.

In the slow exchange limit, the lifetime broadenings are

$$\Delta\nu_A = k_A/\pi = 1/\pi\tau_A \text{ and } \Delta\nu_B = k_B/\pi = 1/\pi\tau_B \tag{14.7}$$

The two resonances are unequally affected due to their different lifetimes; the less concentrated species has a shorter lifetime and a larger broadening. For fast exchange, a single resonance is observed as the weighted average frequency

$$\nu_{av} = p_A\nu_A + p_B\nu_B \tag{14.8}$$

and the line broadening is given by

$$\Delta\nu = \frac{4\pi\, p_A p_B (\delta\nu)^2}{k_A + k_B} \tag{14.9}$$

Equation (14.9) is expected since each nucleus spends a proportion p_A of its time in site A and p_B in site B, giving an average resonance frequency weighted in favour of the more stable site. These expressions reduce to the formulae for symmetrical exchange when $p_A = p_B = 1/2$ and $k_A = k_B = k$. With increasing temperature, the same pattern of broadening, coalescence and subsequent narrowing of the resonance peak as found for symmetrical exchange is observed.

The principles discussed above for exchange between two sites may be extended to multiple-site exchange processes. However, the spectra would be more complicated and need to be analysed by a computer. Processes other than exchange may also lead to averaging of the differences in resonance frequency. The important examples are averaging of dipolar and quadrupolar splittings by molecular tumbling in solution. The proton decoupled ^{13}C spectra in which each inequivalent carbon gives a singlet in the absence of the other magnetic nuclei is well suited to the study of chemical exchange processes (see Chapter 15). The above discussion was focussed on chemical shifts being averaged by fast exchange. The same principles apply for the spin-spin couplings and relaxation times. However, the latter topic is not discussed here.

So far, only conformational equilibria have been dealt with. We shall now deal with other examples of chemical exchange. The variety of processes that can be investigated is best treated with some examples.

RING INVERSION

Among the many types of rate processes, the study of the inversion of cyclic compounds has attracted particular attention. The NMR spectroscopy has contributed significantly to our understanding of the conformational mobility of saturated and unsaturated ring systems. Here, the classical example of the inversion of cyclohexane which has been investigated extensively may be cited. Cyclohexane exists mostly in the chair form at room temperature where the carbon-carbon bond angles are all 109.5° and thereby free of angle strain. It is free from torsional strain also. The observed barrier is 45 kJ mol^{-1} for the inversion of the chair conformation (a) through a high-energy conformation such as the boat or the twist-boat into the equivalent chair conformation(a') (Fig. 14.9). The energy is still low enough to allow rapid rates of conversion from one chair conformation to another. The chair conformation is much more rigid than the boat conformation. The boat conformation can relieve some of its torsional strain by assuming twist conformation. All the C–H bonds in the chair conformation are either axial or equatorial. These interchange when one chair conformation inverts to the other. The twist-boat conformer is 23 kJ mol^{-1} above the chair conformation. Equilibrium exists between the chair and twist-boat forms with the more stable chair form being favoured 10,000 to 1 at room temperature. The transitions between the boat and the twist-boat conformers occur simply by rotation around carbon-carbon single bonds. Such conformational transitions that are generally distinguished by low energy barriers are called pseudo rotations.

The measurement of the rate of inversion of cyclohexane depends on the fact that the equatorial and axial protons have considerably different chemical shifts. The equatorial protons appear at low field. Upon inversion of the ring, they

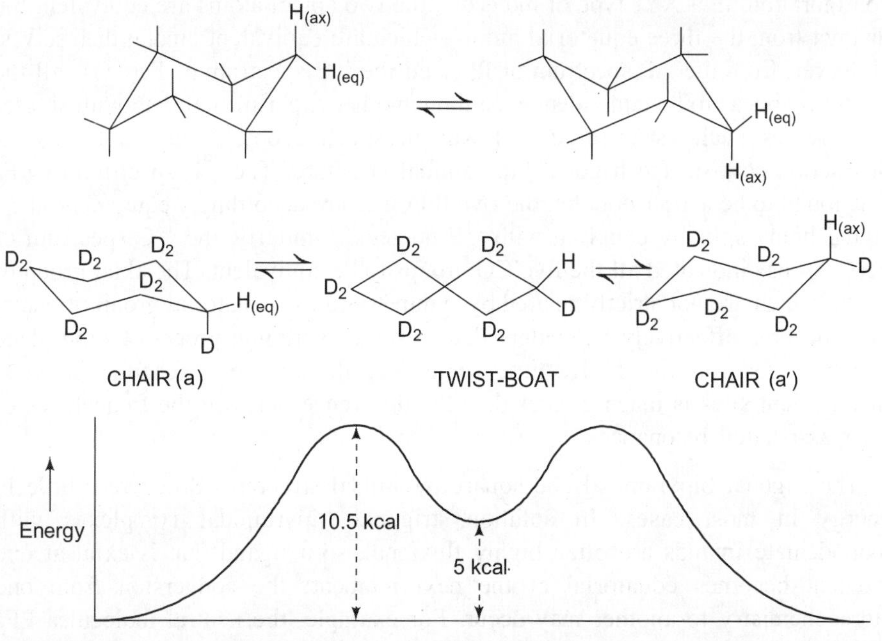

Fig. 14.9 Inversion of cyclohexane-d_{11}

exchange environments. At normal temperature this process is very rapid and all the protons appear as a single narrow peak. Thus at room temperature, cyclohexane exhibits only a singlet in the ^1H NMR spectrum. At low temperatures interconversion between different conformations of cyclohexane becomes slow, the chemical shifts of the axial and equational protons are resolved and complex spin-spin splittings appear and the spectrum becomes rather complex. The spectral complexity can be avoided by replacing all but one of the protons by deuterium, giving C_6HD_{11}. Particularly simple spectra are obtained for cyclohexane-d_{11} if the spin-spin interaction between the proton and the deuterium atoms are eliminated by decoupling. For cyclohexane-d_{11} at −89°C two sharp singlets of equal intensity are seen which coalesce at – 60°C. The two resonances correspond to the axial and equatorial hydrogen atoms of the two chair conformations. At this temperature, τ is nearly 0.008 s; $k = 69\ s^{-1}$ $(1/2\ \tau)$. Several monosubstituted cyclohexanes have also been investigated.

FLUXIONAL MOLECULES

There are a significant number of cases in which the molecular vibrations or intramolecular rearrangements as noted above carry a molecule from one nuclear configuration into another. The resulting two or more configurations are chemically equivalent and this type of stereochemically non-rigid molecule is called a fluxional molecule. Five coordinate compounds are generally stereochemically non-rigid because there are two alterate structures with similar energies. Thus an important class of fluxional molecules are those with a trigonal bipyramidal configuration. In a XY_5 type of molecule, the two apical atoms are equivalent but distinct from the three equatorial atoms, which are equivalent among themselves. However, from the ^{31}P spectrum of PF_5 and the ^{13}C spectrum of $Fe(CO)_5$ all the Y atoms appear to be equivalent even though other experimental data with shorter time scales such as infrared and Raman spectroscopy, X-ray and electron diffraction confirm the trigonal bipyramidal structure. The ^{19}F spectrum of PF_5 was found to be a 1 : 1 doublet, the five fluorines are accordingly equivalent, their signal being split by coupling with ^{31}P nucleus. Similarly, the ^{13}C spectrum of $Fe(CO)_5$ has indicated all the five CO groups to be equivalent. The axial group of nuclei which are not strictly related by symmetry to the equatorial group of nuclei have become effectively equivalent due to some averaging process taking place which is fast on the NMR time scale, i.e. the rate of exchange between inequivalent sites is much greater than the difference between the frequencies of their associated resonances.

The trigonal bipyramidal and square pyramidal structures differ very little in energy in most cases. In solution, trigonal bipyramidal complexes with monodentate ligands are often highly fluxional, so a ligand that is axial at one moment becomes equatorial at the next moment; the conversion from one stereochemistry to another may occur. For example, the neutral molecules PF_5 and $Fe(CO)_5$ have trigonal bipyramidal structure in the crystal; however, in solution the ligands exchange their axial and equatorial positions at a rate that is fast on the NMR time scale but slow on the infrared time scale. Berry suggested that the interconversion takes place by simple angle-deformation motions and in this way the axial and equatorial atoms of the trigonal bipyramid may be interchanged. Berry mechanism for this interchange is shown in Fig. 14.10. The square pyramidal configuration may go back to the trigonal bipyramidal configuration in either of the two ways, one is simply going back to the original and the other places the axial atoms in equatorial positions and two of the erstwhile equatorial atoms in the axial positions. The process can be repeated with other atoms. This rearrangement involves only a vibrational process and no bond breaking. It explains the equivalence of the five fluorines as well as the spin-spin coupling between ^{31}P and ^{19}F in PF_5 molecule. Berry mechanism is often called a pseudo rotation.

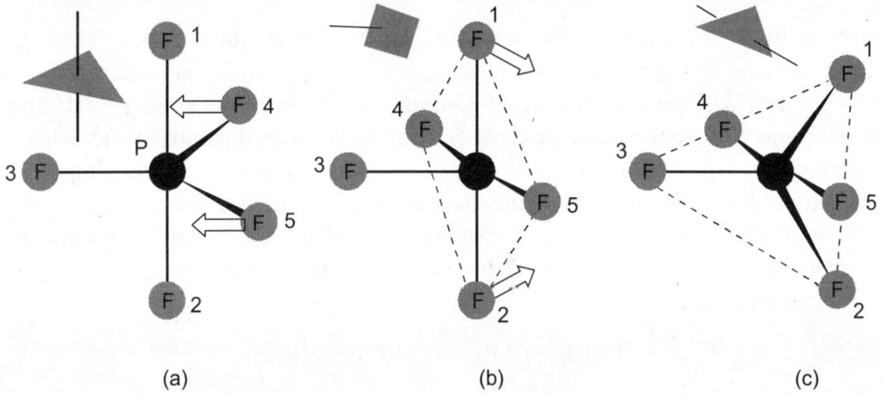

Fig. 14.10 Berry pseudo rotation in which a trigonal bipyramidal PF_5 molecule distorts into a square-pyramidal structure and then to a trigonal bipyramid again.

If these pseudo rotations occur rapidly enough say one in every 10^{-4}s, the F or CO atoms will be moving from one environment to the other so rapidly that they will be seen in an NMR experiment as equivalent since the characteristic time scale for the NMR observation under normal conditions is many times greater than each individual period of residence of a nucleus at a single vertex. In cases such as PF_5 and $Fe(CO)_5$, the energy of activation for the rearragement process is so small that it is not possible to slow down the process sufficiently at low temperature for the symmetry inequivalent environments to be observed in the NMR experiments.

SF_4 molecule: The ^{19}F spectrum of SF_4 measured at –100 °C displays two equally intense triplets separated by 48 ppm. This suggests two pairs of structurally non-equivalent fluorine atoms in the molecule. The coupling constant between the two types of fluorines is about 78 Hz. The structure of SF_4 (Fig. 14.11) may be considered as a distorted trigonal bipyramid with one equatorial position on sulphur being occupied by a lone pair of electrons. On increasing the temperature, the triplet signals of SF_4 become averaged at –94 °C and the two fluorine signals coalesce near –20 °C. The energy of activation for this process is about 17 kJ mol^{-1}.

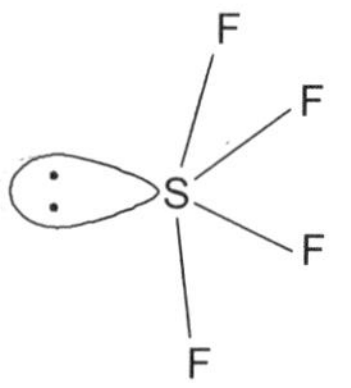

Fig. 14.11 Structure of SF_4.

INTERMOLECULAR EXCHANGE PROCESSES

Proton Exchange

The majority of the exchange processes which have been studied by DNMR are intramolecular. However, the derivations are equally valid for intermolecular exchange provided the first order rate constants are replaced by pseudo first order rate constants. Exchange processes that are intermolecular in nature include almost all proton transfer reactions such as those of ethanol discussed earlier. Important information about exchange mechanisms can also be obtained by considering the coupling constants, as the following examples illustrate. The collapse of spin-spin multiplets is in many cases of diagnostic value. The processes involved are second order reactions such as

$$CH_3OH + {}^{*}HOH \leftrightarrow CH_3OH^{*} + HOH$$

in contrast to the first order reactions considered earlier.

The ^{1}H NMR spectrum of a sample of neat methanol, CH_3OH at 25 °C consists of two singlets with the hydroxyl resonance δ 1.6 ppm to the low field of the methyl resonance (Fig. 14.12). If the temperature is gradually reduced the singlets broaden; structure appears. At -60°C the hydroxyl resonance is a 1 : 3 : 3 : 1 quartet and the methyl resonance is a doublet. In addition the hydroxyl resonance moves downfield, to δ 2.3 ppm from the methyl doublet. The chemical shift difference between the OH and CH_3 resonances decreases, since the hydrogen bonds break with increasing temperature. The OH resonance moves to higher field. This effect arises because of the very small energy difference that exists between magnetic sites of different Larmor frequency. In methanol, the OH proton is bonded to the $-CH_3$ group that has different total spin (+ 3/2, + 1/2, –1/2, –3/2) and different effective Larmor frequencies result. In order to remove the corresponding energy differences the life time of the OH protons at each site must satisfy the equation (14.1) and intermolecular exchange through hydrogen bonds must be slow. Otherwise the differentiation between individual signals disappears and only time-averaged spectrum is obtained. The exchange process affects not only the positions of the bands in a spectrum but also the multiplicity of the band.

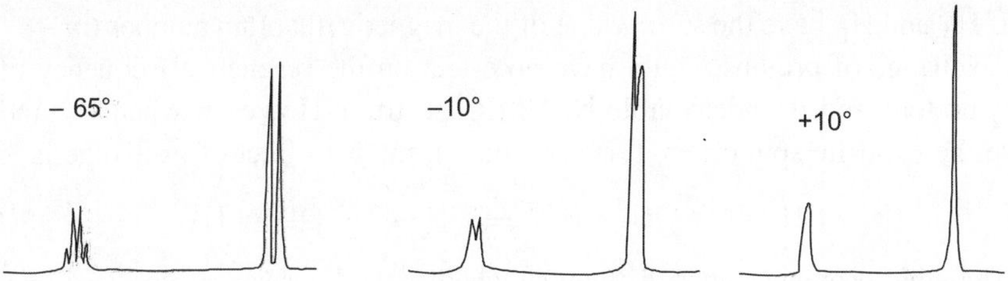

Fig. 14.12 Temperature dependent ^{1}H NMR spectrum of methanol.

Earlier we noted that the splitting due to OH interaction is visible only in the highly purified ethanol, without any traces of acid or base. Under this condition, the OH singlet splits into a 1 : 2 : 1 triplet resulting from the spin-spin coupling with the protons of the CH_2 group (J = 4.8 Hz) and each line of the CH_2 quartet becomes a doublet with the same coupling constant (Fig. 14.13). On addition of as little as 10^{-5} mol dm^{-3} of HCl, these splittings vanish. The addition of acid catalyses rapid exchange of the hydroxyl protons between the neighbouring molecules. This exchange is sufficiently rapid to 'average out' the electron environment of the proton and only an average of the nuclear spins of the OH is felt by the CH_2 protons.

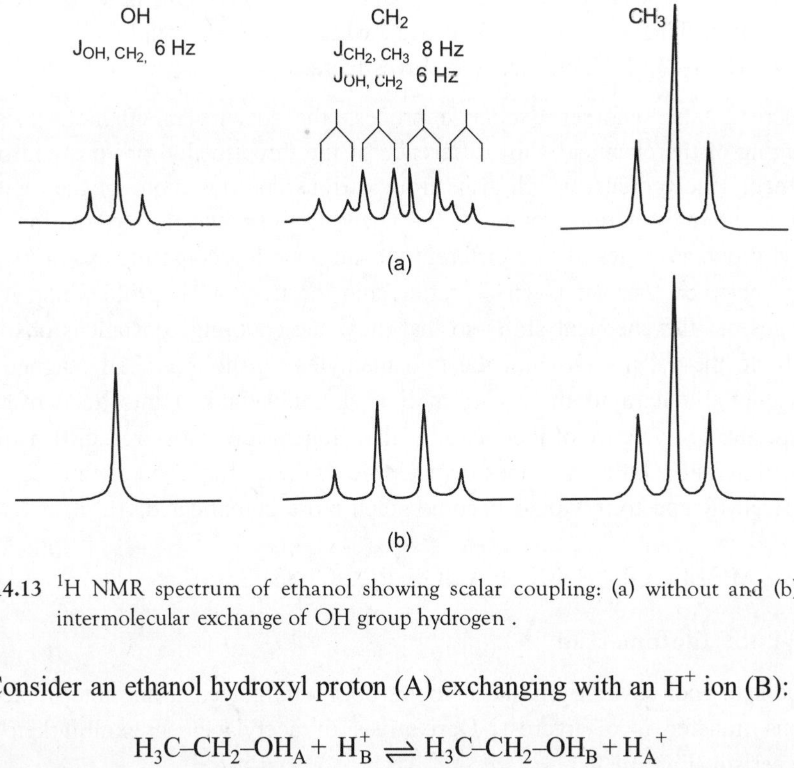

Fig. 14.13 ^{1}H NMR spectrum of ethanol showing scalar coupling: (a) without and (b) with intermolecular exchange of OH group hydrogen .

Consider an ethanol hydroxyl proton (A) exchanging with an H^+ ion (B):

$$H_3C{-}CH_2{-}OH_A + H_B^+ \rightleftharpoons H_3C{-}CH_2{-}OH_B + H_A^+$$

If H_A and H_B have the same value of the magnetic quantum number ($m = \pm 1/2$) the exchange of protons would have no effect on the resonance frequency of the CH_2 protons and is undetectable by NMR spectrum. However, when H_A and H_B have the opposite spin orientations ($\uparrow$ and $\downarrow$), the net effect of exchange is

$$CH_3–CH_2–OH\uparrow + H^+\downarrow \rightleftharpoons CH_3–CH_2–OH\downarrow + H^+\uparrow$$

The ethyl protons cannot distinguish whether the hydroxyl proton is A or B, or whether it is up spin $\uparrow$ or down spin $\downarrow$. Therefore, at each H^+ exchange, the CH_2 resonance frequency changes by $\pm$ J. If this exchange is fast compared to the frequency difference involved i.e. J, the doublet splitting of the CH_2 multiplet collapses to zero and the signal of the CH_2 group becomes a quartet. The triplet structure of the OH resonance also disappears when the proton exchange is faster than about 10 Hz. These coupling constants cannot be determined in the presence of rapid proton exchange.

Because the splitting is usually very small (about 1 Hz), a proton must remain attached to the same molecule for longer than about 0.1 s for the splitting to be observable. In water, the exchange rate is much faster than that and hence alcohols show no splitting from the OH protons. In dry dimethyl sulfoxide (DMSO), the exchange rate may be slow enough for the splitting to be detected. Dimethyl sulfoxide slows down the exchange of protons of OH as well as NH and NH_2 groups. This solvent is hence used when spin-spin couplings of the type H–C–O–H, H–C–N–H, etc. are to be investigated.

There is also another exchange process that occurs in ethanol. It is the averaging of the chemical shift difference of the three methyl protons and of the two methylene protons, which arise from various conformations of the molecule by rapid internal rotation around the C—C and C—O bonds. Rapid rotation of the methyl group averages all the different vicinal coupling constants existing at any instant between the three $-CH_3$ protons and the two $-CH_2$ protons and it also averages out the chemical shifts so that only one coupling constant is observed. The three methyl protons and the two methylene protons are all magnetically equivalent. Such rapid molecular motion should be taken into account while considering equivalence of the nuclei. If the temperature is lowered further till the rate of internal rotation becomes comparable to these frequency differences, then the ^1H NMR spectrum would become much more complicated.

INTRAMOLECULAR EXCHANGE PROCESSES

Keto-Enol Tautomerism

The keto-enol equilibrium that involves both intramolecular and inter-molecular protons transfer is of interest. Derivatives of acetylacetone exhibit keto-enol tautomerism (Fig. 14.14).

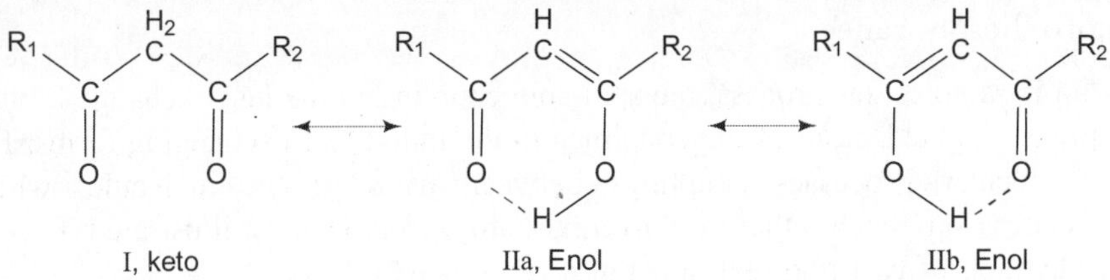

Fig. 14.14 Keto – enol tautomerism in acetylacetone (R_1= R_2= CH_3).

If the rate of interchange between the tautomeric forms is slow compared to the time scale of the NMR process (10^{-3}s) separate signals would be observed for the protons in each of the tautomeric forms. It is expected that the rate of interchange in the equilibrium I ↔ II will be considerably slower than in the equilibrium IIa ↔ IIb for R_1= R_2= CH_3.

In the spectrum of acetylacetone in carbon tetrachloride (Fig. 14.15), the signals of the keto form (δ CH_3 2.2, δ CH_2 3.7 ppm) and the enol form (δ CH_3 2.0, δ = CH 5.7 ppm) are observed. It indicates that the rate of the keto–enol equilibrium reaction is slow on the NMR time scale. From the relative intensities of the signals it is readily seen that at room temperature, the enol form predominates in the equilibrium mixture. Intensities show that the mixture contains about 85% of the enol form and 15% of the keto-form. Strong deshielding of the enol hydroxyl proton (δ 15.2 ppm) is due to the intramolecular hydrogen bond formation. The magnetic equivalence of the methyl groups of the enol form indicates rapid intramolecular exchange of the hydrogen atoms between the two oxygen atoms. By adding a base or by raising the temperature, the reaction rate can be increased leading to line broadening and finally coalescence to a time averaged spectrum.

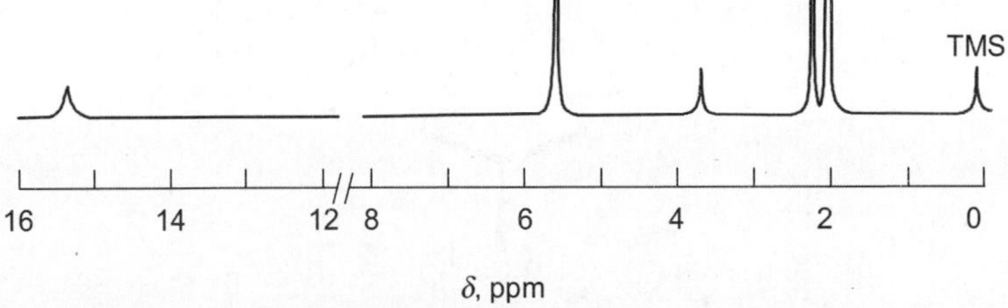

Fig. 14.15 ^{1}H NMR spectrum of acetylacetone at room temperature.

Fluorophosphoranes

For an intramolecular process, the spin-spin coupling in the fast exchange limit is simply a weighted mean of the couplings in the individual exchanging forms. For intermolecular processes couplings between parts of the molecule, which dissociate from each other are averaged to zero. This is illustrated from a consideration of two fluorophosphoranes (Fig. 14.16).

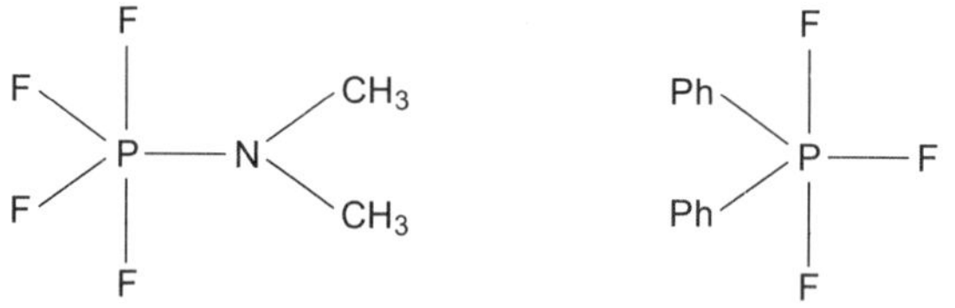

Fig. 14.16 Structure of fluorophosphoranes.

At low temperatures, the ^{31}P spectrum of $PF_4N(CH_3)_2$ is a triplet of triplets with spin-spin couplings to two axial and two equatorial fluorines. On warming, the splitting pattern gradually changes to finally become a regular quintet. This indicates that the fluorine atoms exchange sites. For $PF_3(Ph)_2$, a similar process occurs, all the lines of the ^{31}P splitting pattern changes from a doublet of triplets to a quartet. On warming, all the lines coalesce giving finally a single line. An intermolecular exchange takes place probably involving a catalysed fluoride transfer.

Organometallic Compounds

(*i*) Fluxional behaviour is characteristic of certain classes of organometallic compounds. Of greater interest is the stereochemical non-rigidity that is often observed when a conjugated cyclic polyene is bonded to a metal atom through some but not all of its carbon atoms. In such complexes, the metal-ligand bonding may hop around the ring. For example, in Ge(η^1 – C_5H_5) $(CH_3)_3$ (Fig. 14.17) the single site of attachment of the Ge atom to the cyclopentadienyl ring hops around the ring in which a carbon-metal bond is replaced by a carbon-metal bond to the next C atom around the ring. It occurs on a time scale of 10^{-2} to 10^{-4} s and can be studied by both ^1H and ^{13}C NMR.

Fig. 14.17 η^1-Cyclopentadienyl germyl compound

Many studies have been carried out to determine the rearrangement pathway. In η^1-cyclopentadienyl compounds such as the germyl compound (Fig. 14.17), all the ring protons are not equivalent, and they give a second order $[AB]_2C$ spectrum at low temperature. On warming, an exchange process takes place involving the movement of the germyl group from one carbon atom to another. The spectra indicate that eventually all the hydrogen atoms become equivalent. Detailed analysis has shown that migration occurs by a series of 1, 2-shifts and the activation energy for the fluxional process has been determined.

(*ii*) Another interesting example of a fluxional molecule in which the averaging of two different nuclear environments occurs as a result of intramolecular rearrangement is provided by $(C_5H_5)_4Ti$, which has the structure shown in Fig.14.18. At –30 °C, the five protons in the *b* and *b′* rings are indistinguishable owing to rapid internal rotation which is characteristic of the rings bonded in this way to metal atoms. Thus, at –30 °C, there are effectively two types of protons in the molecule, ten of one kind in the *a* and *a′* rings and ten of another kind in the *b* and *b′* rings, and the rate at which these two ring types interchange their roles is slow enough for two separate sharp signals to be observed. As the temperature is raised, a process in which rings of the types *a* and *b* exchange their roles becomes increasingly rapid and finally rapid enough for the two types of proton environment to be no longer detectable in the NMR spectrum. The energy of activation was found to bc 68 kJ mol^{-1}.

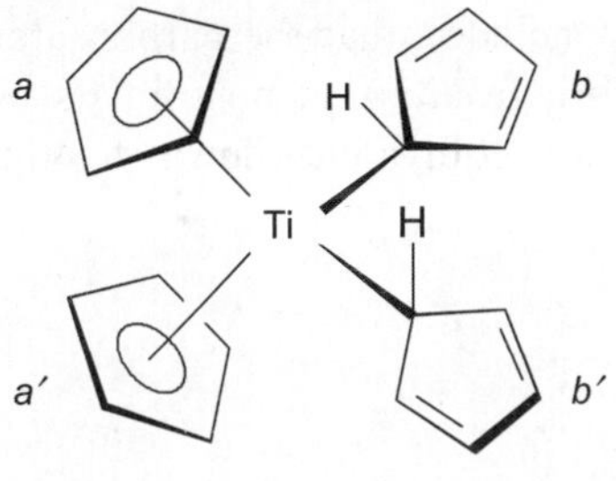

$Ti(\eta^1—C_5H_5)_2(\eta^5—C_5H_5)_2$

Fig. 14.18 Structure of Ti $(C_5H_5)_4$.

(*iii*) A study of interconverting geometrical isomers is provided by the *cis* and *trans*-isomers of bis-(cyclopentadienyl)(tetracarbonyl)di-iron(0) $(\eta^5\text{-}C_5H_5)_2Fe_2(CO)_4$ shown in Fig.14.19. The protons of the cyclopentadienyl rings appear at different positions for the two isomers and the energy of activation for the interconversion is about 50 kJ mol^{-1}. At –70 °C, the rate of interconversion is only about $8 \times 10^{-2}\ s^{-1}$ and each isomer exhibits its own separate proton signal. However, at room temperature the rate of interconversion is about $4 \times 10^3\ s^{-1}$ and the resonances from both the molecules are found at the mean position as one sharp signal. At intermediate temperatures, the signals broaden and collapse into a single peak.

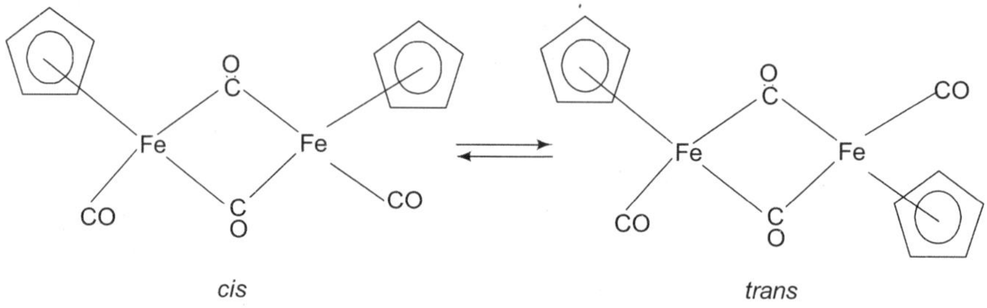

Fig. 14.19 Geometrical isomers of $(\eta^5\text{-}C_5H_5)_2\,Fe_2(CO)_4$.

(*iv*) The temperature dependence of the NMR spectrum of trimethylaluminium in toluene is shown in Fig.14.20. The dimeric $Al_2(CH_3)_6$ has two bridging and four terminal methyl groups (Fig. 14.21). However, at room temperature the proton NMR spectrum is a sharp singlet at – 0.3 δ which suggests that all the six methyl groups are equivalent. It was found that in cyclohexane solution as the temperature is lowered the methyl resonance broadens and separates into two singlets at –50°C, in the intensity ratio 1 : 2 arising from the bridging and terminal methyl groups at δ + 0.5 and –0.6 ppm, respectively. It shows that the exchange between the terminal and bridging methyl groups occurs rapidly on the NMR time scale at room temperature. The activation energy for the exchange calculated from exchange rates obtained from the line shape analysis was found to be 65 kJ/mol. This is similar to the energy of dissociation of the dimer in solution suggesting that the methyl group exchange may proceed via the dissociation of the dimer. From these spectra, it was concluded that the carbon atom of the methyl groups probably forms the bridge and that none of its hydrogens were involved since these were all equivalent down to the lowest temperature.

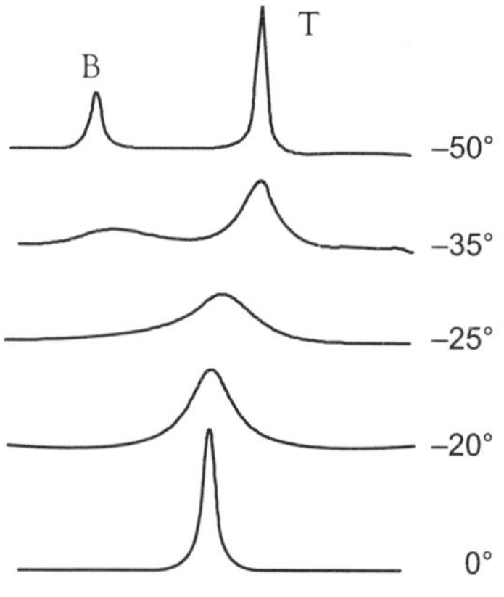

Fig.14.20 ^{1}H NMR spectra of trimethyl aluminium in cyclohexane at different temperatures.

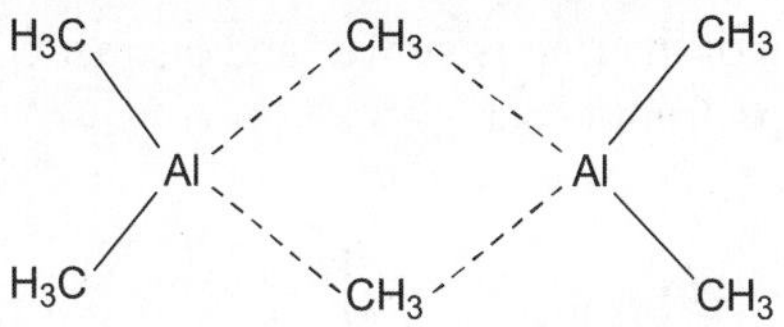

Fig. 14.21 Structure of $Al_2(CH_3)_6$

The equivalence of all of the methyl groups at higher temperature can be rationalized on the basis of the following mechanism. The bridge angle Al-CH_3-Al is approximately 75°. The weakness of the bridge is indicated by the dissociation of dimeric trialkyl aluminium in pure liquid. With every bulky groups, dissociation is virtually complete.

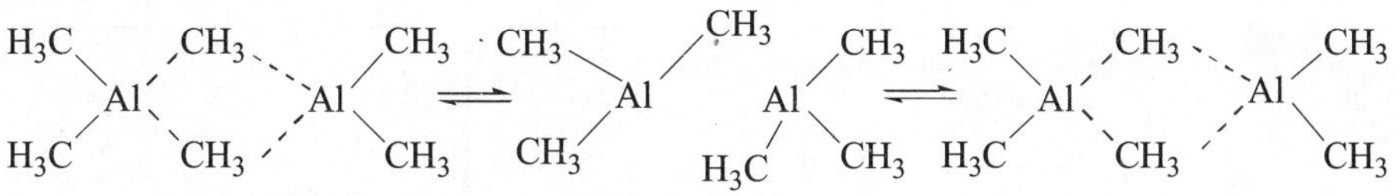

(*v*) The proton decoupled ^{13}C spectrum (see Chapter 15) is particularly suited to the study of chemical exchange processes. There are a number of examples which involve proton decoupled ^{13}C spectra. In the presence of ^{13}C–^{1}H spin-spin couplings, the theoretical analysis of the spectra becomes complicated. For example, the proton-decoupled ^{13}C spectrum of $[Ru(\eta^4\text{-}C_8H_8)(CO)_3]$ below –120 °C exhibits four intense peaks of equal intensity as shown in Fig. 14.22 (a) arising from the eight carbon atoms of cyclooctatetraene ligand. Ruthenium is coordinated to two of the four double bonds of C_8H_8 ring (Fig. 14.22 (b)). Cyclooctatetraene has alternate single and double bonds in a tub-shaped non-aromatic structure. Due to rapid quadrupolar relaxation of ^{99}Ru and ^{101}Ru (both with $I = 5/2$), Ru–^{13}C spin-spin couplings are not observed. With increase in temperature, signals *b* and *c* broaden more rapidly than *a* and *d*. Around –50 °C, a single ^{13}C peak is found at the average chemical shift of the four peaks, due to the rapid rearrangement of the ligand, which renders all the eight carbon atoms of the ring effectively equivalent. A 1,2-rearrangement mechanism [Fig. 14.22 (b)] as in η^1-cyclopentadienyl germyl compound (Fig. 14.17) has been suggested to interpret the observed spectral behaviour.

The X-ray diffraction studies of single crystals have revealed that the ligand is tetrahepto. At lower temperatures, the ^{1}H NMR spectra show the broadening of the signal and then splitting into four peaks. These peaks are expected for the four pairs of protons of a η^4-C_8H_8 ring. At room temperature, the ring moves around the metal atom rapidly compared with the time scale of the NMR experiments, and hence an averaged signal is observed. The ^{1}H NMR spectral studies of $Ru(\eta^4\text{–}C_8H_8)(CO)_3$ are consistent with the ^{13}C spectral results. Accordingly at room temperature, the ^{1}H NMR spectrum exhibits a single sharp peak. At lower temperatures, the motion of the ring is slower and distinct conformations long enough to be recorded in the spectrum occur.

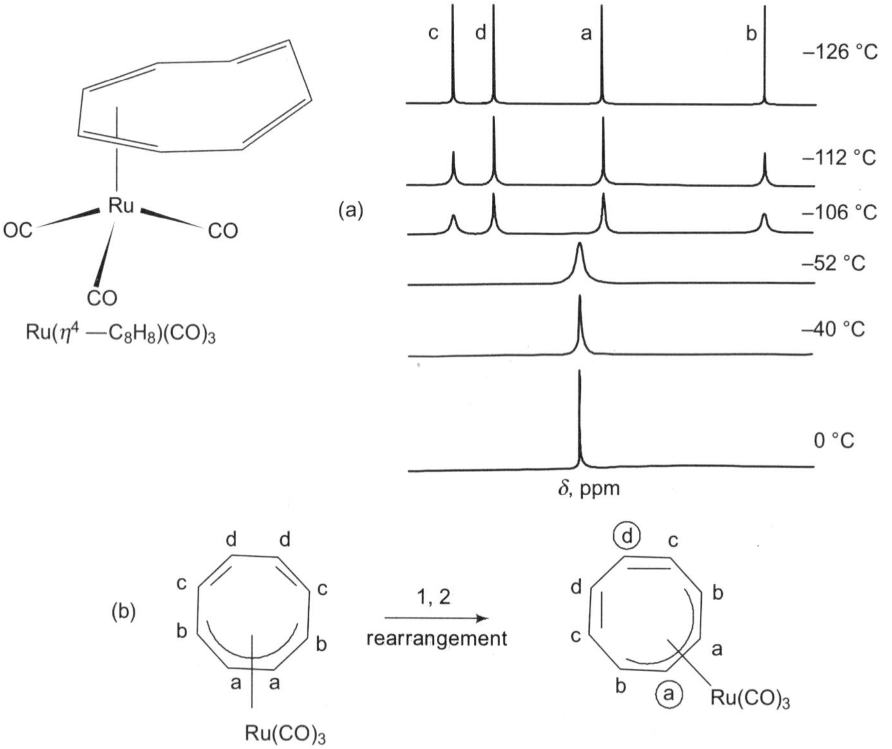

Fig. 14.22 (a) ^{13}C {^1H} NMR spectra of (η^4-C_8H_8)Ru$(CO)_3$ arising from the cyclooctatetraene ligand at different temperatures and (b) 1,2-rearrangement mechanism.

Substituted Ethanes

It is known for a long time that the rotation around the carbon-carbon bond in ethane is hindered. The molecule must therefore overcome an energy barrier in the conversion of the most stable staggered conformation (*a*) into its rotational isomer (*a′*) (Fig.14.23). The barrier to rotation is due to electronic repulsions.

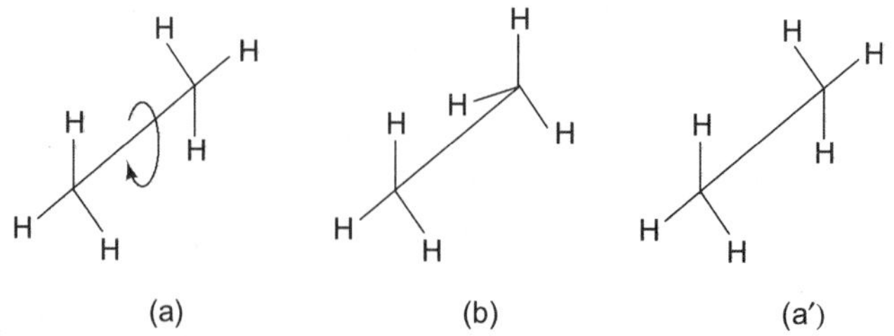

Fig.14.23 Conformation of ethane.

The barrier energy profile for a full rotation as a function of the HCCH torsional angle ϕ is given in Fig. 14.24. The height of the barrier obtained from microwave spectroscopy is about 12 kJ mol^{-1}. The replacement of one or more hydrogen atoms by bulky groups increases the height of the barrier and this shows that steric interactions play a dominant role in restricting the rotation in substituted ethanes.

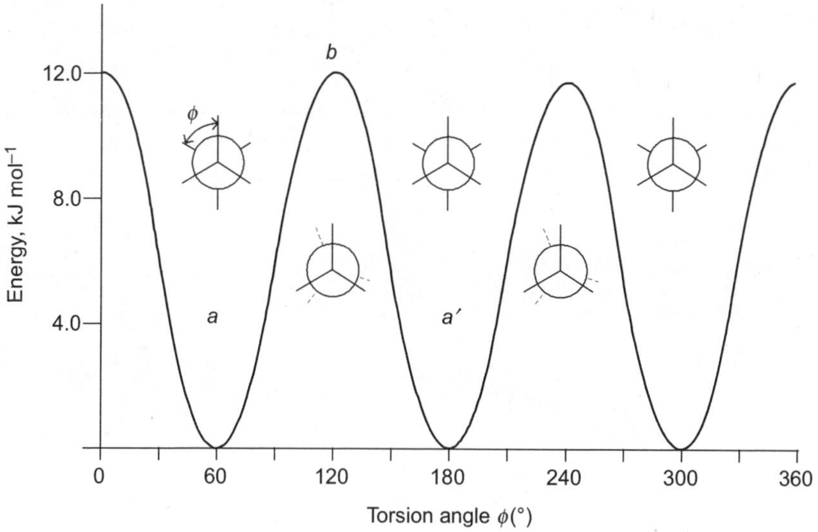

Fig.14.24 Potential energy as a function of torsional angle HCCH in ethane.

The information about the stable conformations of substituted ethanes were obtained from the vicinal coupling constants and their dihedral angle dependence (see Chapter 12). Further the barrier to rotation in several systems has been measured by analysing the temperature dependence of the spectra. For this purpose, ^{19}F NMR spectroscopy was used because of the large differences in the observed ^{19}F chemical shifts.

Rotational isomers of ethane type of molecules such as 1, 2-dichloroethane presents an interesting problem. A typical example of such a system is 1, 2-dichloroethane. The potential is maximum when the two groups are in the eclipsed position and minimum when they are in the staggered position. Figure 14.25 shows the different rotational isomers of 1, 2-dichloroethane. The protons in the three conformations are not magnetically equivalent. At room temperature, the rotation about the C—C bond is very rapid and hence only one signal is obtained. At low temperature, the molecular rotation about the C–C bond is frozen and the isomers become distinct giving rise to a spectrum with multiplet structure.

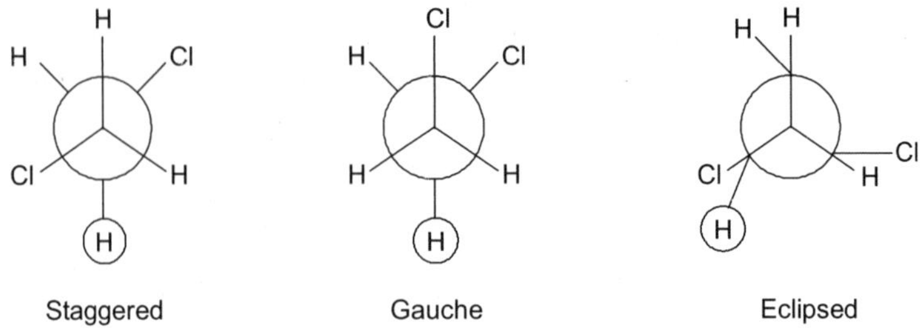

Fig. 14.25 Rotational isomers of 1, 2–dichloroethane (Newman diagram).

SPECTRA OF OTHER NUCLEI: ^{13}C, ^{19}F AND ^{31}P 15

GENERAL REMARKS

The use of NMR spectroscopy of nuclei other than ^1H to characterise and identify organic and inorganic compounds is now common. There are over a hundred different nuclei whose spin is non-zero and, therefore, in principle observable in an NMR experiment. Of these nuclei, nearly 30 of them have their spin $I = 1/2$. The most important nuclei is however the proton. It is not possible to discuss the spectra of all the elements that have been studied by NMR, but a brief account of the NMR spectroscopy of three spin 1/2 nuclei, namely, ^{13}C, ^{19}F and ^{31}P is given here. With the development of Fourier transform methods, extensive studies have been made on the NMR spectra of compounds containing ^{13}C, ^{19}F, ^{31}P and other nuclides. A characteristic feature of the spectra of heavier nuclei is that they have chemical shifts much larger than the proton chemical shifts. The ^{13}C, ^{19}F and ^{31}P spectra when combined with other spectra, especially other forms of NMR, give detailed information on composition, structure and stereochemistry. With these nuclei we can broaden our outlook to other elements of the periodic table.

CARBON-13 NMR

The ^{12}C nucleus is not magnetically "active" (spin $I = 0$) but the ^{13}C nucleus like ^1H has a spin of 1/2. In high resolution NMR, the ^{13}C nucleus is second in popularity to the proton. The ^{13}C NMR spectroscopy is most useful in organic structural analysis. The magnetogyric ratio of ^{13}C is 1.404 as compared to 5.585 for the proton, and hence the ^{13}C signals are inherently weaker. Since the sensitivity of a nucleus to NMR is proportional to γ^3 and $\gamma(^1H)/\gamma(^{13}C) \approx 4$, the ^{13}C nucleus will be less sensitive than the proton by a factor of 64 assuming equal number of nuclei. The low natural abundance (1.1%) of ^{13}C is another factor, which further reduces the sensitivity by nearly two orders of magnitude. The overall sensitivity of ^{13}C compared with ^1H is about 1/5700. As the sensitivity of ^{13}C is very low, usual CW spectrometers are not adequate for recording the spectra. The relatively long spin relaxation time also causes difficulties in recording the ^{13}C NMR spectrum by the CW technique.

The availability of pulsed Fourier transform (FT) instruments has resulted in increased activity in ^{13}C spectroscopy. As stated above, after proton NMR it is the ^{13}C NMR, which is used as a prime tool in structural analysis. The low natural abundance of ^{13}C has an added advantage. The ^{13}C spectra are easily analysed because homonuclear $^{13}C-^{13}C$ spin-spin couplings are absent and the ^{1}H NMR spectra of organic compounds are not affected by heteronuclear $^{13}C-^{1}H$ coupling. The chemical shift of ^{13}C in organic compounds span a range of about 250 ppm. Therefore, the ^{13}C spectrum exhibits a separate and distinct resonance for each chemically different nucleus.

The ^{13}C NMR spectroscopy has been developed into a routine method for structural elucidations. The ^{13}C spin-lattice relaxation measurements which can be carried out routinely provide information on molecular motion and also help in the assignment of ^{13}C spectra. The ^{13}C spectrum of an organic compound yields structural information more directly than a ^{1}H spectrum does. The ^{1}H and ^{13}C NMR spectroscopy are, in fact, complementary. The definition of the chemical shift equivalence given for protons is also applicable to carbon atoms.

For nuclei other than proton the contribution of local diamagnetic term σ_d (due to currents involving local electrons) to shielding is relatively unimportant. Neighbouring group effects are also of minor importance. The paramagnetic term σ_P involving the mixing of the excited electronic states with the ground state in the presence of an applied magnetic field is the dominating term in determining the shielding.

Because of the larger J values for one bond $^{13}C-^{1}H$ (~ 110–320 Hz) and appreciable values of two-bond $^{13}C-C-^{1}H$ and three-bond $^{13}C-C-C-^{1}H$ coupling constants, the proton-coupled ^{13}C spectra usually show complex overlapping multiplets that are difficult to interpret, but some proton-coupled spectra may be simple. The ^{13}C coupling spreads out the intensity of an already weak resonance thus making their observations more difficult. An important development was therefore the use of broad band decoupling of protons. The ^{13}C NMR spectra are usually recorded with protons decoupled (heteronuclear decoupling; see Chapter 17) denoted as $^{13}C\{H\}$ so as to eliminate the $^{13}C-^{1}H$ splittings. Consequently, the spectrum is greatly simplified; no overlapping of the signals occurs; each chemically inequivalent carbon in the molecule gives rise to a singlet in the proton (or broad band) decoupled ^{13}C spectra. Further the ^{1}H-decoupled ^{13}C spectra have higher intensity due to the nuclear Overhauser effect (NOE), which is discussed later and because all the available intensity for each ^{13}C is concentrated in a single peak rather than being spread over several components of the multiplet. The line width of ^{13}C signals is small, 0.5 Hz or less. As a result of the sharpness of the decoupled peaks, impurities are readily detected. The coupling of ^{13}C to other nuclei such as ^{31}P, ^{19}F and ^{2}D may be observed in the proton decoupled spectra.

As a result of broad band proton decoupling, in the absence of other nuclei such as ^{31}P or ^{19}F, each chemically non-equivalent ^{13}C atom shows a single peak. Further, an increase in the signal intensity up to 200% accumulates from the nuclear Overhauser effect. This NOE enhancement results from an increase in the population of the lower energy level of the ^{13}C nuclei concomitant with an increase in the population of the higher energy level of ^1H nuclei on irradiation of the ^1H nuclei. The net effect is a very large reduction in the time needed to obtain a broad band decoupled spectrum as compared to that required for a coupled spectrum. As an example, the ^1H decoupled ^{13}C spectrum of *n*-pentanol is shown in Fig. 15.1.

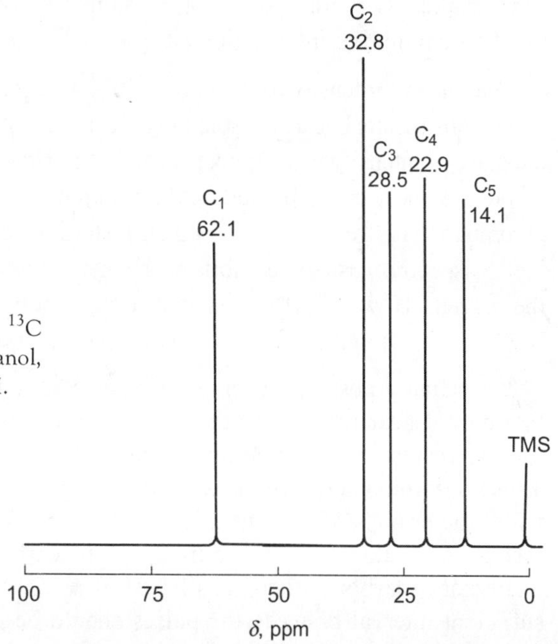

Fig. 15.1 The proton-decoupled ^{13}C NMR spectrum of n-pentanol, $CH_3CH_2CH_2CH_2CH_2OH$.

We noted earlier that the CH_3 protons of N, N-dimethylformamide gave separate peaks for each CH_3 group at room temperature but the chemical shifts of the two CH_3 groups became equivalent at about 123 °C. The ^{13}C peaks show a similar behaviour.

The resonance frequency for ^{13}C is about 1/4 of that for ^1H (the ratio of the magnetogyric ratios for these two nuclei). Thus, for a 7.05 T magnet, the ^{13}C resonance frequency is about 75.5 MHz (300 MHz for ^1H). A ^{13}C spectrum at 75.5 MHz would need about 10 mg of the sample in 0.4 ml of the solvent in a 5-mm outer diameter tube. Samples for ^{13}C spectroscopy are usually dissolved in $CDCl_3$. As in ^1H NMR spectroscopy, the internal reference most commonly employed is TMS and the chemical shift scale is in units of δ, ppm. TMS is arbitrarily assigned the value $\delta = 0.0$. Downfield (deshielding) from TMS is positive and upfield (shielding) from TMS is negative.

Peak Assignments

In FT-^1H NMR spectra, there is usually satisfactory relationship between the integrated peak area and the number of protons because there is sufficient time between the *rf* pulses for the nuclear relaxation to occur. In routine ^{13}C spectra, however, the ^{13}C nuclei, whose relaxation times (T_1) vary over a wide range, are not equally relaxed between the pulses and the peak areas do not integrate for the correct number of nuclei. Thus, the area under the peak in a ^{13}C spectrum is not necessarily proportional to the number of carbon atoms giving rise to the signal. Further, the NOE response is not the same for all ^{13}C nuclei resulting in a further loss of quantitative information.

The relative intensity measurements of ^{13}C resonance peaks often give apparently anomalous results because of the long ^{13}C relaxation times as mentioned above. This anomaly in intensity measurements can be avoided if a much longer time is spent to record the spectrum or by adding the solution of a paramagnetic ion such as iron or chromium. The free electrons in such metal ions can be very efficient relaxers and the ^{13}C spectrum usually exhibits nearly normal intensity pattern after the addition of the reagent. However, if the reagent forms a complex with the given molecule, the NMR signals may become too broad to be observed.

It is usually possible from a ^{13}C spectrum to identify by inspection the nuclei that do not bear protons by their low intensity peaks. The spin-lattice relaxation for ^{13}C results from dipole-dipole interaction with the directly bonded protons. Thus non-protonated carbons will have longer T_1 relaxation times which, together with little or no NOE results in low intensity peaks. It is, hence, usually possible to recognise peaks (except formyl) from carbonyls, nitriles, etc. and other quaternary carbon atoms easily. However, sufficient number of pulses or sufficient interval between the pulses should be given so that these weak signals are not completely lost in the base line noise. The proton bearing carbon atoms have shorter relaxation times as well as enhanced peak intensities caused by the nuclear Overhauser effect (see Chapter 17).

Off-Resonance Decoupling

There are certain disadvantages arising from ^{1}H decoupling. It prevents the measurement of $^{13}C-^{1}H$ coupling constants and thus valuable experimental information is lost. It may render the assignment of signals to specific carbon atoms of the molecule difficult. The nuclear Overhauser effect leads to intensity distortions and the integration of the spectral peaks becomes questionable, as noted above.

There are, however, other techniques which permit spin-spin splitting to be retained; the simplest technique is known as the off-resonance decoupling. The aim of the off-resonance decoupling is to retain spin-spin coupling of the directly bonded protons but eliminate the coupling due to adjacent or further away protons. Thus in the case of ^{13}C NMR, the smaller (geminal, vicinal and long range) ^{13}C, ^{1}H coupling constants are eliminted while the larger splittings due to one bond coupling constants are retained. Thus off-resonance proton decoupled

^{13}C spectrum yields a singlet for a non-protonated carbon, a doublet for CH, a triplet for CH_2 and a quartet for CH_3 group. The proton decoupled and off-resonance decoupled ^{13}C spectra of 1,2,2-trichloropropane are shown in Fig. 15.2. As another example, the off-resonance ^1H decoupled ^{13}C NMR spectrum of vinyl acetate is compared with that of the ^1H decoupled spectrum in Fig.15.3. The resonances of the C=O, =CH, $=CH_2$ and CH_3 carbons occur at δ 167.2, 141.8, 96.8 and 20.2 ppm, respectively. However, in off-resonance decoupling overlap remains a serious problem. The off-resonance decoupling technique is obsolete and has now been completely replaced by a new technique known as the "Distortionless Enhancement by Polarisation Transfer" (DEPT), which is discussed later.

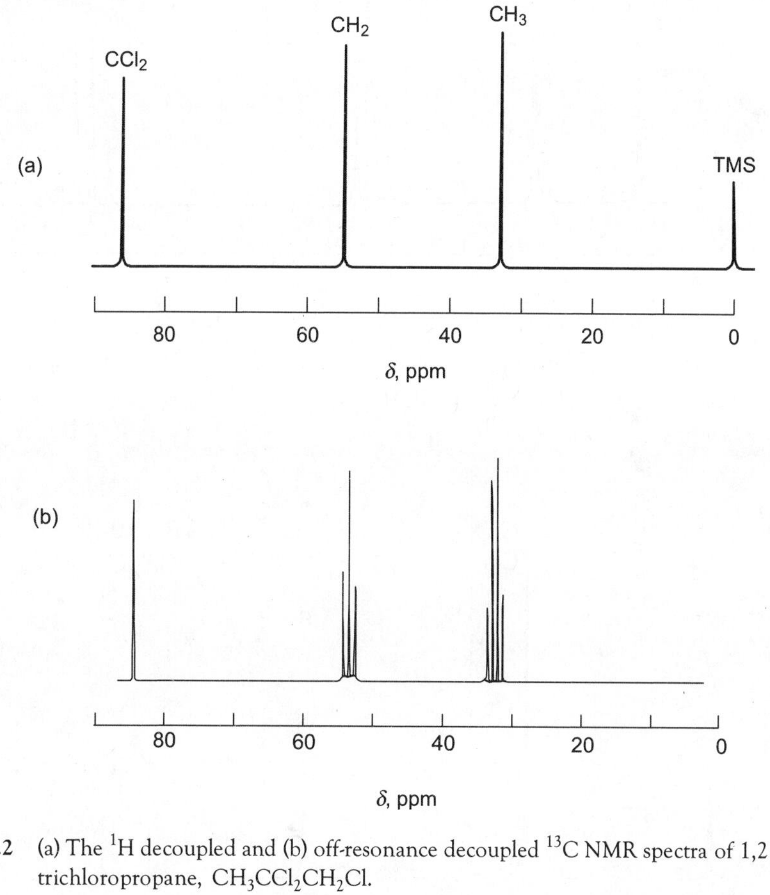

Fig. 15.2 (a) The ^1H decoupled and (b) off-resonance decoupled ^{13}C NMR spectra of 1,2,2-trichloropropane, $CH_3CCl_2CH_2Cl$.

The proton decoupled ^{13}C spectrum (Fig 15.4a) of ethoxy benzaldehyde shows the resonances of two carbons of the ethoxy and one of the aldehyde groups and four of the aromatic ring. Two of the resonances to the low field are rather weak as would be expected for quaternary carbons. Of the remaining signals, two have significantly larger intensity and might arise from two aromatic carbons each. The

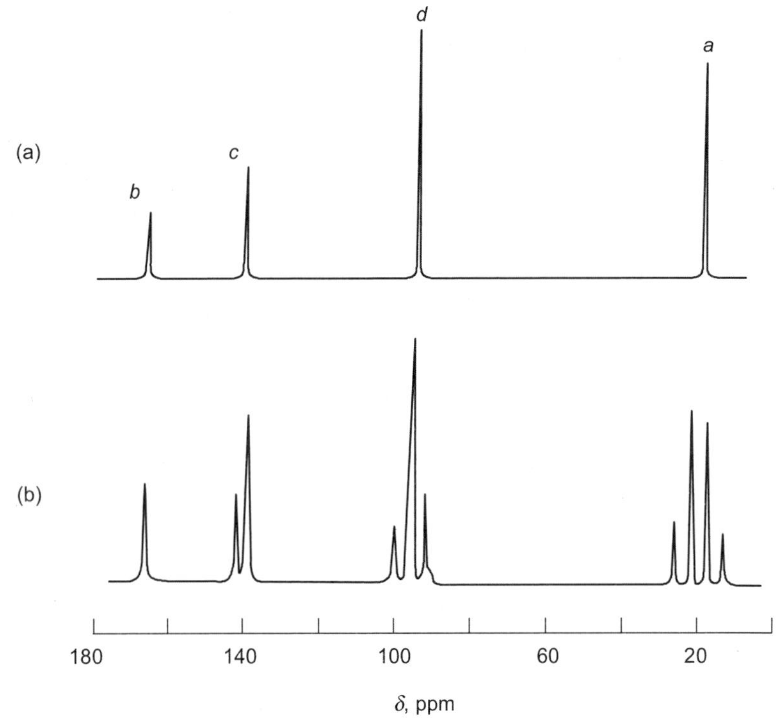

Fig. 15.3 (a) The ^1H decoupled and (b) off-resonance ^1H decoupled ^{13}C NMR spectra of vinyl acetate, $H_2\underset{d}{C}=\underset{c}{C}H-O-\underset{b}{C}(=O)-\underset{a}{C}H_3$.

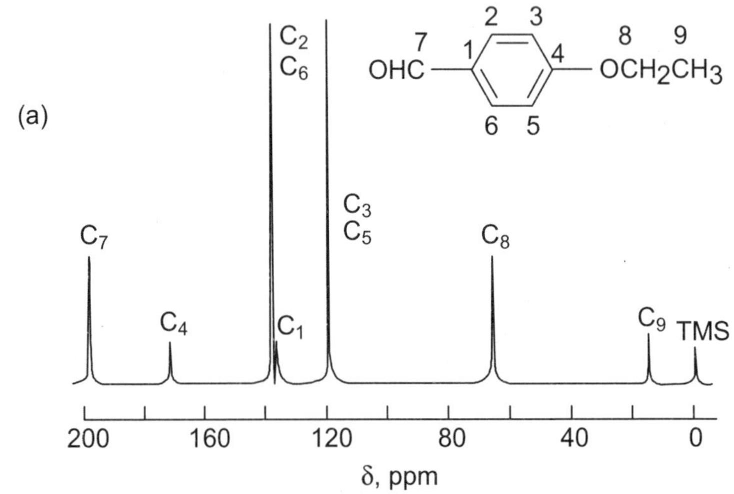

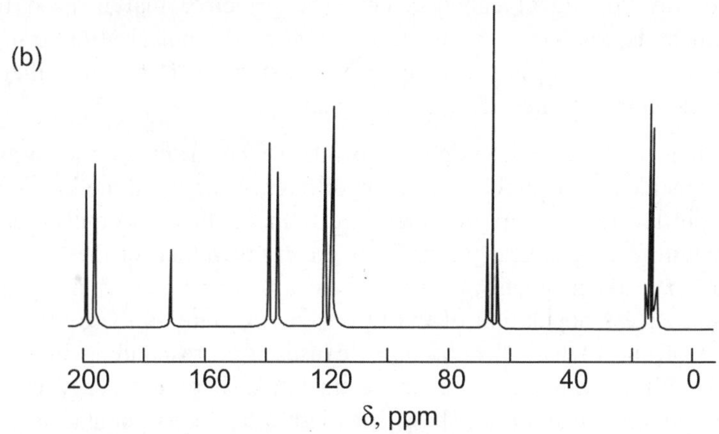

Fig 15.4 The ^{13}C NMR spectra of p–ethoxy benzaldehyde: (a) proton decoupled and (b) off-resonance proton decoupled ^{13}C spectra.

pattern has a form expected for a disubstituted benzene. Thus the spectrm permits us to determine the structure of this molecule. The off-resonance proton decoupled spectrum (Fig. 15.4b) supports the assignments shown in Fig. 15.4a.

Gated Decoupling

Another decoupling technique more often used than the off-resonance technique is gated-decoupling. The broad band decoupler is gated (switched off) during certain pulse NMR experiments an hence spin decoupling occurs only when the decoupler is switched on and not when it is switched off. This technique is used to eliminate either ^{13}C–^{1}H coupling or nuclear Overhauser enhancement in the ^{13}C spectrum. It is employed as a standard procedure in many other ^{1}H NMR experiments such as APT (attached proton test) discussed later. The details of this technique are delineated in Chapter 17.

Other NMR Experiments

There are a very large number of special NMR experiments available and they are usually based on different excitation pulse sequences. An outline of some of these additional experiments is given here since they can greatly simplify the task of assigning the peaks in complicated ^{13}C NMR spectra.

Polarization Transfer Experiment

The lower natural abundance (1.1%) of ^{13}C and its lower γ value lead to a reduction in the strength of the ^{13}C resonance signals, some 6000-fold weaker, than those of the ^{1}H signals as discussed earlier. Some increase in the sensitivity can be achieved through the nuclear Overhauser effect (NOE) (see Chapter 17). The main aspect of heteronuclear NOE effects is the sensitivity gain for the less sensitive nuclei with small γ factors. However, the complications that arise in NOE due to nuclei which have negative γ factors have prompted the search for alternate ways of signal enhancement. The principle that is used for signal enhancement is known as population transfer or polarisation transfer. The techniques rely on the transfer of magnetization from ^{1}H to ^{13}C utilizing scalar spin-spin coupling to intensify the ^{13}C signals.

The intensity of an NMR signal depends on the magnetogyric ratio γ of the nucleus which determines the population difference between the lower and upper energy spin levels in an applied magnetic field. At thermal equilibrium, the ^{1}H population difference is nearly four times greater than that for the ^{13}C nucleus for the same external magnetic field; so there are four times as many protons in Boltzmann excess population of the lower energy state as there are ^{13}C nuclei. Inversion of one of the ^{1}H spin states causes a corresponding increase in the population of the ^{13}C nuclei since the ^{13}C transition shares an energy level common with this proton transition as shown in Fig. 15.5. Such population transfer or polarization transfer procedures are now used extensively in NMR experiments such as INEPT and DEPT discussed later.

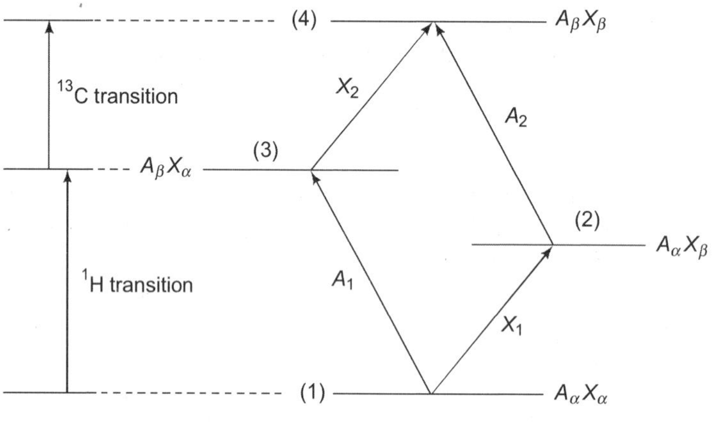

Fig. 15.5 Energy level diagram for a two-spin coupled AX system (^{1}H,^{13}C) showing single quantum transitions. The values in the parenthesis indicate the numbering of the energy levels.

For a two-spin AX system of a sensitive and an insensitive nucleus, for example, ^{1}H and ^{13}C, the equilibrium population of the energy levels and consequently the relative intensities of the A and X lines are governed by the Boltzmann distribution,

$$\frac{N_q}{N_p} = \exp(-\Delta E/kT) = \exp(-\gamma h B_0/kT) = 1 - \frac{\gamma h B_o}{kT} \tag{15.1}$$

The population difference between the two states E_p and E_q are then determined by the magnetogyric ratio of the particular nucleus which changes its spin state during the transition $E_p - E_q$. For energy levels which are connected by the transitions of the sensitive nucleus (A, larger value of γ), a larger population difference results than for those from which the transitions of the less sensitive nucleus (X, smaller value of γ) occur.

The ^{1}H / ^{13}C energy level diagram given in Fig. 15.5 would help to show how polarization transfer, i.e. inversion of one of the ^{1}H spin states enhances the intensity of the two ^{13}C resonances. The energy level diagram for the coupled ^{1}H, ^{13}C has four levels corresponding to (a) $\alpha\,\alpha$, (b) $\alpha\,\beta$ (c) $\beta\,\alpha$ and (d) $\beta\,\beta$ states. The first refers to the proton spin and the second to ^{13}C. The relative energies of the four states are given by the following simple relations, since $E = \pm\, \gamma h B_0 / 4\pi$

$$E_{\beta\beta} = -1/2\,(\gamma_H + \gamma_C)$$
$$E_{\beta\alpha} = -1/2\,(\gamma_H - \gamma_C)$$
$$E_{\alpha\beta} = 1/2\,(\gamma_H - \gamma_C)$$
$$E_{\alpha\alpha} = 1/2\,(\gamma_H + \gamma_C) \text{ (lowest energy state)} \qquad (15.2)$$

For simplicity we add the common factor $1/2\,(\gamma_H + \gamma_C)$ to all the four energy levels. It does not change the energy differences between them. The energy levels will then become γ_0, γ_C, γ_H and $\gamma_H + \gamma_C$. Since $\gamma_H/\gamma_C \sim 4$, the relative population of the four energy levels will then be 0, 1, 4 and 5 (see Fig. 15.6). The signal intensity of ^{13}C nucleus depends on the population difference between its two energy states α and β. Experimentally the population inversion can be achieved by a selective 180°_x pulse applied on one of the lines of the A nucleus. In practice, this is done for a ^{1}H, ^{13}C pair with the proton decoupler applied at the frequency of one of the ^{13}C satellite lines in the ^{1}H spectrum immediately before the ^{13}C pulses. The decoupler pulse then leads to a population inversion between those ^{1}H lines that are connected by the respective satellite lines. On applying the polarization transfer pulse, the population of one of the two ^{1}H states is inverted resulting in a corresponding change in the population of the ^{13}C states. Thus, if the populations 0 and 4 of one of the ^{1}H transitions (A_2 transition) are exchanged, then the ^{13}C population differences become 5 – 0 = 5 and 1 – 4 = –3. (Fig. 15.6(c)) The difference in population between the two ^{13}C spin states before the application of the ^{1}H polarization transfer pulse corresponds to the lower energy state minus the upper energy state, i.e. 1 – 0 = 1 or 5 – 4 = 1 (see Fig. 15.6(a)). Thus, we expect an enhancement of the ^{13}C resonance signal. The net population difference is (5 – 4) + (1 – 0) = 2; after the polarization transfer the difference (5 – 3) is still 2. Thus, there is no net transfer of magnetization. Since the original ^{13}C line intensities before the polarization transfer pulse were +1 and +1, a significant increase in the intensity of the individual antiphase ^{13}C signal occurs. Figure 15.6(b) shows that the alternative exchange of proton populations (5 with 1) through the other ^{1}H transition leads to ^{13}C population differences of 5 – 0 = 5 and 1 – 4 = –3, i.e. the first peak of the ^{13}C doublet has an intensity of +5 and the second peak has an intensity of –3. The two ^{13}C transitions would hence have asymmetric intensities. It shows for the ^{13}C (X) lines increased absorption (X_2) or emission (X_1) and the relative intensities are +5 and –3, respectively. The total intensity found before for the sensitive nucleus is observed now for the absolute intensity of the lines of the less sensitive nucleus. The Boltzmann distribution that determined the spin population for the sensitive nucleus now governs those of the less sensitive nucleus. Contrary to the NOE, the results are independent of the sign of γ_X. For a ^{1}H, ^{13}C pair since $\gamma(^1H)/\gamma(^{13}C) \sim 4$, the relative population numbers 5, 3, –3 and –5 as shown in Fig. 15.6 are obtained. It should be noted that the different natural abundances of A and X nuclei have not been considered here.

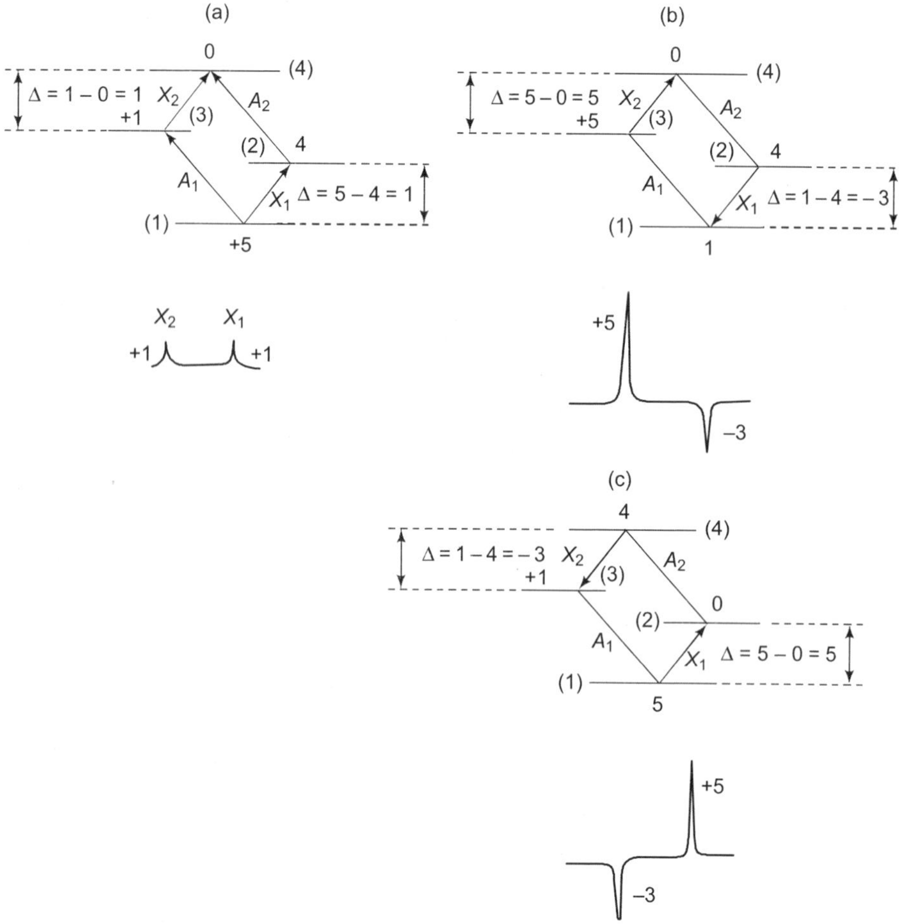

Fig. 15.6 Perturbation of the Boltzmann population in an AX (^{1}H, ^{13}C) spin system through a selective 180° (A) pulse and the resulting X spectra: (a) Boltzmann distribution of spin population at thermal equilibrium determining normal ^{13}C spectral intensities. The expected spectrum is shown. Population inversion of protons through ^{1}H transition : (b) A_2 and (c) A_1 showing perturbed Boltzmann spin population. The expected spectral pattern for the AX (^{13}C–^{1}H) doublet is shown.

Population transfer experiments may be selective or non-selective. If complete population inversion is not achieved, it is termed as selective population transfer (SPT) or selective population inversion (SPI). Selective population transfer experiments have found limited use for signal multiplicity assignments. Non-selective polarization transfer represents a process that allows simultaneous polarization transfer from all the A nuclei (e.g. protons) to all X nuclei (e.g. ^{13}C). Figure 15.7 shows the experimental results for the $^{1}H,^{13}C$ spin system of chloroform which was the first compound to be investigated.

The SPT experiment is useful for signal assignments as well as for the measurement of the spectra of the less sensitive nuclei such as ^{13}C, ^{15}N or ^{29}Si where ^{1}H, ^{19}F or ^{31}P can serve as a sensitive nucleus. The increase in intensity for a first order A_nX spin system ($A = {}^{1}H$, $X = {}^{13}C$) is –3, 5; –7, 2, 9 and –11, –9, 15, 13 for doublet, triplet and quartet splittings respectively as discussed later under INEPT. A limitation of the SPT method is that only one line can be inverted at a time and the sensitivity enhancement is limited to a particular A, X spin pair. For a molecule with several less sensitive nuclei, the experiment should be repeated for each X nucleus. Non-selective population transfer experiments such as INEPT or DEPT discussed later have found extensive applications. These new methods for ^{13}C assignment have the additional advantage of signal enhancement.

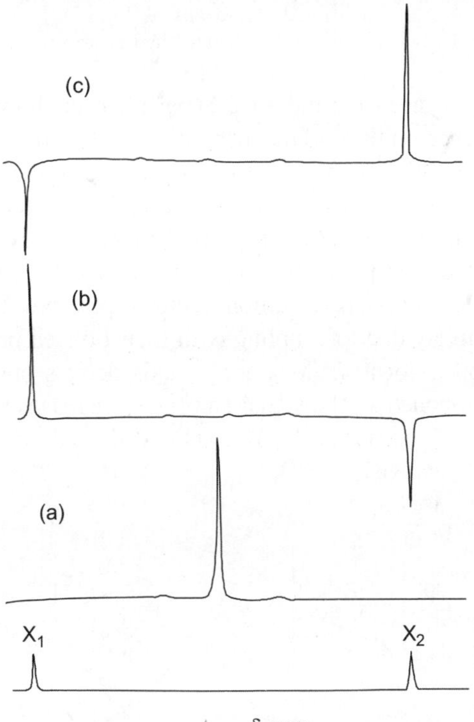

Fig. 15.7 Selective polarization transfer experiment for ^{1}H, ^{13}C spin pair of chloroform. (a) ^{1}H coupled and ^{1}H decoupled ^{13}C NMR spectrum; (b) spectrum obtained after the inversion of ^{1}H line (A_1) (high field ^{13}C satellite peak); and (c) ^{1}H line (A_2) (low field ^{13}C satellite peak).

ATTACHED PROTON TEST (APT)

When a 90° pulse is applied to the ^{13}C nuclei, then during the subsequent interval τ after the pulse, the transverse magnetization vectors of the ^{13}C nuclei of CH_3,

CH_2, CH and quaternary carbons do not rotate synchronously with one another but rotate with characteristically different angular velocities. Characteristic vector orientations result after a certain evolution time. It forms the basis of a method known as the 'Attached Proton Test' (APT) which allows us to differentiate between ^{13}C resonances of CH_3, CH_2, CH and quaternary (C_q) carbons. This is an attractive alternate to the classical method of off-resonance ^1H decoupling.

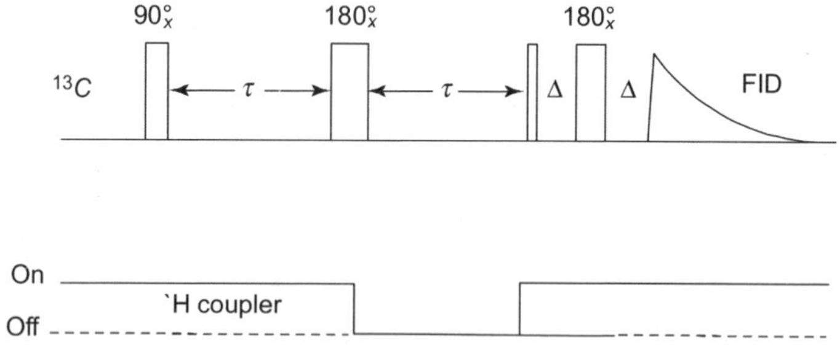

Fig. 15.8 Pulse sequence for the attached proton test experiment.

The pulse sequence used for the APT experiment is shown in Fig. 15.8; the effect of pulse sequence on the ^{13}C magnetization vectors of a CH group is shown in Fig. 15.9. The initial 90° ^{13}C pulse rotates the magnetization vectors of the CH_3, CH_2, CH and quaternary carbons and they come to lie together along the y' axis in the $x'y'$ plane. During the subsequent interval these vectors separate from each other and rotate in the $x'y'$ plane at their characteristic Larmor frequencies. The CH, CH_2 and CH_3 carbon magnetization vectors split into 2-, 3- and 4- vector components, respectively due to coupling with their bonded protons. The detector which by convention is located along the y' axis detects only the sum of these individual vector components. The signal amplitude increases to a maximum value when the vector sum approaches the y'–axis and decreases as they move away from it towards the $-y'$ axis.

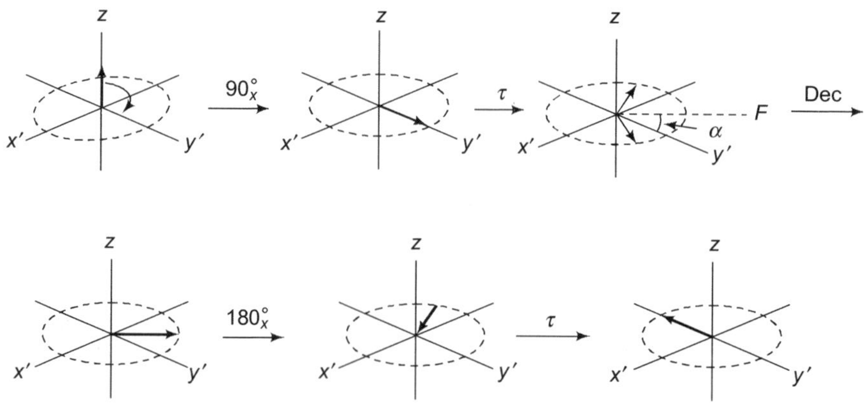

Fig. 15.9 The effect of pulse sequence of the APT experiment on the magnetization vectors of a CH group.

For CH, CH_2 and CH_3 groups (A = ^1H and X = ^{13}C), characteristic vector orientations result after a certain evolution time as stated above. This fact can be utilized for signal selection if $^{13}C,^1H$ coupling is removed during the detection by ^1H decoupling. Switching on the decoupler at the end of the evolution time conserves a particular vector orientation. Because of their larger values (125–250 Hz), this experiment is dominated by one-bond $^1H–^{13}C$ coupling constants and the effect of the geminal and vicinal couplings which are considerably smaller can be neglected.

The decoupler is switched off during the first $1/J$ delay period and it is during this period the J-splitting occurs and the coupling information is retained. The decoupler is switched on at the beginning of the second delay period. It causes the split magnetization vectors to collapse resulting in singlets. At the end of the second $1/J$ interval, the CH_2 and quaternary carbons appear with positive phases while the CH_3 and CH carbons appear with negative phases, thus permitting the CH_3 and CH carbons to be distinguished from the CH_2 and quaternary carbons. A disadvantage of the APT experiment is that it does not allow us to distinguish between CH_3 and CH carbons or between CH_2 and quarternary carbons, although the chemical shifts usually help in the assignment. The signal strength will also be useful since CH_3 carbons are generally more intense than quaternary carbons due to greater nuclear Overhauser enhancement on account of the bonded protons.

The errors may arise in the APT spectrum due to variation in the spin-spin relaxation times T_2 of the different carbons and large variation in one bond and long range $^{13}C—^1H$ coupling constants. The variation in the T_2 relaxation times of the individual carbons can be minimized by keeping the time between excitation and detection constant.

As an example, Fig.15.10 shows the APT spectrum of ethyl acrylate, $H_2C = CH–(C{=}O)–OCH_2–CH_3$. In the APT spectrum, the negative signals represent CH_3 or CH carbons. Therefore, the signal at δ 14.0 ppm of ethyl acrylate is assigned to the methyl carbon and that at δ 128.6 ppm to the olefinic CH carbon. The signals with positive amplitude at δ 60.2 and 128.8 ppm are assigned to the CH_2 carbon of the ethyl group and the terminal CH_2 carbon of the double bond, respectively. The positive signal at δ 167.5 ppm is assigned to the ester carbonyl carbon.

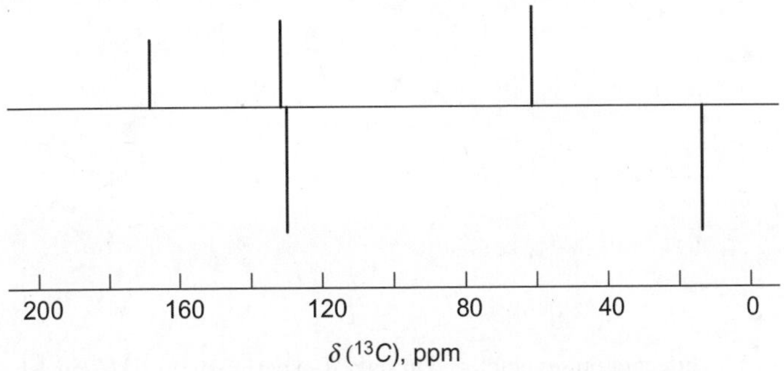

Fig. 15.10 The APT spectrum of ethyl acrylate.

Two intervals are of interest : $\tau = (1/2)J$ which leads to elimination of all ^1H coupled resonances and $\tau = 1/J$ which leads to phase selection with positive amplitude signal for C_q and CH_2 resonances and negative signals for CH and CH_3 resonances. The 180°_x (^{13}C) pulse eliminates the effects of chemical shifts and transverse relaxation and the ^1H-decoupler determines the evolution of ^{13}C magnetization. Additional APT experiments with evolution delays of $1/2J$ are necessary to distinguish between CH and CH_2 resonances. For this reason, nowadays, polarization transfer methods such as DEPT (distortionless enhancement by polarization transfer) and INEPT (insensitive nuclei enhancement by polarization transfer) discussed next have replaced the APT technique for ^{13}C assignments.

INEPT SPECTRA

Another NMR technique which is useful for the determination of carbon

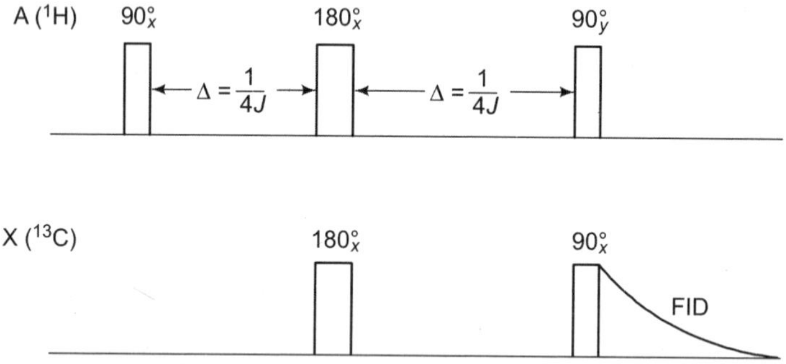

Fig. 15.11 Pulse sequence for INEPT experiment.

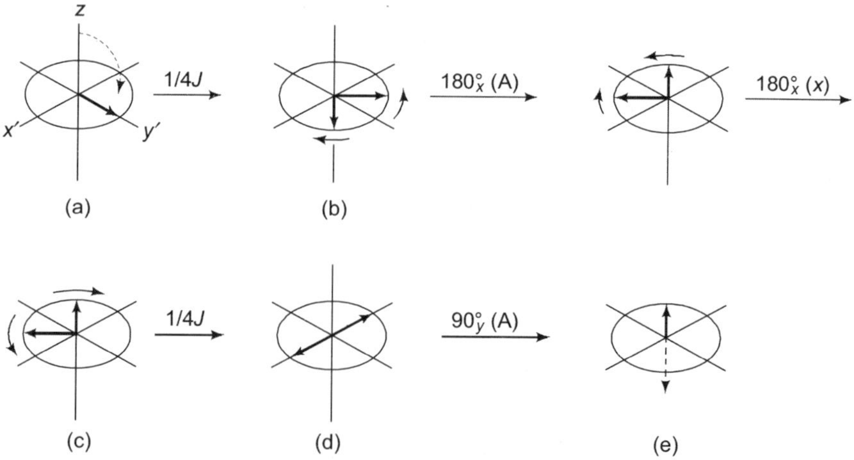

Fig. 15.12 Effect of pulses employed in INEPT experiment on ^1H (A nucleus of an AX spin system, X = ^{13}C) magnetization in the rotating frame ($\nu_0 = \nu_A$).

multiplicities is known as INEPT as mentioned above. In this technique non-selective polarization transfer occurs from all protons to all ^{13}C nuclei having the appropriate $^{1}H-^{13}C$ coupling. The pulse sequence employed for INEPT experiment is shown in Fig. 15.11. The effect of pulse sequence on the ^{1}H magnetization of an AX system is shown in Fig. 15.12.

Application of the 90° ^{1}H pulse aligns the vectors along the y' axis. During the subsequent interval $\tau = 1/4\ J$, the two vectors of the ^{1}H nucleus spread out in the $x'y'$ plane. Applying a 180 ^{1}H pulse causes them to jump across the x' axis and adopt mirror image positions in the $x'y'$ plane. The simultaneous application of the $180_x°$ ^{13}C pulse causes an interchange of the spin labels of the two vectors resulting in an exchange of their "identities" (c). It is accompanied by a corresponding change in the direction of rotation, so that in the subsequent interval $\tau = 1/4\ J$ they move away from each other and align along the x'-axis at the end of the evolution period (d). As in the polarization transfer experiment, a $90°_y$ ^{13}C pulse results in their rotation in the $x'z$ plane so that they point in the opposite directions along the z-axis (e). Thus, while at equilibrium both component vectors of the ^{1}H doublet point towards the $+z$ axis and by applying the INEPT sequence one of these two vector components becomes inverted so that it now points towards the $-z$ axis. This $-z$ magnetization created on one of the ^{1}H vectors causes a corresponding change in the population difference of the ^{13}C vectors. The ^{13}C NMR spectrum now shows signals that are enhanced by a factor of γ_H/γ_C due to greater population difference between the upper and lower energy states of the ^{13}C nuclei. It is read by applying a $90_x°$ ^{13}C pulse that inverts the z-magnetization (longitudinal magnetization) of the ^{13}C nuclei to transverse magnetization in the $x'y'$ - plane.

An important factor in the polarization transfer experiment is the relaxation time of the sensitive nucleus. Since nuclei like ^{1}H or ^{19}F have shorter relaxation times than, for example, ^{13}C the repetition rate for data accumulation can be much higher than for direct measurement of the less sensitive nucleus. This reduces the time required to record the spectra of the less sensitive nuclei. In practice, a five to six-fold enhancement of the sensitivity of INEPT spectra can be obtained whereas in proton-decoupled ^{13}C spectra the repetition rate of successive scans is

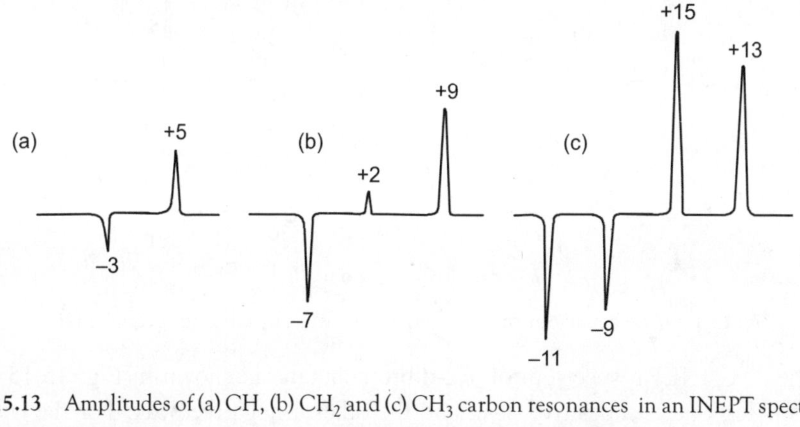

Fig. 15.13 Amplitudes of (a) CH, (b) CH_2 and (c) CH_3 carbon resonances in an INEPT spectrum.

determined by the longer ^{13}C spin-lattice relaxation times. Thus the sensitivity of the INEPT spectra is superior to the four-fold enhancement obtainable by the nuclear Overhauser effect when the proton decoupled ^{13}C NMR spectra are recorded.

In polarization transfer experiments such as INEPT the multiplets do not have the standard binomial intensity distributions. Instead they exhibit intensities of –3, 5 for doublets, –7, 2, 9 for triplets and –11, –9, 15 and 13 for quartets, etc. as shown in Fig. 15.13.

The precession of the ^{13}C multiplet components during the delay Δ by angle θ, where $\theta = \pi J\Delta$ occurs at different angular velocities and it is related to the intensities of CH, CH_2 and CH_3 carbon resonances. Since in the rotating frame, the doublet vectors of the CH carbons precess at frequencies of $+J/2$ and $-J/2$, if the interval Δ is set at $1/2\,J$ (i.e. $\theta = 90°$), then only CH carbons will appear. With Δ set at $1/4\,J$ (i.e. $\theta = \pi/4$), all protonated carbons will appear with positive amplitudes; with Δ set at $3/4\,J$ (i.e. $\theta = 3\pi/4$) CH and CH_3 carbons will show positive signals and the CH_2 carbons will show negative signals. The quaternary carbons do not appear since they have no bonded protons from which polarization transfer can occur. Comparison of the ^{13}C spectra obtained at different intervals Δ readily yields carbon multiplicities. Thus to distinguish between the three kinds of carbon atoms (CH_3, CH_2 and CH), three different spectra should be recorded with suitable delays to make θ equal to $\pi/4$, $\pi/2$ and $3\pi/4$. Appropriate combinations of these spectra allow us to generate subspectra containing CH, CH_2 and CH_3 carbons as shown in Fig.15.14.

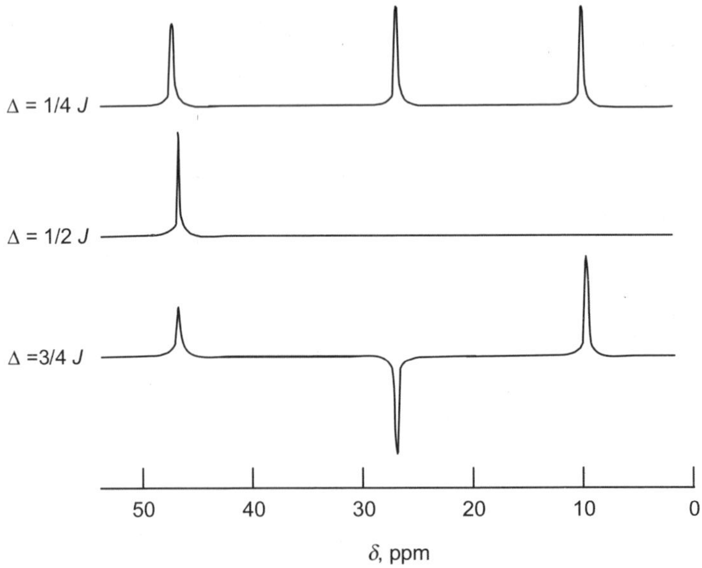

Fig. 15.14 The behaviour of CH, CH_2 and CH_3 groups during spectral editing.

The ^{13}C INEPT spectrum of 1,2-dibromobutane is shown in Fig. 15.15. For the doublet, one is a positive peak and the other a negative peak. For the triplets the outer two peaks appear with positive and negative amplitudes and the central peak with a weak positive amplitude. The quartets have the first two peaks with positive amplitudes and the remaining two peaks with negative amplitudes. Accordingly the assignments have been made for 1,2-dibromobutane.

The INEPT experiment has the following important characteristics.

(i) It enhances the sensitivity of a less sensitive nucleus X such as ^{13}C or ^{15}N by a factor of γ_H/γ_X due to population transfer from an attached sensitive nucleus such as ^1H and by a factor of $(1 + \gamma_H/\gamma_X)$ due to the nuclear Overhauser effect generated by broad-band decoupling. The sensitivity enhancement is particularly marked when nuclei of low magnetogyric ratios are detected. Further, negative γ factor is not a disadvantage.

(ii) The INEPT technique can be used to determine the multiplicity of the carbon signals. It is possible to generate separate subspectra for CH, CH_2 and CH_3 groups.

The evolution time of the INEPT pulse sequence is based on one bond A,X coupling and therefore if J(A, X) is not known it has to be estimated. The APT spectrum does not readily allow us to distinguish between the CH_3 and CH carbons or between CH_2 and quaternary carbons. On the other hand, the INEPT spectrum not only can provide information about the multiplicity of all the carbons but also show sensitivity-enhanced ^{13}C resonances due to polarization transfer.

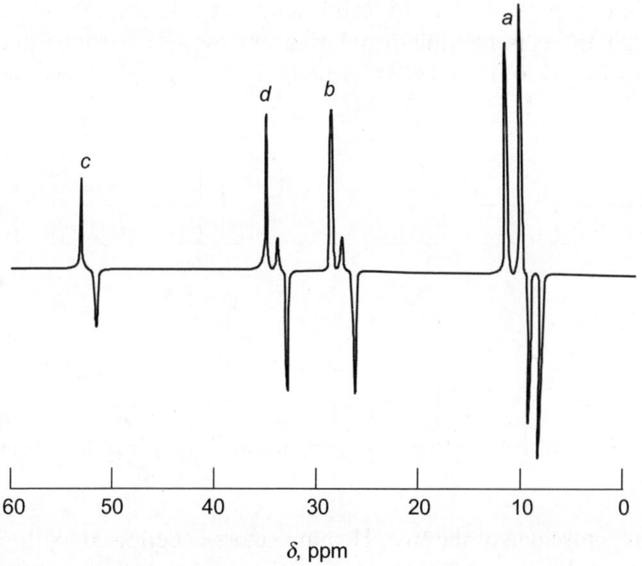

Fig. 15.15 INEPT ^{13}C spectrum of 1, 2-dibromobutane, $Br—{}^dCH_2—{}^cCH(Br)—{}^bCH_2—{}^aCH_3$.

DEPT SPECTRA

In the analysis of the ^{13}C spectrum of a complex molecule, it is very helpful to know *a priori* whether a given ^{13}C resonance arises from a CH_3, CH_2, CH or quaternary carbon. A one-dimensional ^{13}C NMR technique that allows measurement of a ^{13}C spectrum separately for each of the CH_3, CH_2 and CH groups is known as DEPT (Distortionless Enhancement by Polarization Transfer). The DEPT spectrum is most widely used nowadays for determining the multiplicities of the carbon atoms. This has generally replaced the use of off-resonance decoupled ^{13}C spectra from which multiplicities could be read directly.

The pulse sequence used in the DEPT experiment is shown in Fig. 15.16. The effect of pulse sequence on ^1H, ^{13}C magnetization vectors is shown in Fig. 15.17. A 90°_x ^1H pulse rotates the equilibrium magnetization from the z-axis to the y′-axis. During the subsequent evolution period, the transverse ^1H magnetization is modulated by the CH coupling, splitting it into two counter-rotating vectors at the end of the first 1/2J delay, the vectors point along the $+x'$ and $-x'$ axis. A 180°_x ^1H pulse is then applied to remove any field inhomogeneities. Simultaneously a 90°_x ^{13}C pulse is applied which serves to rotate the ^{13}C magnetization from its equilibrium position along the z-axis to the y'-axis. Since there is no z-magnetization for either the ^1H or the ^{13}C nuclei, they are effectively decoupled. Therefore, during the subsequent 1/2J interval, they remain stationary at their original positions and the interval serves to remove the field inhomogeneity effects by the preceding 180°_x ^1H pulse. At the end of the second 1/2J interval a proton pulse of angle θ is applied. Thus to record only CH carbons, the value of θ is kept at 90°, so that there is maximum polarization for CH carbons and negligible x'– magnetization for CH_3 and CH_2 carbons.

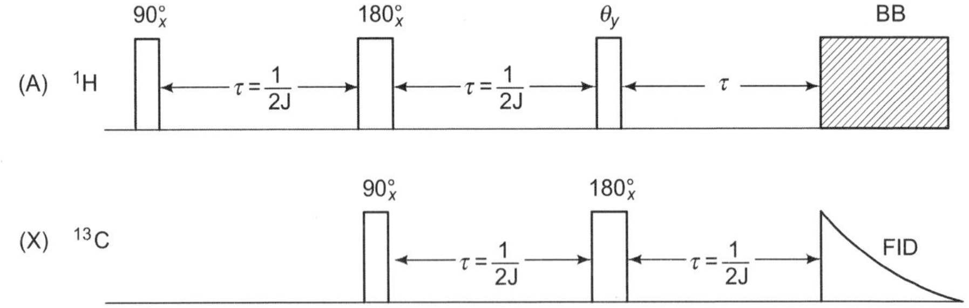

Fig. 15.16 The pulse sequence for the DEPT experiment. The angle of the last pulse θ_y can be varied.

With a 90°_x pulse one of the two ^1H spin vectors is deflected to the $+z$ axis (i.e. to its original equilibrium position) and the other to the $-z$ axis i.e. its population is inverted. As in the APT and INEPT experiments discussed earlier, this inversion of one of the ^1H spin vectors enhances the population difference between the two ^{13}C spin states with which ^1H is coupled, resulting in the intensification of

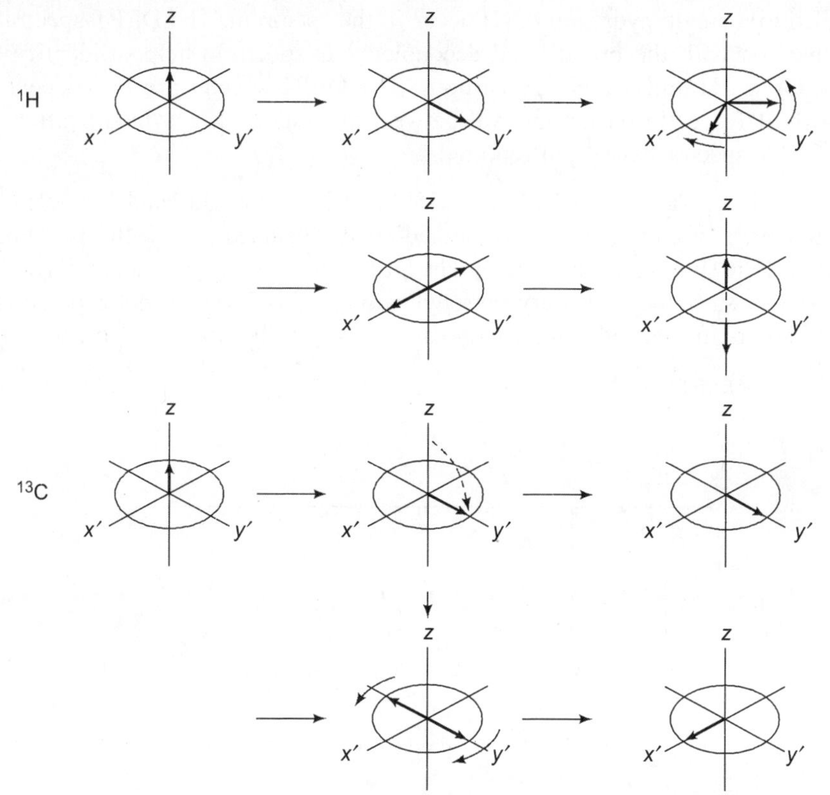

Fig. 15.17 Effect of the DEPT pulse sequence on ^1H and ^{13}C magnetization vectors. The ^{13}C magnetization can be recorded as multiplets or as singlets if broad band decoupling is applied during the acquisition period.

the ^{13}C signal. Simultaneously, a 180° pulse is applied to the ^{13}C nuclei to remove any field inhomogeneity during the third 1/2J delay period. During this last delay, the ^{13}C signals are modulated by ^{13}C–^1H coupling since *z*-magnetization now exists for the ^1H nuclei. The ^{13}C vectors rotate in the x′y′ plane and at the end of the third 1/2J delay, the FID is acquired with or without ^1H decoupling. The intensities of the CH_3, CH_2 and CH carbon signals in the DEPT spectrum depend on the angle θ of the last polarization pulse. A novel feature in the DEPT sequence is a variable proton pulse that is set at 45°, 90° and 135° in three separate experiments. Thus in a DEPT experiment, three spectra can be measured. One is a normal broad band decoupled spectrum which contains all the CH, CH_2 and CH_3 signals. With θ set at 135°, the CH and CH_3 carbons are modulated differently from the CH_2 carbons so that CH and CH_3 carbons give positive (i.e. normal) signals and CH_2 carbons give negative (i.e. inverted) signals and no signal appears for quaternary carbons. With $\theta = 45°$, all protonated carbons (CH_3, CH_2 and CH) show positive signals. The DEPT 90° spectrum is obtained such that only resonances of carbon atoms bonded to a single hydrogen (CH) occur in the spectrum. The DEPT spectra on comparison with the broad band decoupled ^{13}C spectrum help to identify the CH_3, CH_2, CH and quaternary carbons. The DEPT spectra are less dependent than INEPT spectra on the interval between the pulses. Many modifications of the DEPT spectra have been reported.

Figure 15.18 shows the DEPT ($\theta = 135°$ and $90°$) and broad-band decoupled ^{13}C NMR spectra of ethyl acrylate. The broad-bond decoupled ^{13}C NMR spectrum of ethyl acrylate shows five resonances. The DEPT ($\theta = 135°$) spectrum displays only four signals since the quaternary carbonyl carbon (δ 167.5) does not appear in the DEPT spectrum. Thus, the quaternary carbons are easily differentiated from CH,

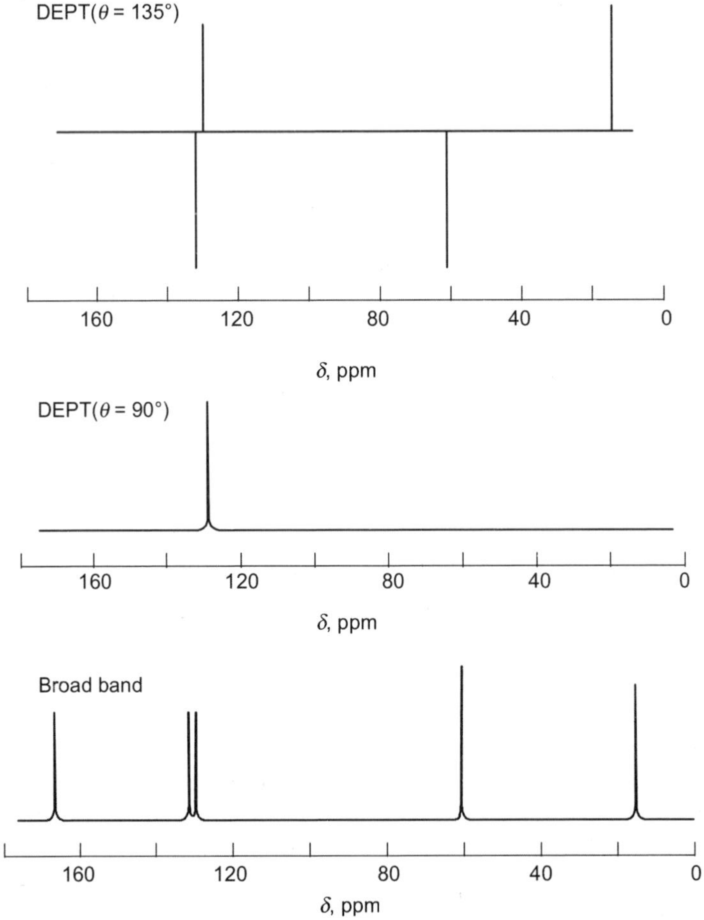

Fig. 15.18 DEPT (θ = 90° and 135°) and broad band ^1H decoupled ^{13}C spectra of ethyl acrylate, 128.8 128.6 167.5 60.2 14.0

$$H_2C = CH - \underset{\underset{O}{\|}}{C} - O - CH_2 - CH_3$$

CH_2 and CH_3 carbons. The CH (δ 128.6) and CH_3 (δ 14.0) carbons appear as normal signals whereas the CH_2 carbons (δ 60.2 and 128.8) appear as inverted signals. The DEPT ($\theta = 90^o$) spectrum exhibits only the methine carbon (δ 128.6) and hence CH_3 carbons are readily distinguished from CH carbons.

In the DEPT experiments, the intensity enhancement by NOE and polarization transfer are conserved. DEPT experiments are therefore superior to APT experiments both in sensitivity and multiplicity assignments. Quaternary carbons which lack any bonded hydrogen atom are not benefited from the polarization transfer and, therefore do not appear in the INEPT and DEPT spectra. Because of the *J*-dependence of the evolution time, in many cases, two different DEPT experiments are necessary since the ^1J (^{13}C, ^1H) values for *sp*, sp^2 and sp^3 carbons differ considerably.

Both DEPT and INEPT experiments are based on polarization transfer from the sensitive nuclei to the less sensitive nuclei and therefore major portions of their pulse sequences are common. The INEPT spectrum taken without refocussing and decoupling exhibits "distorted" multiplets. For example, the two lines of a doublet appear in antiphase with respect to one another. On the other hand, in the DEPT experiments, all the signals of the less sensitive nuclei are in phase at the beginning of the acquisition, so no refocussing period with accompanying loss in sensitivity is required. Since the multiplets appear in phase it is called a distortionless technique. Further, the DEPT spectrum depends on the angle θ of the last polarization transfer pulse, and it is less dependent on the interval 1/2J between the pulses. An error of $\perp$ 20% in the estimation of *J* values still gives acceptable DEPT spectra. This is a distinct advantage over the INEPT technique which requires a more accurate setting of the interval between the pulses. The DEPT experiment is usually preferred in practical applications. It yields on ^1H decoupling singlet signals for all types of ^{13}C resonances. Nowadays, a DEPT spectrum is recorded routinely as a part of the ^{13}C spectrum.

^{13}C CHEMICAL SHIFTS

The range of chemical shifts generally found in routine ^{13}C spectra is about 250 ppm as noted earlier. The trends in the chemical shifts of ^{13}C are somewhat similar to those of ^1H. Solvent effects are important in both spectra. The chemical shift of a given carbon atom in a series of related compounds is sensitive to the influence of the substituents. It is related mainly to the hybridization and electronegativity of the substituent. As in other types of spectroscopy, peak assignments are made on the basis of reference compounds. The spin-lattice relaxation time measurements are also helpful in the assignment of ^{13}C resonances in large organic molecules. This is dealt with later (see Chapter 16).

There is an enormous amount of information on ^{13}C chemical shifts. Typical ranges of ^{13}C chemical shifts are shown in Table 15.1. For saturated carbon atoms the usual range of chemical shifts is δ 0 – 100 ppm relative to TMS and electronegative substituents generally increase the value of the chemical shift. The

Table 15.1 The range of ^{13}C chemical shifts in different organic compounds

Compound	***δ, ppm***
n-Alkanes	5–25
Cycloalkanes	25–40
Alkenes	100–150
Alkynes	75–90
Aromatics and heteroaromatics	120–145
Alkyl halides	15–75
=C=	205–225
Alcohols and ethers	40–90
Alkyl amines	25–80
Alkyl C–S	30–45
Aryl C–S	115–135
R–N≡C	150–165
R–C≡N	105–125
Ketones	195–220
Unsaturated ketones	190–205
Aldehydes	190–210
Carboxylic acids	170–185
Acid chlorides	160–175
Amides	160–180
Esters	155–185
Anhydrides	150–175
Imides	160–180
Lactones	170–178
Ureas	150–170
Carbonates	150–160

chemical shift associated with the methyl group in compounds of the type CH_3X is directly related to the electronegativity of the group X and to the number of lone pair of electrons on the X atom. But the effects are not additive and hence the chemical shifts for multiple substituted compounds such as CH_2XY cannot easily be predicted. Carbonyl carbon atoms are the most strongly deshielded ones and their resonances form a separate region at the lowest field. Chemical shifts for three coordinated carbon atoms usually lie in the range of δ 80 – 240 ppm while chemical shifts for alkynes are smaller, between δ 20 and 100 ppm. Chemical shifts for aromatic compounds are in the range δ 110 – 150 ppm. However, resonances in π-bonded metal alkene and arene complexes may be shifted up to 100 ppm to the low frequency from their usual range depending on the mode of coordination. The chemical shift of CI_4 (δ –293 ppm) represents one extreme. At the high frequency end, metal carbonyls are usually found between 170 and 290 ppm while metal carbene complexes typically have resonances between 250 and 370 ppm. It is often difficult to record the ^{13}C spectra of metal carbonyl complexes since the relaxation times for ^{13}C nuclei in these compounds can be very long because the major relaxation pathway involves the dipolar coupling of bonded protons. This problem can be overcome by the use of paramagnetic relaxation agents such as tris(acetylacetonato)chromium(III), which is discussed later.

The hybridization of the carbon atom has a significant effect on the chemical shift. The sp^3 hybridized carbon shows signals at high field (0 – 60 ppm downfield of TMS), sp^2 carbon at low field (110 – 150 ppm downfield of TMS) and *sp* hybridized carbon absorbs at intermediate values (65 – 90 ppm).

The chemical shifts observed in a particular region of the spectrum provide a good indication of the functional groups present in the molecule as can be seen from the correlation charts. The fact that a given functional group e.g. a heterocyclic compound having ring nitrogen or a substituted benzene shows a very characteristic shift has made the finger print type of application of ^{13}C NMR even more successful than that of ^{1}H NMR. The predictions of ^{13}C chemical shifts based on the tables of empirical data are reliable.

The substituent effects in alkanes, where the replacement of a hydrogen atom by a methyl group leads to deshielding by 9-10 ppm for the α and β carbons and to shielding of 2.5 ppm for the γ carbon have been noted. The values of chemical shift (in δ, ppm) for some alkanes are given below for illustration.

H_3C — CH_2 — CH_2 — CH_3
13.0 23

H_3C — CH_2 — CH_2 — CH_2 — CH_3
13.7 22.7 34.5

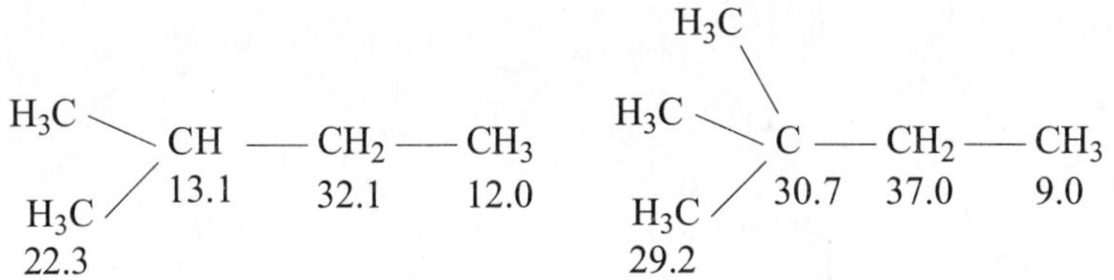

For cyclohexane, the increment for axial and equatorial substitution varies. This difference between the axial and equatorial substituents is important for conformational analysis using their ^{13}C chemical shifts. Tables of substituent induced ^{13}C chemical shifts for different types of compounds have been published. Large ^{13}C chemical shift variations are observed on protonation and deprotonation. For example, ^{13}C resonances of pyridine at C 2,6 at 150.2, C 4 at 135.9 and C 3,5 at 124 ppm occur at 148, 137.5 and 124 ppm respectively on changing from an acidic to basic pH. The values of chemical shift for selected carbon compounds are given in Table 15.2.

Table 15.2 ^{13}C Chemical shifts in δ, ppm for selected compounds

CH_4 — –2.3

$CH_3—CH_2—CH_3$ — 15.9, 15.4

$CH_3—CH_2OH$ — 17.6, 57.0

$CH_3—CHCl—CHCl_2$ — 20.0, 60.0, 75.0

$H_2C{=}CH_2$ — 123.3

$CH_3—CH{=}CH_2$ — 18.7, 135.0, 115.0

$CH_3—CH_2—CH{=}CH_2$ — 13, 28, 140, 113

$R—C{\equiv}N$ — 110-125

$H_2C{=}CH—CH{=}CH_2$ — 116.3, 136.9

$CH_2{=}CH—CH_2—OH$ — 115, 140, 65

$CH_3—CH_2—CH_2—NH_2$ — 11.2, 27.3, 44.9

$CH_3—C({=}O)—O—CH_2—CH_3$ — 30, 170, 70, 20

$CH_3—CH_2—CH_2—CH_2—Cl$ — 13.4, 20.4, 35.0, 44.6

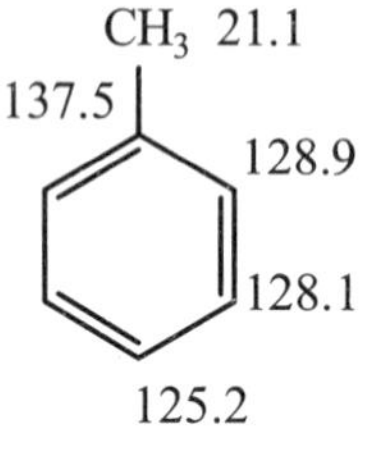

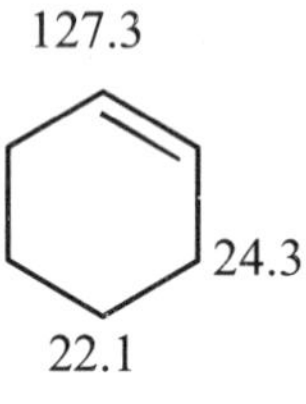

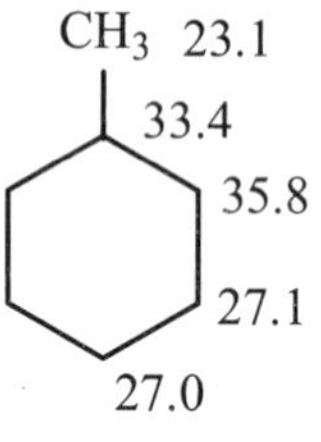

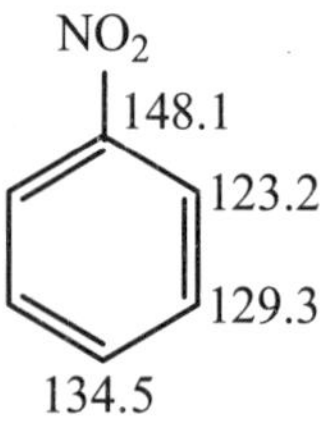

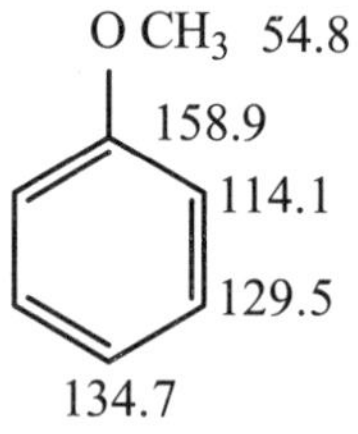

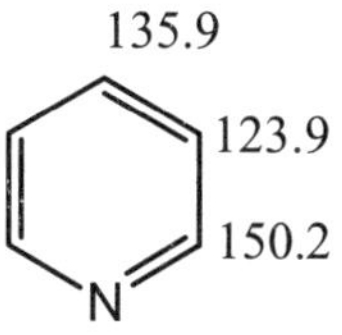

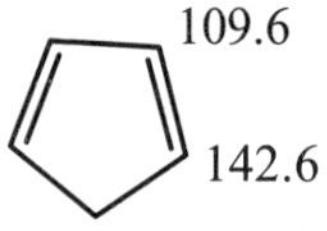

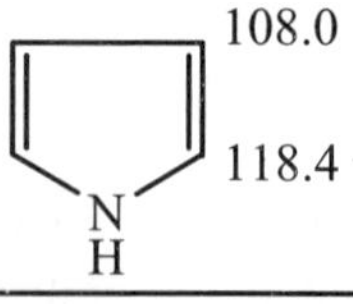

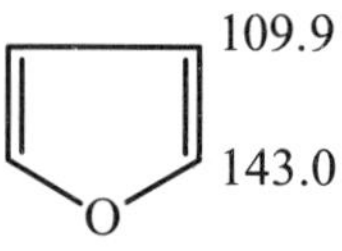

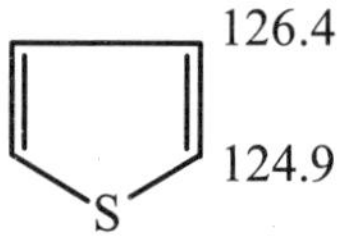

A number of empirical chemical shift–molecular structure correlations have been developed. They have proved useful for the analysis of ^{13}C spectra. The ^{13}C chemical shifts are determined mainly by the varying paramagnetic contribution σ_p. For estimating the contribution to the ^{13}C chemical shifts, the following expression is useful.

$$\sigma_p^i \approx \frac{1}{\Delta E}\left(\frac{1}{r_i^3}\right)_{2p} \sum_{j \neq i} Q_{ij} \tag{15.3}$$

where ΔE is the mean electronic excitation energy, r_i is the average radius of the carbon $2p_z$ orbitals and Q_{ij} is the bond order. Equation (15.3) shows that the largest contributions to σ_p comes from the electronic transitions of lowest energy. For example, the increasing shift of ^{13}C chemical shifts in the series–alkanes, alkenes and carbonyl compounds–is as expected (see Table 15.1). Equation (15.3) can be used for the interpretation of important features observed for carbon shielding. However, nowadays sophisticated quantum chemical methods are available for the calculation of ^{13}C chemical shifts and spin-spin coupling constants.

Partial negative charge leads to an increase in r_i (orbital expansion) and consequently to a decrease in σ_p and increased shielding results. For partial positive charge, on the other hand, a deshielding effect should occur. An empirical relation based on the ^{13}C chemical shift of aromatic ions and the corresponding changes in π charge density $\Delta\rho$

$$\Delta\sigma = 160\,\Delta\rho \tag{15.4}$$

has been proposed. Proportionality constant varying from 60 to 360 has been suggested based on the calculations using π electrons only and those which include both π and σ electrons. Equation (15.4) and similar equations with different proportionality constants are valid to certain classes of compounds. The variation of ^{13}C chemical shifts with charge density also forms a basis for the chemical shift assignments as shown below.

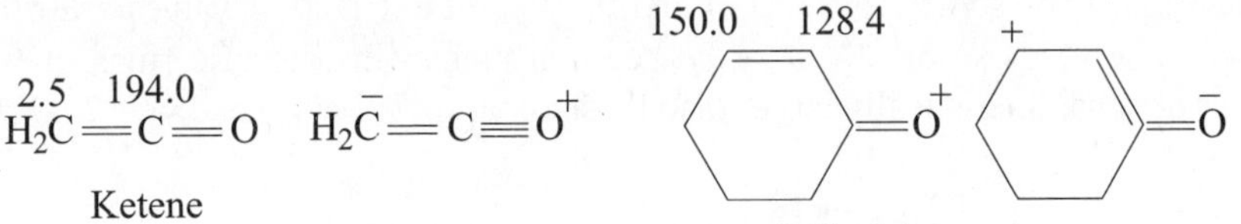

Ketene

2-Cyclohexenone

(Chemical shift in δ, ppm)

The charge density-chemical shift correlations have also been made for ^{1}H and ^{19}F nuclei.

^{13}C–^{1}H COUPLING CONSTANTS

There are three components of a coupling constant arising from the nucleus-electron interactions. First, the magnetic moment of one nucleus interacts with the field produced by the orbital motion of the electrons, which in turn interacts with a second nuclear moment. Secondly, there is the dipolar interaction involving the electron spin magnetic moments. The remaining contribution arises from the Fermi contact interaction. It is the most important factor in the case of proton-proton couplings but for other nuclei, the situation is rather complex. All these terms depend on the magnetogyric ratios of the two nuclei involved.

In ^{13}C spectra, homonuclear ^{13}C, ^{13}C couplings are usually not observed because of the low natural abundance of ^{13}C (1.1%) and its sensitivity is 1.6% of that of ^{1}H. If we take acetaldehyde as an example, among the molecules containing a ^{13}C nucleus at a given position, about only one in a hundred contain a second ^{13}C nucleus. Thus, the spectrum of $^{13}CH_3{}^{13}CHO$ should be nearly hundred times weaker than that of either $^{12}CH_3{}^{13}CHO$ or $^{13}CH_3{}^{12}CHO$. Therefore, the ^{13}C–^{13}C splittings are often not noticed (see Chapter 19). Of course, the most abundant isotopomer $^{12}CH_3{}^{12}CHO$ shows no ^{13}C NMR spectrum.

In organic molecules, the dominant couplings experienced with the ^{13}C nucleus are with their directly bonded protons. There is extensive tabulation of the experimental data concerning ^{13}C–^{1}H coupling constants mostly over one bond. These have been obtained from ^{1}H and ^{13}C NMR spectra. The multiplets produced by these couplings, for example, a quartet for a methyl carbon and a singlet for a quaternary carbon provide valuable guidance for assigning the peaks in the spectrum to particular carbons in the molecule.

The values of spin-spin coupling constants are less important in ^{13}C NMR than in ^{1}H NMR spectra. Since routine ^{13}C spectra are usually proton decoupled, ^{13}C–^{1}H coupling constants are not obtained. One-bond ^{13}C–^{1}H coupling constant ($^{1}J_{CH}$) ranges from about 110 to 300 Hz increasing with increased *s* character of the C–H bond and with substitution of electron withdrawing groups on the carbon atom and angular distortion. Appreciable ^{13}C–^{1}H coupling also extends over two or more bonds. Geminal and vicinal ^{13}C–^{1}H coupling constants are much smaller than the couplings over one bond and generally values between 1 and 12 Hz are observed. Table 15.3 gives some typical $^{1}J_{CH}$ and $^{2}J_{CH}$ values. The $^{3}J_{CH}$ values are roughly comparable to those of $^{2}J_{CH}$ for sp^3 carbon atoms. In aromatic rings, the $^{3}J_{CH}$ values are characteristically larger than those of $^{2}J_{CH}$. In benzene $^{3}J_{CH} = 7.4$ Hz and $^{2}J_{CH} = 1.0$ Hz.

The dependence of ^{1}J (^{13}C–^{1}H) data on the structure has been of considerable interest. An *s*-character dependence has been found. For hydrocarbons, the empirical relation (15.5) relates one bond coupling to the fractional *s*-character of the C–H bond involved.

$$^{1}J\,(^{13}C-{}^{1}H) = 500\ s(i) \qquad (15.5)$$

and for proton s(i) = 1 for the 1*s* orbital. The Fermi contact term is a measure of the *s* character of the bond at the two nuclei. Since the *s* orbital of the hydrogen accommodates all of the proton electron density, the magnitude of $^{1}J(^{13}C—^{1}H)$

Table 15.3 Typical values of $^1J_{CH}$ and $^2J_{CH}$ coupling constants

Compound	*J (Hz)*	*Compound*	*J (Hz)*
$^1J_{CH}$			
CH_3CH_3, CH_4	125	$CH_2=CH_2$	156
CH_3CH_2, CH_3	119	$CH_3\underline{C}H{=}O$	172
$(CH_3)_3CH$	114	$NH_2CH{=}O$	188
CH_3NH_2	133	$HC{\equiv}CH$	250
CH_3OH	141	$HC{\equiv}N$	269
CH_3Cl	151	$^2J_{CH}$	
CH_2Cl_2	178	CH_3CH_3	−4.5
$CHCl_3$	209	$\underline{C}H_3C\underline{H}{=}O$	26.7
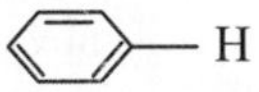	159	$CH_2{=}CH_2$	−2.4
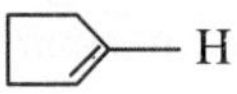	162	$(C\underline{H}_3)_2\underline{C}{=}O$	5.5
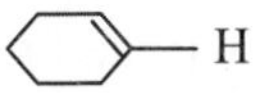	158	$CH_2{=}\underline{C}HC\underline{H}{=}O$	26.9
	220	C_6H_6	1.0
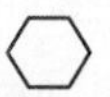	123	$HC{\equiv}CH$	49.3

for directly bonded carbon and hydrogen will depend upon the fraction of the *s* character, in the carbon hybrid orbital bonding with the hydrogen. The following equation could be used to estimate the electron density ρ from ^1J (^{13}C–^1H) data.

$$^1J(^{13}C-^1H)\,Hz = 500\,\rho_{CH} \tag{15.6}$$

It can be seen that ^1J ($^{13}C-^1H$) should be a sensitive measure of the hybridization of carbon. As expected from equation (15.6), the ^{13}C–^1H coupling constant in a saturated hydrocarbon is about 125 Hz, that for ethylenic and acetylenic hydrogens it is around 160 and 250 Hz corresponding to sp^3, sp^2 and sp hybridizations (ρ = 0.25, 0.33 and 0.5), respectively. The $^1J(^{13}C$–$^1H)$ values can be of use in assigning the peaks in a spectrum but electronegative substituents increase the coupling constant greatly, for example, in $CHCl_3$ to 209 Hz. In other cases, the *s* electron contribution from each atom must be considered. Like $^3J(^1H, {}^1H)$ data, vicinal $J(^{13}C, {}^1H)$ couplings are of interest for conformational analysis since they depend on bond lengths and dihedral angles. Equations relating bonding parameters to vicinal ^3J (^{13}C, ^1H) coupling constants have been derived experimentally and theoretically.

^{19}F NMR

Chemical Shifts

Fluorine has only one naturally occurring isotope, ^{19}F, which has a nuclear spin $I = 1/2$. The sensitivity of ^{19}F is about 0.83 of that of ^{1}H and it is an ideal nucleus for studying NMR. The resonance frequency of fluorine is slightly lower than that of proton for a particular applied magnetic field (56.5 MHz as compared to 60 MHz for proton in a magnetic field of 1.4 T). The ^{19}F chemical shift is more influenced by the solvent unlike the proton chemical shift. The ^{19}F NMR was routinely recorded using continuous wave (CW) instruments in exactly the same way as for ^{1}H NMR. Trichlorofluoromethane $CFCl_3$ is used as a standard ^{19}F reference compound (δ 0.0 ppm); it is generally inert, volatile and gives rise to a single ^{19}F resonance.

Fluorine-19 chemical shifts span a wider range of nearly 900 ppm. The paramagnetic shielding is the predominant factor. Less than one percent of the shielding of the ^{19}F nucleus in organic compounds is caused by the diamagnetic shielding. The chemical shifts for ^{19}F are rather difficult to predict. They are sensitive to the electronegativity and the oxidation state of the neighbouring group, stereochemistry and the effect of more distant neighbours. Fluorine resonances tend to be well-separated in the spectrum and usually first order spectra are obtained. For example, the chemical shifts for the axial and equatorial fluorines in ClF_3 differ by 120 ppm, those of XeF_2 and XeF_4 by 180 ppm and those for ethyl and methyl fluorides by 60 ppm. Because of this large variation, ^{19}F NMR can serve as a sensitive method for following the reactions and changes in structure.

Table 15.4 shows the chemical shift ranges for some common organofluorine groups and the chemical shifts for certain fluorine compounds are given in Table 15.5. The variation in chemical shifts between the compounds is rather large and no trend seems to emerge.

Table 15.4 Typical range of ^{19}F chemical shifts

Functional group	*δ, ppm*
$C–CF_3$	68–98
$C=CCF_3$	62–92
$–CF_2–$	90–135
–CHF–	180–225
$–CF_2H–$	122–142
$–CHF_2–$	188–242
–CF=C–	50–140
$CF_2=CF–$	152–205
–CF=CF–	95–185
Ar–F	110–176

Coupling Constants

Some typical values of $^{19}F-^{1}H$ coupling constants are given in Table 15.6. Coupling constants between fluorine nuclei span a wider range than those in

Table 15.5 ^{19}F Chemical shifts of selected compounds (δ, ppm, reference $CFCl_3$). Larger negative number indicates higher field

Compound	*δ, ppm*	*Compound*	*δ, ppm*	*Compound*	*δ, ppm*
F_2	432	BF_3	− 133	CF_4	− 60
HF	− 190	ClF_3	80	SiF_4	− 160
F^-	125	BrF_3	− 38	CH_3F	− 228
NF_3	140	BrF_5	132, 269	CF_3Cl	− 32
PF_3	− 36	IF_5	4	CF_2Cl_2	− 9
AsF_3	− 48	IF_7	238, 274	CF_3I	− 4
NSF_3	66	XeF_2	− 210	CF_3COOH	− 79
SF_6	42	XeF_4	− 23	$(CF_3)_2CO$	− 82
SeF_6	50	XeF_6	50	OF_2	250
SiF_6^{2-}	129	$XeOF_4$	101		

Table 15.6 Typical $^{19}F-^{1}H$ coupling constants

Type	*J (Hz)*	*Type*	*J (Hz)*
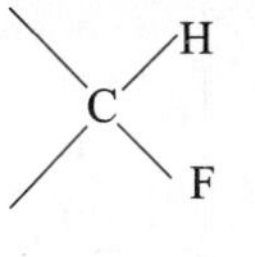	44–80	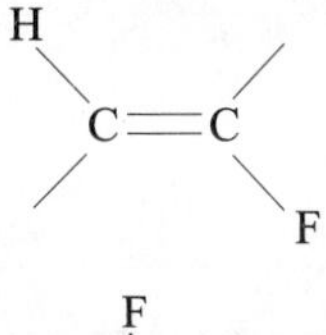	12–40
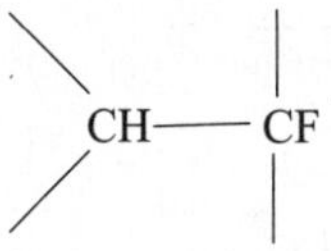	3–25	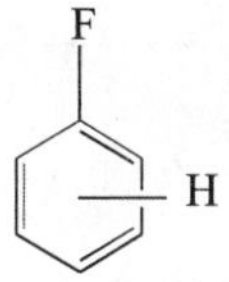	o 6–10 m 5–6 p 2
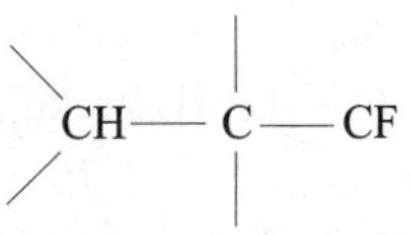	0–4		
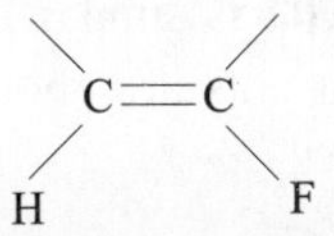	1–8	$^{2}J\ (^{19}F-^{1}H)$	10–50

^{1}H NMR. Geminal F–F coupling ranges from 40 to 370 Hz and vicinal F–F coupling from 0–40 Hz; *cis* fluorines show J values of 105 – 150 Hz. A long range coupling over five bonds through F–C–C–C–C–F varies from 0 to 18 Hz. Coupling between ^{1}H and ^{19}F is also strong. Geminal coupling ranges from 42 – 80 Hz and vicinal H-F coupling constant from 1 – 29 Hz; *cis* H–F shows J value of 0–22 Hz and *trans* H-F shows J value of 11–52 Hz. Fluorine bonded to benzene also couples with protons on the ring: the J(^{1}H–^{19}F) ranges being 7–12 Hz for ortho, 4–8 Hz for meta and 0.2–3 Hz for para hydrogens.

Some Examples

The proton-decoupled ^{19}F NMR spectrum of fluoroacetone CH_3–C(=O)–CH_2F in $CDCl_3$ is given in Fig. 15.19. The spectrum shows a singlet for the fluorine atom. In the proton coupled spectrum also shown in Fig. 15.19, the coupling of fluorine to two sets of hydrogens is observed giving a triplet with a large coupling constant. The triplet is further split into three quartets by a four bond coupling of

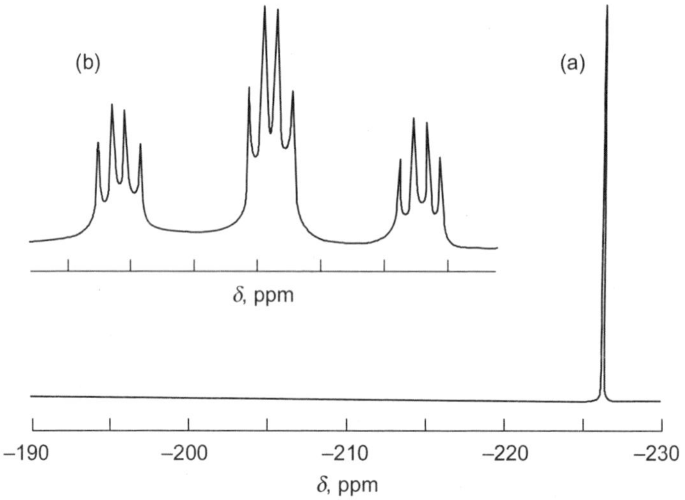

Fig. 15.19 ^{19}F NMR spectrum of fluoroacetone (a) proton-decoupled and (b) proton-coupled.

fluorine with the methyl group. On the other hand, the ^{1}H spectrum of fluoroacetone shows the CH_3 group as a doublet at δ 2.2 ppm (J = 4.3 Hz) resulting from the long-range coupling by the F nucleus. The doublet at δ 4.75 ppm (J = 48 Hz) represents the protons of the CH_2 group coupled to the geminal F nucleus. The splitting of the CH_2 group by CF_3 in CF_3CH_2OH is similar to the splitting of the CH_2 group in CH_3CH_2OH; the coupling constants are about the same, 8 and 11 Hz respectively for H–H and H–F coupling.

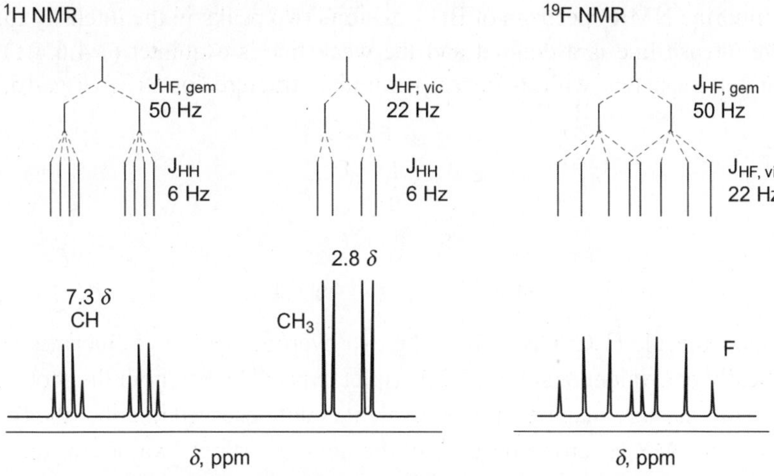

Fig. 15.20 Predicted ^{1}H and ^{19}F NMR spectra of CH_3CHFBr.

The ^{19}F NMR spectrum of 1-bromo-1-fluoroethane (CH_3CHFBr) is given in Fig. 15.20. In the spectrum, the fluorine signal is split into a doublet by the geminal methine proton (J = 50 Hz) and each line is further split into a quartet by the vicinal methyl protons (J = 22 Hz). The fluorine spectrum therefore consists of an overlapping doublet of quartets. In the ^{1}H NMR spectrum, the CH signal is a doublet of quartets and the CH_3 signal is a doublet of doublets.

The fluorine NMR spectra of a large number of compounds of the general formula $R_{5-n}PF_n$, where R is a hydrocarbon, fluorocarbon or halide ion other than fluoride have been reported. The number of peaks in the spectrum and the magnitude of the coupling constants have been utilized to deduce the structure. For a series of compounds of the type R_2PF_3, a trigonal bipyramidal structure is proposed and it was found that the J_{P-F} values for apical fluorines are ~ 170 Hz less than those for the equatorial fluorines. The most electronegative groups are found in the apical position. The spectra obtained for these compounds at low temperatures indicate that the two methyl groups in $(CH_3)_2PF_3$ are equatorial and two trifluoromethyl groups in $(CF_3)_2PF_3$ are axial. At room temperature, rapid intramolecular exchange occurs and the effect of this exchange is to average out the coupling constants.

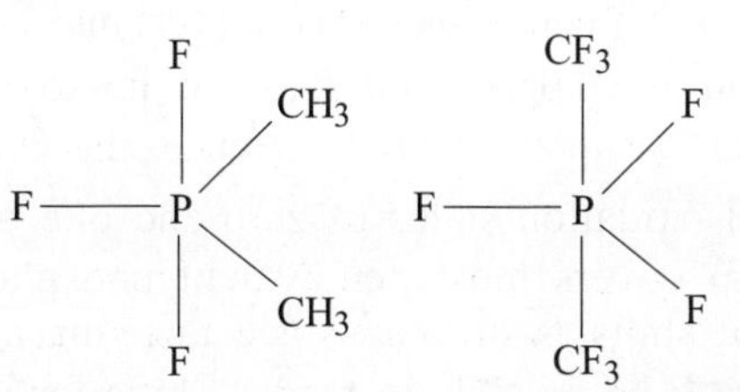

The fluorine NMR spectrum of BrF_5 exhibits two peaks in the intensity ratio of 4:1. The intense line is a doublet and the weak line is a quintet (1:4:6:4:1). The spectrum is consistent with a square pyramidal structure for BrF_5 (Fig. 15.21).

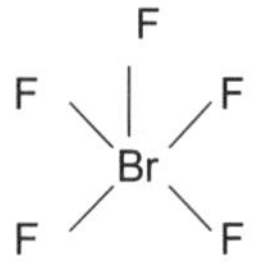

Fig. 15.21 Structure of BrF_5.

For the molecule $H_2C{=}CF_2$, where the two hydrogens and two fluorines are not magnetically equivalent, a simple 1:2:1 triplet expected when both the protons and both the fluorines are magnetically equivalent is not observed in either the fluorine or the proton NMR spectrum because of the non-equivalent J values, as discussed earlier.

^{31}P NMR

Chemical Shifts

The only naturally occurring isotope of phosphorus is ^{31}P. It shows magnetic properties similar to ^1H and ^{19}F. Phosphorus is of great interest to organic, inorganic and biochemists. Phosphorus is of great biochemical interest primarily because of the nucleic acids, which contain phosphorus esters and also other molecules such as ADP, ATP, etc. The reagents containing phosphorus range from various inorganic forms of phosphorus to the organic phosphines, phosphites, phosphorus salts, etc.

^{31}P is a spin 1/2 nucleus, with a positive magnetogyric ratio. There is a rich literature of ^{31}P NMR. The ^{31}P chemical shifts are quoted with reference to 85% H_3PO_4, the only preferred reference compound (external). This is not normally a desirable material to add directly to the solution under investigation but can be inserted in a sealed capillary. The chemical shift range for ^{31}P is rather large and normally lies in the range ± 250 ppm relative to 85 % H_3PO_4 ($\delta = 0$). ^{31}P NOE enhancement from proton decoupling is positive with a maximum of 1.23. Several studies correlating the chemical shift of ^{31}P with structure have been reported. In general, the resonances of phosphorus(III) compounds occur at much lower fields than those of phosphorus(V) compounds. The ^{31}P chemical shifts can be as low as $\delta - 460$ ppm (P_4) and the upper limit may go up to +1362 ppm as in the phosphinidene complex, t-BuP$\{Cr(CO)_5\}_2$. Hence, the extreme limits are found for ^{31}P having formal oxidation states of zero and one. A much smaller range of δ –50 to 100 ppm covers most pentavalent phosphorus compounds. The interpretation of these shifts is difficult since there may be many contributing factors. However, it may be possible to predict shifts for unknown compounds if those for similar species are known. For example, the values of ^{31}P chemical shifts of PX_2Y or PY_3 can be predicted using those of PX_3 and PXY_2. These simple additive relationships may often fail particularly for electronegative atoms such as Cl or F. It is, therefore, advisable to assign the spectra by comparison with literature data. The chemical shifts for certain representative phosphorus compounds are listed in Table 15.7.

Table 15.7 Representative ^{31}P chemical shifts

Compound	*δ, ppm*	*Compound*	*δ, ppm*
PMe_3	– 62	PX_5	– 70 to – 100
PEt_3	– 20	PCl_5	– 80
PPh_3	– 7	$OPMe_3$	36
PF_6^-	– 145	$OPEt_3$	48
PH_3	– 238	$OPPh_3$	29
PX_3	90–185	$P(OMe)_3$	140

The spectra of phosphorus compounds containing organic groups are usually obtained with proton decoupling. In a $^{31}P\{H\}$ spectrum each distinct type of phosphorus atom gives a single signal showing coupling to other ^{31}P atoms, unless other non-hydrogen nuclei with spins are present. Valuable information can often be obtained from the NMR spectra by noting the number of peaks and their relative intensities. For example, phosphorus pentachloride in CS_2 solution exhibits a single ^{31}P resonance at about $\delta \sim 80$ ppm which may be expected from its trigonal bipyramidal structure found in the gas phase. However, in the solid state, phosphorus pentachloride shows two equally intense ^{31}P peaks indicating clearly that the change of phase is accompanied by a change of structure. Actually, PCl_5 undergoes disproportionation in the solid state. It crystallizes as $[PCl_4^+][PCl_6^-]$. The chemical shift of PCl_4^+ and PCl_6^- is nearly δ –95 and 280 ppm, respectively.

Coupling Constants

Some typical proton-phosphorus coupling constants are given in Table 15.8. The coordination number of the phosphorus atom may easily be distinguished by the value of $^1J_{PH}$ which is typically 180 and 400 Hz for three and four coordinated atoms, respectively. The derivatives of phosphorus(III) and phosphorus(V) can also be distinguished using coupling constant data. The value of $^1J_{PH}$ for phosphorus(V) compounds is generally larger, 400-100 Hz, than that for P(III) compounds, 250-100 Hz. Similarly, $^1J_{PF}$ coupling constants can be used to determine the coordination number of phosphorus and in the case of five coordinated species, to distinguish between axial and equatorial fluorine substituents.

Table 15.8 Typical $^{31}P-^{1}H$ coupling constants

Compound	J_{PH} *(Hz)*
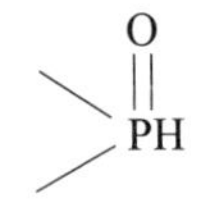	630–700
$(CH_3)_3P$	2.7
$(CH_3)_3P{=}O$	13.4
$(CH_3CH_2)_3P$	0.5 (HCCP) 13.7 (HCP)
$(CH_3CH_2)_3P{=}O$	11.9 (HCCP) 16.3 (HCP)
$CH_3P(=O)(OR)_2$	10–13
$CH_3C{-}P(=O)(OR)_2$	15–20
$CH_3OP(OR)_2$	10.5–12
$P[N(CH_3)_2]_3$	8.8
$O{=}P[N(CH_3)_2]_3$	9.5

Some Examples

The ^{31}P NMR spectrum of HPF_2 is of interest. If $^1J_{PF} > {}^1J_{PH}$, a triplet with each of the lines split into a doublet is expected. On the other hand, if $^1J_{PH} > {}^1J_{PF}$, the expected spectrum consists of two triplets. If the two coupling constants are similar, the spectrum would be a complex pattern, intermediate between the two expected spectral patterns. In most compounds that have been investigated, $^1J_{PF} > {}^1J_{PH}$ (~ 1500 Hz Vs ~ 200 Hz) and as expected the spectrum shown in Fig 15.22 was found experimentally.

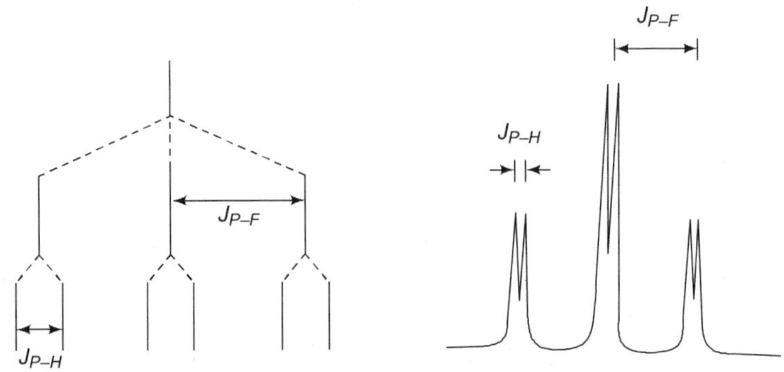

Fig. 15.22 ^{31}P NMR spectrum of HPF_2

The characteristic chemical shift of phosphorus bonded to hydrogen occurs in a limited range and the peaks have fine structures corresponding to $^1J_{PH}$. The ^{31}P resonance of $HPO(OH)_2$ is a doublet and that of $H_2PO(OH)$ is a triplet consistent with the following structures. The coupling of hydroxyl to phosphorus is either too small to be observed or is not observed due to a fast proton exchange process. Similar results have been obtained for $FPO(OH)_2$ and $F_2PO(OH)$.

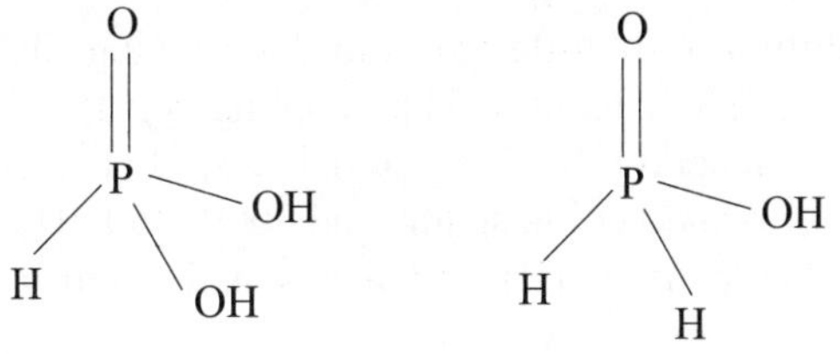

Other useful applications have been made to transition metal complexes where coupling between the metal and ligand atoms depends on the coordination number of the metal and hence on its *s*-orbital contribution to each bond. Thus, values of $^1J_{PtP}$ for the four and six coordinated complexes I and II are in the ratio of 3 : 2 but only complexes that are similar in all other respects should be compared.

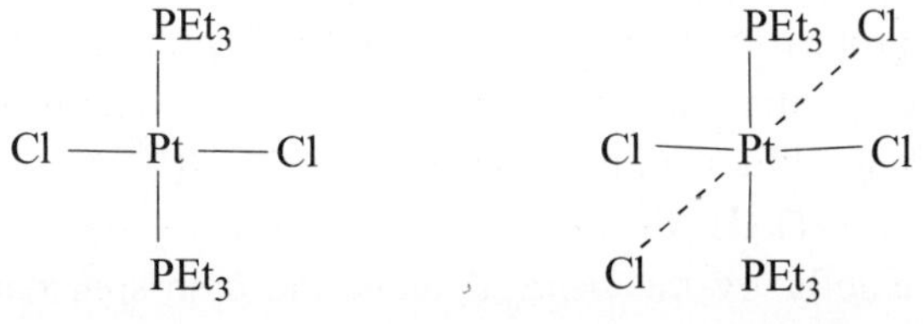

I II

The proton decoupled ^{31}P NMR spectrum of the complex *cis*–$[Pt(PEt_3)_2Cl_2]$ is shown in Fig. 15.23 and the structure of the compound is also given. Only one

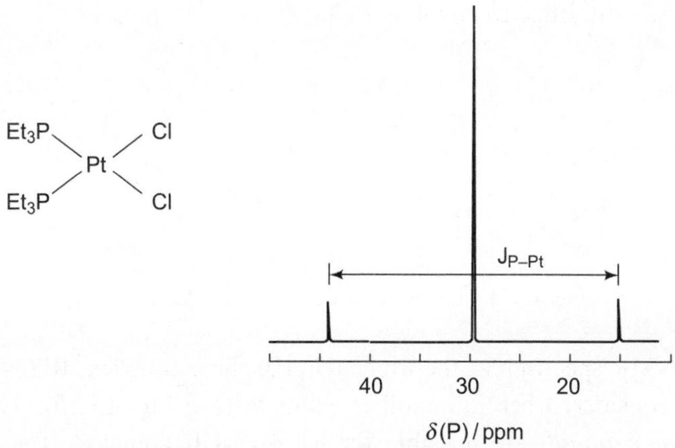

Fig. 15.23 The complex *cis*–Pt $(PEt_3)_2$ Cl_2 and its ^{31}P {H} NMR spectrum.

signal is expected since the two phosphorous atoms are equivalent. However, nearly 34% of Pt atoms are spin active (I = 1/2). The NMR spectrum of those molecules containing the spin active platinum atoms will show a doublet due to ^{1}J (Pt–P) coupling while the other molecules not containing the spin active platinum atoms will display a single peak. The observed spectrum is a superposition of the two spectra. The weaker peaks are usually referred to as satellites to the main peak as discussed earlier (see Chapter 12).

The chemical shifts are of little value in determining the stereochemistry of $PtX_2(PR_3)_2$ complexes (X = halide). The coupling constants, on the other hand, are quite distinctive. Generally for PtX_2 complexes, ^{1}J (P – Pt) is 2000–2500 Hz for a phosphine *trans* to another phosphine and 3000–3500 Hz for phosphine *trans* to a halide. For other platinum(II) complexes, values may vary but the trend remains the same.

Geometrical Isomers

There have been several interesting applications of ^{31}P NMR for the determination of structures of complexes with phosphorus ligands. *Cis-trans* effects in the complexes are also manifested in their NMR spectra. Ligands are affected much more strongly by the groups *trans* to them and are relatively less affected by the groups *cis* to them. Thus, metal-ligand coupling constants are dependent upon the nature of the group or atom *trans* to the ligand. For instance, in the complex $PtHCl(PEt_3)_2$ in benzene (Fig.15.24), $^1J(^{195}Pt–^{31}P)$ is nearly 1400 Hz in the position trans to H while it is nearly 3,500 Hz *trans* to Cl. Chemical shifts are determined chiefly by the *trans* ligands and spin-spin coupling is strong between the *trans* pair of ligands and weak between the *cis* pair. This provides a useful guide for the identification of geometrical isomers. The hydride resonance occurs at δ 16.9 ppm to the high field of TMS; it is split into a triplet by the two equivalent phosphorus atoms, J(Pt–H) = 1276 Hz, and ^{2}J(P–H) = 15 Hz. It has identical Pt satellites due to coupling to 33.7% of ^{195}Pt present.

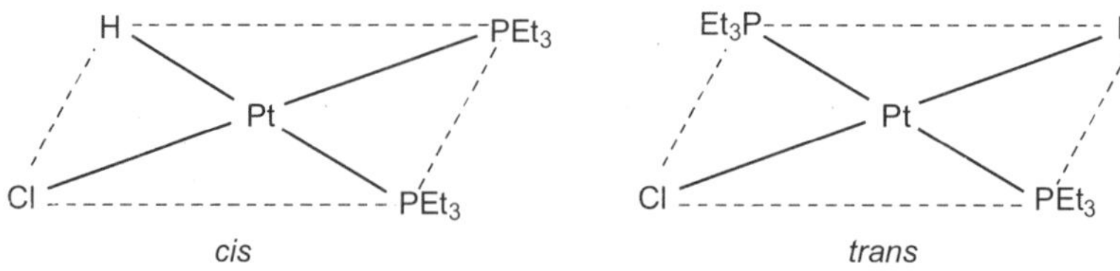

Fig. 15.24 Geometrical isomers of Pt HCl $(PEt_3)_2$.

The ^{31}P NMR spectrum of the triphenyl phosphine complex, $Rh(\phi_3P)_3Cl_3$ (Rh, I = 1/2) is considered here as another example (see Fig. 15.25). Two isomers, *facial* and *meridional*, are possible for $Rh(\phi_3P)_3Cl_3$ complex. The phosphorus ligands are equivalent in the *facial* isomer since all the phosphorus atoms are *trans*

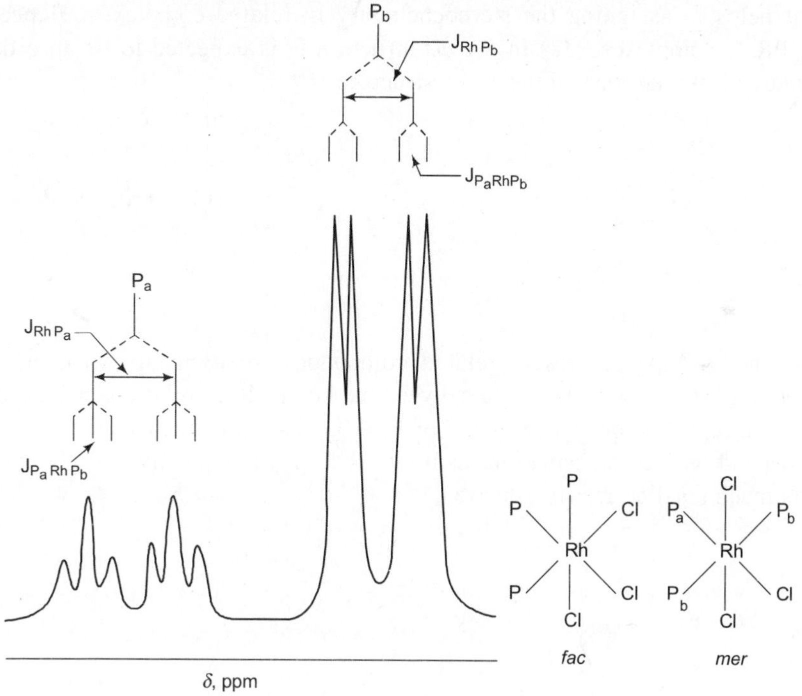

Fig 15.25 ^{31}P NMR spectrum of $Rh(\phi_3P)_3Cl_3$.

to a chloride ligand. The spectrum shown in Fig. 15.25 indicates that the complex has a *meridional* structure. In this isomer, two phosphorus atoms designated as P_b are *trans* to one another and the one *trans* to chloride is labeled as P_a. Thus, in the *mer*-$Rh(\phi_3P)_3Cl_3$, the two ligands are equivalent but different from the third. The expected splittings are shown in the stick diagram (Fig. 15.25). The spectrum shows two doublets for the two equivalent ligands and two triplets of half the intensity for the unique ligand (coupling to the other two). Thus, in principle we can determine whether a complex of the type $M(PR)_3X_3$ adopts the *facial* or *meridional* structure just on the basis of the number of peaks observed in its ^{31}P NMR spectrum.

Two–Bond Coupling

Coupling over two bonds seems to have been ignored, since for many homonuclear systems the nuclei are equivalent and hence this coupling is not observed. However, the magnitude of $^2J_{PP}$ often provides useful information about the stereochemistry of the complexes. The magnitude of this coupling constant is usually much larger when two phosphorous atoms are *trans* (~ 750 Hz) to one another than when they are *cis* (~ 20 Hz) to one another. For example, the *trans* nuclei in heavy metal complexes usually couple more strongly than the *cis* nuclei and it helps in assigning the stereochemistry in related complexes. Hence, for $PtX_2(PR_3)_2$ complexes, $^2J_{PP}$ in the *cis* structure I is expected to be an order of magnitude less than that in the *trans* structure II.

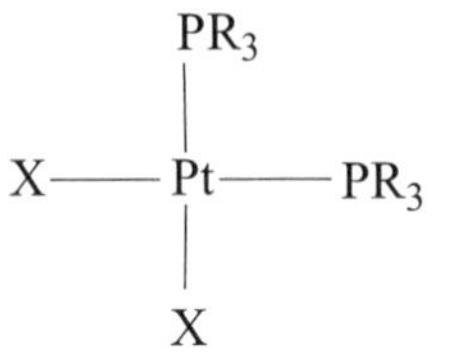

I

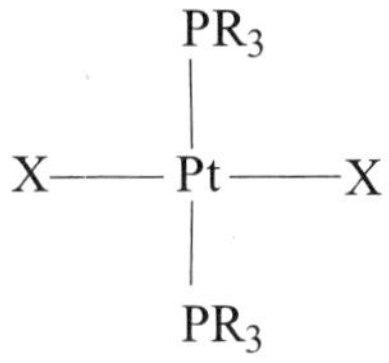

II

In other groups of closely related compounds, relationship exists between two-bond couplings and the angle between the bonds. Such relations, for example, have been reported for $^2J_{FF}$ and FCF or FPF bond angles. The two-bond coupling may depend on the conformation. In structure I, $^2J_{PP}$ is typically 450 Hz while in II it is much smaller, nearly 100 Hz.

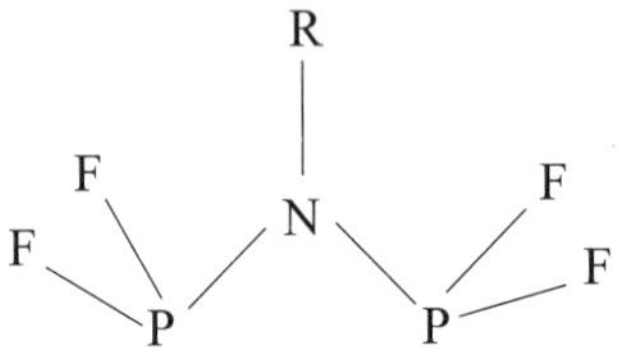

I

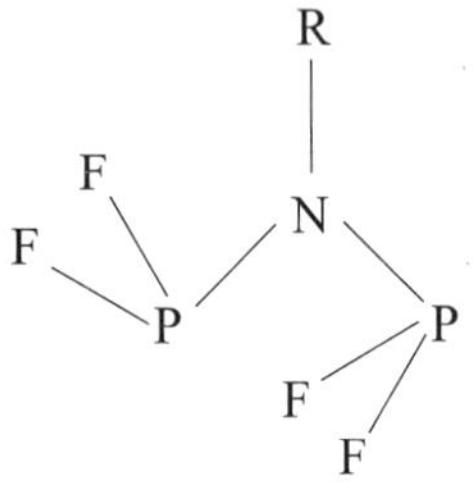

II

When two-bond H-H couplings cannot be observed because of chemical and magnetic equivalence, it is useful to utilize partial deuteration procedure discussed earlier. For example, the ^1H spectra of phosphines PH_2X consist of doublets giving $^1J_{PH}$ but not $^2J_{HH}$. But the spectrum of PDHX shows an extra 1 : 1 : 1 triplet splitting arising from $^2J_{DH}$. Multiplication of the coupling constant $^2J_{DH}$ by γ_H/γ_D where γ is the magnetogyric ratio yields $^2J_{HH}$.

Long Range Coupling

In saturated XCCY units $^3J_{XY}$ depends primarily on the dihedral angle ϕ of X–C–C–Y. The coupling constant is related to the dihedral angle by

$$^3J_{XY} = A + B\cos\phi + C\cos 2\phi \tag{15.7}$$

where A, B and C are empirical constants (see Chapter 12). This equation has been applied to $^3J_{HH}$, $^3J_{PH}$, $^3J_{PC}$ and $^3J_{PP}$.

Smaller couplings over four or more bonds are sometimes observed. Long range couplings may be associated with extended π electron systems. They may be observed for atoms which are physically close in space although separated by several bonds. Such long range couplings give important structural information.

RELAXATION PROCESSES 16

GENERAL REMARKS

The appearance of an NMR spectrum, like the ESR spectrum, is also determined by the two relaxation processes–spin-lattice and spin-spin relaxation–as discussed earlier. The concept of relaxation times appropriate to an assembly of magnetically active nuclei is of great importance to the technique and it helps us to understand a considerable number of NMR phenomena. The spin-lattice relaxation times of ^1H and ^{13}C are very useful in the assignment of chemical shifts and to understand the dynamics of molecules in solution. However, chemical shifts and spin-spin coupling constants are the NMR parameters most commonly utilized.

SPIN–LATTICE RELAXATION

Before the sample is placed in between the pole faces of a magnet, the (2I+1) energy levels of a spin-I nucleus are degenerate and their populations are equal. When an external magnetic field B_0 is applied, the spin states split apart in energy. The spin-lattice relaxation enables the spins to redistribute amongst their energy levels so as to establish the Boltzmann population difference. As the nuclei approach equilibrium, the energy released is dissipated to the surroundings (the lattice). One comes across two macroscopic magnetization in NMR experiments as discussed earlier. They are the longitudinal magnetization along the z-axis and the transverse magnetization in the x,y-plane. Both are subject to relaxation processes.

The build up of the equilibrium magnetization M_0 requires a time T_1 and the variation of the z-component of the macroscopic magnetization obeys the first order equation,

$$\frac{dM_z}{dt} = \frac{M_0 - M_z}{T_1} \tag{16.1}$$

The rate constant for the return of the perturbed system to the equilibrium state is $1/T_1$. During the time T_1 energy is transferred from the spins to the environment referred to as the lattice. This process characterized by the equation (16.1) is known as the longitudinal or spin-lattice relaxation and T_1 as the longitudinal or spin lattice relaxation time.

The spin-lattice relaxation describes the change in the z–component of magnetization (M_z) which is parallel to the applied magnetic field. Relaxation to equilibrium usually occurs exponentially. Hence for a large collection of spin ½ nuclei, the difference in the number of spins of $m = ½$ and $m = -½$ varies according to the relation,

$$\Delta n(t) = \Delta n_{eq} \{1 - \exp(-t/T_1)\} \tag{16.2}$$

where Δn_{eq} is the equilibrium population difference and t is the time as shown in Fig. 16.1. After placing the sample in the magnet, one should therefore wait for sometime which is sufficiently long compared to the spin-lattice relaxation time T_1 before recording a spectrum. Theory predicts that a delay of 5 T_1 will be required for the complete restoration of the equilibrium state. Since typical T_1 values for spin ½ nuclei in liquids are only a few seconds, this presents no difficulty. By the time $t = T_1$, the population difference grows to 63% of its value (Δn_{eq}) at thermal equilibrium. A similar behaviour may be expected whenever the nuclear spin populations are disturbed.

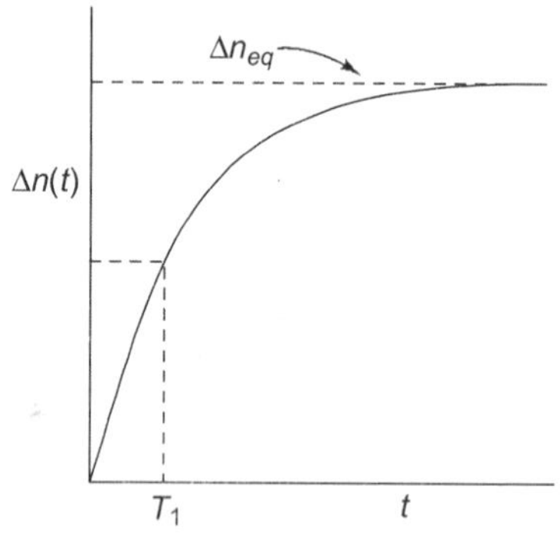

Fig. 16.1. The difference (Δn) in the population of $m = ½$ and $m = -½$ for an assembly of nuclei of spin $I = ½$ as a function of time (t) after inserting the sample in an external magnetic field. T_1 denotes the spin-lattice relaxation time.

By the application of B_1 magnetic fields with small amplitudes, the macroscopic magnetization is deflected from the Z-axis by a small angle θ. Even during resonance, Z-magnetization is maintained since the system tries by relaxation processes to restore the normal Boltzman distribution. A portion of the absorbed energy of the radiofrequency field is transferred to the surroundings. The magnitude of the new equilibrium magnetization M_Z is then a function of the spin-lattice relaxation time and the amplitude of the B_1 field. For strong B_1 fields, the maximum intensity I of the continuous wave signal at $\omega = \omega_0$ is given by

$$I(\omega_0) = \text{constant}/B_1 T_1 \tag{16.3}$$

Long relaxation times (T_1) and high amplitudes (power) of the oscillating magnetic field therefore reduce the signal intensity, i.e. saturate the resonance absorption, while much shorter relaxation times may broaden the resonance lines. When the excited state lifetime of the nuclei is short, it causes an uncertainty in the energy ΔE

according to the uncertainty principle $\Delta E.\Delta t \geq h/4\pi$. Since $\Delta E = h\,\Delta \nu$, we have $\Delta \nu\,\Delta t \geq 1/4\pi$ or $\Delta \nu \geq 1/4\pi\,\Delta t$ for the uncertainty in the determination of the resonance frequency. The line width therefore depends on $1/\Delta t$ or $1/T_1$. In liquids, T_1 for protons is generally of the order of a few seconds or less so that spin-lattice relaxation contributes no more than 0.1 Hz or less to the line width.

There are several mechanisms which contribute to the exchange of energy between the spin system and its surroundings and the rates for these processes are additive. Direct intra- and intermolecular dipole-dipole interactions with other nuclei are important in liquids. Interaction between the nuclear spins and molecular rotation is also important and the energy transfer occurs in this case when the rotation is interrupted by collisions. This mechanism is particularly important for liquids of low viscosity at high temperatures and for gases. The relaxation rate is proportional to the square of the applied magnetic field.

Relaxation mechanisms that operate in other forms of spectroscopy are ineffective for NMR. Spontaneous emission is extremely slow at NMR frequencies as discussed earlier (Chapter 1). Deactivation of the excited states by molecular collisions is also negligible because nuclear spins interact so weakly with the rest of the system that they are effectively decoupled from the motions of the molecules containing them. The mechanism of nuclear spin relaxation lies in magnetic interactions, the most important interaction being the dipolar interaction. The dipole dipole interaction, as noted earlier, between two nuclei depends on their separation r and on the angle θ between the internuclear distance vector and the applied magnetic field. This anisotropic coupling does not give rise to splitting in NMR spectra. As the molecules tumble in solution, r and θ vary in a complicated manner causing the interaction to fluctuate rapidly. The dipolar coupling modulated by the translational and rotational motions of the molecules causes the nuclear spins to experience fluctuating local magnetic fields which if they contain a component at the NMR frequency ($= \gamma B_0/2\pi$) can induce radiationless transitions which return the spins to equilibrium. The magnetic energy received by the lattice is then transformed into thermal energy. Since the spin energies are minute compared to the rotational, vibrational and translational energies of the molecules in solution, nuclear spins are relaxed with a negligible change in the temperature of the sample. Nuclear spin relaxations are slow since the local magnetic fields are rather very weak and they interact rather weakly with the spins. The longitudinal relaxation is particularly effective if paramagnetic materials are present in solution. This is because the relaxation time T_1 is inversely proportional to the square of the magnetic moment that gives rise to the fluctuating magnetic field. The magnetic moment of an unpaired electron is larger than the nuclear magnetic moment by 10^3 times. Therefore, T_1 becomes less than 10^{-1}s and the resonance lines become very much broadened. Therefore the presence of even trace amounts of atmospheric oxygen in the sample accelerates the spin-lattice relaxation process.

The spin-lattice relaxation time of protons is chiefly determined by the intermolecular interactions, whereas for ^{13}C nuclei the relaxation processes are determined by intramolecular interactions. For heavy nuclei with $I = 1/2$ additional contributions to spin-lattice relaxation comes from the anisotropy of the chemical shifts, intramolecular processes like rotation of methyl groups or scalar coupling to quadrupolar nuclei such as ^{35}Cl, ^{79}Br etc., which are discussed later. Quadrupolar relaxation is less effective for ^{2}H and ^{14}N.

The origin of other relaxation mechanisms is also nearly the same; an intramolecular or intermolecular magnetic (or for $I > ½$ nuclei, electronic) interaction, which becomes time-dependent by random molecular motion. Rotational motions occur at approximately the right frequency and modulate intramolecular interactions and hence are most favourable for relaxation. The molecular vibrations are too fast (10^{12} to 10^{14} Hz) and so do not contribute. Translational motions are also relatively less efficient because couplings are generally weaker than in the intramolecular case.

Relaxation to the Boltzmann distribution of nuclei with spin ½ can only occur by the interactions of its magnetic moment with the random magnetic fields in the sample. These random fields originate in the rotational and translational motions of the molecules as stated before. The overall relaxation rate therefore depends on the correlation times characterizing these motions and on the energies of interaction between the spin system and the molecular motions. Correlation time τ_c characterizes the dynamic behaviour of the molecules and is regarded as the average time which a molecule in solution takes to rotate through an angle of one radian (60°). In liquids of low viscosity τ_c is typically 10^{-11}s. This corresponds to a frequency of 10^5 MHz. Nuclei in small molecules which tumble faster generally have relatively shorter relaxation times than the nuclei in large molecules. Thus the ends of long chains are more mobile than the middle ones. The ^{13}C T_1 measurements have been used to study the mobility of the different parts of the polymer chains.

The spin-lattice relaxation rate of a pair of interacting nuclei is proportional to $(r^{-3})^2$ i.e. r^{-6}. Therefore, the spin-lattice relaxation time T_1 sensitively depends on the intermolecular separations and hence on the molecular structure. Further the mean square dipolar field produced by a spin X is proportional to γ_X^3 so that the spin-lattice relaxation rate of a nearby spin (B) should be proportional to $\gamma_X^2\gamma_B^2$, where γ denotes the magnetogyric ratio. A proton hence relaxes a nearby ^{13}C nucleus much less efficiently than another proton at the same distance.

SPIN-SPIN RELAXATION

Spin-lattice relaxation causes broadening of the order of $1/\pi T_1$ as noted above. It usually contributes less than 0.1 Hz to the line width. However, observed line widths are larger and may amount to several kHz in the case of solids. Thus T_1 is not the only relaxation process that affects the NMR line widths and hence it is useful to define a new parameter, namely, the spin-spin relaxation time T_2,

$$\frac{1}{\pi T_2} = \Delta\nu \tag{16.4}$$

where $\Delta\nu$ is the line width associated with the relaxation processes.

Relaxation in a plane perpendicular to the applied magnetic field is described by the spin-spin relaxation time T_2 and it could be regarded as a line width parameter. Both the relaxation times depend on the tumbling motion of the molecules and the two relaxation times are identical in the extreme narrowing limit when the tumbling is fast compared to the resonance frequency ω_0.

In the classical description of the NMR experiment, in addition to the Z–magnetization, a second magnetization in the x,y–plane termed as the transverse magnetization ($M_{x,y}$) exists. The transverse relaxation time T_2 is introduced since the time dependence of $M_{x,y}$ usually differs from that observed for M_z. In the simplest case, for liquids $T_2 = T_1$ since after resonance the x, y–component of magnetization vanishes as the longitudinal magnetization attains its previous value M_0 along the z–axis. However, the transverse magnetization can also be reduced without simultaneous increase in the Z-component ($T_2 < T_1$). As in the case of spin-lattice relaxation, fluctuating fields can interact with the transverse component $M_{x,y}$, thereby reducing its magnitude. The transverse relaxation can also occur from presence of static dipolar fields and the component B_z is static in the laboratory frame.

An important mechanism for spin-spin relaxation is the energy transfer within the spin system. Any transition between the spin states of a nucleus changes the local field at nearby nuclei to stimulate a transition in the opposite direction. The life-time of the spin states will decrease by this process and it therefore contributes to the NMR line width. The total energy of the spin system does not change. The random local magnetic fields produce small time-dependent variations in the Larmor frequencies of individual spins leading to loss of phase coherence in the sample as a whole.

In solids transverse relaxation is strongly affected by the presence of static dipolar fields. In the absence of molecular motion, each spin experiences a slightly different local field as a result of dipolar interactions with its neighbours. Since the transverse magnetization $M_{x,y}$ is a macroscopic quantity that can arise only if the individual magnetic moments in the sample have the same Larmor frequency and the same phase angle with respect to an arbitrary origin, it is clear that the spread in Larmor frequencies resulting from different local fields will destroy $M_{x,y}$. This process can be described as fanning out of the vector $M_{x,y}$ as shown in Fig.16.2. Since M_z is not affected by this process, the total energy of the spin system in the applied magnetic field B_0 does not change.

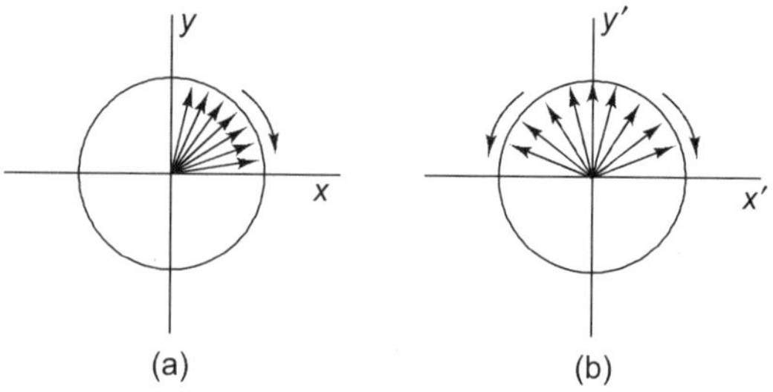

Fig. 16.2. Fanning out of $M_{X,Y}$: (a) in the laboratory frame and (b) in the rotating frame.

In liquids, the inhomogenity ΔB_0 of the magnetic field B_0 is the most important factor for the time dependence of $M_{x,y}$. Exposure of the individual nuclear spins to different external fields $B_0 \pm \Delta B_0$ will result in a spread of their Larmor frequencies and in the fanning out process for $M_{x,y}$ analogous to that shown in Fig.16.2.

The width at half the signal height is given by

$$\Delta \nu_{1/2} = 2/T_2 \tag{16.5}$$

where $\Delta\nu_{1/2}$ is the line width of the resonance signal at half-height (see Fig. 14.3). The NMR signal has a Lorentzian line shape. Since the decay of $M_{x,y}$ is caused both by field inhomogeneity and spin-spin relaxation, we can write

$$\Delta \nu_{1/2} = 2/T_2'$$

$$\text{with} \quad \frac{1}{T_2'} = \frac{\gamma \Delta B_0}{2} + \frac{1}{T_2} \tag{16.6}$$

where the first term is the contribution of field inhomogeneity to the line width. In hertz, we have,

$$\Delta \nu_{1/2} = \frac{1}{\pi T_2'} \tag{16.7}$$

if T_2' is measured in seconds. For example, if a line in the spectrum has a width of 10 Hz, the effective transverse relaxation is $T_2' = 1/\pi\ 10\ s^{-1} = 32\ ms$.

For viscous liquids and solids, T_2 is much smaller than T_1. As the line widths increase when T_2 decreases, difficulties arise in obtaining the spectra. A very effective reduction of the transverse relaxation time occurs when the molecules undergo intramolecular or intermolecular dynamic processes such as proton transfer, conformational equilibria, etc. Thus scalar interactions with another spinning nucleus can contribute to relaxation, if the coupling constant for the two nuclei is time-dependent, as a result of chemical exchange (see Chapter 14) or if the second nucleus is itself relaxing rapidly and there is coupling between the two. The second situation occurs with quadrupolar nuclei which is discussed later. In cases where this is the dominant mechanism for transverse relaxation, reaction rates can be derived from the temperature dependent T_2 values. Thus the kinetics of rapid reversible reactions can be studied by NMR spectroscopy as discussed earlier (Chapter 14).

MEASUREMENT OF RELAXATION TIMES

Several techniques are available for the measurement of spin-lattice and spin-spin relaxation times, T_1 and T_2, respectively. The two relaxation times can be measured independently using the Fourier transform techniques involving two or more radiofrequency pulses. The spin-lattice relaxation times are now routinely measured for ^1H and ^{13}C nuclei. Inversion recovery and spin-echo methods are two of the simplest experiments commonly used for the measurement of T_1 and T_2 respectively and they are briefly described here.

Measurement of T_1 : Inversion Recovery Method

Measurement of T_1 by population inversion is the most popular of the several available methods. This method for the determination of longitudinal relaxation time T_1 utilizes a preparation, evolution and detection sequence as in a two-dimensional NMR experiment (see Chapter 19). Here the nuclei are allowed to equilibrate such that their net magnetization is directed along the direction of the static magnetic field (z). This magnetization is then perturbed by an inverting *rf* pulse (preparation) and allowed to evolve during a time t (evolution) before their z magnetization is monitored by a 90° pulse (detection). The spectra are taken for various values of t and the peak heights are analysed as a functions of t. Since the recovery is described by a simple exponential function, it is necessary to have the computer fit the data for each peak to simple exponentials and obtain the relevant relaxation times (T_1).

In an applied magnetic field, the nuclear magnetic moments precess about the field direction B_0 as discussed earlier. Consider the macroscopic magnetization M_z in the rotating coordinate system. In a pulse experiment, it is possible to deflect the magnetization vector M_z by 90°, 180° or any other angle depending on the duration of the pulse as described in Chapter 13.

The pulse sequence denoted as $180° - \tau - 90°$ shown in Fig. 16.3 is utilized. The first step in the inversion recovery method is irradiation of the sample with a 180° pulse. So the magnetization vector is rotated through 180° and it points in the $-z$ direction. No signal can be seen at this stage since there is no component of magnetization in the x, y – plane i.e. M_{xy} where detection coils are sensitive. After

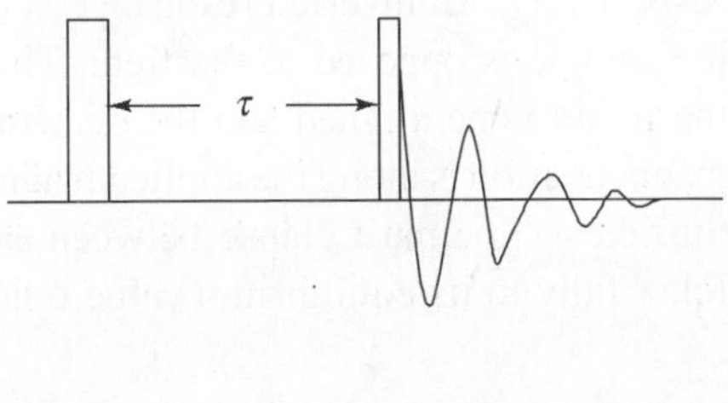

Fig. 16.3. The pulse sequence for the inversion recovery method.

the 180° pulse is turned off, during the delay decay occurs and the magnitude of M_z decreases at a rate determined by the longitudinal relaxation time T_1. The β spins begin to relax back into α spins and the magnetization vector decreases exponentially back to its thermal equilibrium value M_z. The arrows in Fig. 16.4 depict the exponential decay of the M_z vector with time to its thermal equilibrium value, the magnitude of M_z decreases, passes through zero and then increases in the positive z-direction to finally attain its initial value. Since the individual moments relax in a purely random manner, no x, y–component results.

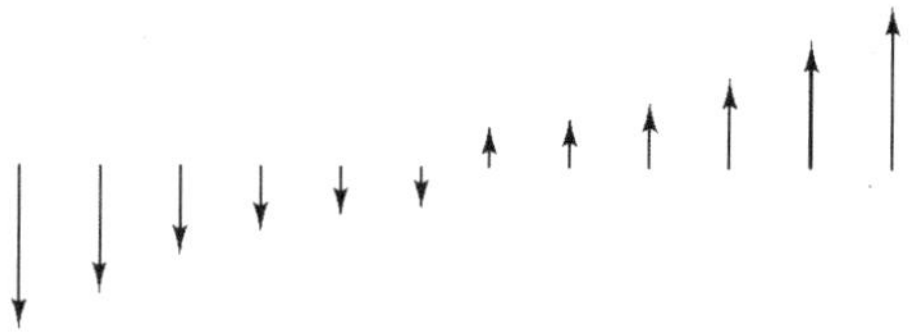

Fig. 16.4. Decay of the M_Z component after a 180° pulse with time.

Consider now irradiation of the sample after an interval with a 90° monitoring pulse that rotates the magnetization vector into the x,y plane. We can then detect the magnetization. After waiting for thermal equilibrium to be reestablished (usually a time corresponding to 5 to 10 times T_1 is employed), the sample is again subjected to the same 180° pulse, wait a while and then follow it with a 90° monitoring pulse that rotates the magnetization vector into the x,y plane. As shown in Fig. 16.5, during the delay τ, M undergoes partial spin-lattice relaxation (c) to give z magnetization (M_z (τ)) which the 90° pulse rotates on to the y-axis (d), where it produces an FID signal. The frequency domain spectrum is then obtained by Fourier transformation. The process is repeated by waiting longer for different values of τ before applying the 90° monitoring pulse to determine the recovery of the inverted magnetization. The intensity of the spectrum obtained in this manner depends on the magnitude of the magnetization vector (M_{xy}) that is deflected onto the x,y-plane. The magnetization vector returns exponentially to its thermal equilibrium value, as the interval between the two pulse is gradually increased so that the intensity of the spectrum also returns to its equilibrium value with increasing τ. The value of T_1 can therefore be obtained by fitting an exponential curve to the series of spectra obtained after different values of the interval τ.

The situation (c) to (d) of Fig. 16.5 is characterized by the time delay τ_1, τ_2 to τ_4 after the 180° pulse, the magnetization is detected by 90° pulses at τ_1, τ_2, and τ_3, τ_4 which align M along the negative and positive y- directions, respectively. These signals differ in phase by 180° and thus lead to an emission and absorption line respectively. An inverted resonance is obtained whenever the net magnetization of the sample is opposed to the field. The transition then is from a state opposed to the field to one aligned and the detector senses this as an emission. When the net magnetization is along the applied magnetic field direction, it leads to an absorption. Sufficient time must elapse between each 180°/90° pair of pulses to allow M_Z to relax fully to its equilibrium value otherwise incorrect results may be obtained.

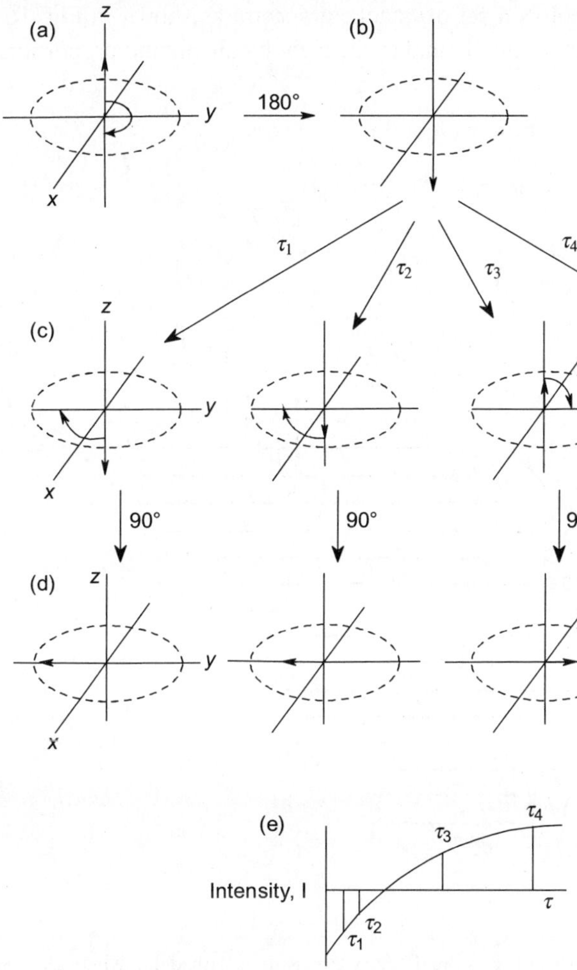

Fig. 16.5. Inversion recovery method for the measurement of T_1 : (a) thermal equilibrium, (b) after the 180° pulse, (c) after four different delays $\tau_1 < \tau_2 < \tau_3 < \tau_4$, (d) after the 90° pulse, and (e) the observed NMR signal intensity I as a function of the delay τ, $I(\tau)$.

The computer connected to the spectrometer is utilized to carry out the experiment for the measurement of T_1. The sequence of successive pulses is repeated several times with the computer executing the pulses and adjusting automatically the values of varying delay periods between 180° and 90° pulses as well as fixed relaxation time intervals between the successive pulses. The free induction decay is recorded and Fourier transformed to obtain a spectrum containing peaks whose NMR intensities $I(t)$ are proportional to $M_z(t)$. An application of such experiments is shown in Fig. 16.6 for the ^{13}C NMR spectrum of a sample showing a single resonance. Each spectrum is the result of the pulse sequence 180°– τ – 90° applied to the sample; varying the delay time from 2.0 to 40 s so as to monitor the recovery of the inverted magnetization shown in

Fig. 16.5 (e). The computer plots a set of 'stacked' spectra as shown in Fig. 16.6 where each spectrum is offset vertically and to the right by an amount proportional to the delay time.

Assuming exponential relaxation, we obtain

$$M_z(\tau) = M_0[1 - 2 \exp(-\tau/T_1)] \tag{16.8}$$

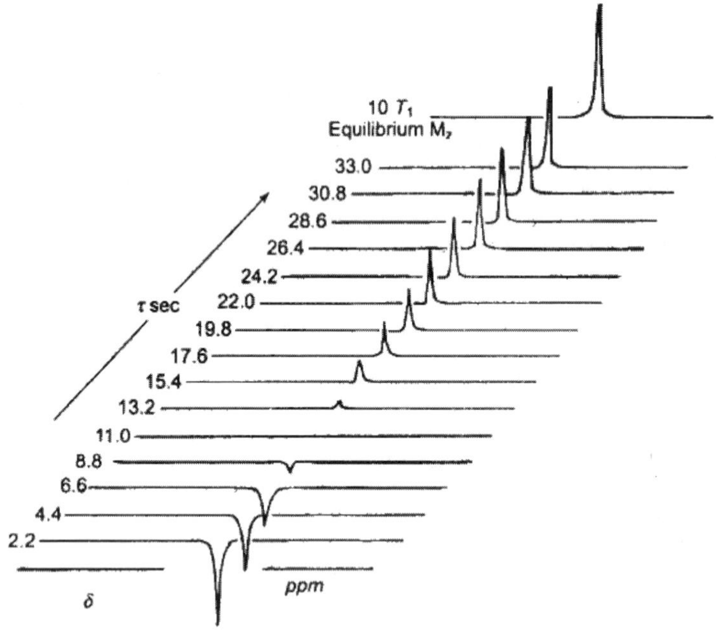

Fig. 16.6. Spectra obtained from the inversion recovery experiment with delay τ between 180° and 90° pulses increasing as we move from down to up.

The equation (16.8) can be rewritten as

$$A(\tau) = A_0 [I–2 \exp(-\tau/T_1)] \tag{16.9}$$

where A_0 is the equilibrium amplitude measured in a normal FT NMR spectrum and τ is the interval between 180° and 90° pulses. The T_1 of each peak can be obtained by plotting ln [1(∞)–I(t)] against τ where I(∞) is the fully relaxed NMR intensity for $\tau >> T_1$. A plot of ln I against τ will give a straight line from which T_1 can be calculated.

The assumption of exponential recovery in the equation (16.9) is justified by the good linearity in the plot of ln[A_0–A]/2A_0 versus τ. If one plots the area of the band A in Fig. 16.6, against interval τ between the pulses,·a first order decay curve is obtained as shown in Fig. 16.7 from which T_1 can be calculated.

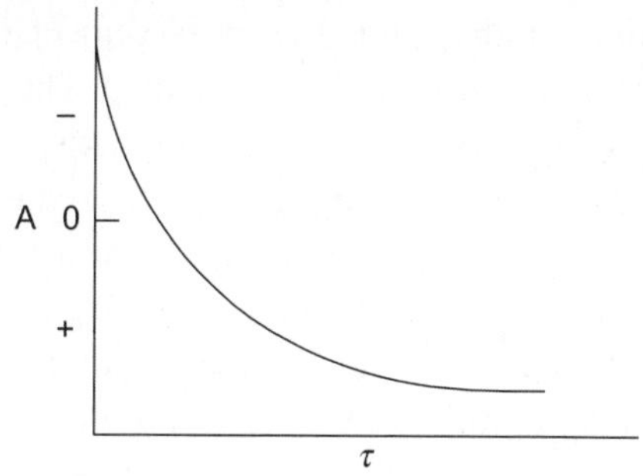

Fig. 16.7. First order decay curve of the M_z component after the 180° pulse.

Alternatively T_1 can be obtained more accurately from a semilogarithmic plot of the intensity changes M_0–M_z against τ since from equation (16.1), by integration, we get

$$\ln (M_0 - M_z) = \ln 2\, M_o - \tau / T_1 \qquad (16.10)$$

An approximate value of the spin lattice relaxation time T_1 can simply be found from the spectral plot (Fig. 16.6). The spectrum which has the signal of any particular ^{13}C or proton passing from negative to positive amplitude is determined. This zero transition (null point) will vary for different protons (or ^{13}C) in a molecule, depending on their respective spin-lattice relaxation time. At time τ_o no signal is observed since here the sample is not magnetised. For this situation the relation

$$\tau_o = T_1 \ln 2 = 0.693\, T_1 \qquad (16.11)$$

holds good and from it the relaxation time T_1 can be determined. Determining τ_o in seconds at which the amplitude of the signal from a particular nucleus is zero and dividing it by 0.693 yields its spin-lattice relaxation time T_1. In this particular example (Fig. 16.6), the signal becomes zero at about 11 seconds and hence relaxation time is 11/0.693 = 15.8 seconds. Thus the T_1 measurements are relatively easy to perform and fairly accurate T_1 values are obtained. There are other ways of determining T_1 but this is the most convenient one.

If there are several resonances in the spectrum and each relaxes at its own rate, it is possible by this method to measure the relaxation times of all the nuclei of the same isotope in a molecule and obtain detailed information about molecular motion. If the nuclei are spin coupled, then the apparent relaxation times may differ from the real ones. Carbon–13 relaxation times are usually measured with ^1H decoupling and this yields satisfactory results.

Measurement of T_2: Spin Echoes Method

In principle, the spin-spin relaxation time T_2 could be determined directly from the NMR line width. However, in practice inhomogeneities in the magnetic field usually dominate the line widths. Thus the measurement of T_2 depends on the elimination of the effects of inhomogeneous broadening. The NMR line broadening and the free induction decay arise from the interactions that cause the spins to precess at slightly different Larmor frequencies, consequently destroying their phase coherence i.e. fanning out magnetization in the transverse plane. The factors responsible for the loss of phase coherence are spin-spin relaxation induced by inter- or intramolecular magnetic fields and the spatial inhomogeneity of the external magnetic field B_o as discussed above. As the width of an NMR line depends on both these factors, spin-spin relaxation cannot be obtained easily from line widths. The spin-echo method is an elegant method for the accurate measurement of transverse relaxation time T_2. It allows one to separate the two processes. It is required that $T_2 < T_1$.

The salient feature of the spin-echo method is that it can suppress the effects of both field inhomogeneities and chemical shifts. The principle of spin-echo method can be understood from the pulse sequence given in Fig. 16.8 which can be formulated as $90°_x..\tau...180°_x....\tau$ ....FID.

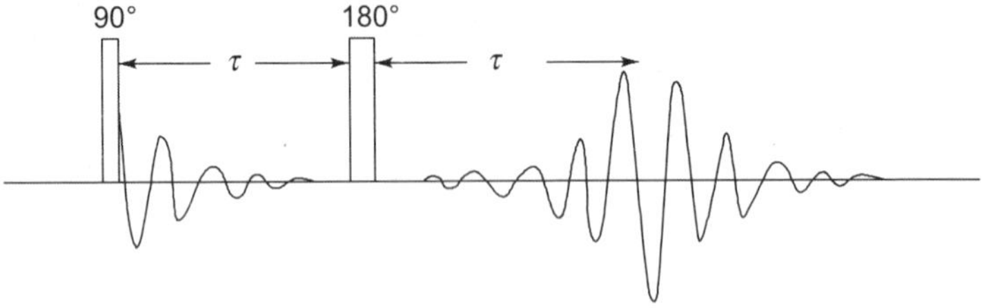

Fig. 16.8. The pulse sequence for the spin echoes method.

The overall magnetization of the sample could be considered to be made up of a number of small regions, each region experiencing a slightly different magnetization due to the applied field inhomogeneity, neglecting for the present spin-spin relaxation. The macroscopic magnetization vector M along the z-axis in the laboratory frame is shown in Fig. 16.9 (a). First a $90_x°$ pulse in the x-direction of the rotating frame is applied to the sample; it causes deflection of the vector along the y' axis in the x,y–plane creating M_{xy} (Fig. 16.9 (b)). The magnetization of each region precesses at slightly different frequencies some below the radiofrequency and some above because of the slight inhomogeneity of the applied field so that the magnetization vectors begin to fan out in the x,y–plane (Fig. 16.9 (d)). The precession frequencies also differ if there is more than chemical shift present. This dephasing reduces the transverse magnetization as a whole finally to zero. We wait for a time τ s during which the nuclei in different parts of the sample experience different fields and precess at different frequencies. After the delay period τ, a 180° pulse is applied along the x' axis of the rotating frame which flips the magnetization of each region around the x' axis to the symmetrical positions in the x,y plane in the negative y

direction (Fig. 16.9 (e)) i.e. across to the other side of the x, y–plane when the precession continues for a further time τ. The advantage of the 180° refocussing pulse is that it compensates for any errors due to the field inhomogeneities. It should be noted that a 180° rotation about x' is different from a 180° rotation about the z-axis. The $180°_x$ pulse rotates the magnetization vectors of the faster spins packets into positions previously occupied by the slower spin packets and vice versa. Thus the vectors continue to precess, the faster vectors now lag behind the slower ones; the fan begins to close up again. The second delay period is kept identical to the first delay period τ. After time 2τ, all the vectors will once again be aligned along the y-axis to reach a maximum M_{xy} and the fanning out caused by field inhomogeneity is said to have been refocussed. Irrespective of the precessional frequency and the value of τ, all regions come back into phase perfectly at the end of this second delay (Fig. 16.9 (f)). This is known as spin-echo; the dephasing caused by field inhomogeneity is refocussed by the $180°_x$ pulse in the negative y direction and the NMR signal which decays rapidly after the 90° pulse reappears as 2τ. Nuclei resonating at different chemical shifts will also experience similar refocussing effects. A spin-echo is the magnetic analogue of an audible echo. The transverse magnetization which is created by an *rf* pulse decays away and it grows to form an echo.

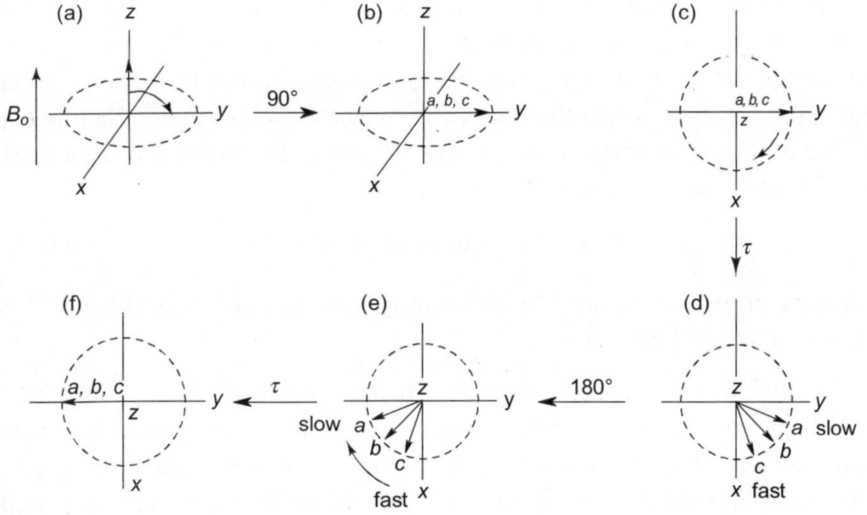

Fig. 16.9. Spin echo experiment: (a) thermal equilibrium, (b) after the 90° pulse, (c) same as (b) viewed vertically from the z-axis, (d) after a delay τ, (e) after the 180° pulse and (f) after the second delay τ. The three magnetization vectors a,b,c arise from nuclei experiencing slightly different magnetic fields B_0.

The intensity of the maxima or echoes falls at a rate determined by the real T_2. There are actually two effects that reduce the amplitude. (i) T_2 of the nucleus which fans out the magnetization randomly as opposed to the systematic nature of the applied field and (ii) diffusion of the molecules to different parts of the sample in the tube in the time 2τ. The first effect varies as $\exp(-\tau)$ and the second effect varies as $\exp(-\tau^3)$ and it is hence possible to obtain molecular diffusion times as well as spin-spin relaxation time T_2 by measuring the signal amplitude as a function of τ.

Because the effects of field inhomogeneities have been suppressed by the refocussing, the echo signal will be attenuated by a factor of $\exp(-2\tau/T_2)$ caused by spin-spin relaxation alone. After the time 2τ, the magnetization will continue to precess fanning out once again giving a resultant that decays with time constant T_2'. The true transverse relaxation arises from fields that fluctuate on a molecular time scale and there is no guarantee that an individual fast spin will remain fast in the refocussing phase; the spins within the packets therefore spread with a time constant T_2. Hence, the effects of the true relaxation are not refocussed and the size of the echo decays with time $T_{2.}$ The resultant transverse magnetization can now be detected in the receiver coil as a signal, the so-called spin-echo. The second half of the echo which has the same form as the free induction decay is detected and Fourier transformed to obtain the NMR spectrum, and the amplitudes of the spectral lines are now independent of the field inhomogeneity.

Consider now the effect of relaxation on the echo amplitude. During both τ delays, spin-spin relaxation destroys the phase coherence created by the 90° pulse and causes the transverse magnetization to decay at the rate of T_2^{-1}. This dephasing, which is due to the fluctuating magnetic fields arising from random molecular motions, is not refocused by the 180° pulse. The intensity of each line in the spin echo spectrum is therefore given by

$$I(2\tau) = I(0)\exp(-2\tau/T_2) \tag{16.12}$$

The experiment is repeated for different delay time τ and T_2 is obtained from a plot of ln $[I(2\tau)]$ against τ.

From the above discussion, it is clear that the intensity of spin echo should depend only on the transverse relaxation rate, i.e. the irreversible loss of transverse magnetization during the period 2τ, since contributions of the field inhomogeneity to the spreading out of the spins have been eliminated by the refocusing process. In such a case, the echo amplitude should be proportional to $\exp(-2\tau/T_2)$. In practice, diffusion processes complicate the situation by changing the position of the spins in the magnetic field and consequently increasing the spread of the Larmor frequencies. However, to eliminate the effect of diffusion, instead of using a single 180° pulse at interval τ, the spin-echo experiment should be repeated with pulses at different intervals τ, 3τ, 5τ .... The decrease in the amplitude of the spin echo that is measured at 2τ, 4τ, etc. is now proportional to $\exp(-\tau/T_2)$ and the effect of diffusion is negligible if the interval between the pulses is small. The intensities of the echoes (signal intensities) are then plotted as a function of delay. This yields a straight line whose slope is $-1/T_2$.

The important feature of the spin-echo method is that the size of the echo is independent of the inhomogeneities in the magnetic field and local fields that remain constant during the two τ intervals. The spin echoes method gives accurate and precise results. The T_2 measurements are somewhat difficult to carry out and T_2 values have not been useful in structural studies.

QUADRUPOLAR RELAXATION

The dipolar mechanism was considered above as the source of relaxation. For nuclei with $I > ½$, there is an additional dominant relaxation mechanism. A nucleus with $I = 1$ possesses a non-zero nuclear quadrupole moment eQ where e is the unit of electrostatic charge and Q is a measure of the deviation of the charge distribution from spherical symmetry (see Chapter 20). The electric field gradient at a nucleus arises from the asymmetries in the spatial arrangement of the bonding electrons. In highly symmetrical environments such as spherical, cubic octahedral or tetrahedral environments, the electric field gradients generated by the surrounding charges cancel out and no net quadrupolar interaction occurs.

Nuclei possessing quadrupole moments are usually very efficiently relaxed by the fluctuating electric fields that arise from the dipolar solvent and solute molecules. The interaction of the nuclear quadrupole moment with the electric field gradient provides an extremely efficient mechanism for nuclear spin relaxation through molecular tumbling. For example, for ^{14}N in CH_3CN, T_1 is 22 *ms* but for aqueous ammonium ion, which has no electric field gradient, it is more than 50 *s*. The quadrupolar interactions are manifested in large line widths (small T_2) generally observed for nuclei such as ^{14}N, ^{17}O, ^{35}Cl, and ^{37}Cl. The other consequence of quadrupolar relaxation is the loss of spin-spin splitting for spins that are scalar-coupled to the quadrupolar nuclei. If T_1 is very short, the nuclear spins interchange energy and change their orientation so rapidly that a coupled nucleus interacts with all possible spin states in a short time. It can distinguish only an average value of the interaction and a singlet resonance results. For instance, chlorinated hydrocarbons show no evidence of coupling of proton spin to chlorine nuclei, ^{35}Cl and ^{37}Cl both with $I = 3/2$. At intermediate relaxation, a broad line is observed. Efficient spin-lattice relaxation of, for example, an $I = 1$ nucleus causes it to flip rapidly between its three spin orientations ($m = +1, 0, -1$) so that a coupled spin (chemical shift frequency ν_o, coupling constant J) has a resonance frequency that changes rapidly between $\nu_o + J$, ν_o and $\nu_o - J$. If this occurs at a rate faster compared to the frequency differences involved i.e. J and 2J then a single line at the mean frequency ν_o is observed.

EFFECT OF QUADRUPOLAR RELAXATION ON THE SPECTRUM

Rapid nuclear quadrupole relaxation has a dominant effect on the line width in the NMR spectra of protons or other nuclei bonded to the quadrupolar nucleus as mentioned above. The splitting of the proton resonance from the quadrupolar nucleus may not be observed or the proton signal may be too broad to be observed. This is the case for most halogen (except fluorine) compounds. Sharp signals have been obtained for symmetrical compounds of the halogens (e.g. ClO_4^-) where the spherical charge distribution gives rise to a small field gradient at the nucleus, leading to large values of T_1. It is of interest to note that chlorine chemical shifts have been observed for the compounds $SiCl_4$, CrO_2Cl_2, $VOCl_3$ and $TiCl_4$ where the chlorine is in an environment of lower than cubic symmetry. Like rapid chemical exchange, rapid nuclear quadrupole relaxation in effect places the proton on a nucleus (or, nuclei, for chemical exchange) whose spin state changes rapidly. Nuclear quadrupole relaxation rates often correspond to an intermediate rate of chemical exchange and hence extensive broadening is usually observed (see Chapter 14). As a result of quadrupole relaxation, the proton NMR spectrum of $^{14}NH_3$ (^{14}N, $I = 1$) consists of three very broad signals, while in the absence of this effect, the spectrum of $^{15}NH_3$ (^{15}N, $I = ½$) consists of a sharp doublet. On the other hand, in $^{14}NH_4^+$, where a spherical distribution of electron density gives rise to a zero-field gradient, a sharp three line spectrum is observed. Similarly, the ^{14}N resonance of the pyramidal $N(CH_3)_3$ is about 100 Hz wide while $^{15}N(CH_3)_3$, has nitrogen NMR line width of less than 1 Hz. Such line broadenings severely reduce spectral resolution. It accounts for the relatively less amount of work carried out on quadrupolar nuclei in NMR studies of liquids. Since quadrupolar relaxation is sensitive to temperature and viscosity, the line shapes observed for coupled nuclei are altered by viscosity and temperature changes. An increase in temperature leads to slower relaxation and a better resolved multiplet.

An interesting effect has been observed for the fluorine NMR spectrum of NF_3 shown in Fig. 16.10. At –205°C, a sharp singlet is obtained for NF_3; as the temperature is increased the line broadens and finally a spectrum consisting of a sharp triplet appears at 20°C. These changes observed as a function of temperature are opposite to those normally expected for the exchange process (see Chapter 14). At higher temperature, the rate of quadrapole relaxation is slow. At low temperature, the slow molecular motions are most effective for quadrupole relaxation of ^{14}N and consequently, the rate of quadrapole relaxation is very short and a single peak is obtained. At higher temperatures, relaxation is not as effective and the life time of the given state for the ^{14}N nucleus is sufficient to cause spin-spin splitting.

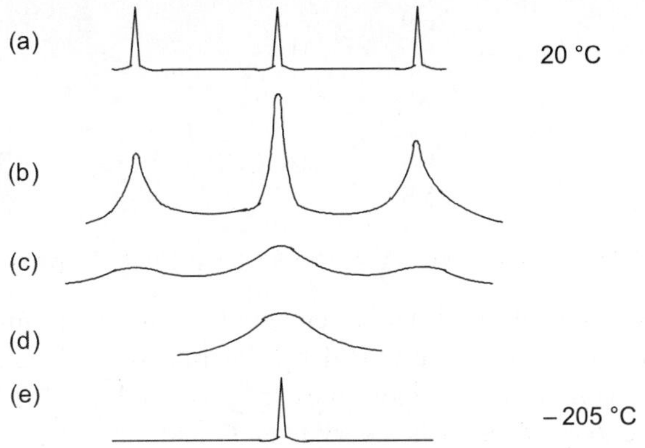

Fig. 16.10. The ^{19}F spectrum of NF_3 at different temperatures.

APPLICATIONS OF RELAXATION TIMES

The applications of relaxation times in NMR spectroscopy generally are in the areas of structure and dynamics. The T_1 measurements are more useful for dynamical studies. For structural investigations, some knowledge of the dynamical properties of the molecule is useful. The T_1 values for carbon atoms and protons can be utilized to describe certain aspects of the dynamical behaviour of polymeric chains. The T_1 measurements are increasingly utilized in spectral assignments and structural elucidations.

The ^{13}C spin lattice relaxation times of organic molecules are dominated by ^{13}C–^{1}H dipolar interactions, the closest protons having the greatest effect because of the dependence on r_{CH}^{-6} distance. Quaternary carbons which have no directly bonded protons have relatively long relaxation times, typically ranging from 15 to 100 s, than the CH and CH_2 carbons which have a relaxation time typically ranging from 2 to 20 s in the same molecule. The quarternary carbon gives rise to a low intensity peak. The smaller nuclear Overhauser enhancements (NOE) for the quaternary carbons are consistent with slower dipolar relaxation. The most useful relaxation parameter is the NOE. The r^{-6} dependence of the dipolar relaxation mechanism on a single internuclear distance is a valuable source of information on the structure and conformations of molecules (see Chapter 17). Spin lattice relaxation times are rather less informative, since they are determined by the dipolar interactions to all nearby nuclei as well as by other relaxation mechanisms.

The T_1 values of both ^{13}C and ^{1}H are useful in assigning NMR peaks. The T_1 values for the protons of vinyl acetate are shown in Fig. 16.11 as an example.

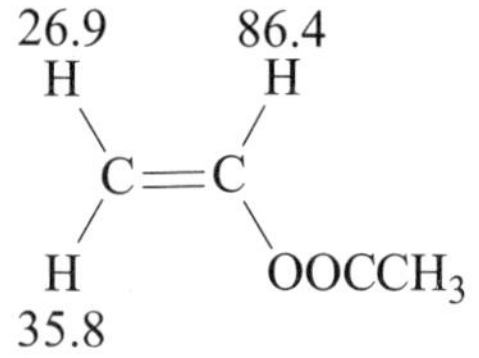

Fig. 16.11. The T_1 values in seconds for the protons of vinyl acetate.

The T_1 values give us useful information about the molecular motion in solution since the relaxation is also determined by the rate of molecular motion. The individual ^{13}C T_1 values in some molecules are given in Fig. 16.12. In the case of toluene, the quaternary carbon has a very long T_1 and the CH carbons relax more efficiently and at different rates. This occurs because the molecular reorientation is anisotropic and the 2 and 3 ring carbons reorient more often than does the 4 carbon. The rotation of the methyl group about C–CH_3 bond is however so fast that effectively it averages the dipolar interactions leading to slower relaxation. Typical relaxation times for free –CH_3 groups are 10–20 seconds. Relaxation times for propylamine show that the segmental motion increases along the alkyl chain especially if the T_1 values in *s* are multiplied by N_h, the number of hydrogen atoms bonded directly to each carbon. The effect increases dramatically if the amino group is protonated to give an end group which is a solvated ion interacting with the solvent.

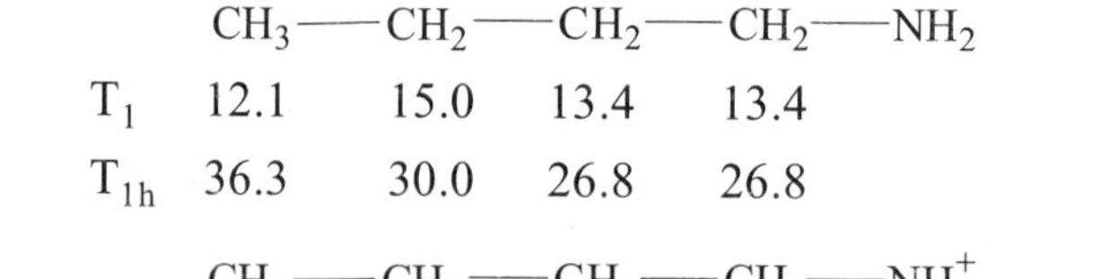

(in acetone $-d_6$)

	CH_3—	CH_2—	CH_2—	CH_2—	NH_2
T_1	12.1	15.0	13.4	13.4	
T_{1h}	36.3	30.0	26.8	26.8	

	CH_3—	CH_2—	CH_2—	CH_2—	NH_3^+
T_1	4.0	3.1	2.3	1.5	
T_{1h}	12.0	6.2	4.6	3.0	

Fig. 16.12. Carbon-13 spin-lattice relaxation times in some molecules at 38 °C (in D_2O). T_{1h} denotes the product of T_1 in seconds and N_h, the number of directly bonded hydrogen atoms to each carbon.

MULTIPLE RESONANCE TECHNIQUES 17

HOMONUCLEAR DOUBLE RESONANCE

An important technique for simplifying a complex spectrum is the double resonance method. The aim of the technique is to remove the effect of scalar spin-spin coupling between nuclei or groups of nuclei and the procedure may hence be referred to as the 'spin-decoupling' technique. Spin decoupling is one of the whole family of multiple irradiation experiments that have been developed over the years. Spin decoupling is accomplished by irradiating one group of nuclei with a strong *rf* field B_2 while simultaneously observing another group with usual *rf* field B_1.

The double resonance or spin-decoupling has proved to be most valuable for the simplification of complex spectra and for determining which nuclei are coupled in such spectra. Decoupling can be particularly useful in cases where the origin of one particular multiplet which lends itself to decoupling is known unambiguously. If two nuclei A and X are coupled by spin-spin coupling giving two doublets while observing X with an usual weak *rf* field B_1, we simultaneously use a second much stronger *rf* field B_2 at the resonance frequency of A. Irradiation at the A frequency results in the loss of structure due to coupling from this nucleus (A) at the X resonance. The nucleus X is said to be decoupled from A nucleus. Similarly, irradiation at the X frequency leads to collapse of the multiplet structure of the A resonance. The effect of this stronger B_2 field is to cause nuclear spin transitions in say X of the AX system, leading to saturation of X nuclei so that they become effectively decoupled from the A nuclei, that is, X makes no contribution to the A spectrum. The A doublet collapses to a singlet which occurs at its mid point. When the spectrum of A is measured with X being irradiated, this is denoted as $A\{X\}$. In practice, decoupling can be achieved if $\nu_A - \nu_X > 5\,|\,J_{AX}\,|$. Spin decoupling is not viable when the signal to be decoupled is close to the signal to which it is coupled, that is, the difference between the two signals is less than about 1 ppm. It is not necessary that spectrum as a whole should be first order. The value of the method is two-fold. Firstly, the complex pattern of the NMR spectrum which results particularly in intermediate coupling cases is simplified which aids in spectral analysis. Secondly, the spin-spin splitting parameters can be assigned unambiguously. The proton sequence in a molecule can be established from double resonance experiments. If we observe the nucleus A while irradiating transitions associated with the nucleus X, it may be possible to find out a great deal about X

without observing it directly. This is an inexpensive and versatile method of extending the usefulness of the spectrometer.

In the normal single resonance CW NMR experiments, an observing radiofrequency field, B_1 cos $\omega_1 t$, where ω_1 denotes the radiofrequency is applied at right angles to the static magnetic field B_0. In a double resonance experiment, a second radiofrequency, B_2 cos $\omega_2 t$ is also applied at right angles to B_0. It is possible to use for irradiation either a single frequency or a band of frequencies. The frequency and the amplitude (power level) of B_2 are important, since they determine the form of the observed spectrum. Further, by controlling the timing of switching B_2 on and off it is possible to separate the effect of intensity and transition frequency(*vide infra*).

There are a number of examples of the application of double resonance in spectral analysis. The ^{1}H NMR spectrum of *n*-propyl bromide is shown in Fig. 17.1(a). Now consider the CH_3 resonance of *n*-propyl bromide $CH_3CH_2CH_2Br$ and simultaneously irradiate the sample with a second intense radiofrequency field at exactly the resonance frequency of the β–CH_2 group. This will induce rapid transitions between the spin-states of the protons of the β–CH_2 group. As a result, the coupling of the spin states of the β–CH_2 group to the CH_3 group will be lost and the CH_3 resonance will appear as a singlet. The same argument applies to the resonance of α–CH_2 group when β–CH_2 is strongly irradiated; it will also collapse to a singlet. The decoupled spectrum will then be as shown in Fig. 17.1(b). It is not possible to record the β–CH_2 resonance signal under these conditions. With these double resonance experiments, one can only determine which proton or group of protons couples with the nucleus that is being irradiated. Another decoupling experiment that can be performed upon this molecule is illustrated in Fig. 17.1(c). Here the second irradiation is carried out at the frequency of CH_3 group. Of course, the NMR spectrum of propyl bromide can be interpreted easily without decoupling experiments. However, decoupling experiments are valuable in more complicated cases.

In a FT double resonance experiment, a time sharing procedure is employed to obtain the spectrum. For recording data points of free induction decay, only a fraction of the dwell time is needed. Dwell time is the time used to produce a particular data point. It is given by the reciprocal of the sampling time. For a spectral width of 5 kHz, dwell time is 10^{-4} *s* or 100 μs. The receiver and the decoupler channel can be synchronized in such a way that decoupling is applied between data collection. In both cases, the spin-spin splitting due to coupling with the irradiated nucleus is removed. The splitting patterns observed in spin-decoupling experiments depend on the frequency of irradiation (ideally, it should be exactly at the centre of the doublet in the X spectrum) and on the power level. For the AX spin system, irradiation of X transitions at moderate power levels can cause splitting of lines in the A spectrum. If the power level is increased further ($B_2 >> 2\pi J_{AX}/\gamma_X$), the coupling between A and X is lost completely.

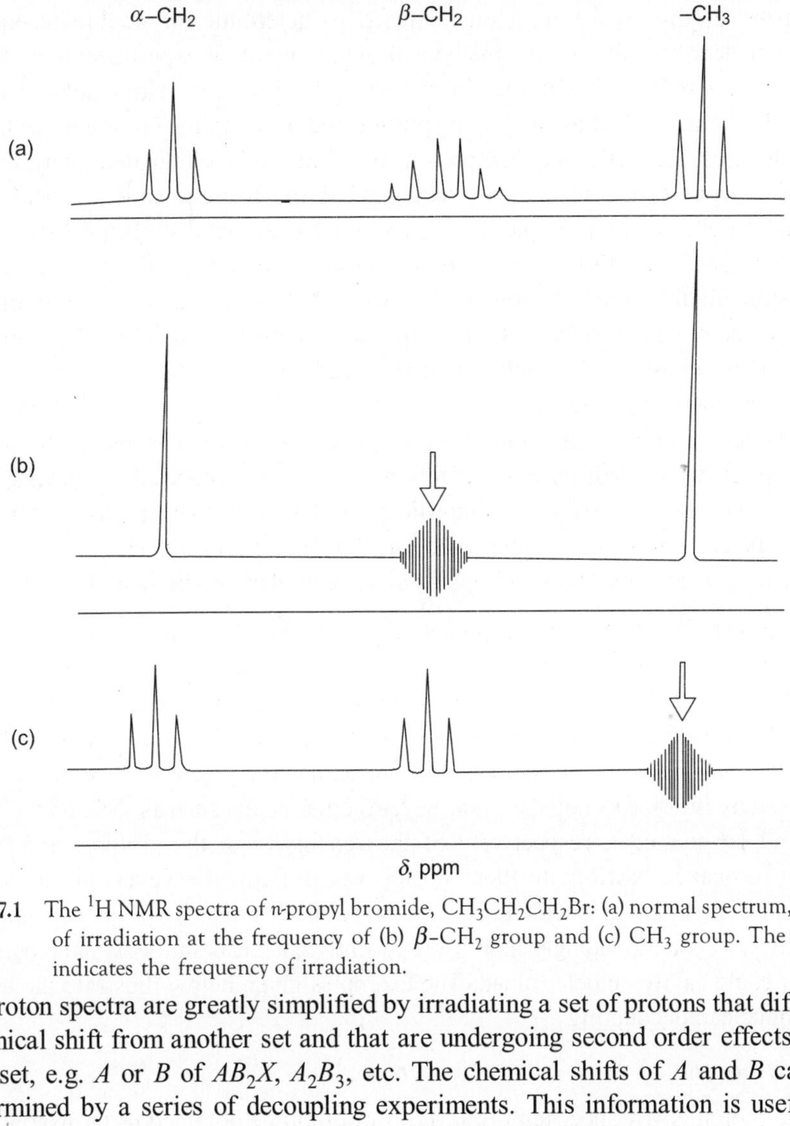

Fig. 17.1 The ^{1}H NMR spectra of *n*-propyl bromide, $CH_3CH_2CH_2Br$: (a) normal spectrum, effect of irradiation at the frequency of (b) β–CH_2 group and (c) CH_3 group. The arrow indicates the frequency of irradiation.

Proton spectra are greatly simplified by irradiating a set of protons that differ in chemical shift from another set and that are undergoing second order effects with this set, e.g. A or B of AB_2X, A_2B_3, etc. The chemical shifts of A and B can be determined by a series of decoupling experiments. This information is useful in the analysis of a second order spectrum. In an $A\{B\}$ spectrum, only A is detected, and hence its chemical shift is accurately obtained. If the chemical shifts of A and B are known, the only unknown J_{AB} could easily be determined. This information is more valuable in complex systems where many coupling constants and chemical shifts have to be determined. By double resonance experiments the chemical shift of a nucleus can be determined even when its resonance is not observed because of overlapping with the signals of other nuclei.

HETERONUCLEAR DOUBLE RESONANCE

In a decoupling experiment if the irradiated and the observed nucleus are from the same nuclide, the experiment is known as homonuclear double resonance. The extension of this to different nuclei leads to heteronuclear double resonance. Heteronuclear spin decoupling methods provide important experimental evidences for the assignment of NMR spectra of various heteronuclei. Many nuclei such as ^{13}C, ^{19}F, ^{31}P and ^{11}B cause spin-spin splitting and the coupling constants are larger. In such cases, their effects on the proton spectrum can be eliminated by irradiation of the sample at the resonance frequency of the nucleus causing them. In a magnetic field of, for example, 9.4 *T*, carbon-13 resonates at 100.6 MHz while proton resonates at 400 MHz. To obtain proton decoupled ^{13}C spectrum at this field strength, the sample is irradiated at two radiofrequencies, one radiofrequency field is employed to produce ^{13}C resonance and another strong *rf* field to cause all the protons to undergo rapid transitions leading to saturation. However, the frequency difference due to coupling here lies in the megahertz range. The decoupling frequency is conveniently provided nowadays by a frequency synthesizer, which will give accurately any frequency required for NMR. Such experiments are useful for the simplification of spectra that are complicated by heteronuclear spin-spin coupling such as J (^{1}H, ^{19}F) or J (^{1}H, ^{31}P). The line broadening caused by ^{14}N quadrupolar effect can also be eliminated.

Heteronuclear decoupling is potentially an important method for determining the chemical shift of a range of nuclei such as ^{13}C, ^{14}N, ^{19}F, ^{195}Pt, ^{199}Hg .... Not only because of the versatility with which it invests a spectrometer with a single probe but also because of high sensitivity. The sensitivity of a nucleus to NMR depends upon its magnetic moment and proton has the greatest sensitivity closely followed by fluorine as noted earlier. Several other nuclei such as ^{14}N and ^{13}C have 0.1 and 1.6 per cent, respectively of the sensitivity of the protons, and hence single resonance NMR with these nuclei was difficult. However, in the double resonance method, the detection is made at the ^{1}H or ^{19}F frequency and the sensitivity is correspondingly high. Consequently chemical shifts of 'unfavourable' nuclei could easily be determined. The examples given below illustrate the use of heteronuclear decoupling.

(i) Diborane

The proton NMR spectrum (100 MHz) of diborane obtained using isotopically pure $^{11}B_2H_6$ is given in Fig. 17.2. This spectrum results from two sets of non-equivalent protons (bridged and terminal) being split by the ^{11}B nuclei. The terminal proton 1 : 1 : 1: 1 quartet is to the low field of TMS and the bridged proton septet in the intensity ratio of 1 : 2 : 3 : 4 : 3 : 2 : 1 is to the high field of TMS. Each of the terminal protons has major spin coupling to one boron atom. The bridge protons are equally coupled to both the boron atoms. The spectrum contains second order splittings, since both the ^{11}B atoms and the terminal protons are not magnetically equivalent. The splitting caused by ^{11}B can be removed by saturation of the boron nuclei by the spin decoupling technique. Two peaks in the intensity ratio of 2:1 are then obtained corresponding to the four terminal and two bridging protons.

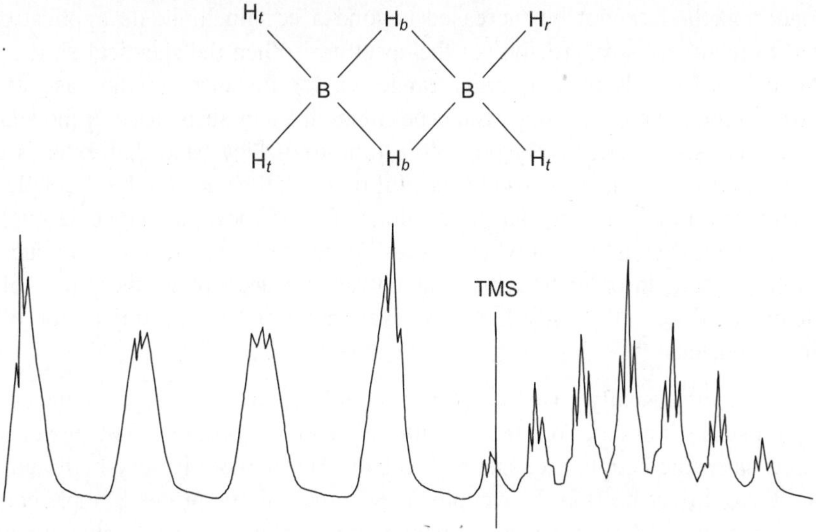

Fig. 17.2 The ^1H NMR spectrum of B_2H_6 nearly pure in ^{11}B isotope.

(ii) N_2F_2

Two isomers have been obtained during the synthesis of N_2F_2. One has a *trans* structure with one fluorine on each nitrogen as shown in Fig.17.3. The structure of the second isomer was reported to be a *cis* isomer and also $F_2N{=}N$. Analysis of the results obtained by several physical methods along with ^{19}F NMR spectrum including from a double resonance experiment has been made to resolve this problem. Saturation of the ^{14}N nucleus in the second isomer with a strong radiofrequency field causes collapse of all nitrogen splitting, indicating that the chemical shift of the two nitrogens must be equivalent, and this excludes the $F_2N{=}N$ structure. The value for $J_{N\text{-}F}$ calculated for a *cis* structure was in fair agreement with the $J_{N\text{-}F}$ for NF_3. Additional evidence for the *cis* structure has been obtained from infrared and Raman spectra.

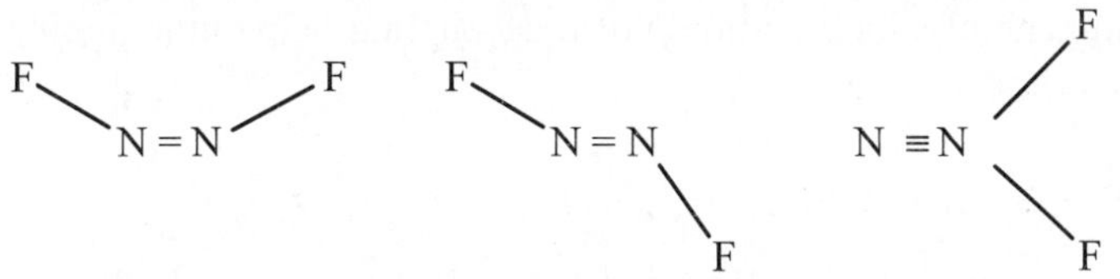

ci *tran*

Fig. 17.3 Structure of N_2F_2.

BROAD BAND DECOUPLING

An important development in the area of heteronuclear decoupling is the technique of broad band decoupling. Since the amplitude of the B_2 field in the double resonance method cannot be increased beyond a certain limit, its application is limited to relatively small regions of the spectrum. When the chemical shift of the heteronucleus extends over a larger range as, for instance, in the case of ^{31}P resonances complete decoupling would be impossible. In such cases, a modulation technique is used to effectively produce a radiofrequency band that extends over several kilohertz and span the whole spectral range of the nuclei to be decoupled by saturating transitions that are causing coupling. This is known as 'noise decoupling' or 'broad band decoupling' in which a noise generator is employed to produce the decoupling effect. In order to achieve complete spin decoupling, the power of the B_2 field should be sufficiently large to cover the complete spectral region of the particular nucleus.

One difficulty usually encountered in decoupling experiments particularly with inorganic systems arises from the size of the coupling constants. The power level required when the coupling constants are 1000 Hz or more is very high and it is difficult avoid over heating the sample or even melting the probe. Sample cooling is one measure. In such cases, a new technique known as composite pulse decoupling based on the spin-echo experiment (see Chapter 16) appears promising. It is similar to noise decoupling but can deal with coupling constants of several kHz. It should prove useful to inorganic chemists.

Complete removal of the effects of the irradiated nuclei from the observed spectra by noise decoupling is routinely used. Without broad band proton decoupling, ^{13}C spectra of organic compounds would be very complicated with each type of carbon coupling to several different sets of protons. Broad band decoupling is employed routinely in ^{13}C NMR spectroscopy. With broad band decoupling, the ^{13}C spectrum usually consists of a single resonance for each chemically distinct carbon and thus easier to interpret (see Chapter 15). The same principles apply for many other nuclei which couple to the abundant nuclei (hydrogen). An added advantage of ^{1}H decoupling method is an intensity enhancement of resonances from (i) collapse of the multiplets into singlets improving the signal to noise ratio and (ii) an intensity increase for the ^{13}C resonance by about 200% due to nuclear Overhauser enhancement discussed later. Apart from general simplification, noise decoupling helps in assigning particular coupling constants.

OFF-RESONANCE DECOUPLING

As the name suggests, it is a partial decoupling technique. In this technique, an intense radiofrequency radiation close to but not coinciding with the frequency of the protons i.e. at a frequency, a few hundred hertz away from the resonances involved, is employed. The effect is to remove all long-range couplings but not quite remove the largest (normally one bond) couplings so that it is possible to observe multiplet structure and hence determine the number of neighbouring spinning nuclei. In the case of ^{13}C spectrum, the smaller the $^{13}C-^{1}H$ coupling constants (geminal, vicinal and long range) are eliminated whereas the spin-spin splittings caused by the larger one-bond coupling constants are retained. Consequently, in an off-resonance ^1H decoupled ^{13}C spectrum first order multiplicity is observed. Quartets, triplets, doublets and singlets are observed for primary(CH_3), secondary(CH_2), tertiary(CH) and quaternary carbon atoms, respectively. The off-resonance ^1H decoupled ^{13}C NMR spectra of vinyl acetate and p-ethoxy benzaldehyde were discussed earlier (Chapter 15). The residual coupling observed in off-resonance decoupled spectra is given by

$$J' = 2\pi J \Delta\nu / \gamma_H B_2 \tag{17.1}$$

where J is the usual coupling constant, $\Delta\nu$ is the difference between the resonance frequency of the coupled proton and the frequency of the decoupling signal and B_2 is the decoupler power. Thus the coupling constant J is not easily obtained from the off-resonance decoupled spectra.

This method is very often used for ^{13}C spectra, which are extremely complicated when all the proton couplings are retained. With total ^1H decoupling, the spectrum becomes too simple with series of single lines but assignment becomes difficult. With off-resonance decoupling, the single lines are replaced by narrow multiplets and it is hence easier to identify primary, secondary, tertiary and quaternary carbon atoms as mentioned above. The method is general, although mainly used for ^{13}C spectra but to some extent it has been superseded by multi-pulse techniques. For example, the technique of DEPT (distortionless enhancement by polarization transfer) discussed earlier is used nowadays to obtain the multiplicities of carbon (Chapter 15).

GATED DECOUPLING

There are some disadvantages arising from ^1H decoupling while measuring the ^{13}C NMR spectra. (i) The elimination of spin-spin splittings prevents the measurement of ^{13}C, ^1H coupling constants and thus valuable experimental information is lost. (ii) The nuclear Overhauser enhancement leads to intensity distortions and the integration of the spectral peaks becomes questionable. We might also want to avoid the nuclear Overhauser effect where it could lead to negative or zero intensities, or when the intensity enhancement is required while retaining a fully coupled spectrum, and (iii) the assignment of the resonances to specific carbon atoms of the compound is not unambiguous.

The technique of 'gated decoupling' successfully overcomes the limitations of spin-decoupling and it is often used. Different time scales are characteristic of the spin decoupling experiment and the nuclear Overhauser effect. While the NOE is governed by the spin-lattice relaxation and thus evolves and decays within seconds, decoupling sets in or vanishes almost immediately after the B_2 irradiation field has been switched on or off.

This difference is made use of in gated decoupling. The pulse sequence given in Fig. 17.4 illustrates this difference. In the first, the decoupler is gated so that it is on during the interval between the pulses and switched off during the acquisition of data. The spectra are therefore, not decoupled but the NOE enhancement is retained. Thus ^{13}C,^{1}H coupling constants can be obtained without losing NOE completely. If correct intensities are of interest, a second set of experiments is carried out. Here, the decoupler is switched on during the acquisition of data but switched off during the remaining period before the next pulse. The ^{13}C, ^{1}H spin-spin coupling is thus eliminated and the NOE enhancement does not occur. The NOE builds up slowly and hence does not influence the free induction decay (FID). There must be a delay after acquisition of the FID to allow the NOE to decay before the next pulse sequence starts. In this way decoupling occurs without NOE. Thus the effect of intensity-change and decoupling may be realised separately. The gated decoupling experiments are more time consuming than the simple population inversion technique discussed earlier. (Chapter 15)

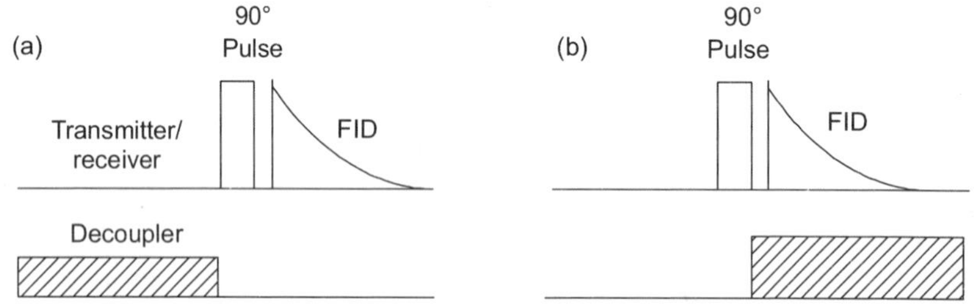

Fig. 17.4 Pulse sequence for gated decoupling. (a) The decoupler is switched off during acquisition of data, part of NOE is retained, ^{1}H-coupled ^{13}C spectra are obtained. (b) The decoupler is switched on during acquisition of data ; NOE is eliminated; ^{13}C spectra obtained are proton decoupled.

To deal with the NOE variations, 'inverse gated decoupling' is useful. The ^{1}H broad band decoupler is 'gated' (switched on) only during the ^{13}C pulse and the acquisition period; it is switched off during the delay period. The NOE builds up only slightly during the pulse and acquisition period. Decoupling is established almost immediately on irradiation with the ^{1}H broad band decoupler so that the end result is a number of singlets whose intensities are proportional to the number of carbon atoms they represent.

SPIN TICKLING

Spin tickling is a subtle variation of the spin decoupling experiment and it can be regarded as an incomplete spin decoupling technique. Instead of irradiating the whole multiplet to be decoupled with a powerful decoupler signal to eliminate the coupling, here a much less powerful decoupler signal, about one-tenth as powerful, is employed and it is centred just on one line of the multiplet. This has the effect of perturbing the energy levels of the spin system. Very small but distinct changes in the appearance of the resonances coupled to the signal being "tickled" are observed. A series of heteronuclear spin tickling experiments can provide more information than that obtainable from a normal spectrum of the other nucleus. Spin tickling experiment which is essentially a CW one can also be performed with pulse FT technique.

Spin tickling can be used to determine the positions of the individual lines more accurately. For example, consider the NMR spectra of phosphite anion, HPO_3^{2-}. The ^{1}H and ^{31}P spectra are identical and each is a doublet with a splitting $J(^{31}P-^{1}H)$ of 690 Hz (Fig. 17.5 (a)). If sufficiently large power is applied at the ^{31}P resonance frequency as in a double resonance experiment, the proton doublet collapses to a single line. (Fig. 17.5 (b)). This is rather difficult to carry out since J is very large. At very low power levels, when the second frequency is set on or

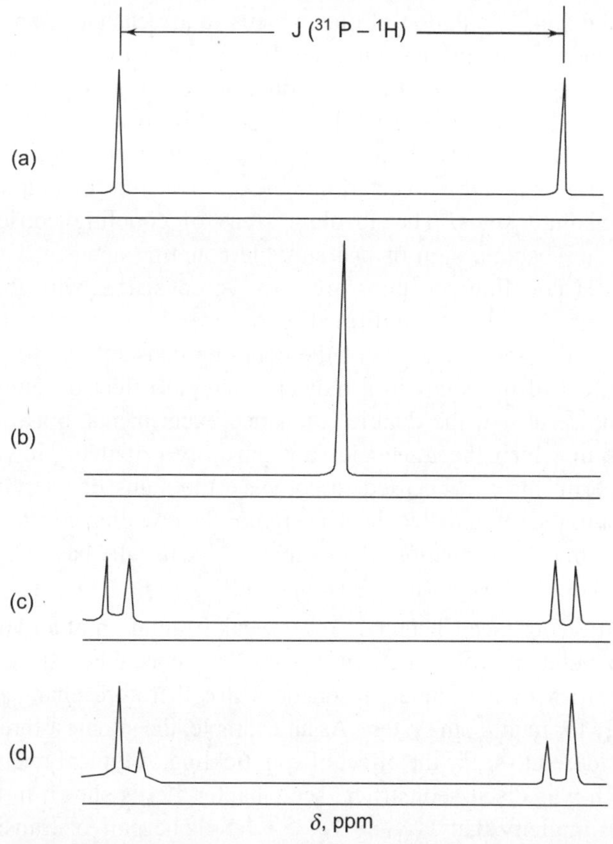

Fig. 17.5 The ^{1}H NMR spectra of the phosphite anion HPO_3^{2-}: (a) normal spectrum, (b) ^{31}P decoupled, (c) spin tickling of a ^{31}P line exactly and (d) tickling close to a ^{31}P line.

very close to one of the individual lines in the ^{31}P spectrum, each of the lines in the proton spectrum is split into a doublet as shown in Figs. 17.5 (c) and (d) which is symmetrical when the second radiofrequency field is set exactly on either ^{31}P line. If the molecule under investigation is more complicated say $(CH_3O)_3P$, a symmetric triplet will occur in which the relative intensities depend on the nature of the system.

SIGN OF THE COUPLING CONSTANTS

Spin tickling is a valuable technique for assigning the spectral lines in a complicated spectrum. Thus by "tickling" a nucleus with a weak field one can identify the other nuclei to which it is coupled. It allows energy level diagram to be set up for the system. The spin tickling utilizes much less power than the usual decoupling experiments. This is an advantage in cases where a conventional decoupler signal would perturb the other signals in the spectrum.

Spin tickling is useful in the determination of the relative signs of the coupling constants. The determination of the sign of the coupling constants was first demonstrated for $(C_2H_5)_2{}^{205}Tl^+$ by Evans. The coupling constant is a measure of the ability of the spin orientation of one nucleus to affect that of another through the electron cloud. If this influence is such that the lowest energy level is the one with two nuclear spins parallel, then the coupling constant is said to be negative and it is positive otherwise. For example, $J(^{13}C\text{-}^{1}H)$ is positive while $J(^{13}C\text{-}^{19}F)$ is negative in $^{13}CFHCl_2$. For any spin system with three or more sets of spins coupled together, it is possible to determine the relative signs of coupling constants but not the absolute signs. The absolute signs of coupling constants can be determined if the absolute sign of one suitable coupling constant is known. The one-bond $^{13}C\text{-}^{1}H$ coupling constants are positive consistent with the theoretical predictions. This has been confirmed experimentally in some cases. The determination of the absolute signs of the coupling constants is difficult because the reversal of all of the signs in a molecule has no effect on either the NMR spectrum or the results of the double resonance experiments, but it can be done using samples in which the molecules are partially oriented. However, double resonance experiments can be used to compare the signs of coupling constants between nuclei in the same molecule. The sign of the coupling constant is valuable and it yields chemical information. For example, $^{1}J_{PN}$ usually has different signs in phosphorus(III) and phosphorus(V) compounds.

It was noted earlier (see Chapter 15) that weak irradiation of a two-spin system perturbs the populations of the two energy levels connected by the transition and that the intensities of associated resonances are therefore changed. The same principles apply for multi-spin system. As an example, the simplest three-spin AMX system is considered to study the effect of spin tickling. A typical twelve line AMX spectrum which was discussed earlier (see Chapter 12) is shown in Fig. 17.6 (a) assuming for simplicity that $J_{AM} >> J_{AX} >> J_{MX}$. The pair of transitions A_1, A_2 may lie to the high or low frequency side of A_3, A_4 depending on the sign of J_{AM}. If $J_{AM} > 0$, then A_3, A_4 lie on the low frequency side of A_1, A_2. Similarly, the sign of J_{MX} determines the positions of X_{11} and X_{12} relative to X_9 and X_{10}. Decoupling at the resonance position of proton A eliminates the coupling of A with M and X and only the small coupling J_{MX} remains and two doublets are observed in Fig. 17.6 (b). If spin tickling is accomplished by weakly irradiating one of the high frequency lines of the A spectrum, lines 6 and 8 of M resonance and similarly lines 10 and 12 of the X resonance are further split (Fig. 17.6(c)). Thus if a low field line of a multiplet is tickled and if it leads to splitting in the high field part of one associated multiplet and the low field part of another, then the two coupling constants are of opposite sign. From the coupling constant J_{AM}, it is clear that the lines 5 and 6 are associated with one particular spin orientation of A while lines 7 and 8 are associated with the opposite spin orientation of A. Similarly, from the coupling constant J_{MX}, it is evident that the lines 5 and 7 are associated with the one spin orientation of X while lines 6 and 8 are associated with the opposite spin orientation of X. The spectral appearance would change slightly if the decoupler is centred on a different line of the A multiplet. This slight change in appearance is sufficient to establish the connection between the coupled signals.

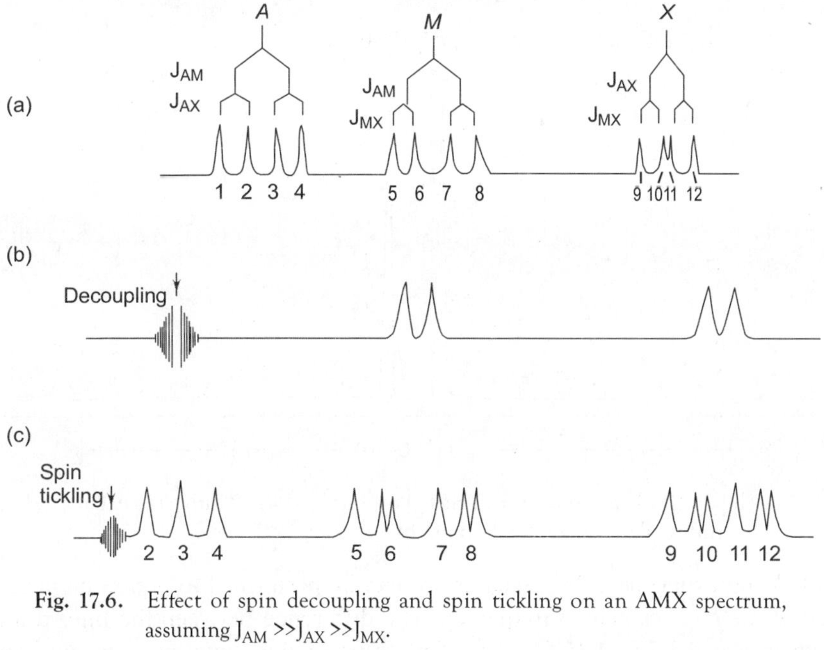

Fig. 17.6. Effect of spin decoupling and spin tickling on an AMX spectrum, assuming $J_{AM} >> J_{AX} >> J_{MX}$.

On the basis of above arguments, the lines 9 and 10 in the X part of the spectrum arise from one particular spin orientation of A while the lines 11 and 12 arise from the opposite spin orientation. Similarly, lines 9 and 11 are associated one particular spin orientation of M while lines 10 and 12 are associated with the opposite spin orientation of M. We have to choose between the two alternative spin orientations. (i) If the + orientation of A gives rise to the set of lines 5, 6 and 9, 10, then the – orientation would give rise to lines 7, 8 and 11,12. (ii) The other alternative is that the + orientation of A would give rise to the set of lines 5, 6 and 11, 12, in which case the lines 7,8 and 9,10 arise from the – orientation. Spin tickling of line 1 has perturbed the lines 6, 8 and 10,12, so that these lines must all arise from the same spin orientation of A; the signs of the coupling constants J_{AM} and J_{AX} must be the same and opposite to that of J_{MX}. Absolute signs can only be found if the absolute sign of one of the sets is known. This is often possible as for example $^1J_{PF}$ is always negative and $^1J_{CH}$ and $^1J_{Pt-P}$ are always positive.

COUPLING WITH LOW–ABUNDANT NUCLEI

Many magnetic nuclei that are of interest to chemists have low natural abundance e.g. ^{13}C (1.1 per cent), ^{29}Si (4.6 per cent), etc. In these cases, only a small proportion of the molecules of the sample will show the effect of spin-spin coupling between the proton and the other nuclei. The proton spectrum will then be a strong central signal with satellite lines on either side. Thus the ^1H spectrum of trimethyl phosphite $(CH_3O)_3P$ is a strong doublet due to splitting by ^{31}P which is 100 per cent abundant with a pair of weak symmetrically placed doublets arising from the molecules containing ^{13}C nucleus as shown in Fig. 17.7. A heteronuclear

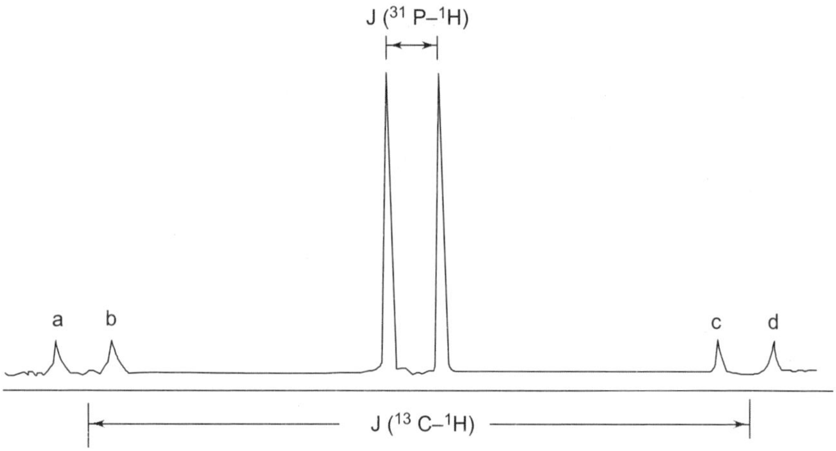

Fig. 17.7. The ^1H NMR spectrum of trimethyl phosphite $(CH_3O)_3P$ with ^{13}C satellites (a, b, c, d). Intensity scale is exaggerated for clarity.

spin tickling experiment discussed above can be performed to obtain an accurate value of the ^{13}C resonance frequency. The difference between the lines a and b shown in Fig. 17.7 is J ($^{13}C-^{31}P$). The value of this coupling constant can be confirmed by measuring the difference between the ^{31}P frequencies with lines a and c. These experiments also provide the signs of J($^{31}P-^1H$) and J($^{31}P-^{13}C$) relative to J($^{13}C-^1H$). This method has advantage over the measurement of J($^{13}C-^{31}P$) from the ^{31}P or ^{13}C single resonance spectrum. Thus, a significant feature of heteronuclear double resonance experiments is their ability to provide signs and magnitudes of the coupling constants between pairs of nuclei that are not observed directly.

THE NUCLEAR OVERHAUSER EFFECT

We had discussed the NMR spectra of molecules involving the coupling of nuclei through the chemical bonds (i.e. scalar coupling). Nuclei can also interact with each other directly through space (i.e. dipolar interaction). This latter form of interaction accounts for an interesting phenomenon known as the 'nuclear Overhauser effect' (NOE), which is the change in the intensity of the resonance of one nucleus when the transitions of another nucleus that lies spatially close to the first nucleus are perturbed by irradiation. It occurs in all double resonance experiments.

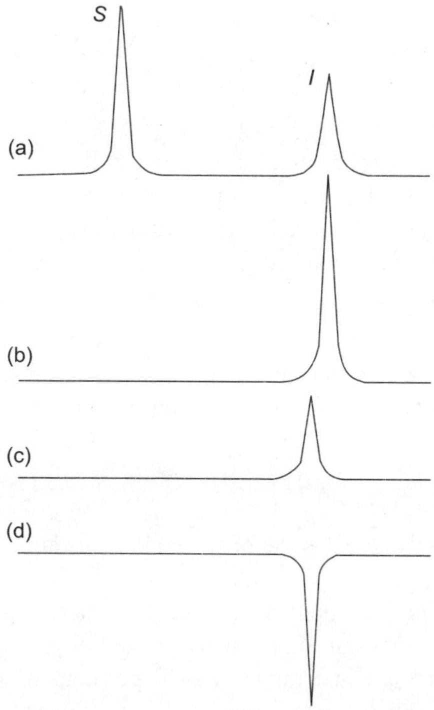

Fig. 17.8. Schematic diagram of the NMR spectra showing different nuclear Overhauser enhancements. (a) Normal spectrum of two neighbouring spins *I* and *S*. The *I* signal becomes (b) stronger, (c) weaker or (d) inverted due to saturation of the *S* resonance.

Consider a simple AX type of system containing two different protons (or groups of chemically inequivalent protons or, a ^{13}C nucleus and a proton) *S* and *I* with no scalar coupling between them so that the ^{1}H spectrum consists of a singlet at each of the chemical shifts. The two protons *S* and *I* should be close enough to allow dipolar spin-spin interactions but not the scalar spin-spin coupling. Suppose that the *S* spins are saturated at their resonance frequency by applying an intense radiofrequency field while recording the spectrum. Evidently this eliminates the NMR signal of *S* but it can also affect the intensity of *I* resonance if the two spins have an appreciable dipolar interaction. As shown schematically in Fig. 17.8, the *I* resonance peak may get stronger, weaker, or even invert. Thus, NOE as stated above is the change in intensity of the resonance of one nucleus when the transitions of another nucleus lying spatially close to the first nucleus are perturbed by irradiation. The perturbation is usually carried out with a weak *rf* field and it eliminates the population differences for the transitions of the saturated nucleus. This brings about a compensatory response from the system through a readjustment of the population differences for the transitions of a nearby nuclei. The resulting change in intensity of the NMR peak may correspond to an increase, as in small rapidly tumbling molecules or to a decrease as in large, slowly tumbling molecules. One advantage of protons is their high magnetogyric ratio which results in relatively large Boltzmann population differences. In NOE, relaxation processes involving spin dipole-dipole interactions aid the transfer of Boltzmann population advantage of the proton to another nucleus (^1H or ^{13}C) so that the other non irradiated nucleus resonances are enhanced. The NOE theory for strongly coupled systems is rather complex.

The spectrum is recorded with a relatively low power decoupler signal irradiating H_S. If H_I and H_S are separated by more than about 4.5 Å, then no enhancement may occur. If H_I is relatively nearer to H_S, say less than 4.5 Å, then the intensity of H_I may be enhanced in the second spectrum.

If we assume that the relaxation of nucleus *I* occurs only through a pure dipolar interaction with nucleus *S* and that a steady state equilibrium has been reached, then the nuclear Overhauser enhancement η of nucleus *I* is given in the absence of other relaxation mechanisms by the following expression assuming that extreme

$$\eta = \frac{\gamma_S}{2\gamma_I} \tag{17.2}$$

narrowing limit exists, where *I* is the observed nucleus and *S* is the nucleus that is irradiated, γ_I and γ_S are the magnetogyric ratios. For the homonuclear case, $\gamma_S = \gamma_I$ and according to equation (17.2) the maximum NOE enhancement expected is nearly ½ (50%) of the intensity of the normal signal, and the effect is quite small. However the usual range is 1-20 per cent whatever be the distance between the nuclei *S* and *I* due to a number of interfering factors. The NOEs are further weakened when the proton that is observed is also relaxed by protons other than the irradiated protons. Thus, a methyl group in which each proton already has two nearby protons to speed up relaxation often shows very little NOE when a nearby proton is irradiated. Therefore, NOEs are more easily detected in methine groups than in methyl groups. The NOE results must therefore be interpreted with caution.

While the final magnitude of NOE depends on the relaxation pathways, the initial rate of build up of NOE (transient NOE) depends only on the cross-relaxation between the nuclei and this can provide valuable information about the distance between the nuclei. Cross relaxation is the mutual intermolecular or intramolecular relaxation of magnetically equivalent nuclei e.g. dipolar relaxation. The dipolar field is proportional to r^{-3}, where r is the internuclear distance, and the relaxation effect is proportional to the square of the field and therefore to r^{-6}. The rate of build up of NOE is proportional to r^{-6}, where r is the distance between the nuclei. Thus, if the proportionality constant is determined, we can calculate approximately the distance between the two nuclei.

The NOE is noticeable over short distances, generally 2-4 Å, falling off rapidly as the inverse sixth power of the distance between the nuclei. The best results are obtained in rigid molecules when the nuclei are less than 3 Å apart. It is possible to measure NOEs by integrating the signals with the irradiation on and off and noting the difference. The accuracy of the integration is such that this method can be used reliably if the NOE is 10 per cent or more. Even then NOE has been a useful method for detecting groups which are close to each other in space. Since the magnetogyric ratio of the proton is about four times greater than that of carbon, i.e. $\gamma_H/\gamma_C \approx 4$ if ^{13}C is observed and ^{1}H is irradiated at the extreme narrowing limit, $\eta \approx 2$ that is, 200%. Thus the ^{13}C signal should appear with a four-fold enhancement of intensity due to the NOE effects. The NOEs of this order are found for the ^{13}C atoms directly bonded to protons in noise-decoupled ^{13}C spectra where the ^{13}C signals appear with enhanced intensities. This is a useful feature. For $^{13}C\{^{1}H\}$, NOE enhancements are usually observed for all carbons bound to hydrogen but not for quaternary carbons as discussed earlier (Chapter 15). For example, for ^{13}C nuclei in biphenyl, Overhauser enhancement factors of 2.7 to 2.8 have been found for o-, m- and p-carbon resonances. Quaternary carbon shows an NOE of 0.61 due to the long range interactions with the protons in the molecule. An improved procedure for recording heteronuclear NOEs is to irradiate individual lines of the multiplet with a low decoupling power instead of exciting the entire multiplet with a higher *rf* power. This results in greater selectivity of the protons being irradiated and higher sensitivity of the carbon signals.

Dipolar relaxation is particularly important for spin ½ nuclei. In addition to the dipole-dipole relaxation processes, which depend on the strength and frequency of the fluctuating magnetic fields around the nuclei, there are other factors that affect NOE. (a) The intrinsic nature of the interacting nuclei I and S, (b) the internuclear distance r between them, and (c) the rate of tumbling of the relevant segment of the molecule in which the nuclei I and S are present i.e. the effective molecular correlation time τ_c. Molecular correlation time τ_c represents the time taken for the relevant portion of the molecule containing I and S to change from one orientation to another. It is so chosen as to be nearly equal to the minimum time interval between various orientations rather than as the average time interval between the orientations. In small molecules with molecular weights upto about 300, the tumbling frequency in common NMR solvents is nearly 10^{10}–10^{11} s^{-1} and NOE is an enhancement (positive NOE). In large molecules with molecular weights greater than 600 in relatively viscous solvents such as DMSO, the NOE corresponds to a reduction in the observed signal (negative NOE).

The dipole-dipole relaxation as discussed above depends on the molecular correlation time τ_c, the internuclear distance r and the magnetogyric ratios of the two nuclei γ_S and γ_I. For fast molecular motion in liquids, the following relation hold good for the dipolar relaxation rate,

$$R_i = k\,\gamma_I^2\,\gamma_S^2\,(r)^{-6}\,\tau_c \tag{17.3}$$

where k is a constant. While the rate of change of dipolar interaction depends on τ_c, its magnitude depends only on the internuclear distance and is independent of τ_c. The equation (17.3) forms the basis for distance measurements using homonuclear NOE experiments. The r^{-6} distance dependence of the dipolar relaxation is an invaluable source of information on the structure and conformation of molecules. The efficiency of the relaxation process is therefore governed by the r^{-6} relationship. The strength of the dipole-dipole interaction depends on the square of the product of the magnetogyric ratios of the two nuclei and hence proton-proton interactions are the main source of relaxation processes. Carbon atoms relax mainly through the bonded protons and hence methyl carbons containing three hydrogen atoms will normally relax faster than methine carbons bearing only one hydrogen. The low intensities of non-protonated carbons is usually due to their long relaxation times. The addition of a paramagnetic substance such as tris(acetylacetonato)chromium(III), $Cr(acac)_3$ can produce a hundred-fold increase in the relaxation rate with the rate varying linearly with the concentration of $Cr(acac)_3$; so the NOE is also reduced a hundred-fold. Protonated carbons with faster relaxation rates are less affected (about ten-fold increase in relaxation rates) and the reduction of NOE is correspondingly less. The NOE is quenched by all mechanisms that allow relaxation other than through intramolecular dipole-dipole interaction. Consequently NOE can be used to study relaxation mechanisms (see Chapter 16).

The NOE is an important tool for improving the sensitivity of the NMR measurements with less sensitive nuclei i.e. nuclei with smaller magnetogyric ratios. This is the standard practice for example in ^{13}C NMR spectroscopy where ^{1}H broad band decoupling is applied as discussed earlier. Problems arise if the X nucleus has a negative magnetogyric ratio, e.g. ^{15}N or ^{29}Si. Equation(17.2) predicts negative enhancements if either γ_I or γ_S is negative. For ^{29}Si–^{1}H, the maximum effect is –1.5 and for ^{15}N{^{1}H}, it is – 4. The NOE is then negative and the observed I signal will be inverted. Some complications may arise in the application of NOE measurements, since in molecules, a single proton generally has a large number of different neighbours and not isolated two-spin systems. Scalar spin-spin coupling is another factor that can complicate NOE. With larger molecules the dependence of NOE on the correlation time is also important (equation (17.3)).

Experimentally, NOE measurements for the estimation of relative proton distances are performed easily with the help of FT NMR NOE difference spectroscopy. Two different sets of experiments are recorded, one in which certain protons are irradiated and enhancements are obtained of other nearby protons and the other without such irradiation so that a normal spectrum is obtained. Subtraction from the perturbed FID that of the unperturbed one, followed by Fourier transformation of the difference FID gives the NOE difference spectrum, which contains only the NOE enhancements. This form of NOE spectroscopy is very sensitive and allows the detection of very small intensity differences, one per cent enhancements or even less since the signals from experiments with and without irradiation are recorded under identical conditions. By a series of experiments, a map of the relative proton positions can be constructed.

The NOE experiment is one of the most powerful and widely utilized methods for structure determination. The NOE difference and the two dimensional nuclear Overhauser spectroscopy (NOESY) (see Chapter 19) are utilized extensively to obtain useful structural information especially for determining the stereochemistry. They provide an indirect way to obtain information about internuclear distances. Consequently, NOE measurements are valuable aids in structural elucidation and conformational analysis when it is needed to decide which of the two nuclei say X or Y is closer in space to a third nucleus A in the same molecule. For example, in the compound (I) shown below, on irradiation of the methyl group *cis* to the –CH group, the –CH proton signal registers 17 per cent increase in intensity. The irradiation of the *trans* methyl group does not affect the intensity of the –CH signal. It was possible using NOE experiments to assign unambiguously the chemical shifts of the two methyl groups of N,N-dimethyl formamide. The methyl group at the lower field was identified as the one that is *trans* to the carbonyl group.

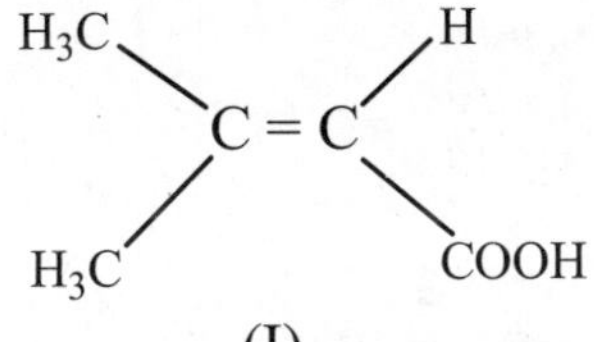

(I)

In many cases, the structure of the compound may be known except for some small important details like the stereochemistry or the positioning of a substituent. For example, the following two isomers will give very similar NMR spectra. Without an authentic spectrum, it would be rather difficult to distinguish between the two and NOE could be utilized to identify the two isomers unambiguously.

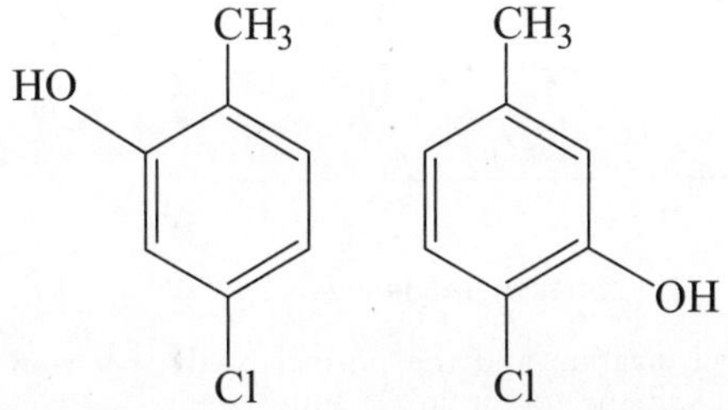

INTERNUCLEAR DOUBLE RESONANCE

The internuclear double resonance (INDOR) technique is another double resonance method of establishing spin-spin couplings. It is potentially a very useful technique but it is rather difficult to carry out the experiment. In the INDOR method, the intensity of a single line is observed while a perturbing field is swept through the remaining part of the spectrum. A decrease or an increase in the amplitude of the observed signal occurs when the perturbing field irradiates a line which has an energy level in common with the observed signal.

A simple INDOR experiment is illustrated using the energy level diagram for an AX spin system given in Fig. 17.9 (a). The terms $\alpha\alpha$, $\alpha\beta$, etc. refer to the spin quantum numbers (m_I) of the nuclei A and X. At thermal equilibrium, the population differences are proportional to the energy differences between the nuclear spin levels in an applied magnetic field. The intensity of A signal (peak height) is monitored as a weak radiofrequency source is swept through the (X_1 and X_2) transitions of the nucleus X. If the transition say X_1 is saturated by a low power radiofrequency radiation, the population of the levels connected by this transition becomes equal. It causes a decrease in the intensity of the line A_1 and an increase in the intensity of the line A_2. On the other hand, irradiation at X_2 transition enhances the intensity of A_1 but decreases that of A_2. The intensities of the lines A_1 and A_2 depend on the population differences of their associated energy levels. A plot of the intensity of A_1 (or A_2) against irradiation (X) frequency as shown in Fig.17.9 (b) is known as the INDOR spectrum. Thus INDOR measurement provides information about the frequencies of the X spectrum while

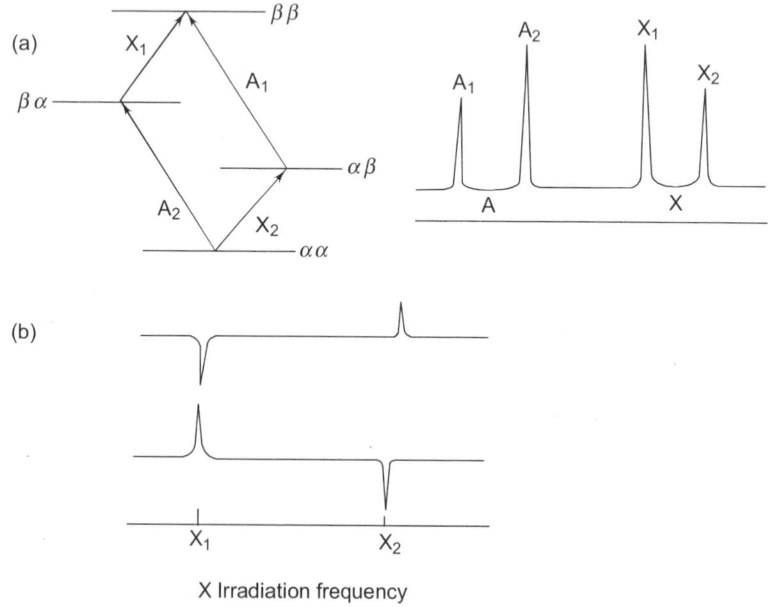

Fig. 17.9. (a) Energy level diagram and the normal NMR spectrum for an AX spin system. (b) Typical INDOR spectra for an AX spin system.

observing only the A spectrum. The INDOR can readily be accomplished using continuous wave spectrometers. Since multinuclear NMR spectrometers are widely available, the INDOR experiment is not important nowadays. However, it may still be useful if X is an insensitive nucleus for NMR study or if only a proton spectrometer is available. The INDOR technique is useful to obtain the energy levels of the spin system and to determine the relative signs of the coupling constants in the first order spectra.

The energy level diagram for an AX spin system given in Fig. 17.9 (a) helps us to understand the changes in the intensities of the A signal during irradiation of the X signal. Each line in the spectrum represents a transition between the spin states $\alpha\alpha$, $\alpha\beta$, $\beta\alpha$ and $\beta\beta$ assuming that energetically $A_1 > A_2 > X_1 > X_2$. Irradiation at X_1 will produce an INDOR signal from A_1 since they have an energy level ($\beta\beta$) in common. Irradiation at X_1 induces the transition $\beta\alpha \rightarrow \beta\beta$ and the population of spin state $\beta\beta$ is increased. Therefore, the probability of the transition $\alpha\beta \rightarrow \beta\beta$ (the A_1 line) becomes less favoured and the intensity of A_1 signal decreases (negative INDOR signal). On the other hand, when the transition $\beta\alpha \rightarrow \beta\beta$ takes place, the population of spin state $\beta\alpha$ diminishes and therefore the transition $\alpha\alpha \rightarrow \beta\alpha$ (the A_2 line) becomes more probable; the intensity of the A_2 signal increases (positive INDOR signal). Similarly, irradiation at X_2 leads to the transition $\alpha\alpha \rightarrow \alpha\beta$; population of the spin state $\alpha\alpha$ decreases while that of the state $\alpha\beta$ increases. Consequently the intensity of the A_2 line will decrease (negative INDOR peak) while that of the line A_1 will increase (positive INDOR signal). The intensity effects in INDOR take time of the order of spin-lattice relaxation time T_1 to build up and decay.

The term progressive transition is used if there is a progressive increase in energy from the lower end of one transition to the far end of the connected one and the term regressive transition is used if the two levels have the same common highest or lowest energy level. Positive signals are produced when the irradiated and monitored transitions are progressively connected and regressively related transitions give rise to negative signals. The populations of the energy levels become modified. The relationship between the transitions X_2 and A_1, and between X_1 and A_2 is said to be progressive and the relationship between the transitions X_2 and A_2, and between X_1 and A_1 is regressive. The three levels characterising the connected transitions change their energy steadily in progressive transitions. On the other hand, for regressive transitions, the value of the central energy level is larger or smaller than the energy of the initial and final states as shown in Fig. 17.10.

The INDOR technique was originally used for studying nuclei such as ^{13}C. Scanning through the ^{13}C resonance frequency using a low intensity *rf* radiation for decoupling produces changes in the associated ^{1}H spectra. Without directly observing the ^{13}C resonance much information about it could be obtained and hence the name internuclear double resonance. A similar technique is used for proton-proton INDOR experiments. The perturbed signals and the monitored signals can be much closer in the spectrum than in the case of decoupling or spin tickling experiments. The INDOR experiment can be carried out in either the CW or the FT mode.

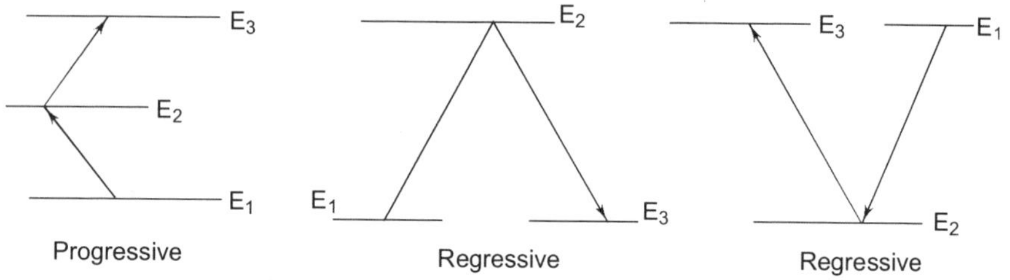

Fig. 17.10. Progressive and regressive spin transitions.

The INDOR spectra can be recorded directly on to the same chart paper as the normal spectrum; the INDOR peaks show up with the original transitions that are being affected. It allows us to observe readily only the INDOR spectrum for the system of interest. It is usually difficult to record ^{14}N spectrum of NH_4^+. The INDOR spectrum of the ^{14}N resonance of NH_4^+ in acidic solution exhibits a quintet with $J(^{14}N-^{1}H)$ = 52.6 Hz. The intensities of homonuclear INDOR spectrum are only half as intense as the ordinary spectrum and hence relatively concentrated solutions are needed. This factor is important in a complex spectrum such as say that of styrene oxide. If one line of a complex spectrum is monitored and an unresolvable multiplet swept by the perturbing radiation, INDOR peaks will be recorded at those points within the complex multiplet where there are signals coupled to the monitored line. In most cases, the coupling constants can be deduced, since the spacing between INDOR signals will be the same as J (unless other couplings are also present). The intensity of irradiating *rf* radiation is very low in INDOR, and hence much less disturbance occurs to the adjacent signals than in other double resonance techniques. Therefore, more precise and selective information about closely spaced multiplets can be obtained.

The power level of irradiation field used in INDOR is similar to or less than that required for spin tickling. The perturbing rf field need not be applied continuously. In INDOR only the populations of the spin states are varied and no change in the energy levels occur. With Fourier transform spectrometers, it is possible to obtain equalization of population using a 90° pulse and the effect on intensities is the same as for continuous wave irradiation. However, a 180° pulse inverts the X spins and hence the population of the associated levels. The effect of this is to change the intensities of the A transitions even more dramatically and this forms the basis of the polarization transfer techniques (See Chapter 15). The information obtained from spin tickling and INDOR spectroscopy which are typical CW techniques are readily obtained nowadays from 2-D NMR experiments.

SELECTED TOPICS 18

SPECTRA OF PARAMAGNETIC MATERIALS

CONTACT SHIFT

It is usually difficult to obtain the NMR spectra of paramagnetic compounds because of the very effective spin lattice relaxation and the consequent line broadening caused by the unpaired electrons. For organic free radicals, this broadening obliterates the spectrum completely. For certain complexes of transition metals however informative spectra can be obtained partly due to the fact that extremely large chemical shifts may occur which far exceed the line broadening in magnitude. Thus paramagnetic materials can be investigated by NMR spectroscopy if the paramagnetic species is 'diamagnetically diluted' as in a transition metal complex or as a free radical occurring in low concentration in the diamagnetic species. The spectrum of a diamagnetic complex may also show the effects in the presence of a paramagnetic species even at very low concentrations. The NMR spectrum of a ligand coordinated to a metal is usually that expected from its structure. However, the chemical shifts and the coupling constants of the free and complexed molecules will differ.

The magnetic nuclei in a ligand or a diamagnetic complex of the ligand exhibit normal chemical shifts due to shielding by the bonding electrons. For diamagnetic complexes, the variation in the chemical shifts of the ligands is small, usually less than 1 to 2 ppm, from that of the uncoordinated molecule. However, paramagnetic complexes may show shifts at least an order of magnitude larger. For example, the ^1H NMR spectrum of methyl amine shows resonances at δ 3.08 and 1.92 ppm due to the methyl and amine protons, respectively. In the spectrum of the paramagnetic nickel(II) complex, $Ni(CH_3NH_2)_6^{2+}$, the NH resonance moves upfield to about $\delta - 98$ ppm while that of the CH_3 protons moves downfield to nearly δ 80 ppm with respect to TMS as shown in Fig. 18.1. Thus the interaction of the protons with the unpaired electrons of the paramagnetic metal ion leads to large up field or down field shift of the NMR signals, relative to the signal of the same protons in the corresponding free ligand or its diamagnetic complex. Unpaired electrons give rise to large dipolar magnetic fields since the gyromagnetic ratio of the electron is 660 times that of the proton. It can result in large nuclear shielding or deshielding. The difference in the chemical shift between a paramagnetic compound and its diamagnetic analogue is referred to as the isotropic shift. A detailed NMR study in terms of the electronic and geometrical structure of the metal complex is usually possible.

We need to understand the relaxation processes occurring in the paramagnetic systems. There are two allowed spin states $m_s = +\frac{1}{2}$ and $m_s = -\frac{1}{2}$ for an electron in a magnetic field. When an electron is delocalized on to a proton, we expect two NMR peaks arising from the coupling of the nuclear spin to the two spin states of the electron as in the case of the ESR spectrum of an unpaired electron interacting with a nucleus of $I = \frac{1}{2}$ (see Chapter 3). A large separation between the two peaks corresponding to the value of the hyperfine coupling constant found in the ESR spectrum should result. However, this is not the case.

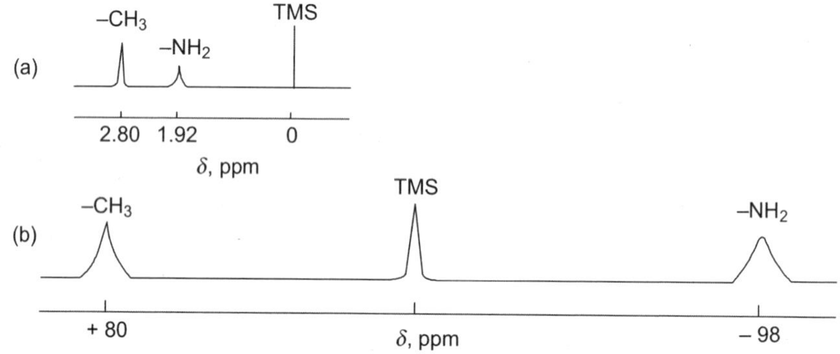

Fig. 18.1. The ^{1}H NMR spectra of methylamine and its nickel(II) complex, $Ni(CH_3NH_2)_6^{2+}$ in solution.

The unpaired electron changes its spin state $m_s = \pm \frac{1}{2}$ with time. For a transition metal complex in solution, the lifetime of the electron spin state can be related to either the electron spin relaxation time τ_e or the ligand exchange time (T_{le}) (whichever is faster); τ_e is a measure of the time taken for the transfer of spin energy to other degrees of freedom, while T_{le} denotes the time spent by the ligand at a particular metal ion. The rate of molecular tumbling in solution is characterized by a correlation time τ_c, which is of the order of 10^{-11} s. The electron spin-nuclear spin coupling constant A_i is typically 10^8 Hz and for NMR, we shall use the symbol A_i. If $1/T_{le} >> A_i$ or $1/\tau_e >> A_i$, the nucleus will sense only a population-weighted average of the two electron spin states; the nucleus will be coupled weakly to the electron spin and hence undergo much less efficient relaxation under these conditions. In order to observe the NMR spectrum in a paramagnetic molecule, it is necessary that one or both of the conditions,

$$1/\tau_e >> A_i \text{ or } 1/T_{le} >> A_i$$

is fulfilled. If neither of these conditions are met T_2 for the nuclear spin will be very short and the resonance signal correspondingly broad and probably not observable. It means that slow electron relaxation will produce fluctuating magnetic fields by flipping of electron spins which have components with frequencies suitable for inducing nuclear relaxation, whereas fast electron spin relaxation will not. In the latter case, the nuclear spins will sense only the average magnetic field produced by the electron spins. Very rapid intermolecular electron exchange or ligand exchange would have the same effect since it would place electrons with different m_s values at the proton. This is similar to decoupling of the proton in NMR spectroscopy and the latter like exchanging of the OH protons of ethanol. Electrons relaxing at exactly the same frequency as A_i are most efficient in broadening the NMR spectrum. Thus the lifetime of the electron spin state in a transition metal complex necessary for the observation of NMR spectra ranges from 10^{-9} to 10^{-13} s. Interestingly, a rapid nuclear spin relaxation results in a broad NMR signal but a rapid electron spin relaxation decreases the efficiency of the spin-lattice relaxation time T_1 for the proton leading to sharp NMR signals. However, the observed spectra may not be as sharp as those of the diamagnetic molecules. High power radiofrequency radiation is employed for recording the NMR spectra of paramagnetic materials since there is little likelihood of saturating the resonances.

Some general observations on the electron spin lifetime of the first row transition metal ion complexes can be made. It will help us to predict when to expect sharp signals (see Chapter 8). Complexes with triply degenerate (T) ground electronic state generally have short electron spin lifetimes, leading to 'sharp' NMR spectra. These include octahedral complexes of d^1, d^2, low spin d^4 and d^5, high spin d^6 and d^7 configurations. When large distortion from octahedral symmetry occurs the degenerate state splits and the NMR spectrum usually becomes broader. Particularly broad peaks are observed for low symmetry complexes of d^1 configuration where there is no zero-field splitting to cause rapid relaxation. Broad NMR signals are also obtained for octahedral complexes with d^3, high spin d^5 and d^9 configurations. There may be a rapid exchange between a free ligand and a ligand coordinated to a paramagnetic metal ion i.e. an equilibrium of the type

$$M^+ + L \rightleftharpoons (ML)^+ \tag{18.1}$$

may exist. In such cases, the observed contact shifts will be smaller by a factor equal to the fraction of the coordinated molecule. Maganese complexes are example of systems with slowly relaxing electrons ($1/T_e \approx A_i$). A single very broad NMR peak may be observed. However, the ESR spectrum of maganese(II) complexes show narrow lines. Sharp lines are observed for octahedral Ni^{2+} complexes (having a non-degenerate ground state) because of zero-field splitting. Tetrahedral complexes with a triply degenerate ground state, for example, tetrahedral Co^{2+} complexes also give 'sharp' NMR peaks.

The hyperfine splitting of the NMR peak due to electron spin is not observed for two reasons. (i) The fast electron spin relaxation and (ii) the rapid exchange of electrons between the metal ions and the diamagnetic molecules in solution. Time averaging results and the hyperfine splitting vanishes, since the electron interacts with a large number of nuclei in different spin states. The averaged NMR signal does not occur in the same position as the corresponding resonance signal of the diamagnetic compound due to the different populations of the two electron spin states. The energy difference between $m_s = +1/2$ and $m_s = -1/2$ is substantially larger for an unpaired electron and the lower energy level $m_s = +1/2$ is more populated than the energy level $m_s = -1/2$ that is $N_{+1/2} > N_{-1/2}$ and contributes more to the time averaging of the signal. That is, the position of the signal ν is given by

$$\nu = N_{+1/2}\,\nu_{+1/2} + N_{-1/2}\,\nu_{-1/2} \tag{18.2}$$

The observed isotropic shift in a paramagnetic compound is made up of two components. The first component is the contact shift and the other component termed as the pseudo contact shift is due to a through-space dipolar interaction between the magnetic dipoles of the unpaired electron and those of the nuclear spins.

The magnitude of the contact shift which is field-dependent is given by the equation,

$$\frac{\Delta B}{B_0} = \frac{A_i \gamma_e^2 h}{8\pi\gamma_p\, kT} \tag{18.3}$$

where A_i denotes the coupling constant, γ_e and γ_p are the magnetogyric ratio of the electron and the proton, respectively, k is the Boltzmann constant and T is the absolute temperature. There is also a dependence on the concentration of the paramagnetic compound. The above equation permits us to determine the sign of the hyperfine coupling constant A_i. This information is not available from the ESR spectrum alone.

ORIGIN OF CONTACT SHIFT

The origin of the contact shift may be understood by considering the interaction between the electron spin and the nuclear spin. When a complex containing an unpaired electron is placed in a magnetic field, the two spin states $m_s = +1/2$ and $m_s = -1/2$ are not equally populated as noted before. A mole fraction weighted average of the chemical shift for the complex with m_s=+1/2 and that with m_s=–1/2 would be observed when the spins rapidly undergo exchange or relaxation. The difference between N_β and N_α i.e. the total number of electron spins in the α ($m_S = +\frac{1}{2}$) and β ($m_S = -\frac{1}{2}$) states gives the excess unpaired spin along the magnetic field direction. It represents the net magnetic effect induced in the system by the unpaired electrons due to the external magnetic field. The weighted average value of m_s for the system in a magnetic field is referred to as the average electron spin polarization (S_q), which can be related to the observed shift. The shift results from the additional field that the electron moments produce in the vicinity of the nucleus. The magnitude of this additional field at the nucleus will depend on the type of electron-nuclear coupling (scalar or dipolar) which will, in turn, depend upon the bonding and geometry of the molecule. Under fast relaxation conditions, the collapsed singlet may occur some thousands of Hz away from the expected position for the diamagnetic complex and this shift of the NMR signal is the isotropic shift, as stated earlier. When more than one electron spin is present, the value of $\overline{S}_q$ is given by

$$\overline{S}_q = \frac{g_{av}\, \beta\, B\, S(S+1)}{3\, kT} \tag{18.4}$$

where g_{av} is the average value of g and $\overline{S}_q$ denotes the average electron spin polarization defined earlier. Equation (18.4) holds satisfactorily for systems obeying Curie law. Since the excess of α spins will be greater at low temperatures, it leads to a $1/T$ (Curie law) dependence for isotropic hyperfine shifts. When the expression given above for $\overline{S}_q$ is not valid, we can write,

$$\frac{\Delta B}{B} = \frac{\Delta \nu}{\nu} = \frac{A_i}{\hbar g_N \beta_N B} < \hat{S}_Z > \tag{18.5}$$

$< \hat{S}_Z >$ should be properly expressed.

The theory involved in the interpretation of the contact shifts is nearly the same as that discussed earlier to interpret the hyperfine splittings in the ESR spectra of free radicals. It is usual to interpret the hyperfine coupling constants in terms of the spin densities (refer Chapter 3). Contact interactions have been observed for a large number of metal complexes. For example, first row transition metal acetylacetonates (Fig. 18.2) which have only two distinct types of protons, CH_3 and CH, have been studied. The observed shifts of CH_3 and CH for V(III) are δ – 45.7 and – 40.7 ppm, respectively and the corresponding values for Mn(II) are δ –25.9 and –18.8 ppm, respectively. The shifts which are predominantly contact in origin are interpreted in terms of the delocalization of the electron spin on to the ligand π-molecular orbitals. Pseudo contact shifts play a negligible role in the shifts observed for the first row transition metal tris(acetylacetonates).

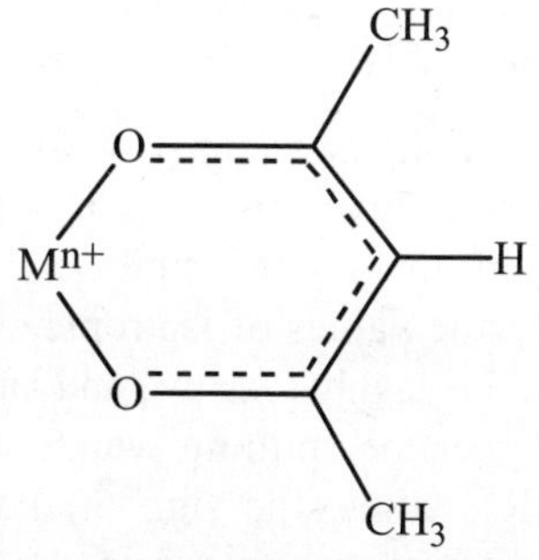

Fig. 18.2 Structure of the metal acetylacetonate complex.

The isotropic electron spin-nuclear spin hyperfine coupling constant, A, is the same as the hyperfine splitting constant 'a' discussed earlier under ESR spectroscopy except that the capital letter A is used here to indicate the possibility of more than one unpaired electron in the molecule or ion. It is given by the relation,

$$A = \frac{8\pi}{3} g_e g_N \beta_e \beta_N |\psi(0)|^2 \tag{18.6}$$

where g_e is the isotropic g-factor and the other symbols have been defined earlier (see Chapter 3). A normalization factor of $1/2S$ is added when there are more than one unpaired electron spins, and g_e is replaced by g_{av} (average value of g) for the complex giving

$$A = \frac{8\pi}{6S} g_{av} g_N \beta_e \beta_N |\psi(0)|^2 \tag{18.7}$$

with A in ergs. The value of A in gauss is given by the relation,

$$A(\text{ergs}) = g\,\beta\,A\ (\text{gauss}) \tag{18.8}$$

The contact interaction constant A_i can be related to the measurable parameters by the equation,

$$\frac{\Delta B}{B} = \frac{\Delta \nu}{\nu} = \frac{A_i\, g_{av} \beta S(S+1)}{3kTg_N \beta_N} \tag{18.9}$$

where k is the Boltzmann constant and T is the temperature in degrees Kelvin. Here B represents the magnetic field at which resonance of the ith nucleus occurs, ΔB_i is the change in the field required to make that particular nucleus resonate when it is changed from the pure ligand or the zinc complex (both diamagnetic) to a paramagnetic complex; ν and $\Delta\nu$ are the corresponding frequencies, A_i is the contact interaction constant of the ith nucleus, β is the Bohr magneton and S is the total electron spin. Equation (18.9) shows that a very small value of A_i can lead to a large isotropic shift. The equality $\Delta B/B = \Delta\nu/\nu$, where ν is the fixed probe frequency follows from the relation, $h\nu = g\beta B$ for the nuclear spins. If measurements are made at different temperatures for systems exhibiting Curie law behaviour, a plot of $\Delta\nu$ *vs* $1/T$ is a straight line with a slope proportional to A_i.

The isotropic hyperfine coupling constant A obtained from the NMR spectrum arises from the same effect that gives rise to the hyperfine splitting constant a obtained from the ESR spectrum as stated before. If the same system could be studied by both the methods, the values of isotropic shift a and A obtained should be identical. The NMR method is highly sensitive and large proton shifts (e.g. 50 Hz) would give rise to proton hyperfine splitting which are not resolved in the ESR spectra. Further from NMR studies, the sign of the coupling constant can be obtained from the direction of the shift, while the sign has no effect on the observed ESR spectrum.

PSEUDO CONTACT SHIFT

Whenever there is an unpaired electron in the molecule, dipolar interaction with the nucleus may occur through space. With an isotropic *g*-value, this dipolar effect is averaged to zero by rapid tumbling of the molecules in the applied magnetic field. With anisotropic *g*-tensor, the magnitude of the dipolar contribution to the magnetic field at the nucleus of interest from the unpaired electron density on the metal will depend on the orientation of the molecule with respect to the field. Since *g* has different values for different orientations, this through space contribution does not average to zero by rapid tumbling of the molecules. Thus, the same effects that give rise to anisotropy in *g* will also give rise to pseudo contact interaction. Pseudo contact interaction also leads to shifts in the values of the chemical shift and it operates when a strong anisotropic paramagnetic centre is present in the molecule. The unpaired electrons of the lanthanide metals, for example, possess such anisotropic properties (*vide infra*). Their influence on the ^1H resonance is a result of the dipolar interactions between the magnetic moments through space. Its magnitude is proportional to $(3\cos^2\theta - 1)/r^3$, where r is the distance between the nucleus under consideration and the paramagnetic centre and θ is the angle between the symmetry axis of the paramagnetic moment and the distance vector r to the nucleus. The through-space nature of the pseudo-contact effects is comparable to the neighbouring group anisotropic contribution to the proton chemical shifts discussed earlier (see Chapter 11), which depends upon the differences in the diamagnetic susceptibility, χ_{dia} for different orientations of the molecule. The same is true for the paramagnetic susceptibility χ_{para}. The pseudo contact interaction may be observed both in crystals and in solutions.

To understand the origin of the isotropic shift in the NMR spectrum, several approaches have been followed. Owing to the complications that arise when both contact and dipolar shifts occur, systems can be selected in which the dipolar contribution is negligible compared to that of the contact shift contribution. Therefore, much of the work on the NMR spectroscopy of paramagnetic complexes have been carried out on systems in which either contact or pseudo contact interaction is dominant. Octahedral complexes of the form ML_6^{n+}, where L is a monodentate ligand ($g_\parallel = g_\perp$) will have no contribution to pseudo contact shifts. If the ML_6^{n+} complex is Jahn-Teller distorted, the distortion would be expected to be dynamic in solution on the NMR time scale. Even when the distortion is not dynamic, rapid ligand exchange may average the pseudo contact shift to zero.

APPLICATION OF CONTACT SHIFTS

Contact shifts are useful in the study of the electronic configuration of the metal ion in transition metal complexes. Contribution to the contact shift due to the bonding of the ligand to a metal can result only if there is covalency in the metal-ligand bond. Accordingly, many attempts have been made to study the trends in covalent bond strength as well as metal-ligand pi-back bonding in a series of complexes from such shifts. The principle involved in the qualitative interpretation of the contact shifts is illustrated using as an example the nickel(II) complex of pyridine, where the magnitudes of the shifts follow the order H(2) > H(3) > H(4). Further, when 4-methylpyridine is coordinated to a metal, the methyl proton resonance shifts in a direction opposite to that of H(4) on pyridine. The pattern of contact shifts observed is characteristic of the molecular orbitals of the ligand that are involved in the spin delocalization. The molecular orbital of the nitrogen lone pair in pyridine has decreasing contributions in the order H(2) > H(3) > H(4). Thus a characteristic sigma delocalization pattern occurs through the mixing of the $e_g(d_{x^2-y^2}, d_{z^2})$ metal orbitals with the pyridine sigma molecular orbitals. It results in a large shift for H(2), a smaller shift for H(3) and the smallest one for H(4). The interpretation of the spin delocalization mechanism is thus closely related to the molecular orbital description of the complex and the ligand. Contact shifts as discussed above serve as a measure of the unpaired spin density at the resonating nucleus and are hence useful in studying the spin distribution in organic radicals and in organometallic and inorganic complexes. In an octahedral complex that contains unpaired electrons only in e_g orbitals, evidence for the existence of a π-delocalization shift pattern is found, showing that spin polarization mechanism is operative.

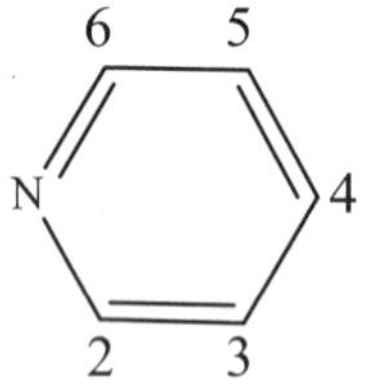

Both ^1H and ^{13}C contact shifts have been determined for bis(cyclopentadienyl) complexes of vanadium, chromium, cobalt and nickel. Information about the orbitals involved in metal-ring bonding and the mechanism of electron delocalization have been obtained and compared with various theoretical models.

The shifts in the proton resonances of methylamine in $Ni(CH_3NH_2)_6^{2+}$ referred to in the beginning of this chapter can now be understood. Most of the unpaired electron density delocalized on to the ligand is on the nitrogen, and little is delocalized directly onto the N–H proton. This leads to large spin polarization of the N–H bond which dominates the shift at this proton, leading to an upfield shift. This appears to be the case for all N–H protons when the nitrogen is directly bonded to nickel. The methyl proton resonances shift down field due to delocalization of the unpaired electron density.

One of the advantages of studying the NMR spectra of paramagnetic complexes is that the separation of resonances arising from non-equivalent atoms in a molecule is enhanced. For example, two isomers, *fac* and *mer,* are possible with asymmetric chelating ligands such as 4,6-dimethyl phenanthrolene as shown in Fig.18.3. The ^1H NMR spectrum of the cobalt(II) complex of this ligand shows four sets of resonances, one from the symmetrical *fac* isomer and three from the asymmetrical *mer* isomer. The intensities showed the presence of the two isomers in the ratio of three *mer* to one *fac*.

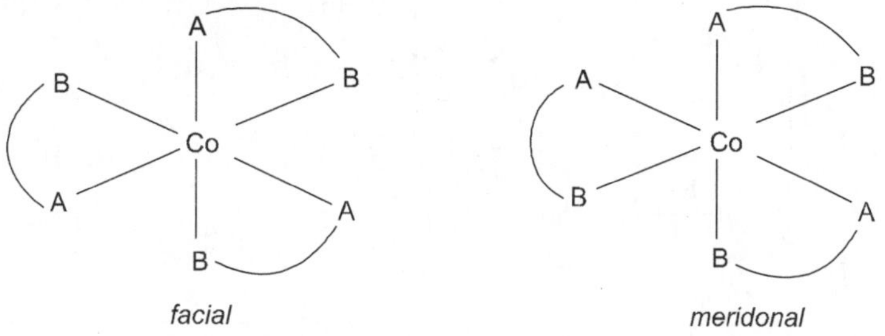

Fig. 18.3 Geometrical isomers of the cobalt(II) complex of 4,6-dimethyl phenanthrolene.

The isotropic shift is also useful for determining the nature of the second coordination sphere around a transition metal complex in solution. Solvation of the metal ions could be investigated. There have been several applications of the NMR spectroscopy of paramagnetic complexes to equilibrium and kinetic studies. The larger range of the shifts enables one to determine the equilibrium constants more accurately and to investigate faster chemical exchange processes.

DIAMAGNETIC COMPLEXES

For diamagnetic transition metal complexes, a quantity known as the 'coordination chemical shift' can be defined. It normally shows systematic variations. For example, in Pt(0) (d^{10} configuration) complexes, the coordination chemical shift decreases as the coordination number increases. For higher coordination number, the chemical shift becomes closer to that of the free ligand. The shift may be positive or negative e.g. for *mer*–$IrCl_3(PMeph_2)_3$, the two types of phosphorus atoms show coordination chemical shifts of -2.6 and $+6.5$ ppm *trans* to Cl^- and $PMeph_2$ respectively. *Cis-trans* effects in complexes are also manifested in their NMR spectra (refer Chapter 15).

SPECTRA OF FREE RADICALS

An added advantage of the NMR method is its sensitivity. It enables us to determine very small hyperfine coupling constants that cannot be measured from ESR spectra directly. In favourable cases, it is also possible to observe the NMR spectra of free radicals and to determine the ESR hyperfine coupling constants from the magnitude and signs of the contact shifts. As an example, Fig.18.4 shows the NMR spectra of 1-propylnaphthalene for different concentrations of the corresponding radical anion. It is to be noted that the hyperfine coupling constant for the protons of the methyl group and the β-methylene group have opposite signs since the CH_3 protons are shifted to lower field while the CH_2 protons are shifted to higher field. The magnitude of the contact shift calculated from the spectra is as expected from equation (18.2) and it is proportional to the magnitude of the hyperfine splitting constant a. Further, the line width is also proportional to the magnitude of a and it is a function of $1/r^6$, where r is the distance of the

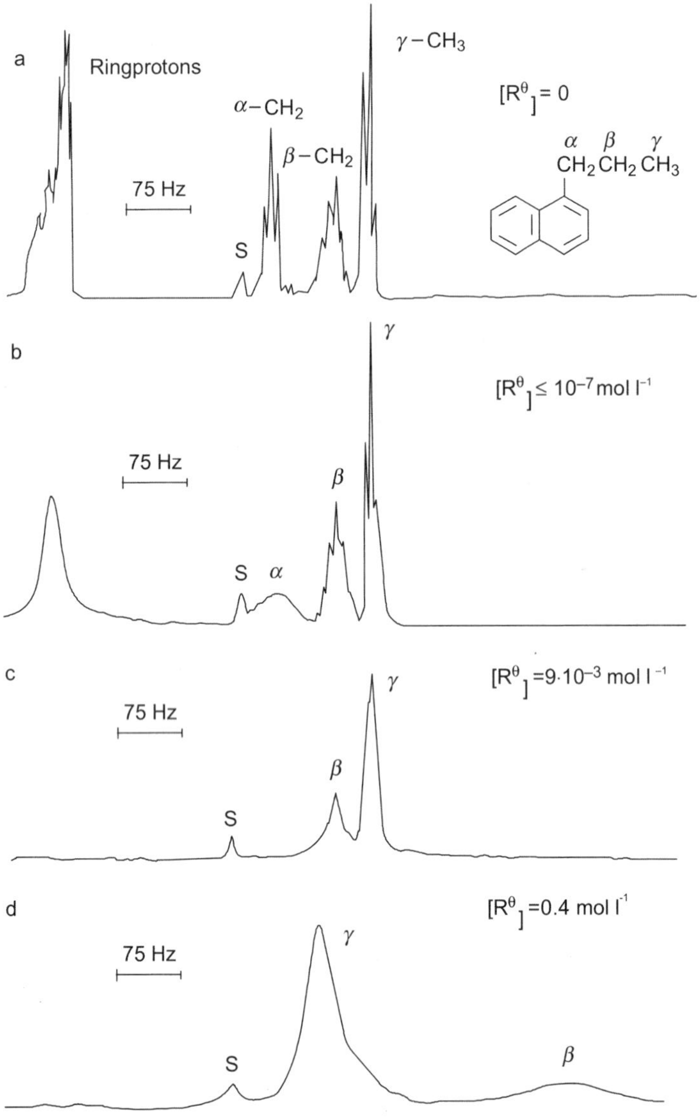

Fig. 18.4 The NMR spectra of 1-propylnaphthalene and its radical anion at different concentrations.

nucleus under consideration from the radical centre. This is the $2p_z$ orbital of the corresponding carbon atom of the naphthalene ring. The resonances of the ring protons and the α–methylene protons are highly shifted and broadened so that they can no longer be observed in the spectrum at higher concentrations of the radical anion (Figs.18.4 (c) and (d)). The values of the hyperfine coupling constants determined from the spectra are given below.

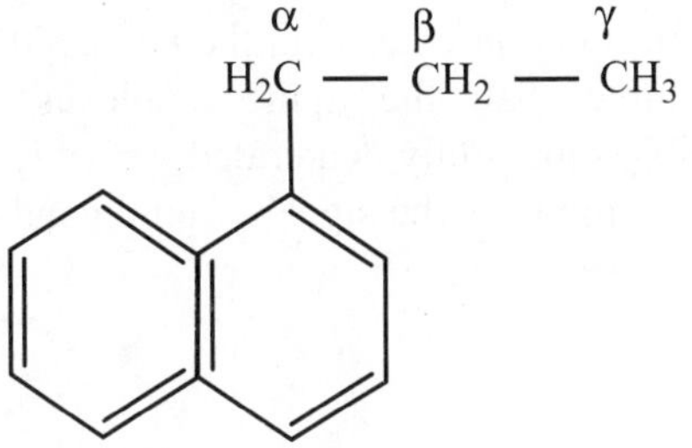

$a_\alpha = 267.0\ \mu T$

$a_\beta = 21.2\ \mu T$

$a_\gamma = 6.4\ \mu T$

LANTHANIDE SHIFT REAGENTS

The paramagnetic species are commonly used to simplify the NMR spectra of organic compounds. Many rare earth complexes that exhibit this behaviour are called shift reagents. Lanthanide shift reagents are mild Lewis acids which bond to sites which have lone pairs. The site of complexation of the substrate, for example, NH_2, NHR, –OH, C=O, etc. should not be sterically hindered which prevents the close approach of the bulky reagent to the lanthanide ion. Shift reagents significantly simplify complex spectra in which numerous proton resonance signals overlap. The reagent stretches out the spectrum, thus separating the overlapping signals. Fortunately the line broadening caused by the paramagnetic moments of the lanthanides is relatively small. The ^1H spectral shifts induced by the shift reagents are assumed to be dipolar in origin. This produces pseudo contact shift of the protons in the organic molecule which can be very large.

Lanthanide shift reagents (LSR) are complexes of *β*-diketones and have the general formula $Ln(R_1COCHCOR_2)_3$, where *Ln* stands for a paramagnetic lanthanide ion, *Eu*, *Pr*, *Dy* or *Yb*, the former two being the most common ones. Complexes of various lanthanides with different ligands are commercially available. Two ligands, dipivaloylmethane, *dpm*, $R_1COCH_2COR_2$ ($R_1 = R_2 = t\text{–}Bu$) and the fluorinated chelate 1,1,1,2,2,3,3-heptafluoro-7,7-dimethyl-4,6-octanedione, $(CH_3)_3CC(O)CH_2C(O)CF_2CF_2CF_3$, abbreviated as fod ($R_1 = C_3F_7$, $R_2 = t\text{–}Bu$) anions have been found to be particularly useful as they are chemically inert, impart good solubility and possess favorable association properties for complex formation. The best of these lanthanide shift reagents for general purpose is *Eu*(fod)$_3$, the structure of which is shown in Fig. 18.5. Rare earth trischelates of *β*-diketonate derivatives are stronger Lewis acids and good complex formation occurs even with weak bases and cause large pseudo contact shifts in Lewis bases that are added to the solution containing the lanthanide complex. The resulting resonances for several of these complexes (for example, of europium and praseodymium) are very sharp. As is usually the case with these systems, the shift reagent complex is in fast exchange with excess of the Lewis base.

The lanthanide complexes are octahedral and have excellent solubility in organic solvents. Lanthanides are capable of having coordination numbers higher than six so that free coordination sites still exist. If the organic molecule possesses a suitable donor site (e.g. O or N), it can interact with the lanthanide complex. The fod complexes are more soluble than the dpm complexes although their hygroscopic nature can be troublesome. Fully deuterated derivative of fod is also commercially available and this eliminates the signals from ligand protons.

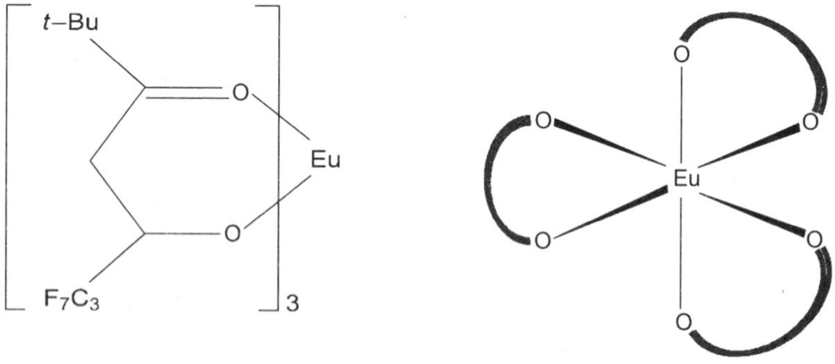

Fig. 18.5 Structure of $Eu(fod)_3$.

There are many applications of the shift reagents. Upon complexation, different protons in the molecule are shifted to different extents. On the basis of the angle factor $3\cos^2\theta - 1$ shifts both to lower and higher fields may occur. The observed spectra are time averaged between free and complexed substrate. Individual lanthanides differ with respect to the sign of their dipole field. The magnitude of the resulting shift is strongly dependent on the basicity of the complexing group in the substrate, which decreases in the following sequence,

$$-NH_2 > OH > C{=}O > COOR > CN$$

As the concentration of the shift reagent is increased the chemical shift difference $\Delta\nu$ becomes greater than J, leading to first order spectra. Thus, the shift reagents can be used to simplify the second order (i.e. $J \sim \Delta\nu$) spectra. Acidic groups usually lead to decomposition of the lanthanide complex and hence cannot be studied. Hydrocarbons which do not form complexes with lanthanide shift reagents cannot also be studied.

The use of $Eu(fod)_3$ is illustrated using *n*-pentanol. Figure 18.6 shows the ^{1}H NMR spectrum of *n*-pentanol in $CDCl_3$. The alkyl signals denoted as *a,b,c* are not well-separated in Fig. 18.6(i). The normal spectrum of pentanol consists of a triplet due to the methylene protons adjacent to the alcohol group and two groups containing a series of overlapping lines due to the remaining protons which all have similar chemical shifts. Figure 18.6(ii) shows the spectrum taken after adding 10 mgs of $Eu(fod)_3$. The signals are beginning to stretch out downfield.

The closer the protons are to the site of complexation, the greater is their downfield shift. On adding 20 mg of $Eu(fod)_3$, all the alkyl signals are separated out sufficiently to give a first order spectrum. In cases where the complexation is relatively weak, more amounts of $Eu(fod)_3$ will be required to obtain the desired separation. In general for the europium complexes, the resonances all move downfield of TMS. If a praseodymium complex is added all the resonances shift upfield of TMS.

The change in the chemical shift $\Delta\delta$ of the proton is inversely proportional to the cube of its average distance from the europium atom and is given by the relation

$$\Delta\delta = \text{constant } (3\cos^2\theta - 1)/r^3 \tag{18.10}$$

where θ is the angle between r_1, the distance vector which joins the proton and the europium atom and the line which joins europium to the atom bearing lone pair to which it is bonded. Distinguishing two very similar isomers becomes relatively simple with $Eu(fod)_3$.

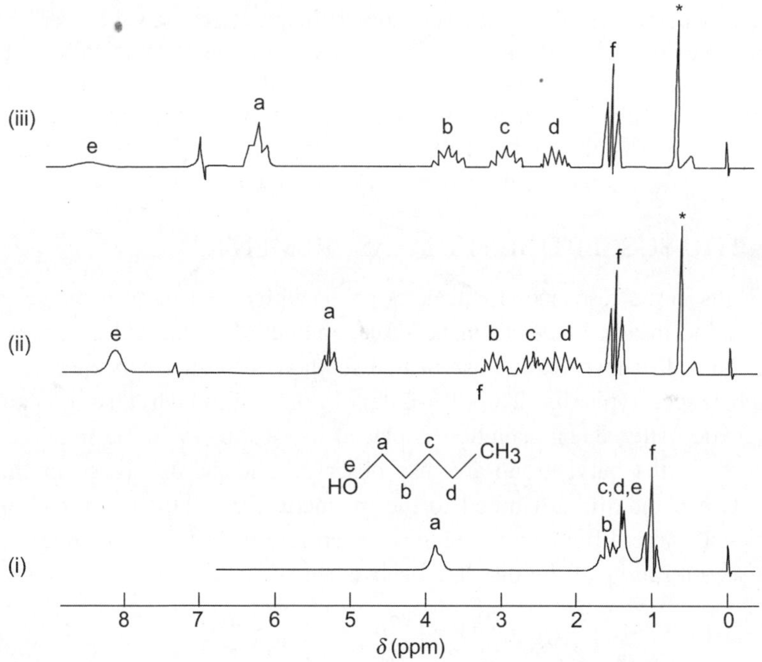

Fig. 18.6 ^1H NMR spectra of *n*-pentanol in $CDCl_3$ containing various amounts of $Eu(fod)_3$: (i) 0.0 mg ,(ii) 10.0 mg and (iii) 20.0 mg. * 1– Butyl signal from $Eu(fod)_3$.

When the solvent has lone pairs $Eu(fod)_3$ is not used. For $Eu(fod)_3$, paramagnetic relaxation is rather inefficient and hence does not cause line broadening even at high concentrations. Erbium and holmium compounds generally produce greater shifts but some line broadening also occurs which can complicate the analysis of the multiplets. The line broadening is due to the paramagnetic ion which decreases the relaxation time. The lanthanide shift reagents do not work well at high magnetic fields.

Another application of the shift reagents involves their use in determining the geometries of molecules in solution. This experiment is usually carried out in the fast exchange region. Initially, a structure is assumed for the molecule whose structure is to be determined and the expected dipolar shifts are calculated at various nuclei in the molecule. The assumed structure of the molecule is varied to obtain a best fit of the experimental shifts to the calculated ones using computer programs available for this purpose. The effect of the shift reagents on ^{13}C resonances provides additional support for ascertaining the solution structures. There are some limitations regarding the applicability and accuracy of the method. There is some evidence that Fermi contact interaction may contribute to the shift and hence considerable caution should be exercised in utilizing these data. The coordination of the shift reagent may affect the conformation of the biomolecule.

Quantitative analysis of lanthanide induced shifts is of considerable interest for conformational analysis as discussed above. In practice, however very often a quantitative empirical analysis is sufficent to solve a stereochemical problem by the application of shift reagents. Lanthanide complexes with chiral ligands are used quite frequently to determine the optical parity of the enantiomeric mixtures. Different NMR signals for diastereomeric complexes of *D*– and *L*–species are observed if a shift reagent with a unique stereochemistry of the ligand is used.

MAGNETIC SUSCEPTIBILITY MEASUREMENT

The shifts in the resonance frequencies produced by the paramagnetic substances can be used to measure their magnetic susceptibilities. This method is known as the 'Evans' method. It involves the use of two solutions of equal concentration of an inert substance, typically 2% of *t*-butanol, one of which also contains the paramagnetic material; one solution is placed in a capillary inside the other. Two resonances for the butyl group are then observed and the difference in chemical shifts between them is attributed to the magnetic material; the shift is directly proportional to the difference in volume susceptibilities for the two solutions. The mass susceptibility χ of the dissolved substance is given by the equation,

$$\chi = \frac{3\Delta\nu}{2\pi\nu m} + \chi_0 + \frac{\chi_o(d_o - d_s)}{m} \tag{18.11}$$

where $\Delta\nu$ is the shift in the resonance frequency ν, m is the mass of the material per millilitre of the solution, χ_o is the mass susceptibility of the solvent (-0.72×10^{-6} for dilute aqueous *t*-butanol) and d_o and d_s are the densities of the solvent and the solution, respectively. For strongly paramagnetic substances, such as many transition metal complexes, the last term of equation (18.11) is small and could be neglected.

SOLID STATE NMR

WIDE LINE NMR

Every nucleus with a nonzero spin quantum number has a magnetic dipole. A magnetic nucleus therefore behaves like a tiny bar magnet. We had mentioned earlier about the direct interaction of nuclear magnetic moments through space which was referred to as dipole-dipole interaction or dipolar coupling. This interaction should result in the splitting of the resonance signal. The magnetic field produced by a nucleus of magnetic moment μ at a second nucleus is given by

$$B_b = K\mu\left(\frac{3\cos^2\theta - 1}{r^3}\right)\left(\frac{\mu_o}{4\pi}\right) \tag{18.12}$$

where μ_o is the permeability of vacuum, K is a constant, r is the distance between the two nuclei and θ is the angle between the line joining the nuclei and the direction of the applied magnetic field B_o as shown below.

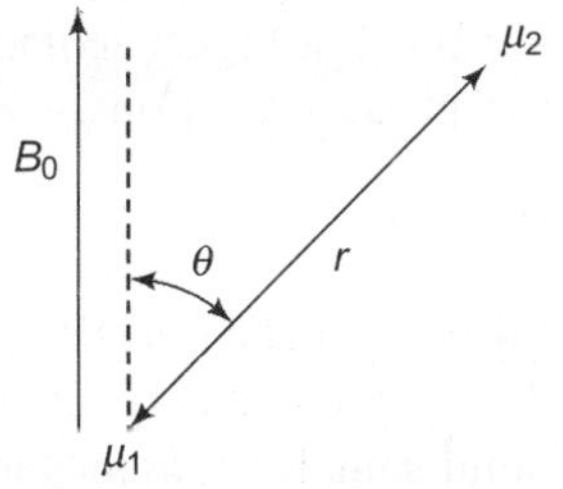

Equation (18.12) applies to a spin system of protons (nuclei) with same resonance frequency such as those in methylene bromide. Line splitting due to dipolar interaction is not observed in solution. Because of the rapid tumbling motion of the molecules in solution, the angle θ varies with time and consequently $3\cos^2\theta - 1$ vanishes. However, in a solid, the molecules have fixed orientation with respect to the direction of the external magnetic field B_o and with respect to each other. Thus, the intramolecular dipolar couplings are much more pronounced in solids, where intermolecular dipolar interactions are also important. In solids, besides dipole-dipole interactions, scalar coupling, quadrupole interactions, and the anisotropy of nuclear shielding lead to severe line broadening. These factors lead to spectra with structureless broad peaks with half width of several kilohertz. The information obtainable about the nuclei through chemical shifts and spin-spin coupling so useful in high resolution NMR is lost. However, equation (18.12) indicates that distances between nuclei say protons can be determined from the NMR spectra of solids, which is also known as wide-line NMR spectroscopy.

Solid state NMR spectroscopy can be applied to microcrystalline and amorphous materials. The high resolution solid state NMR can give us information on bonding, structure and dynamical behaviour of solid state structures. Since the line width is a function of the internuclear distance, it has been possible in simple cases to measure proton-proton distances, which are difficult to measure by other techniques. The line shape also depends upon the number of spins. In many crystalline hydrates, it has been possible to determine whether the water of crystallization is present as H_2O, H_3O^+ or OH^-. For example, borax Na_2O. 2 $B_2O_3.10H_2O$ has been shown to exist as $Na_2[B_4O_5(OH)_4]$. $8H_2O$.

If the interacting spins move so that their relative orientations change, for example, by the rotation of a group of atoms around some axis, then their through space coupling and the line width of their signal is reduced. By measuring the line widths at different temperatures, information about the motion of various groups in the crystal can be derived. For the adduct, $(CH_3)_3N—BF_3$, at 68 K, there is no rotation around the C—N bonds and the line is very broad. When the temperature is increased to 100 K, a significant narrowing occurs due to the onset of rotation of the methyl groups around the C—N bond. Above 100 K, further narrowing occurs due to the onset of rotation of the $N(CH_3)_3$ group as a whole around the B—N bond. Finally near 400 K, the line width narrows to a fraction of a millitesla since the whole molecule begins to rotate isotropically and diffuse within the solid crystal. The fluorine resonance of the BF_3 moiety is broad below 77 K and the BF_3 group rotates around the B—N bond at higher temperatures. Changes in the line width can often be attributed to phase transitions.

MAGIC ANGLE SPINNING

In recent years, high resolution solid state NMR spectroscopy of the powdered solids with relatively narrow lines (line width 10–50 Hz) has been developed. For the measurement of spectra, solid samples present three problems, which have to be overcome.

(i) The peaks are broadened due to dipolar and long range couplings which have to be reduced to zero. If the given nucleus has a quadrupole moment, it gives rise to further broadening due to its interaction with the electric fields of the sample.

(ii) The chemical shift anisotropy occurs in the absence of molecular tumbling and it adds to the line broadening since the molecules take all directions for a powdered sample.

(iii) The spin-lattice relaxation time T_1 is extremely long in solids because of the absence of tumbling. It restricts the rate at which spectral data can be acquired. This has to be overcome if less receptive spin 1/2 nuclei are to be studied.

To achieve in the solid state spectra the high resolution that is obtained from the solution spectrum, the technique of magic angle spinning (MAS) has been developed. In a typical solid, all values of θ of equation (18.12) will occur. When θ is at an angle of 54° 44′ to the field, the value of B_b is zero and the nuclei do not perturb B_b at each other locations and hence do not affect their resonance frequencies. If the sample filled in a small capsule is mounted as a cylinder with its axis at an angle of 54° 44′ to the main magnetic field direction and spin it rapidly around this axis as shown in Fig. 18.7, its average B_b will be zero and all the broadening should disappear and result in narrow NMR lines. The angle 54° 44′ is known as the magic angle. The quadrupolar interaction and the chemical shift anisotropies also follow the relation $3\cos^2\theta - 1$ and they are also eliminated by the magic angle spinning (Fig. 18.7). Figure 18.8 compares the ^{31}P NMR spectrum of $(NH_4)H_2PO_4$ (ammonium dihydrogen phosphate) obtained in the solid state with and without magic angle spinning.

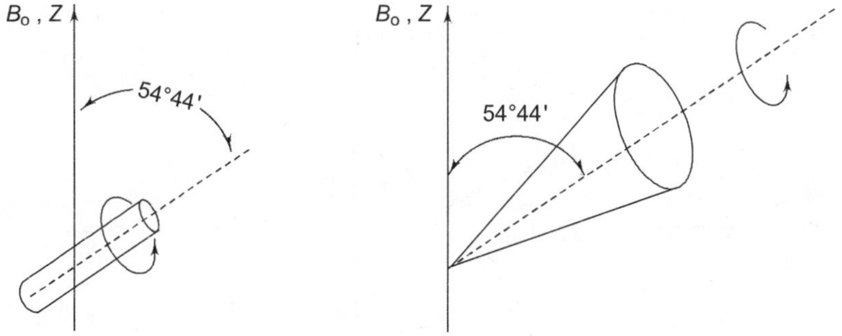

Fig. 18.7 Magic angle spinning.

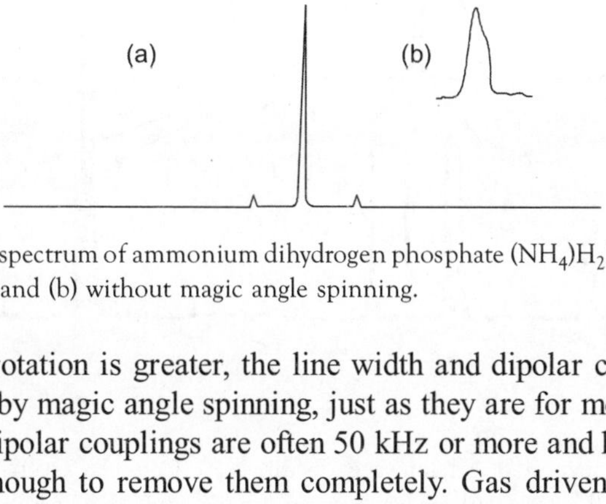

Fig. 18.8 ^{31}P NMR spectrum of ammonium dihydrogen phosphate $(NH_4)H_2PO_4$ in the solid state : (a) with and (b) without magic angle spinning.

If the rate of rotation is greater, the line width and dipolar couplings are also averaged to zero by magic angle spinning, just as they are for molecules tumbling in solution. But dipolar couplings are often 50 kHz or more and hence the rotation cannot be fast enough to remove them completely. Gas driven sample spinners that can be rotated upto 25 kHz are now available. Thus it may not be possible to remove all effects of directly bonded protons which give maximum line widths. These dipolar effects can be removed for heteronuclear systems by decoupling in the normal way but very high decoupling powers are needed. Hence, for nuclei such as ^{13}C or ^{29}Si for which only heteronuclear couplings are important, high-resolution spectra comparable with those obtained for solutions can be obtained. Figure 18.9 shows the dramatic narrowing effects of decoupling and magic angle spinning on the ^{13}C spectra of calcium acetate dihydrate, which permit us to resolve fairly small splittings. This fine structure arises from four molecules of acetate, which are unrelated by symmetry in the crystal. Therefore, information on the crystal structure can be obtained from NMR spectra. The same splittings are also observed in the spectrum of the analogous crystalline barium acetate for which the crystal structure is known. While recording the ^{13}C

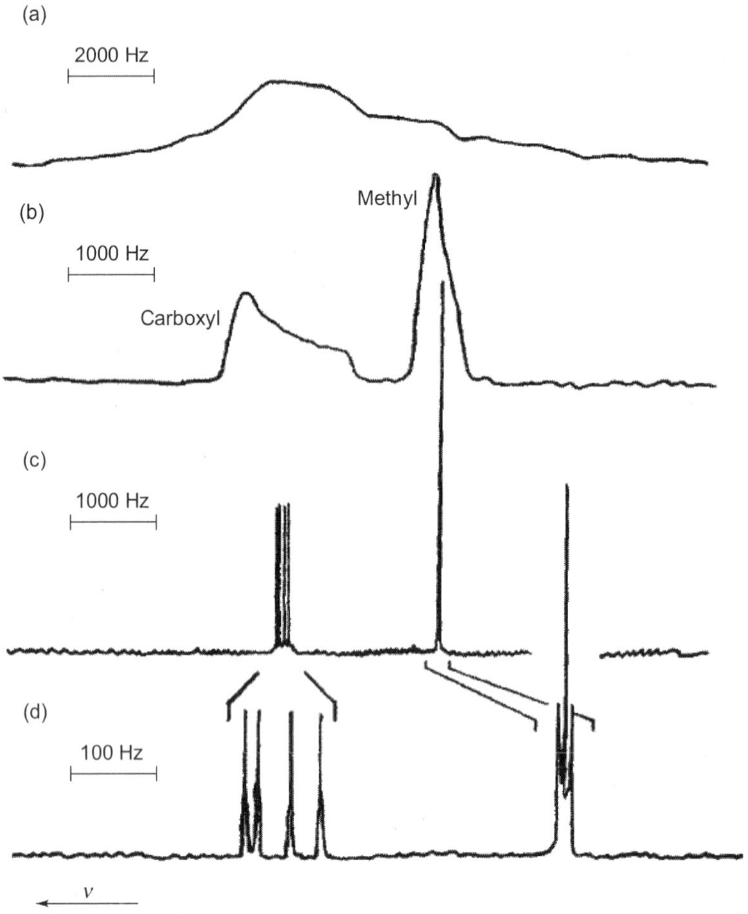

Fig. 18.9 The ^{13}C spectra of calcium acetate dihydrate $Ca(CH_3COO)_2\ 2H_2O$ at 22.6 MHz: (a) static, (b) static with high power proton decoupling, (c) with decoupling and magic angle spinning and (d) expansion of (c) to show the fine structure due to the presence of different molecules in the unit cell unrelated by symmetry.

NMR spectra, great improvement in the resolution can be obtained by combining double resonance with magic angle spinning. Using appropriate spinning speeds, it is possible to narrow the resonance lines to an extent comparable to that in the high resolution NMR.

The problem of long relaxation times for dilute nuclei with spin 1/2 in the presence of abundant spins can be circumvented by a technique known as cross-polarization, which will not be discussed here. Using one or more of these methods, high-resolution spectra of solids, including crystalline powders, gels and composite materials could be obtained.

The ^{31}P spectrum of the diphosphine platinum complex shown below exhibits two sets of resonances with slightly different chemical shifts in the solid state. Both centre resonances have ^{195}Pt satellites and the couplings are not equal, 1980 and 1820 Hz. Therefore, there must be two different phosphorous environments in the crystal. The ^{195}Pt spectrum showed that there is only one type of platinum atom in the crystal, coupling to two different ^{31}P nuclei. Therefore, there must be one complete molecule per asymmetric unit and the molecules themselves have no symmetry. This was later confirmed by X-ray crystal structure analysis. The observations of these site splittings indicate that chemical shifts are very sensitive to the environment.

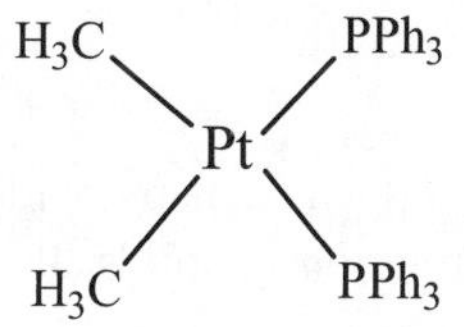

Solid state NMR spectra of nuclei of spin 1/2 having high natural abundance (e.g., ^{1}H, ^{1}F) are most difficult to obtain since their high natural abundance leads to strong homonuclear dipolar interactions and homonuclear decoupling is not possible. Due to quadrupolar interactions, nuclei with spin $I \geq 1$ e.g. ^{2}H, ^{11}B, ^{17}O, etc. are also difficult to study. With quadrupolar nuclei, the splitting of the resonance by quadrupolar coupling occurs in the solid state which in general cannot be eliminated by magic angle spinning. The degeneracy of individual transitions is lost and multiplet structures (doublets, triplets) arise. Therefore, spectra of molecules with nuclei such as ^{13}C, ^{15}N, ^{29}Si and ^{31}P have been widely studied. It is advantageous to study these nuclei since they are diluted either because of their low natural abundance (^{13}C, ^{29}Si) or as in the case of ^{31}P the compound of interest generally contains only one atom of this kind per molecule. Homonuclear coupling is absent and heteronuclear interactions can be eliminated through spin decoupling. For insensitive nuclei, magnetization transfer could be brought about from the abundant protons by cross polarization. This technique has helped in the development of high resolution ^{13}C NMR spectroscopy of solids. For obtaining high resolution solid state ^{13}C NMR spectra, the ^{13}C nuclei are excited by ^{1}H,^{13}C cross polarization (CP). During signal detection, high power ^{1}H decoupling is applied to eliminate the dipolar and scalar heteronuclear spin-spin coupling. However, the signals are broadened by the anisotropy of the chemical shift which results from the anisotropy of the shielding constants σ in the applied magnetic field. In the case of powder spectra different σ-values are observed depending on the orientation of the individual microcrystals relative to the applied magnetic field direction. This distribution is reflected in the different line shapes of the experimental spectra. From these spectra, the anisotropy of the chemical shift can be determined. It provides information about electron distribution around the nucleus which is directly related to chemical bonding. The anisotropy of the shielding constants can be obtained from single crystal studies, where the sample is measured at various orientations with respect to the external magnetic field.

When MAS is combined with cross polarization, CP/MAS solid state spectrum is obtained. The CP/MAS technique has been used for the detection of a large number of spin 1/2 nuclei, for example ^{29}Si, ^{33}S, etc. Without cross polarization one can obtain only MAS spectra if the sample does not contain protons as for example, the ^{29}Si spectra of silicates. For complex spectra such as those observed for ^{13}C spectra of organic solids, unambiguous assignment becomes difficult.

APPLICATIONS

Solid state NMR spectroscopy has many wide-ranging applications. It is a powerful tool for studying interaction in solids. It can also be used in the study of amorphous materials. Applications of the MAS or CP/MAS technique for structural studies are numerous. It is useful in following solid state reactions, phase changes and polymorphism. Structural information on zeolites and aluminosilicates have been obtained from ^{29}Si and ^{27}Al spectra. The chemical shift differences are sufficiently large to distinguish the environment of different ^{29}Si nuclei. It has even been possible to obtain information on $^{29}Si - {}^{29}Si$ bonding. Internuclear distances could be determined from single crystal NMR. Information about the electronic structure of solids can be obtained through chemical shifts data.

Significant differences between the structure of a compound in the solid state and in solution are reflected in the larger differences in chemical shifts. Solvation effects which are absent in the solid may stabilize a certain structure in solution when differences between the liquid and solid state spectra may be found. ^{31}P MAS NMR spectrum of PCl_5 clearly reveals the existence of ionic structure PCl_4^+ PCl_6^- because two ^{31}P resonances separated by δ 377 ppm are found.

The study of dynamic processes in solids by high resolution solid state NMR spectroscopy has attracted much attention. The basis for such investigations is the same as in liquids. Temperature-dependent studies reveal the mechanisms and the thermodynamic parameters for conformational changes and other intramolecular processes by line shape analysis. Interestingly, the ring inversion in cyclohexane has been found to have nearly the same barrier in the solid state as in solution. It shows that packing in the molecular crystals of organic compounds is not very different from solvation in solution.

MAGNETIC RESONANCE IMAGING

Magnetic resonance imaging (MRI) is concerned with the application of nuclear magnetic resonance to heterogeneous specimens such as parts of the human body. It is a novel diagnostic technique that has revolutionalised medical and biomedical fields. The basic principles of this development are outlined here.

The biological samples, tissues, etc. contain water and other fluids which contain protons. Due to the high concentration of hydrogen in living systems and their very favourable NMR characteristics protons have been the favourite nuclei for magnetic resonance imaging. Other nuclei such as ^{2}H, ^{13}C, ^{14}N, ^{23}Na, ^{19}F and ^{31}P could also be used. Since these nuclei have much weaker NMR sensitivity in living systems, a significantly lower image resolution will be realised when compared with proton imaging. One important area of medical research is based on the observation of signals from ^{31}P compounds containing phosphorus esters such as adenosine triphosphate (ATP) and adenosine diphosphate (ADP) which are involved in most metabolic processes.

One of the most important applications of NMR spectroscopy is its use in medical field for creating images of sections of human body to detect tumours and other growths. Magnetic resonance imaging is non-invasive and hazardless to living cells for clinical diagnosis unlike X-rays used in conventional scanning techniques. The parameters normally used for MRI are chemical shift, spin density, relaxation times T_1 and T_2 and the fluid flow velocity of the object. Thus, MRI besides providing morphological information yields additional diagnostic insights through relaxation parameters. For example, the longitudinal relaxation time T_1 for proton varies over a range of 0.3 to 1.2 s at 100 MHz and is dependent on the pathological state of the tissues. Therefore, relaxation times can often be utilized. This makes MRI an important tool for biology and medicine.

The basic phenomena of conventional NMR and magnetic resonance imaging remain the same in the experimental procedure and applications. The main differences in instrumentation are the much larger aperture magnets that are required for examining human body and the addition of coils to generate field gradients. In 1973, Lauterbur first produced an NMR image by applying a field gradient ΔB over the sample cell. One of the first NMR images produced by this technique was that of two capillaries of 1 mm diameter each filled with H_2O and fitted into a 5 mm sample tube containing D_2O/H_2O. The NMR data were subjected to computer image processing that allowed the printing of a spin density map for the protons in the sample.

The basic principle of MRI is rather simple. Consider a body cavity containing 1-mL specimen of pure water placed in a uniform magnetic field of a conventional NMR spectrometer. The proton NMR spectrum will exhibit a single sharp line. If now a gradient of the order of 1 G/m (10^{-2} T/m) is applied across the tube of water, different parts of the specimen will experience different field strengths. Since the NMR frequency is directly proportional to the field strength, the

NMR spectrum is a one-dimensional (1D) projection of proton density along the direction of the field gradient.

Gradients give rise to a spatially dependent distribution of NMR frequencies. The field gradient introduces a chemical shift between nuclei that would be isochronous in the conventional NMR experiment. Gradients used for imaging are generally about 0.1 to 0.2 T/m or less thereby producing magnetic fields considerably smaller than that of the main field. MRI is thus produced by making the field inhomogeneous while the high resolution NMR spectroscopy utilizes highly homogeneous magnetic fields. In MRI, a portion of the patients body is placed in a powerful magnetic field and irradiated with *rf* radiation. When the magnetic field varies systematically across the object, the resonance frequency of a normal nucleus and that affected by disease appear at different magnetic fields. For example, regions of increased iron disposition in the brain such as the globus pallidus, substantia nigra and red nucleus show a different signal from that of the remainder of the brain. Magnetic resonance image of a section of the brain is given in Fig. 18.10. By taking different cross-sections at different depths, it is easy to determine the location, shape and size of, say, a tumour in the brain or the spinal cord. The main observed species is water with the image intensity being dependent on the concentration of water in the sample and also transverse and longitudinal relaxation times or its diffusion coefficient as mentioned earlier.

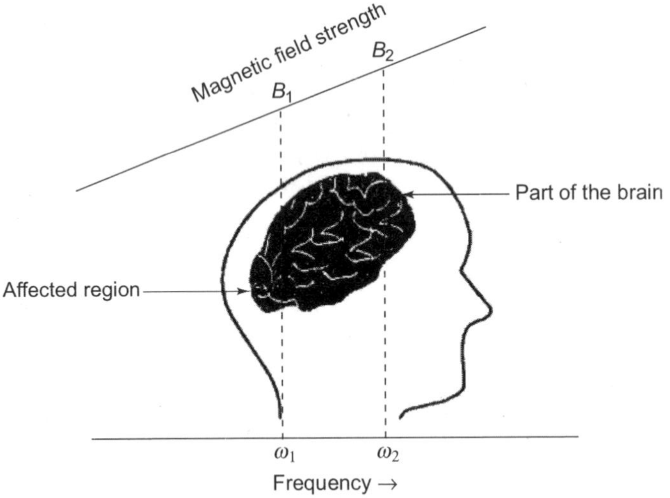

Fig. 18.10 Magnetic resonance imaging. B_1 and B_2 denote the signal from the regions of affected and normal brain, respectively.

One-dimensional projections give some structural information but for an image at least a two-dimensional representation of the NMR response is required . In order to obtain 2D or 3D information by using this essentially 1D probe, a time dependence is introduced by modulating, switching or rotating the gradient. The

NMR response must be restricted to a plane or slice of the object. A set of 2D images of a coplanar thin slice of the object provides a complete 3D representation of the object. Further, suitably placed surface coils allow more detailed investigations of certain regions of the human body.

There are several different methods of magnetic resonance imaging. They are classified according to the spatial element that is detected. The commonly used ones are, (i) sequential point, (ii) sequential line, (iii) sequential plane, and (iv) simultaneous volume method. Considerations regarding the signal to noise ratio, distortionless image, minimum imaging time, etc. decide the selection of the technique. The simplest procedure consists of isolating and directly measuring the NMR response A_{pq} from one element P_q at a time moving sequentially from one to the next until all n^2 elements over the entire region of interest have been scanned. Such a procedure is called a sequential point method. In the sequential line method, a line-by-line scan is performed through the region of interest. This method is more efficient and sophisticated than the previous one. In a similar way, an image can be produced by selecting sequential planes or by simultaneously imaging the whole volume. If the line width of the individual resonances is small compared to the field gradient, signals from different parts of the object – more precisely those from nuclei residing in various planes perpendicular to the gradient – can be detected. An intensity plot will then give information about the spatial distribution. The most successful human whole body images to date have been obtained by planar imaging.

To create a two-dimensional image, two gradients are applied along the *X*- and *Y*–directions and a series of one-dimensional images recorded in different directions in the *XY* plane. A technique known as back projection can then be used to combine the various one-dimensional images of the object. Especially adopted surface will allow more detailed investigations of certain regions of the human body. The majority of the imaging experiments used presently employ multi-dimensional Fourier transformation rather than the back projection to create the two-dimensional images. The technique can be extended to three dimensions. MRI continues to attract much attention for clinical applications. Its potential for non-clinical applications has also been recognised.

TWO-DIMENSIONAL NMR SPECTROSCOPY 19

INTRODUCTION

Over the last twenty five years, two-dimensional (2D) methods have revolutionalized the practice of NMR spectroscopy and it has resulted in the development of several techniques for the elucidation of structures of complex organic molecules. A detailed discussion of all the theoretical and practical aspects of two-dimensional NMR spectroscopy is beyond the scope of the present text. We restrict ourselves to a brief introduction to the basic principles and a discussion of the most popular 2D experiments. There are basically two main types of 2D NMR experiments: shift correlation through bonds (e.g. COSY and HETCOR) and shift correlation through space (e.g. NOESY). Interestingly all 2D NMR techniques are referred by suitable abbreviations. These spectra may be of homonuclear or heteronuclear type involving interactions between similar nuclei (e.g. protons) or between different nuclei (e.g. ^{1}H and ^{13}C). These simplest and the most powerful NMR experiments reveal the scalar and dipolar couplings that exist between the nuclei in a much 'cleaner' way than the conventional one-dimensional experiments. The COSY, HETCOR, NOSEY and INADEQUATE techniques described in this chapter are examples of a large number of powerful 2D NMR experiments available for probing the molecular structure and motion. There are also three- and four- dimensional NMR methods for studying larger molecules.

PRINCIPLES OF 2D NMR

The procedure followed to record the one-dimensional (1D) Fourier transform ^{1}H NMR spectrum consists of a 90^0_x pulse following thermal equilibrium, delay and signal acquisition for a few seconds (Fig. 19.1). The decay of oscillating transverse magnetization is detected by the NMR spectrometer as free induction decay. Thus in a normal one-dimensional pulse sequence experiment, the receiver signal is a function only of the detection time t_2. This sequence is repeated until NMR signals of satisfactory intensity are obtained. Fourier transformation of the free induction decay (FID) provides the required frequency-domain spectrum. The ^{13}C channel is kept inactive during proton NMR experiment. The channels denote the hardware associated with the irradiation of each nucleus of interest and signal acquisition.

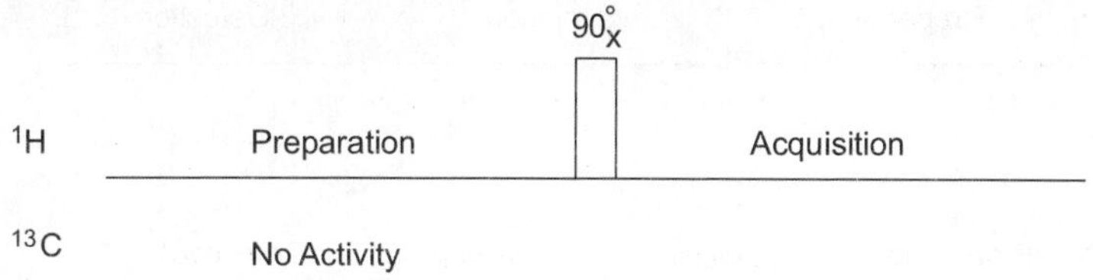

Fig. 19.1 Pulse sequence for 1D ^{1}H NMR; 90° pulse is applied along the *x*–axis.

The spectrum is obtained as a plot of the chemical shift in δ units (frequency) against intensity in arbitrary units. It is referred to as an 1-dimensional NMR spectrum, the dimension that is referred to is the frequency axis that is derived from the acquisition time of the FID through Fourier transformation. The chemical shifts and coupling constants are displayed on the same axis with intensity plotted in second dimension (the omitted dimension). In one-dimensional NMR, there is only one variable, i.e. the acquisition time, t_2 that affects the spectrum.

In one-dimensional NMR, only a single response is recorded and only a single Fourier transformation is required to obtain the desired frequency domain data. The FID signal is the result of phase modulation arising from precessing nuclear magnetization with different chemical shifts and spin-spin coupling constants and consequently different resonance frequencies.

In 1D NMR experiments, the FID is recorded immediately after the pulse. However, if the signal is not recorded immediately after the 90° pulse and a certain interval, t_1 is allowed to elapse before detection followed by a 90° acquisition pulse, then during this interval t_1 known as the 'evolution period' the nuclei interact with each other in different ways depending on the pulse sequence applied. The nuclei precess under the influence of chemical shifts and/or spin spin coupling. The resulting signal in the NMR spectrum is then a function of two variables $S(t_1, t_2)$, the evolution time t_1 and the acquisition time t_2. Introduction of this second dimension in time has resulted in the development of several powerful 2D NMR techniques for probing the structure and dynamics of complex organic molecules. All 2D NMR experiments involve the same basic procedural building blocks and data processing methods with the aim of unraveling obscured or hidden information.

It is convenient to partition the sequence of time of a typical 2D NMR experiment into three periods: (a) preparatory period, (b) an evolution period and (c) a detection period as outlined in Fig. 19.2. The duration of the evolution period is denoted by t_1 while t_2 represents the duration of the detection period (acquisition time). In several 2D NMR experiments, another interval, known as the mixing time (t_m), is added before detection.

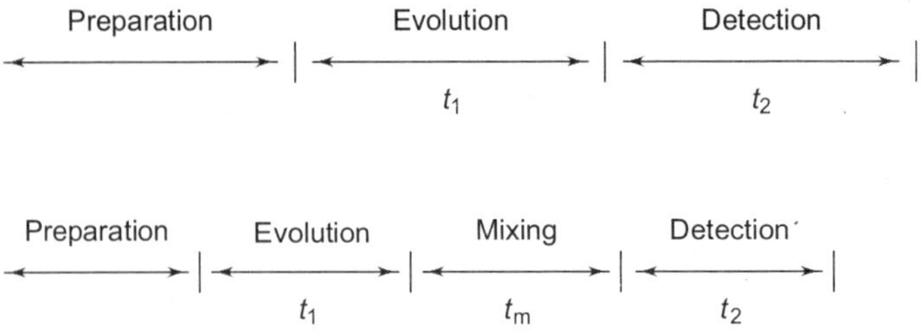

Fig. 19.2 Time scales of 2D NMR experiments.

Preparation

During this period, the spin system of interest is prepared for various pulses or a pulse to which it is subjected to. During this phase, nothing actually happens to the sample. This state can be thermal equilibrium where all the spins have their natural magnetization governed by the Boltzmann population. In most experiments, this period consists of a delay sufficient to give equilibrium magnetization for all nuclei. For this reason, the actual time is usually set to about five times the average relaxation time of the nuclei.

Evolution

The magnetization induced in response to a pulse or more than one pulse in the preparation period are allowed to evolve over a fixed interval. In the evolution time t_1 nuclei develop under the influence of different factors, as for example Larmor precession or scalar spin-spin coupling, before a signal is detected. What the nuclei 'feel' during the evolution time is what makes one 2D experiment different from another. Since the FIDs obtained give the positions and the status of the magnetization vectors, we can map the behaviour of the nuclei during the evolution time from the FIDs. A Fourier transformation with respect to t_1 reveals the natural frequencies of the nuclei determined by their environment during the evolution period. To obtain new frequencies and thus new information, we must establish and control the environment of the nuclei during t_1.

Mixing

At the end of the evolution time we can redistribute nuclear magnetizations among the spins. The redistribution may involve the use of pulses and/or further interval. The aim here is to allow spin communication for a fixed period of time. The mechanism used to allow this communication will determine the way the interactions between the nuclei take place. Two examples are *J*-coupling and dipolar relaxation discussed later.

Detection

The interval during which a free induction decay is measured is known as the acquisition time. Finally the NMR spectrum of the nuclei is recorded in the usual chemical shift pattern.The appearance of the spectrum will generally differ in intensity or phase from the normal 1D spectrum but the features are still similar. These phase and/or intensity variations can be investigated in a detailed manner by systematically varying the evolution time from zero to some upper limit, and collecting spectra for each new value of the evolution time used in the experiment. These variations can reveal details about the chemical and magnetic environments of the nuclei and provide information that might otherwise be not observable.

The common 2D NMR technique is $^1H—^1H$ shift correlation spectroscopy (COSY) which identifies pairs of protons coupled to each other. The other common two-dimensional NMR technique which show $^1H—^{13}C$ shift correlation is termed as HETCOR (heteronuclear correlation) or $^1H—^{13}C$ COSY. These are used to detect directly bonded $^{13}C—^1H$ atoms. The NOSEY spectra show shift correlation between nuclei that undergo mutual dipolar relaxation. The $^{13}C—^{13}C$ shift correlations are revealed by the INADEQUATE spectra.

2D NMR EXPERIMENT

A typical pulse sequence employed for obtaining a 2D-NMR spectrum is shown in Fig. 19.3. It consists of a 90° pulse, an interval (t_1, the evolution period), a second 90° acquisition pulse and acquisition time (t_2). This pulse sequence is repeated many times with incremented t_1 interval. Each time a separate FID results.

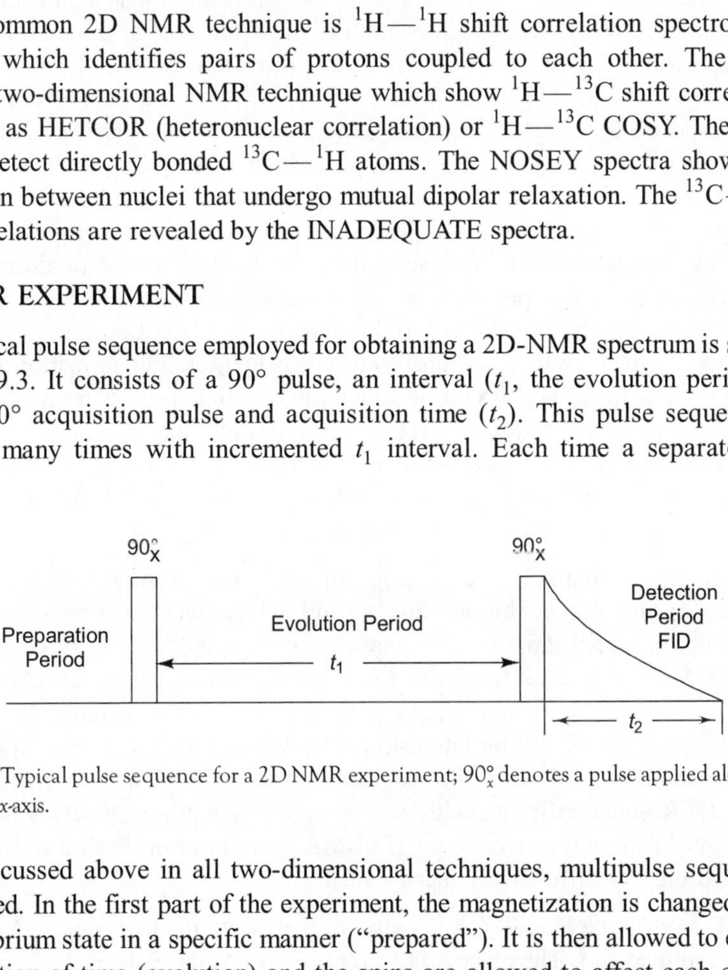

Fig. 19.3 Typical pulse sequence for a 2D NMR experiment; 90°_x denotes a pulse applied along the *x*-axis.

As discussed above in all two-dimensional techniques, multipulse sequences are applied. In the first part of the experiment, the magnetization is changed from its equilibrium state in a specific manner ("prepared"). It is then allowed to evolve as a function of time (evolution) and the spins are allowed to affect each other's behaviour (mixing) e.g. according to whether they are spin-spin coupled or dipolar coupled through space. Finally, the resulting magnetization is detected by collecting FIDs (acquisition). Several experiments are carried out. The resulting signals (FIDs) detected depend on two independent time variables t_1 and t_2, and this means that the FID data can be Fourier transformed twice – once with respect to t_2 and then with respect to t_1. As a consequence, two frequency–variables, F_1 and F_2 are obtained. The entire data matrix after Fourier transformation as described above is plotted so that the F_2 data lies on the x–axis and the F_1 data of

t_1 origin lies on the y–axis. All the spectra would of course be slightly different from one another because of the t_1 increments Δt_1 that separate them all.

The evolution time t_1 is of the order of microseconds and the detection time t_2 is of the order of seconds. The evolution time t_1 is varied systematically by adding small increments Δt_1 and separate FIDs are obtained for each value of t_1 as noted above. The second interval t_2 representing the detection time is kept constant. It is the increment Δt_1 in time which is the origin of the second dimension in a 2D NMR spectrum. The first set of Fourier transformations of these FIDs yields a set of frequency domain 1D NMR spectra (rows). When these 1D NMR frequency-domain spectra are stacked together (data transposition), a new data matrix or "pseudo-FID" is obtained, $S(t_1, F_2)$ in which absorption-mode signals are modulated in amplitude as a function of evolution time t_1. It is hence necessary to carry out second set of Fourier transformations to convert this 'pseudo FIDS' to frequency domain spectra. The second set of Fourier transformations (across columns) on $S(t_1, F_2)$ produces a two-dimensional spectrum $S(F_1, F_2)$. This represents the general procedure for obtaining the 2D NMR spectra (ignoring the dimension of intensity). It is immaterial which Fourier transformation is carried out first and which second (i.e. whether the function is first transformed with respect to the variable t_1 or the t_2). However, it is advisable to transform individual FIDs with respect to t_2 first and then after data transposition to transform the resulting interferogram with respect to t_1 to produce the 2D spectra (or more precisely into sections of 2D NMR spectrum).

The spectra contain the frequencies of those mechanisms which have been effective during the evolution time t_1 and which have caused the observed modulation of signal amplitude or phase. If for example spin-spin coupling was effective during evolution time t_1 and Larmor precession during detection time t_2, the frequency axis F_1 contains coupling constants while the chemical shifts appear on the frequency axis F_2. Our interest in 2D NMR lies in experiments in which the magnetization evolves with one frequency during t_1 and a different frequency during t_2. Resonance frequencies and spin-spin coupling constants which are indistinguishable in a conventional 1D NMR spectrum can thus be separated and presented on two different frequency axes.

In 2D NMR, one axis F_2 (or ν_2) always represents the nucleus detected during acquisition time (t_2). The other axis F_1 (or ν_1) which depends on evolution time t_1 can represent the same nucleus (e.g. ^1H–^1H COSY), a different nucleus (e.g. ^1H–^{13}C COSY), or a coupling constant (e.g. *J*-resolved spectroscopy) which are discussed later. We need to know how the magnetizations are related during the period t_1 and t_2 to assign the cross peaks.

Normally long measuring times are required for 2D NMR spectroscopy. This is also due to the fact that nearly always 90° excitation pulses are used. Therefore, after each individual experiment a delay at least five times the relaxation time is necessary in order to provide uniform starting conditions. Further, large data accumulation has to be made in 2D NMR experiments in order to get sufficiently intense signals. Finally the amount of data which has to be mathematically processed is much larger than that in an one-dimensional experiment.

PRESENTATION OF 2D NMR SPECTRA

There are two basic methods of presentation of the 2D NMR spectrum, either as a 'stack-plot' as already known from T_1 experiments or as a contour plot as shown in Fig. 19.4. Stack plot is impressive and viewed from the top it gives the look of a mountain range. It provides a pseudo 3D representation of the spectrum. However, it may be difficult to analyse a stack plot. It has proved most useful to plot the view that would be obtained if we looked down on such a stacked plot along the axis of peak intensity. The intensities of the peaks are indicated by contours exactly as are the heights of mountains on a map. The contour plot is a

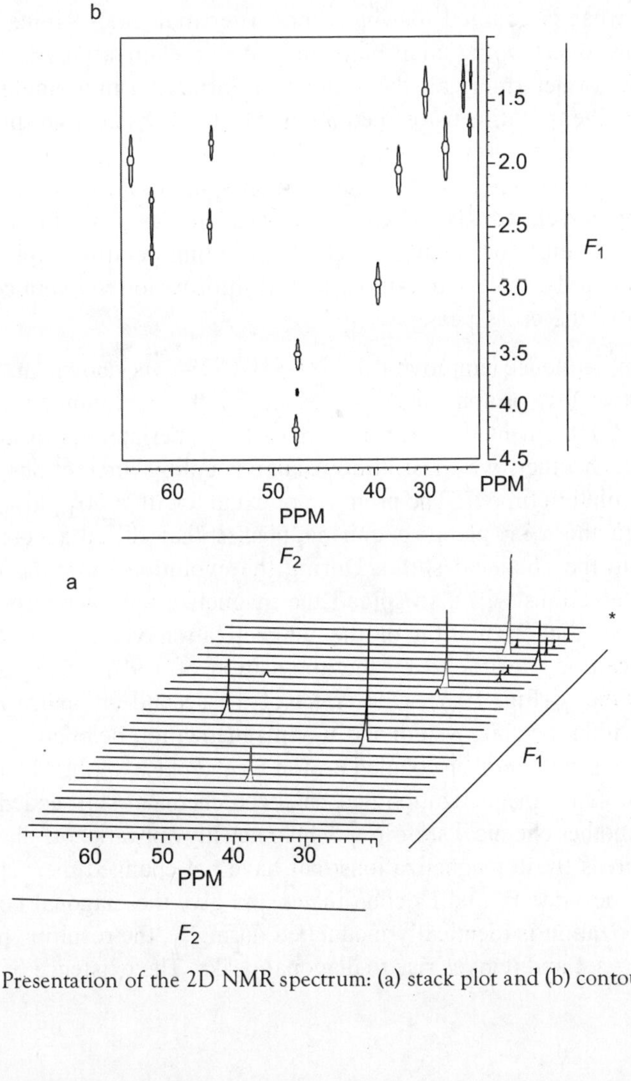

Fig. 19.4 Presentation of the 2D NMR spectrum: (a) stack plot and (b) contour diagram.

cross section through a stack-plot at a chosen height. Thus in contour plots, the spectral peaks are shown as slices of each mountain. Contour plots are better suited for representation of spectra; they are convenient and informative. This method of displaying the spectrum is quicker to plot and easier to analyse. However, the relative intensities of the signals and the structure of the multiplets are less easily observed.

^{1}H-^{1}H COSY

Selective spin decoupling experiments to determine the protons giving rise to the observed couplings in the ^{1}H NMR spectrum to obtain the connectivity of CH_3, CH_2 and CH groups can become tedious for a large molecule giving a complex spectrum. Further, when two or more signals have very similar chemical shifts, irradiation of one of the signals may inevitably hit the other signal at the same time and we are then unable to identify from the decoupling experiment what is coupled to what. For larger molecules giving complicated spectra many overlapping multiplets may occur even at higher fields. It is advantageous to determine all the coupling information in a single experiment. The COSY is the two-dimensional equivalent of the 1D selective spin decoupling and it is the most important one·of the 2D NMR techniques. It could be more aptly referred to as ^{1}H-^{1}H COSY (corelation spectroscopy) to clearly indicate what is being correlated. ^{1}H–^{1}H COSY experiment has proved to be simple and exceptionally useful for deducing ^{1}H–^{1}H coupling relationships. The COSY technique is chiefly used to obtain structural information via spin connectivities as revealed by the cross peaks.

The pulse sequence employed for ^{1}H–^{1}H COSY is shown in Fig. 19.5. It consists of two 90° proton pulses separated by the evolution time t_1 which is systematically incremented. Since the aim of the experiment is to identify nuclei coupled to each other, we must allow scalar coupling interactions to be active during the evolution time t_1. The protons are excited with a 90°_x pulse and allowed to precess in the x, y plane, acquiring phases that differ according to their differences in the chemical shifts. During the evolution time, the homonuclear spin-spin interactions will also spread the magnetization vectors encoding their own values in the orientation of the magnetization vectors of protons. This modulation can be transferred by the application of a 90° pulse just before the acquisition time t_2 (Fig. 19.5). One effect of the second 90° pulse is to mix the spin states within the spin system and x, y magnetization belonging to one proton can be transferred to another coupled proton by the pulse applied along an axis in the x, y plane. Thus magnetization that has been chemical shift coded in t_1 can be detected at another chemical shift in t_2. This intensity appears "off diagonal" in the 2D plot whereas those magnetizations that have not changed their chemical shift values have the same F_1 and F_2 coordinates and give the diagonal peaks. That is, when magnetization is identically modulated during t_1, the resulting peaks will be such that $\nu_1 = \nu_2$; and it gives rise to diagonal peaks. The existence of off-diagonal

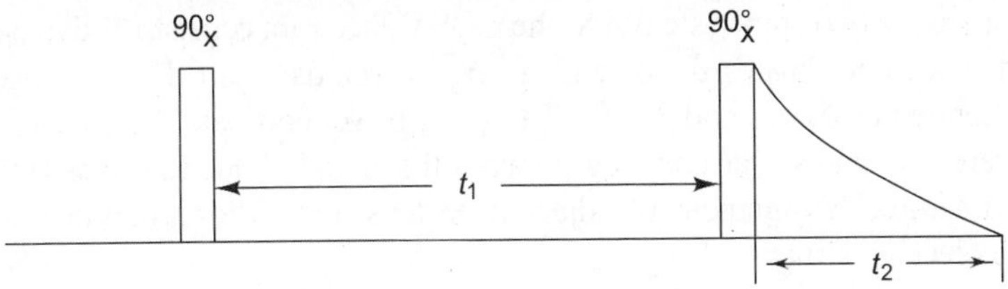

Fig. 19.5 Pulse sequence for a COSY spectrum.

intensity is a proof of spin-spin coupling between the protons. Since the 90° pulses are short and of high amplitude, all possible off-diagonal intensities will be generated leading to a complete description of the coupling constants. Normally at least 100-1000 increments in t_1 are used to record a COSY spectrum. As long as the signals are resolved, the cross-peaks can be associated reliably with a specific pair of coupled protons.

In the presence of spin-spin coupling, the second 90° pulse not only affects transverse magnetization, but also leads to population changes for the various transitions in the spin-system. In this way magnetization is exchanged between all nuclei which are mutually coupled and their signals are amplitude-modulated in a series of t_1 experiments by the frequencies of the neighbouring nuclei. This mechanism leads to the cross peaks of the 2D-spectrum at ν_i, ν_j and ν_j, ν_i.

As discussed above, in the COSY spectra the signals of the observed nuclei are modulated by the resonance frequencies of the partner nuclei to which they are coupled. The result is that after the second Fourier transformation a 2D plot is obtained in which the unmodulated signals lie along the diagonal while cross peaks lying on either side of the diagonal occur at the chemical shifts of the coupled nuclei.

In a COSY spectrum, the chemical shifts are measured along F_1 and F_2 axes and hence COSY spectra can also be called as δ, δ-spectra. Off-diagonal peaks indicate homonuclear scalar spin spin coupling between the protons found on the diagonal; they are located by extrapolation horizontally and vertically from the cross peak. The F_1 and F_2 axes of a COSY spectrum contain the frequencies of the homonuclear scalar spin-spin interactions since during the evolution time both chemical shifts and scalar coupling parameters operate. A fine structure is therefore observed for the diagonal peaks as well as for the cross peaks in a high resolution COSY spectrum. The rules governing the intensity distribution within the cross peaks however differ from those that determine the intensities of the multiplets in the one-dimensional spectra. There is no simple relation between the intensity of the cross peak and the magnitude of the scalar spin-spin coupling. For complex spectra, overlapping signals can lead to elimination of cross peak components. The assignment of certain spin-spin couplings in more complex 2D spectra is therefore not simple. One of the most valuable aspects of the homonuclear shift correlation experiments is that cross peaks appear for all ranges of coupling constants. Couplings constants ranging from 2 to 20 Hz could be detected.

For a simple two spin system AX, the COSY spectrum consists of diagonal signals, centred at the coordinates δ_A, δ_A and δ_X, δ_X as well as of off-diagonal signals (cross peaks) centred at δ_A, δ_X and δ_X, δ_A. The cross peaks correlate the chemical shifts δ_A and δ_X and indicate scalar coupling between the *A* and *X* nuclear spins (Fig. 19.6). It leads to a direct assignment of adjacent protons and yields important information about molecular structure.

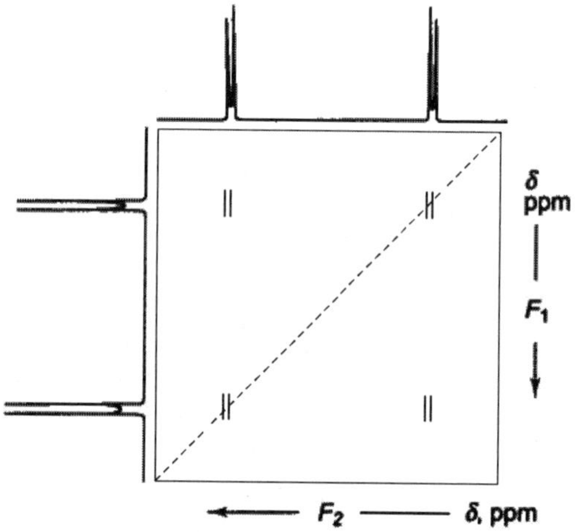

Fig. 19.6 Schematic representation of the COSY spectrum of a pair of scalar coupled nuclei A and X.

The 2D spectrum yields all information about coupled nuclei of interest with only one experiment and is thus superior to 1D decoupling experiments. It is indispensible not only for complicated spectra where it takes less time for a complete spectral analysis but also for routine applications. The time needed for a routine COSY spectrum is in the case of modern high field instruments relatively short and is usually less than one hour with standard samples. Even shorter measuring times are possible by the use of gradient enhanced COSY spectroscopy which is not discussed here.

For illustration, the ^1H–^1H COSY spectrum of m-dinitrobenzene is considered. The contour display of this spectrum is given in Fig. 19.7. In the COSY spectrum, two almost identical chemical shift axes (labelled as F_1 and F_2) are plotted orthogonally. However, the resolution on each chemical shift axis is usually different. The F_2 axis is found on the top or bottom with the proton chemical shift scale as usual; F_1 is displayed on the right with the proton chemical shift scale running from top to bottom. The COSY spectrum is symmetrical about the 'diagonal' which runs from the top right to the bottom left. A series of absorptions in which ν_1 equals ν_2 are found on the diagonal as contours. The diagonal peaks provide no new information beyond the simple 1D ^1H spectrum. All the peaks that are mutually spin-spin coupled are shown by the cross peaks, which are symmetrically placed on either side of the diagonal. So, we need to consider cross peaks only on one side of the diagonal. For convenience of interpretation, a better resolved one-dimensional spectrum is plotted on one of the orthogonal chemical shift axes, usually on the F_2 axis. However, 1D NMR spectrum is not a part of the COSY spectrum. To determine which signal is coupled to which other one(s), one starts with a given cross-peak and projects two perpendicular (dotted) lines, parallel to the F_1 and F_2 axes as shown in Fig. 19.7. The protons corresponding to two signals in the 1D NMR spectrum connected in this way to a cross peak are coupled to each other. Clearly all the connectivities of the signals can be quite easily seen at a glance.

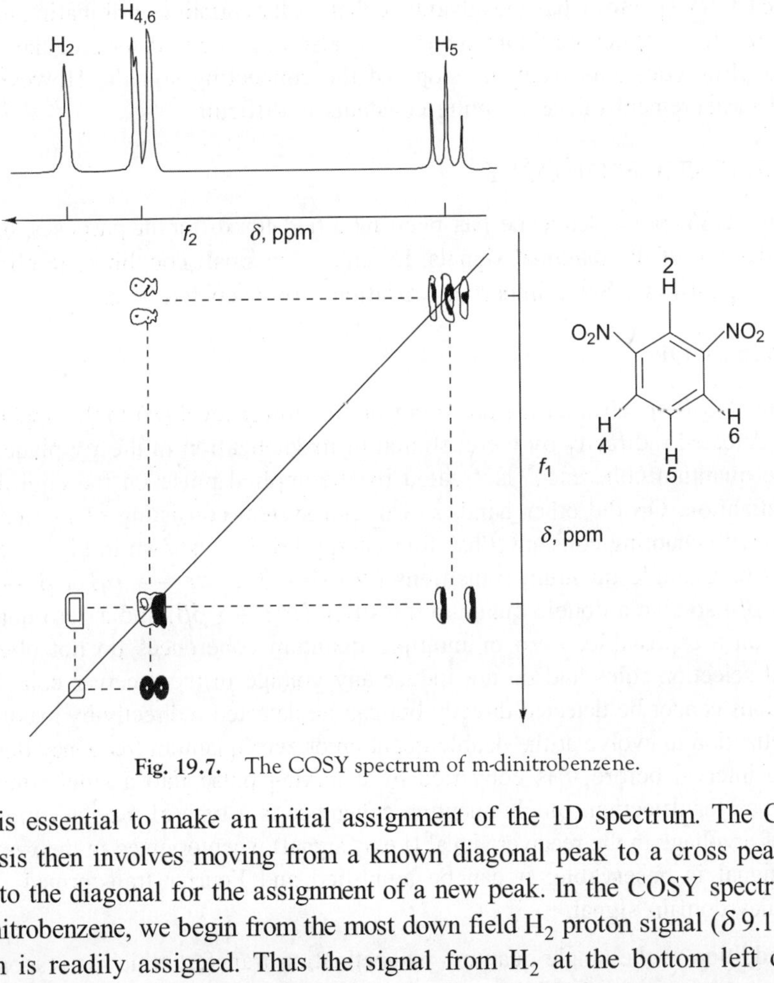

Fig. 19.7. The COSY spectrum of m-dinitrobenzene.

It is essential to make an initial assignment of the 1D spectrum. The COSY analysis then involves moving from a known diagonal peak to a cross peak and back to the diagonal for the assignment of a new peak. In the COSY spectrum of *m*-dinitrobenzene, we begin from the most down field H_2 proton signal (δ 9.1 ppm) which is readily assigned. Thus the signal from H_2 at the bottom left of the diagonal has a cross peak connecting it to the $H_{4,6}$ signal (δ 8.6 ppm). Thus the H_2 proton is coupled to the H_4, H_6 protons. Similarly it is readily seen that the signal from $H_{4,6}$ is connected by a cross-peak to the H_5 signal showing coupling between $H_{4,6}$ and H_5.

The cross peaks provide the coupling constants but not the full multiplicity. The H_2 signal on the diagonal is a triplet. However, the cross peaks lack the centre line of the triplet. The loss of central line also occurs with quintets. However, the doublets and quartets occur as doublets and quartets in the cross peaks. The presence of a cross-peak normally indicates that the protons giving the connected resonances on the diagonal are geminally or vicinally coupled. Long range couplings do not usually give significant cross peaks except when their values are large. By introducing extra delays in the pulse sequence, COSY spectra can be obtained to emphasize long range couplings when coupling constants of the order of 1 Hz could be obtained.

The COSY spectrum has the advantage that scalar couplings originating in large multiplet structure can be found clearly. It is also possible to determine the sign of the coupling constants from the slope of the connecting signals. However, the actual measurement of the coupling constants is difficult.

MODIFICATION OF COSY

The COSY pulse sequence has been modified for different purposes such as the reduction of the diagonal signals, to emphasize small couplings, to eliminate singlet signals, etc. Some important variations are discussed here.

(a) COSY-DQF

In the NMR spectrum, the transition from the lower level (α) to the upper level (β) is detected indirectly by the evolution of magnetization in the *x,y* plane. This 'single quantum coherence' is created by the applied pulse on the equilibrium magnetization. On the other hand, a two-spin system consisting of two coupled nuclei, with coupling constant J has four energy levels as shown in Fig. 19.8. For this system, single-quantum transitions ($\alpha\alpha \longrightarrow \beta\alpha$, $\alpha\alpha \longrightarrow \alpha\beta$, $\alpha\beta \longrightarrow \beta\beta$, $\beta\alpha \rightarrow \beta\beta$) so also a double quantum transition ($\alpha\alpha \longrightarrow \beta\beta$) and a zero quantum transition are possible. Zero or multiple quantum coherences do not obey the normal selection rules and do not induce any voltage in the receiver coil. These transitions cannot be detected directly but can be detected indirectly by making the magnetization to evolve at the double-quantum or zero quantum frequency during a certain interval before it is converted by a mixing pulse into a single quantum coherence for detection. Single-quantum coherence is a type of magnetization that induces a voltage in the receiver signal (i.e. *rf* signal) when oriented in the *xy*-plane. This signal is observable; it can be amplified and Fourier transformed into a frequency-domain signal.

In addition to the single quantum magnetization transfer, which is responsible for cross peaks in the COSY spectrum, there is also zero and multiple quantum magnetization (or coherence) produced by the second 90° pulse. These are not detected in the receiver coil and are thus not recorded in a normal COSY spectrum. Double quantum magnetization can occur in any coupled spin pair. Double

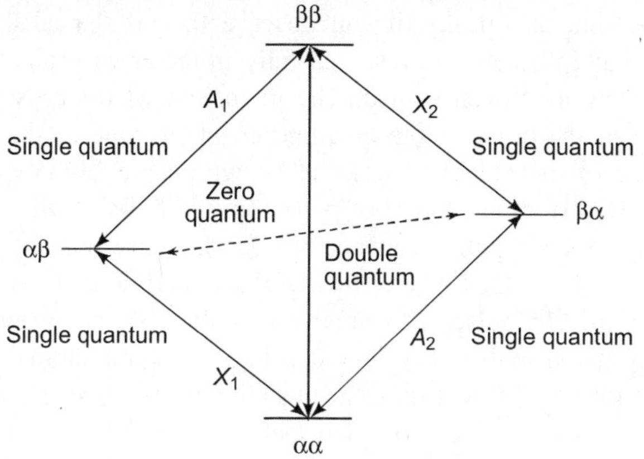

Fig. 19.8 Energy level diagram for a coupled two-spin system.

quantum magnetization can be observed indirectly by reconverting it into single quantum magnetization via the third pulse.

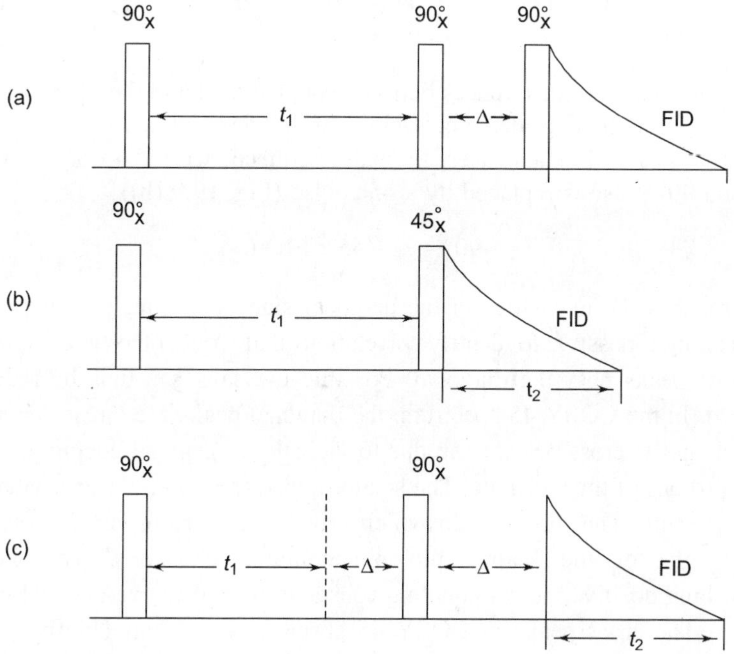

Fig. 19.9 Modification of the pulse sequence used for obtaining the COSY spectrum: (a) COSY with double quantum filtering (COSY-DQF), (b) COSY 45 and (c) COSY long range (COSY-LR).

Nuclei with similar chemical shifts will exhibit cross peaks near the diagonal which will be somewhat difficult to identify and interpret. Because of the dispersive character of the diagonal peaks, it can obscure the cross peaks lying near the diagonal. If the multiplets are not resolved fully in the cross peaks their intensities may reduce or they may even vanish. The intensities of the cross peaks provide some indication of the order of the coupling constant. One of the techniques for reducing the size of the diagonal peaks is known as the COSY-double quantum filtering (DQF). By inserting a third 90° pulse immediately following the second 90° pulse in the simple COSY pulse sequence, we get the pulse sequence for the double quantum filtered ^1H–^1H COSY (COSY–DQF) experiment (Fig. 19.9 (a)). The purpose of the third 90° pulse is to remove or 'filter' signals arising from single quantum magnetization so that only double or higher quantum transitions occur. The double quantum filter will select spin systems with at least two spins (minimum AX or AB) and thus, for example, uncoupled methyl singlets will be greatly reduced. Only in a spin system that consists of two or more scalar coupled nuclei with different chemical shifts can detectable double quantum phenomena arise. Normal COSY spectrum but with diagonal signals of uncoupled nuclei severely reduced occur making the interpretation of the spectrum easier.

(b) COSY 45

If the chemical shift differences between coupled nuclei in the 1D spectrum are small, the COSY cross peaks appear close to the diagonal and may overlap with the diagonal peaks. In such cases, the pulse sequence COSY 45 is useful where the second 90° pulse is replaced by a 45° pulse (Fig. 19.9 (b))

$$90^\circ_x - t_1 - 45^\circ_x;\ \text{FID}\ (t_2)$$

This reduces the intensities of the diagonal signals. It simplifies the diagonal region making it possible to identify correlations that would otherwise be hidden in a group of peaks near the diagonal. Possible overlaps can thus be reduced or eliminated. In the COSY 45 spectrum, the diagonal peaks are thus not important. The off-diagonal cross peaks arise due to vicinal and geminal couplings between coupled protons of the molecule. The smaller pulse angle introduces asymmetry in the cross peaks. The direction in which cross peaks are oriented provides the relative signs of the scalar spin-spin coupling constants. This helps in distinguishing positive vicinal coupling constants from negative geminal coupling constants. The cross peaks in COSY 45 spectrum represents chiefly couplings between directly connected transitions.

(c) COSY-LR (Long Range COSY)

The proton spectra of organic compounds are usually dominated by geminal and vicinal coupling constants. These interactions with magnitude between 2 and 20 Hz also determine the COSY spectra. If the long range couplings are of interest, special provisions have to be made. This can be achieved by the introduction of a fixed delay Δ in the evolution and detection time that is before and after the second pulse of the COSY sequence: $90^\circ_x - t_1 - \Delta, 90^\circ_x, \Delta, t_2$ (Fig.19.9(c)). Thus inserting a significant delay of the order of 1/J (e.g. 1 s) after the first pulse allows t_1 to span a region more favourable to the detection of small J-cross peaks. It is also useful to delay the start of acquisition using the same delay time as that used for t_1. It results in the weakening and often disappearance of cross peaks arising from the stronger direct coupling interactions and intensification of the weaker long-range interactions. This permits long range couplings to be observed clearly. Thus COSY-LR gives connectivities over 4 or 5 bonds. Using this pulse sequence, cross peaks corresponding to ^4J coupling constants of *o*-nitroaniline between meta protons have been observed; 125 ms were used for the delay Δ. With unknown compounds, it is therefore advisable to measure first a COSY 90 or COSY 45 spectrum in order to identify the vicinal and geminal couplings before other sequences can be used to obtain the long-range couplings by a careful comparison of these spectra.

HETCOR (^1H-^{13}C COSY)

Due to the lack of multiplet structure in the carbon resonance signals in the conventional ^1H decoupled ^{13}C spectra, it may not always be easy to assign the carbon resonances since it is difficult to distinguish between resonances of CH_3, CH_2, CH and quaternary carbon atoms. Both ^1H and ^{13}C signals are almost equally affected in their chemical shifts by their surroundings. It would be desirable to know with certainty the multiplicity of the carbon atoms. One such technique to obtain the multiplicity of the carbon atoms is the DEPT spectrum (discussed earlier) and the other is the ^1H–^{13}C correlated spectrum. In the latter technique, a signal from the ^{13}C spectrum is correlated with the signals of the protons bonded to that carbon. The 2D NMR spectrum that displays ^{13}C–^1H chemical shift correlations is known as the HETCOR (Heteronuclear chemical shift Correlation) spectrum; sometimes referred to as ^1H–^{13}C COSY. Correlation signals show which protons are bonded to which carbon atom of the molecule.

As in the case of COSY, in HETCOR, resonance frequencies of scalar coupled ^1H and ^{13}C nuclei are correlated and cross peaks with the coordinates δ (^1H) δ (^{13}C) are obtained. These correlations are generally based on heteronuclear ^{13}C, ^1H couplings over one bond ($^1J_{CH}$). The nuclei which show cross peaks are therefore direct neighbours in a given molecule and the assignment of the proton spectrum can be transferred directly to the ^{13}C spectrum and vice versa. If the proton spectrum can be correlated with the ^{13}C spectrum then the complete assignment of both the spectra and hence the elucidation of structure may be facilitated.

The pulse sequence for 2D HETCOR experiment shown in Fig. 19.10 makes the heteronuclear correlation a highly sensitive technique. The effect of pulse sequence on ^1H magnetization vectors of CH is shown in Fig. 19.11. The first pulse creates initial proton magnetization, the time governing the return of the spin system to equilibrium is the proton spin-lattice relaxation time T_1 which is usually shorter than that of the attached carbon. Therefore the experiment could be repeated after waiting only for a couple of seconds. Since magnetization transfer through a pair of 90° pulses is used, the amount of carbon magnetization obtained is governed by the Boltzmann distribution for the protons and not for the carbons. This enhances the carbon signal by a factor of four relative to the magnetization for carbon in the absence of nuclear Overhauser effect. The rapid repetition rate and the gain in sensitivity due to polarization transfer overcomes the lower sensitivity of nuclei such as ^{13}C, ^{15}N, etc.

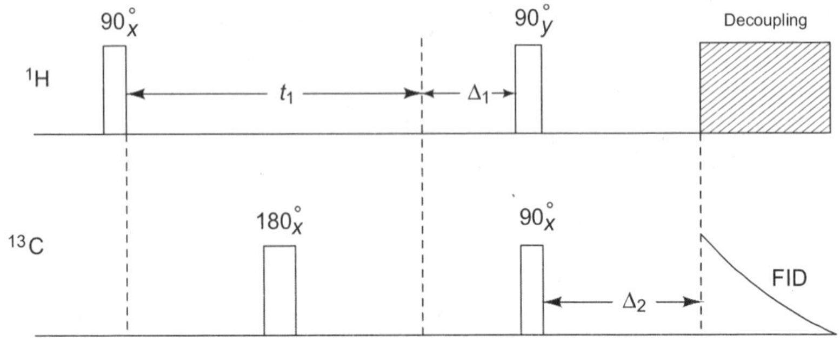

Fig. 19.10 Pulse sequence for obtaining heteronuclear chemical shift correlation (HETCOR) spectrum.

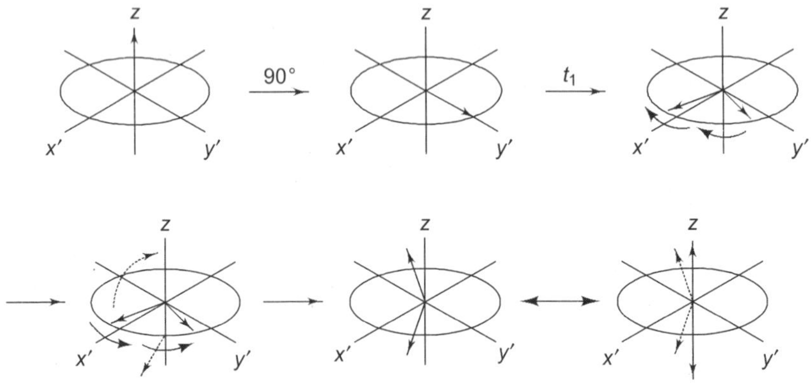

Fig. 19.11 Effect of pulse sequence for HETCOR experiment on ^1H magnetization vectors of CH.

The first 90° ^1H pulse deflects the ^1H magnetization on to the y-axis. During the evolution period, this magnetization precesses in the $x'y'$ plane. It may be regarded as being made up of two vectors corresponding to the lower (α) and higher (β) spin states of carbon to which ^1H is coupled. These two vectors diverge in the first half of the evolution period since their angular velocities differ. The 180° ^{13}C pulse applied at the centre of the evolution period reverses their direction of rotation, so that during the second half of the evolution period they converge and are refocussed at the end of the evolution period t_1. The angle by which the vector diverges from the y' axis corresponds to its chemical shift, the downfield protons diverge larger than the upfield protons. A constant mixing delay time Δ_1 is given after the variable evolution period t_1 and the duration of the mixing delay is set equal to the average value of J_{CH}. After the first 1/2 J interval, the vectors diverge and align in opposite directions. The second 90° ^1H mixing pulse is then applied; it causes rotation of the two vectors towards the $+z$ and $-z$ axes. One vector now points towards the $+z$ axis and the other towards the $-z$ axis. Thus one of the proton lines of the CH doublet is inverted which corresponds to the vector pointing towards the $-z$ axis. It means that a selective population inversion has occurred as in the INEPT experiment described earlier (see Chapter 15). It leads to intensification of the ^{13}C signals due to polarization transfer. The extent of polarization transfer will depend on the disposition of the proton multiplet components, when the second 90° mixing pulse is applied, which depends on the ^1H chemical shifts as well as on the duration of the evolution period t_1. The amplitude of the ^{13}C signals detected during t_2 is therefore modulated as a function of t_1 by the frequencies of ^1H spins giving rise to ^1H, ^{13}C shift correlation spectra. A $90°_x$ ^{13}C pulse is then applied that monitors the evolved spin states. Lastly, broad band decoupling is applied in the proton channel during acquisition so that ^{13}C signals obtained from each FID are singlets.

In a 2-D experiment correlation occurs during t_1 and hence the proton decoupler is not turned on during this period. If broad-band decoupling is not applied during acquisition, then a coupled spectrum is obtained in both dimensions giving a rather complicated spectrum. The splitting of the signals into multiplets results in lowering of intensity. Therefore, it is usual to eliminate the heteronuclear couplings. For inequivalent geminal protons of a CH_2 group, two different cross peaks will occur at the same ^{13}C chemical shift but at two different ^1H chemical shifts, which aids in their identification.

The F_1 axis (ν_1) which is derived from the incremented delay t_1 is the proton chemical shift axis. The F_2 axis (ν_2) obtained during t_2 is the carbon-13 chemical shift axis. Thus the $\pi/2$ pulse is in the ^{13}C channel and the FID acquired during t_2 represents the ^{13}C nuclei.

A HETCOR spectrum, like all 2D NMR spectra, describes the environment of the nuclei during t_1. A feature of the HETCOR pulse sequence is that only interactions responsible for modulating the proton magnetizations that is the proton chemical shifts and heteronuclear couplings occur. One may assign carbon resonances based on known proton chemical shifts or proton resonances based on known carbon chemical shifts. For example, protons that yield a complex spectrum may be bonded to carbons whose spectrum is well resolved or the carbon signals are very similar but the protons are well assigned. Thus HETCOR spectrum helps to clarify the overlapping resonances occurring in a proton spectrum. This complementary nature of the HETCOR spectrum to the ^1H–^1H COSY spectrum makes it extremely useful for larger molecules. Together with a COSY spectrum that identifies neighbouring protons, the HETCOR spectrum allows the sequencing of molecular chains and helps in confirming the proposed structure.

The HETCOR experiment can be utilised for any heteronuclear system in which there is resolvable spin-spin coupling. Of course, high abundance nuclei such as phosphorus and boron give very good HETCOR spectra. Quadrupolar nuclei in asymmetric environments can have unfavourable relaxation properties since their relaxation times may be very short.

Consider the HETCOR spectrum of 1-chloro-2-propanol given in Fig. 19.12.

(i) In this Figure, the ^{13}C chemical shifts of the compound are given along the F_2 axis (X axis) and the ^1H chemical shifts along the F_1 axis (Y axis). Opposite these axes, the corresponding 1-D spectrum is given for convenience and these spectra are not a part of the actual 2D spectrum.

(ii) The 2D spectrum contains only the cross peaks each one relating a carbon to its directly bonded proton(s) in the molecule. Assignment of the proton spectrum is desirable. There are no diagonal peaks in the HETCOR spectrum. Quaternary carbons are also not observed.

(iii) For the interpretation of the spectrum we begin with any carbon resonance and imagine a line being drawn vertically until a cross peak is met. Another line is drawn perpendicular to the first one to find the proton or protons with which it correlates. If we start with a proton resonance we would obtain exactly the same result. It readily reveals which hydrogens are bonded to which carbons.

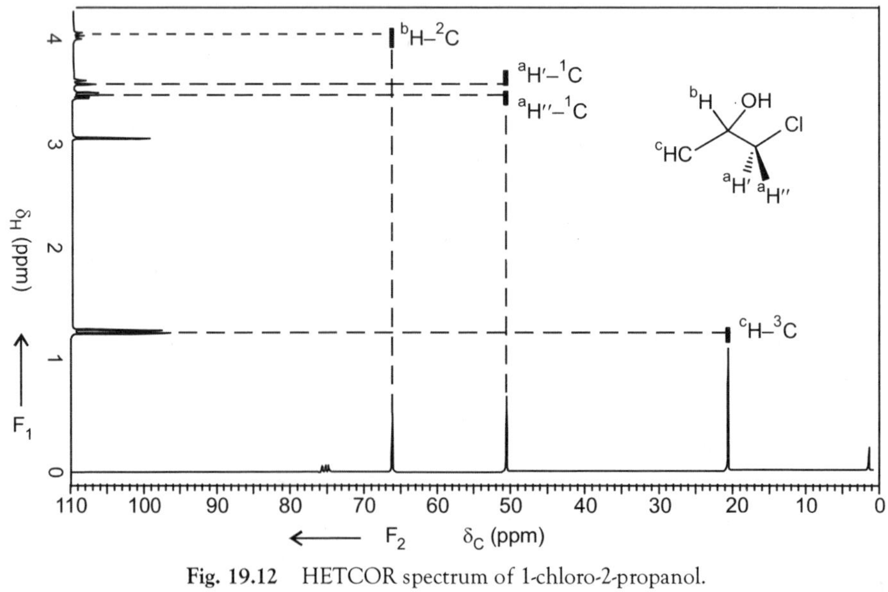

Fig. 19.12 HETCOR spectrum of 1-chloro-2-propanol.

(iv) Broad band ^1H decoupled ^{13}C NMR spectrum of 1-chloro 2-propanol shows bands at δ 20, 51 and 67 ppm assignable to CH_3, CH_2 and CH carbons respectively. Following the above procedure, we find that there is a cross peak that correlates wth the chemical shift of the methyl doublet at δ 1.2 ppm and of C_3 carbon signal of 1-chloro 2-propanol at δ 20 ppm. Similarly the ^1H NMR peak at δ 3.9 due to CH proton of CHOH is connected by (dashed lines) a cross peak to the ^{13}C signal at δ 67 ppm arising from the alcohol carbon C_2. Finally the ^1H NMR peak at δ 3.4 – 3.5 ppm for the two protons of CH_2Cl is connected by a cross peak to the ^{13}C peak (C_1) at δ 51 ppm.

(v) If a line drawn from a ^{13}C chemical shift encounters no cross peaks, then that carbon has no bonded hydrogens. If it encounters only one cross peak, then that carbon may have either 1, 2 or 3 protons attached to it and if two protons are attached then they have either equivalent chemical shifts or they fortuitously overlap. If the line drawn from a ^{13}C chemical shift encounters two cross peaks then we have the special case of diastereotopic protons attached to a methylene group. Much of this information can be obtained from the DEPT spectra. The HETCOR should, whenever possible, be carried out along with DEPT.

Owing to the general importance of heteronuclear shift correlations, extensive research for improvements with respect to sensitivity and selectivity has been initiated and various modifications of HETCOR have been proposed.

Two different pulse sequences could be used for the two dimensional heteronuclear shift correlation between a sensitive nucleus A (e.g. ^1H, ^{19}F, ^{31}P) and an insensitive nucleus X (e.g. ^{13}C, ^{15}N). The more recent techniques of heteronuclear multiple quantum coherence (HMQC) and heteronuclear multiple bond connectivity (HMBC) involve inverse polarization transfer, i.e. the magnetization is transferred from the sensitive nucleus, in general from the proton, to the insensitive nucleus, for example ^{13}C (or ^{15}N).

(a) Heteronuclear Multiple Quantum Coherence (HMQC)

The HMQC is a sensitive technique for the determination of one-bond ^1H–^{13}C shift correlations. The pulse sequence for obtaining a HMQC spectrum is shown in Fig.19.13. Protons not coupled to ^{13}C are suppressed by a bilinear (BIRD) pulse so that only protons directly bonded to ^{13}C give cross peaks in the 2D spectrum. The BIRD pulse is followed by a delay τ whose magnitude is so adjusted that inverted magnetization of protons not bound to ^{13}C nuclei passes through zero amplitude, when the first 90°_x pulse is applied. The interval in the pulse sequence is set at 1/2 J where J is the value of one-bond ^{13}C–^1H coupling constant (100 to 200 Hz).

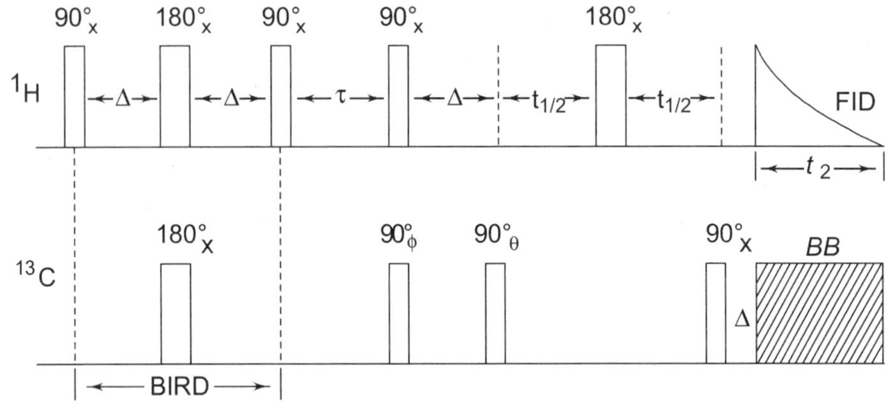

Fig. 19.13 Pulse sequence for HMQC experiment.

In HMQC, magnetization which develops during the evolution time is finally transferred into detectable A (say ^{1}H) magnetization. The sensitive A nucleus is used for signal detection. The relaxation time T_1 of protons directly bonded to ^{13}C is short and hence the interval between repeat scans can be kept very short. Because of the sensitive A nucleus detection, the HMQC spectrum is more sensitive than the spectra based on polarization transfer for the determination of one bond ^{1}H–^{13}C correlations. The HMQC technique is about 16 times more sensitive for ^{13}C detections than the conventional HETCOR technique. The sensitivity advantage of HMQC and HMBC methods is due to the fact that they rely on the equilibrium magnetization derived from protons. Further, since the strength of the NMR signal increases with the frequency of detection, a more intense signal will be obtained at the higher proton observation frequencies than at the lower ^{13}C observation frequencies. If there are two cross peaks at any particular ^{13}C chemical shift in the HMQC spectrum, then the two protons must be the two geminal protons attached to that particular carbon.

(b) Heteronuclear Multiple Bond Connectivity (HMBC)

The HMBC is a sensitive technique for obtaining information about long-range ^{1}H–^{13}C couplings. In the HMBC technique, like the HMQC, the ^{13}C population differences are measured indirectly through their effects on ^{1}H and hence it is a more sensitive technique. The pulse sequence employed for HMBC is shown in Fig. 19.14. The first 90° ^{13}C pulse serves to eliminate one bond $^1J_{CH}$ couplings and hence the cross peaks due to one-bond ^{1}H–^{13}C couplings do not occur and only long range ^{1}H–^{13}C couplings occur. The second 90° ^{13}C pulse creates zero and double quantum coherence which are interchanged by the 180° ^{1}H pulse. After the final 90° ^{13}C pulse, the ^{1}H resonances resulting from ^{1}H–^{13}C multiple-quantum coherence are modulated by ^{13}C chemical shifts and homonuclear ^{1}H couplings.

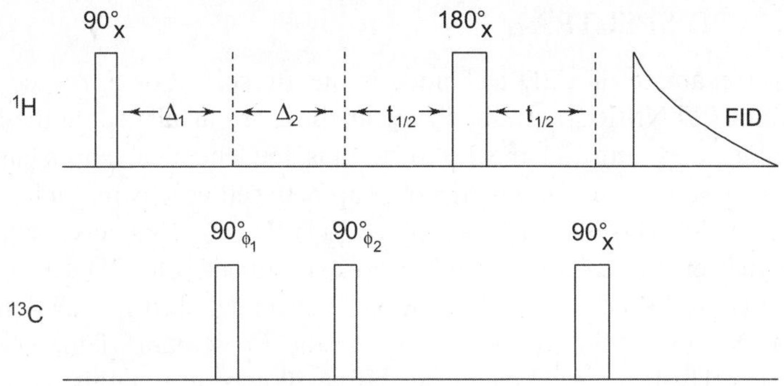

Fig. 19.14 Pulse sequence for obtaining HMBC spectrum.

The long-range ^{1}H–^{13}C COSY pulse sequence also gives a two-dimensional spectrum with ^{13}C chemical shifts on one axis and ^{1}H chemical shifts on the other. However, in this sequence, after the 'preparation' of the nuclei, the delay in the pulse sequence is set equal to ½ J which is about 10 Hz, i.e. a delay of about 50 ms. Since the value of ^{1}H–C–^{13}C (two bond) and ^{1}H–C–C–^{13}C (three bond) coupling constants are generally similar and lie in the range 2 to 20 Hz, the ^{13}C chemical shifts are now correlated with the chemical shifts of those protons that are separated from them by two and three bonds. The HMBC method is particularly useful for locating quaternary carbons through two-bond ($^2J_{CH}$), three bond ($^3J_{CH}$) and some times four-bond ($^4J_{CH}$) couplings. Although the values of the two– and three–bond couplings may overlap, the technique is important since it is possible to detect the presence of a methylene group and a trans double bond in a structure. The HMBC experiment permits us to deduce the skeleton of an organic compound indirectly through carbon–carbon connectivities but the procedure is laborious since it is not known whether correlations result from two– or three–bond couplings.

By a combination of COSY and HETCOR spectra, all ^{13}C and ^{1}H peaks can be unambiguously assigned to their respective carbon and hydrogen atoms. For many compounds assignments are quite difficult to make without the aid of these 2D techniques. The COSY and HETCOR experiments are extremely useful in the elucidation of structures of complex organic molecules. The geminal and vicinal protons and their one-bond connectivities are identified from the HETCOR spectrum. By careful analysis of the COSY and HETCOR spectra, it is possible to obtain information on carbon carbon connectivities of the protonated carbons. Interaction of protons with quaternary carbons across two, three or four bonds through long range carbon proton couplings is important in joining the various fragments together.

J-RESOLVED SPECTRA

Another example of a 2D technique is the *J*-resolved or *J*, δ-spectroscopy. In the resulting 2D NMR spectrum, chemical shifts are displayed on the F_2 axis and the coupling constants on the F_1 axis, thus the chemical shifts and coupling constants are separated avoiding the overlap between nearby multiplets. The pulse sequence employed is $90^\circ_x - t_{1/2} - 180^\circ_x - t_{1/2}$ –FID (t_2). This pulse sequence is the same as that employed for the spin echo experiment. The 2D J-resolved NMR experiments are based on the elimination of chemical shift modulations during t_1 allowing the observation of *J*-couplings in the F_1 domain. If the nucleus being examined is ^1H, then the ^1H chemical shifts will appear along the F_2 axis and the couplings of ^1H with other neighbouring ^1H nuclei will occur along the F_1 axis. A typical ^{13}C 2 D J,δ NMR spectrum as contour plot is shown in Fig.19.15. Broad band ^1H decoupled ^{13}C spectrum is plotted along one side. This permits multiplicities of each carbon to be read easily. It is a good technique but gives rise to potentially confusing artefacts for systems which are second order.

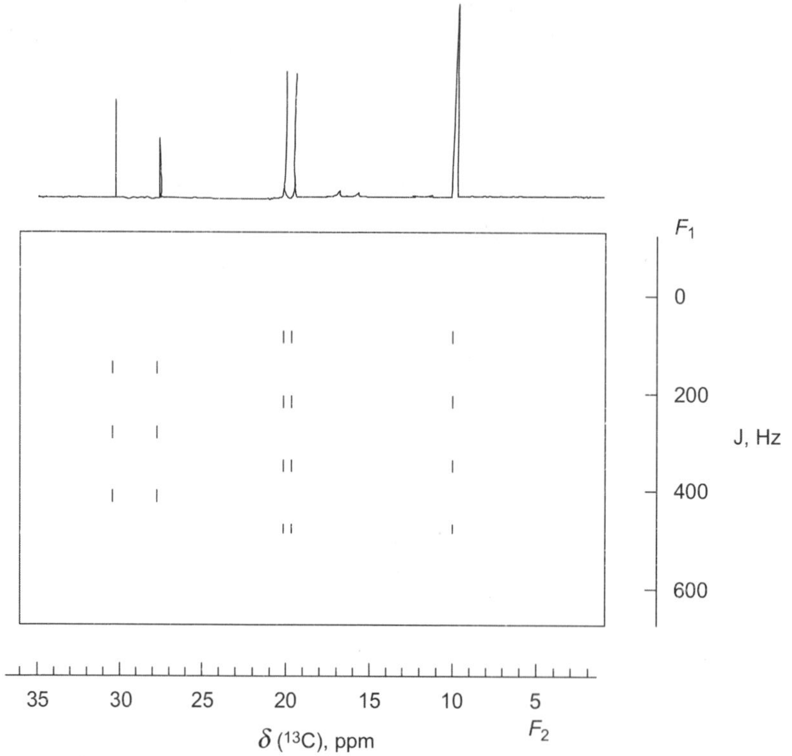

Fig. 19.15 Contour diagram of a typical 2D J, δ NMR spectrum.

In homo- and heteronuclear 2D *J*-resolved experiments, the chemical shifts and coupling constants are separated along the two axes. Heteronuclear 2D *J*-resolved spectra contain the chemical shifts of one nuclear species e.g. ^{13}C

along one axis and its coupling data with another type of nucleus e.g. ^1H along the other axis. 2D *J*-resolved spectra are therefore often referred to as *J*, *δ*-spectra. Homonuclear 2D *J*-resolved spectroscopy is very useful for unraveling complex overlapping multiplets. The main application of 2D *J*-resolved spectroscopy is in the analysis of complex spectra where spin multiplets of different protons strongly overlap as in biomolecules. *J*, *δ*–NMR spectroscopy requires relatively longer measuring time. For strongly coupled spin systems, a simple analysis of the spin multiplets is not possible. Nowadays the use of *J*, *δ*-spectroscopy is less common.

2D NUCLEAR OVERHAUSER SPECTROSCOPY (NOESY)

The perturbation of the longitudinal magnetization of one nucleus can result in changes in the equilibrium magnetization of another nucleus through mutual relaxation. The nuclear Overhauser effect (NOE) is a well known example of this. The NOE experiments in which one nucleus is saturated or its magnetization is inverted by a selective 180° pulse serve as sensitive molecular probes. The rate of transfer of magnetization to a second nucleus during the time following the saturation or inversion is a function of the proximity of the nuclei, other relaxation path ways and their relative motions. By varying the time during which the interaction takes place, the strength of the interaction can be monitored.

A large number of 1D NOE experiments may have to be carried out if the spatial relationships among many protons in a molecule are to be established. In such cases, one can undertake two-dimensional homonuclear Overhauser correlation spectroscopy (NOESY). A two-dimensional spectrum which records all the magnetization transfer arising from mutual dipolar relaxation (NOEs) occurring in a molecule is called a NOESY spectrum. By the 2D version of the 1D NOE experiment (see Chapter 17), i.e. using the NOESY spectrum the entire spatial relationships among many protons could be examined at once. If these protons have closer chemical shifts, then NOESY may be particularly advantageous, since it is difficult to carry out irradiation in 1D experiments without affecting other nearby protons. In a single NOESY spectrum, all interproton NOE effects appear simultaneously and the spectral overlap is minimized due to the spread of the spectrum in two dimensions.

The NOESY is indispensible for the conformational analysis of biomolecules. It is useful for the determination of three-dimensional structures in solution because in addition to correlations based on spin-spin coupling through chemical bonds, direct information about distances between nuclei in space can be obtained. Taking known distances as reference, proton distances of the order of 0.2 to 1.0 nm can be determined fairly accurately. For larger molecules NOESY spectra are very useful for the assignment of at least partial spectra. The NOESY technique can also be employed to study chemical exchange.

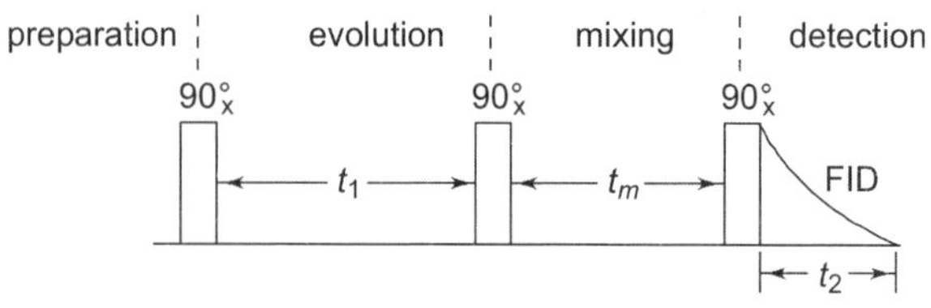

Fig. 19.16 Pulse sequence for recording NOESY spectra.

The pulse sequence used for obtaining NOESY spectrum is shown in Fig. 19.16 and the effect of pulse sequence on the magnetization vectors of the nuclei is shown in Fig. 19.17. The pulse sequence consists of three 90° pulses, the second and third pulses are separated by a fixed interval τ_m during which z-magnetization is transferred between neighbouring spins through NOE. Let us assume that there are two nearby protons S and I that are not involved in scalar coupling. The first $90°_x$ pulse applied to I spins rotates the equilibrium longitudinal z magnetization of I nuclei to the y axis and in the subsequent interval t_1, this magnetization precesses in the x y plane so that after time t_1 it would have undergone precession by an angle $\Omega_I t_1$ where Ω_I is the precessional frequency of the nucleus I. The second 90° pulse will rotate this magnetization from the xy plane to the xz plane. At this time, the x-component of magnetization will be proportional to sin $(\Omega_I t_1)$ and the z-component will be proportional to –cos $(\Omega_I t_1)$.

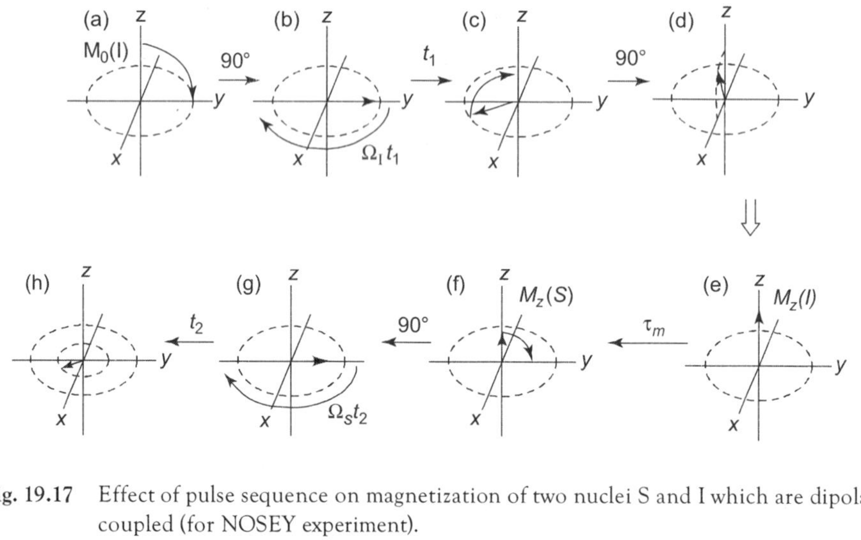

Fig. 19.17 Effect of pulse sequence on magnetization of two nuclei S and I which are dipolar coupled (for NOSEY experiment).

During the interval τ_m, the spins I and S will undergo cross relaxation and some of the non-equilibrium z–magnetization of I spins is transferred to S through the NOE effects. The z-magnetization of S spins is rotated onto the y axis by the third $90°_x$ pulse when it undergoes precession at frequency Ω_s during the detection period t_2. The pulse sequence (Fig. 19.16), which is used for recording the two-dimensional homonuclear Overhauser spectrum (NOESY) is also used for 2D exchange spectroscopy (EXSY) to investigate chemical exchange. In chemical exchange, magnetization is transferred from one site to another. The exchange of magnetization during the mixing time is based on the nuclear Overhauser effect. Figure 19.18 shows as an example the schematic representation of a 2D NOESY spectrum.

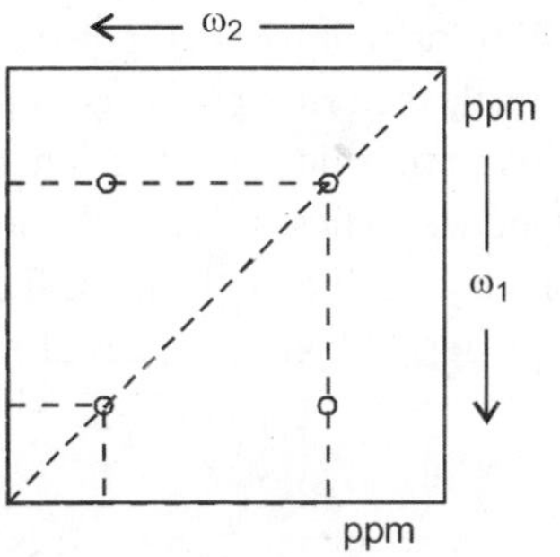

Fig. 19.18 A schematic representation of the NOESY spectrum; the off-diagonal cross peaks denote NOE interactions between various nuclei.

As in the COSY spectrum, two 90° pulses separated by an evolution time separate the nuclei by their relative chemical shifts. It creates non-equilibrium longitudinal magnetization for the spins. This will lead to polarization which will cause redistribution of population among mutually relaxing spins as in an Overhauser effect. Since this imbalance is created for a regularly incremented series of evolution time, all spins regardless of their chemical shifts will experience periods in which they have a non-equilibrium distribution of population. The regular variation in t_1 increments ensures that all spins are sampled equally and that all possible population transfers occur. This population transfer occurs during the mixing period, whose duration should be chosen appropriately. The extent of magnetization transfer is monitored with a third 90° pulse followed by the usual t_2 detection. The magnetization detected at spin S in t_2 can originate in t_1 from spin S itself (leading to a diagonal peak) or from a mutually interacting spin leading to a cross peak. The detection of a cross peak indicates that there is either cross-relaxation between the two spins and therefore they are in spatial proximity or that the two spins are involved in chemical exchange. The latter situation reveals that the parent spin need not be in the same molecule. In this case, two separate spins, say A and X precess following the first pulse and dephase because of the difference in their chemical shifts. At the end of the evolution period they are antiphase in *xy* plane. This difference is reflected in their *z*-magnetization to propagate to other spins following the second 90° pulse. This is then monitored by the final pulse. All possible orientations of A and X magnetization are sampled by regular incrementing of t_1. Thus all possible frequencies in F_1 corresponding to the relaxation rates and chemical exchange rates are monitored. When both exchange and NOE effects operate, ROESY (rotating frame NOSEY) experiments can be used to differentiate between the two effects. In several cases it is possible to eliminate exchange signals by varying the temperature. COSY signals which are produced in most NOESY experiments because of the presence of geminal, vicinal or long range ^1H, ^1H couplings must be eliminated through the phase cycle.

A typical NOESY spectrum is shown in Fig. 19.19. A comparison of the Fig. 19.18 with Fig. 19.8 reveals that a NOESY spectrum has the same structure as a COSY spectrum with each orthogonal axis of proton chemical shifts; the 1D spectrum occurs on the diagonal, the cross peaks as non-diagonal elements indicate those protons that are in close proximity in space indicating interactions through space, rather than through bonds. Thus a NOESY spectrum provides important information about the geometry of the molecules. The 1D spectrum is usually reproduced along one axis of the 2D contour spectral plot for clarity.

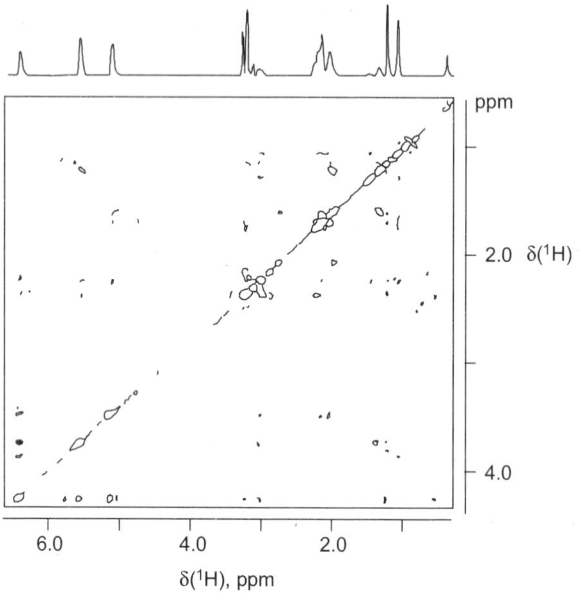

Fig. 19.19 A typical NOESY spectrum.

The main advantages of the NOESY spectrum are that

(i) no initial assumptions are made and the complete spectrum can be examined at once,

(ii) crowded spectra are considerably simplified by spreading resonances into the second dimension,

(iii) selective excitation of the individual resonances as in NOE which is difficult for overlapping lines is avoided and

(iv) measurement of all NOEs simultaneously is more efficient than a series of 1D experiments.

The positive NOE enhancements from small molecules (molecular mass 100–400) are rather weak and these are often best examined by NOE difference spectroscopy. Larger molecules (molecular mass 1000 or more in nonviscous solvents) normally give more intense NOE cross peaks. Since the NOE effect falls off with r^{-6} (where r is the distance between the protons giving rise to the cross peak), the intensity of NOE cross peak falls off rapidly with increasing distance between the protons but it is also affected by other variables. The overlapping resonances in the spectrum may be removed by measuring the spectrum at different temperatures, pH or in different solvents. It is possible from complete COSY and NOSEY spectra to solve the structure of complex molecules; the advantage over X-ray crystallography is that NOSEY applies to molecules in solution and the compound does not have to be crystalline.

2D INADEQUATE SPECTROSCOPY

The most important question in structural organic chemistry is the nature and stereochemistry of the carbon skeleton of the molecule. X-ray crystallography is a powerful tool for obtaining the three-dimensional structure of the molecule. However, the X-ray method is applicable only to those molecules that can be crystallised and the structure in the solid state is obtained. Nuclear magnetic resonance is well suited for studying the molecules in solution and it can provide similar structural information.

The NMR techniques discussed earlier (e.g. COSY, NOESY, HETCOR) provide information about proton-proton interactions and they allow the correlation of protons with carbon or other heteroatoms (e.g. hetero COSY). The information derived from these techniques is very useful for structural elucidation but it does not reveal the carbon framework of the organic molecule directly. The 2D INADEQUATE (Incredible Natural Abundance Double Quantum Transfer Experiment) allows the entire carbon skeleton to be deduced directly via the measurement of $^{13}C–^{13}C$ coupling constants. The determination of ^{1}J ($^{13}C–^{13}C$) coupling constants permits the assignment of neighbouring carbon atoms of interest.

Spin-spin splittings as a result of $^{13}C–^{13}C$ couplings are present in the ^{13}C NMR spectrum only as weak satellites accompanying the intense signals of the molecule. Since ^{13}C enrichment is possible only in a few cases experimental methods which facilitate the measurement of $^{1}J(^{13}C–^{13}C)$ coupling constants are of interest. For compounds of low molecular mass (say less than 500), about 500 mg of the sample dissolved in 0.5 ml of a deuterated solvent and scanning for 24–36 h is needed to obtain a 2 D INADEQUATE spectrum with a good signal to noise ratio.

The INADEQUATE experiment is a homonuclear NMR technique in which $^{13}C–^{13}C$ coupling satellites are detected after suppressing the stronger $^{13}C–^{1}H$ couplings. The presence of two anisochronous ^{13}C atoms in a molecule is necessary to observe $^{13}C–^{13}C$ coupling constants. Because of the low natural abundance of ^{13}C (1.1%), it is very rare for one ^{13}C to be bonded to another. Two adjacent ^{13}C–^{13}C bonded carbons would be about 1 in 10,000 molecules. The experimental determination of ^{13}C–^{13}C coupling constants is rendered difficult because of the inherently low intensity of ^{13}C–^{13}C nuclei in natural abundance as well as the similarity of many ^{13}C–^{13}C coupling constants. The INADEQUATE experiment for the detection of ^{13}C–^{13}C couplings in such molecules is therefore very insensitive.

The $^1J(^{13}C–^{13}C)$ coupling constant falls in a narrow region and it is sensitive to the nature of the carbon-carbon bond. Geometry is the main factor affecting $^1J(^{13}C–^{13}C)$. In the case of hydrocarbons, dependence on the product of *s*-character of the carbon orbitals ϕ_i and ϕ_j forming the sigma bond $C_i - C_j$ has been observed. The ^{13}C –^{13}C coupling constant varies from 35 Hz for $CH_3 - CH_3$ ($sp^3 - sp^3$), 68 Hz for $H_2C = CH_2$ ($sp^2 - sp^2$) to 172 Hz for $HC \equiv CH$ ($sp - sp$). For the methyl-phenyl group in toluene ($sp^3 - sp^2$), the observed coupling constant is 44 Hz. The neighbouring electronegative elements raise the value of the coupling constant, e.g. $^1J(^{13}C–^{13}C)$ for the methyl group in acetates is 59 Hz.

The INADEQUATE spectrum is helpful for establishing the carbon skeleton, while vicinal proton-proton coupling constants help in elucidating the components which make up the chain of carbon atoms. Difficulties arise for example when quaternary carbon intervenes since it has no protons to couple with adjacent protons, the chain is interrupted. Starting at a known carbon, one first determines which carbon(*s*) are connected to that carbon. Then by analyzing further connectivities strings of carbon atoms can be established. For an unknown hydrocarbon this process can result in the complete determination of the molecular structure. Even if the unknown molecule contains heteroatoms nearly all of the structure can be deduced.

The INADEQUATE technique was originally developed for ^{13}C NMR spectroscopy and it was carried out as an one-dimensional experiment in order to determine neighbouring ^{13}C pairs. The 1D INADEQUATE experiment has not been widely used because of its very low sensitivity. The detection of ^{13}C–^{13}C coupling is accomplished with the aid of double-quantum filtering. The phenomenon of double quantum filtering which removes all single-spin transitions, which correspond presently to isolated ^{13}C atoms was discussed earlier under COSY-DQF. In a chain of NMR active ^{13}C nuclei each spin excepting those at the chain ends has two neighbours and therefore participates in two different AX spin systems. Only those transitions with two spins (AB and AX systems) are detected during acquisition time and these reveal the spin connectivities in a particular molecular structure.

A powerful extension of the 1D INADEQUATE experiment was introduced by adding as usual the evolution time t_1 which gives rise to a second dimension in frequency and thus to a 2D experiment. The aim of the experiment is the analysis of the carbon skeleton of the molecule through their $^1J(^{13}C - ^{13}C)$ coupling as mentioned earlier. Using 2D-INADEQUATE spectra, the analysis of the molecular structure of many liquids and highly concentrated solutions of solids have been made.

In principle, the 2D INADEQUATE experiment can be performed using the pulse sequence given in Fig. 19.20. This pulse sequence is used for ^{13}C observation with ^{1}H broad band decoupling. During t_1 it is the sum of the chemical shifts of the coupled nuclei which evolve and since it is double quantum filtered, only two types of ^{13}C–^{13}C spin systems AB and AX contribute significantly to the intensity in the INADEQUATE spectrum.

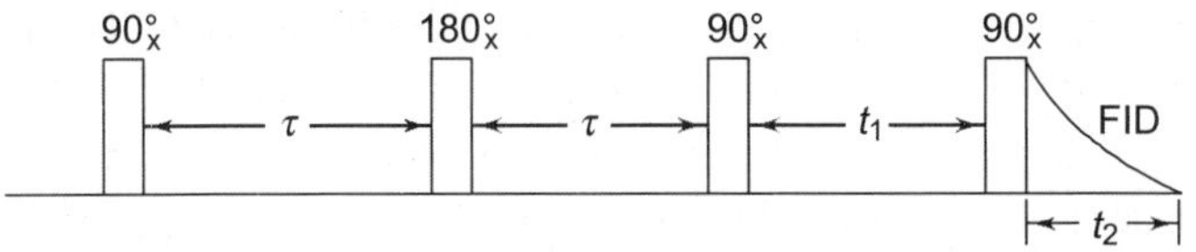

Fig. 19.20 Pulse sequence for 2D INADEQUATE experiment.

The interval τ is generally kept at 1/4 J_{CC}. The coupling constant J_{CC} for directly bonded carbons generally lies between 30 Hz – 70 Hz. The first two pulses and delays (90° – τ – 180°– τ) create a spin echo which is subjected to a second 90°_x pulse which then creates a double quantum coherence for all directly bonded ^{13}C nuclei. The double quantum coherence evolved during incremented evolution time t_1 is then converted into detectable magnetization by a third pulse and the resulting FID is recorded. The most efficient conversion of double quantum coherence is achieved through a flip angle of 90°. Proper choice of the delays (τ) in the pulse sequence allows us to select larger one bond couplings($^{1}J_{CC}$).

Each two spin AX system is characterized by its Larmor frequencies ν_A and ν_X as well as by the double quantum frequency $\nu_{DQ} = \nu_A + \nu_X - 2\nu_o$ (ν_o = transmitter frequency). The frequency axis F_1 contains the double quantum frequencies ν_{DQ} of the AX systems. On the frequency axis F_1, it can thus be distinguished from all the other systems with different ν_{DQ} values, if in a two-dimensional spectrum the F_2–axis contains the Larmor frequencies ν_A and ν_X. Here advantage is taken of the fact that the double quantum frequency ν_{DQ} of an AX system depends on the transmitter frequency ν_o.

In the 2D INADEQUATE ^{13}C,^{13}C spectrum, the F_2 axis is the usual carbon-13 chemical shift axis which of course is related to acquisition time. The F_1 axis which gives the double quantum frequencies is usually given in Hz and it is two times the range in F_2. Since the ^{13}C satellite spectrum is detected at the natural abundance level, long acquisition times are necessary. It is therefore important to keep the two dimensional matrix as small as possible. For practical applications, a number of modifications are necessary. For this purpose, several versions of the INADEQUATE pulse sequence have been developed. In the 2D INADEQUATE spectrum, the cross peaks or correlations for two adjacent ^{13}C nuclei say A and X appear at the coordinates δ_A, $\delta_A + \delta_X$ and δ_X, $\delta_A + \delta_X$ in the F_1, F_2 coordinate system. The actual cross peaks themselves are doublets with spacing equal to the coupling constant $^1J_{CC}$. The mid point of the line connecting the two sets of doublets is $(\delta_A + \delta_X)$, $(\delta_A + \delta_X)/2$. The mid points for all of the pairs of doublets lies on the line running along the diagonal. This can be used to distinguish genuine cross peaks from the spurious ones. Further even when J_{CC} values are identical, overlapping does not occur since each pair of coupled ^{13}C nuclei are found on different horizontal rows.

The 2D experiment requires long acquisition time compared to the 1D experiment particularly because of the low natural abundance of carbon – 13. It also requires fairly large amount of the sample. Alternate 1D experiments such as obtaining ^{13}C–^1H coupling constants and use of shift reagents are therefore of interest if selective excitation is possible. However, the requirement of strong coupling in the satellite spectra is less stringent in the 2D INADEQUATE technique. Further, the analysis of carbon connectivities can be made without any prior knowledge even if $^1J_{CC}$ varies between 30 and 70 Hz. The 2D INADEQUATE has the advantage that the diagonal signals are absent. As in COSY-DQF experiment the solvent signal arising from single quantum magnetization does not occur.

The INADEQUATE spectrum of 2-butanol along with its 1D ^{13}C NMR spectrum is presented in Fig. 19.21. The dotted diagonal line passes through the midpoints between each of the pairs of cross-peaks and it is helpful in identifying the cross-peaks from the noise. Initially a connectivity is established by making a horizontal correlation between cross peaks which are symmetrically situated with respect to the diagonal line. The best procedure for the interpretation of an INADEQUATE spectrum is to begin with an unambigously assigned ^{13}C signal and trace the ^{13}C–^{13}C connectivities. In the present example, the most down field signal could be assigned to the C–O carbon (C–2) since it is bonded to an OH group. The cross peaks denoted as a and a_1 are connected by a horizontal line so also the cross peaks labeled as b and b_1. The cross peaks a and a_1 correspond to the ^{13}C resonances of C_2 and C_3 atoms respectively while the cross peaks b and b_1 correspond to the ^{13}C signals of C_2 and C_1 respectively. The cross-peaks denoted as a and a_1 , and b and b_1 show the connectivities between C_2 and C_3 and between C_2 and C_1 carbons respectively. The remaining connectivity between C_3 and C_4 carbons is shown by the horizontal line connecting the cross peaks c and c_1. The termini of carbon chains are easily found since they have only vertical correlation at each end. Further connectivity is made by the vertical correlations at either end to other cross peaks. Figure 19.21 shows that the upfield cross peaks b_1 and c_1 have no cross peaks above or below them demonstrating that these atoms are bonded to only one carbon atom each.

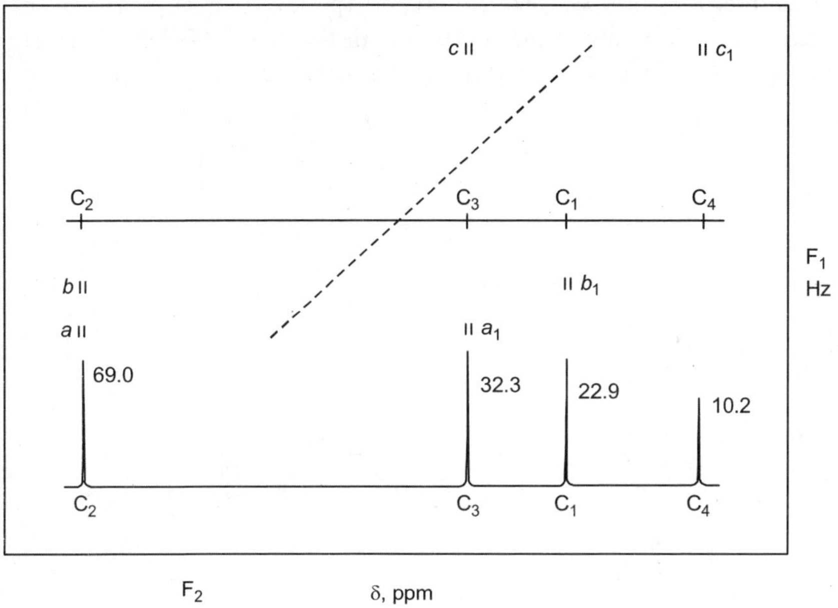

Fig. 19.21 2D INADEQUATE spectrum of 2-butanol, $H_3C\text{-}CH(OH)\text{-}CH_2\text{-}CH_3$ (carbons numbered 1 2 3 4).

As another example the 2D INADEQUATE spectrum of methyl tetrahydrofuran is considered (see Fig. 19.22). A straight line drawn diagonally from the top right to the bottom left as shown in Fig. 19.24 passes through the mid point between each of the pairs of cross peaks. This helps to identify the cross peaks from the noise as stated above. To determine the carbon-carbon connectivities, a horizontal line is drawn between the satellite peaks that are situated at equal distances on both sides of the diagonal. We start with the unambiguously assigned methyl carbon signal observed at δ 17.9 ppm. A horizontal line is drawn from the cross peak A which corresponds to the methyl carbon resonating at δ 17.9 ppm (peak a) connecting it to another cross peak B corresponding to the signal at δ 34.0 ppm (signal *b*). It shows connectivity between these two carbon atoms ($C_a - C_b$). A vertical line is drawn from the cross peak B to the cross peak B_1. The B_1 has another pair of cross peaks C on the same horizontal axis showing the connectivity between the carbons resonating at δ 34.0 (peak *b*) and δ 34.7 ppm (peak *c*). In a similar manner we can start the analysis from the other side of the spectrum. The cross peak E is connected to another cross peak B_2, which reveals the connectivity between the carbons resonating at δ 34.0 and 74.7 ppm (peak *e*), i.e. ($C_b - C_a$). Similarly the cross peak D is connected horizontally to the cross peak C_1 which correspond to ^{13}C resonances at 67.6 and 34.7 ppm respectively showing the existence of $C_d - C_c$ bond. By drawing a horizontal line from the cross peak C to B_1 we show the connectivity between the carbons resonating at δ 34.7 and δ 34.0 ppm (peak δ) (C_b – C_c). These $^{13}C-^{13}C$ connectivities thus establish the structure of methyl tetrahydrofuran. The terminal carbon atom (C_a) is readily found since it has only vertical correlation at one end.

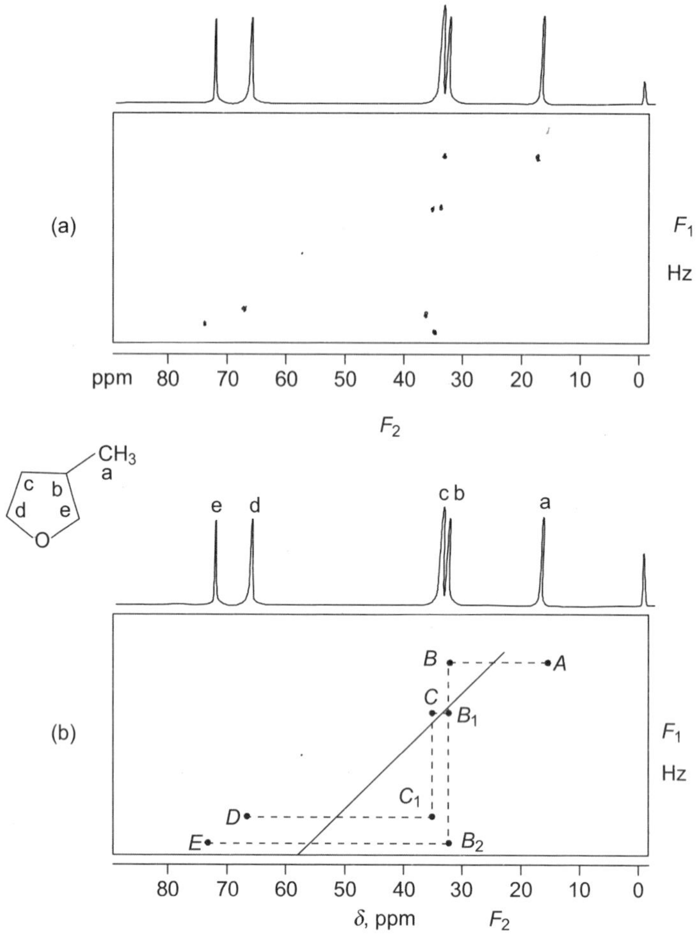

Fig. 19.22 2D INADEQUATE spectrum of methyl tetrahydrofuran: (a) without the lines drawn and (b) with lines drawn for interpretation.

The 2D INADEQUATE technique has limited applicability because of extremely low sensitivity. The INADEQUATE spectra find applications when extremely concentrated solution of the sample can be prepared or the molecule is enriched in ^{13}C isotope.

APPLICATIONS OF 2D NMR

The capabilities of various 2D techniques have greatly enhanced the applications of NMR spectroscopy.

(i) For complex molecules, the 1D spectra obtained even at the highest magnetic fields may still be obscured by extensive overlapping of the signals. Two dimensional NMR in the form of *J*–resolved spectroscopy provides a new method to resolve even the highly overlapping resonances into readily recognizable multiplets and permits the assignment of chemical shifts.

(ii) Complex proton spectra can be analysed by the decoupling techniques. This method becomes impractical in poorly resolved spectra or when a large number of decoupling experiments have to be carried out. Two-dimensional ^{1}H–^{1}H COSY spectroscopy can be used to identify spin-coupled pairs of nuclei even when structural information for the given molecule is not available.

(iii) Two dimensional ^{1}H–^{13}C COSY spectra can be utilized to identify all directly bonded carbon-proton pairs in a molecule. It provides data even without the previous knowledge of specific proton assignments and thus information obtained from one spectrum (e.g. ^{13}C) can help in the assignment of the other spectrum (e.g. ^{1}H). The COSY and HETCOR experiments are extremely useful in the elucidation of the structures of complex organic molecules.

(iv) Nuclear Overhauser effect has been utilized to obtain distance information and to infer about the local neighbourhood of the nucleus. These measurcmcnts arc difficult and require long acquisition times. Two dimensional NMR in the form of NOESY (nuclear Overhauser and exchange spectroscopy) provides a new method to obtain the same Overhauser effect information for all the nuclei in a molecule and without the previous knowledge of the spectral assignments or molecular structure. NOESY is used extensively for stereochemical assignments. It provides an indirect way to extract information about internuclear distances.

(v) Often repetitive variable temperature NMR experiments have to be carried out to study the dynamic processes. Two-dimensional NMR in the form of magnetization-transfer correlation experiments reveal the existence of chemical exchange processes in a sample and help in the measurement of associated rates of exchange without prior knowledge of the coalescence temperatures. The 2D measurements can be carried out simultaneously for all the lines in a spectrum.

(vi) NMR spectroscopy generally provides the structure of a molecule through indirect means. However, two-dimensional NMR methods such as 2D INADEQUATE provide direct evidence for the carbon skeleton of the molecule even without any prior knowledge of the structure of that molecule.

GENERAL REFERENCES

1. R.J. Abraham, J. Fisher and P. Loftus, Proton and Carbon-13 NMR Spectroscopy: An Integrated Approach, Heyden and Son, London, 1978.
2. J.W. Akitt, NMR and Chemistry: An Introduction to the FT Multinuclear Era, Methuen, 1983.
3. R.K. Harris, Nuclear Magnetic Resonance Spectroscopy, Longman, Harlow, 1986.
4. W. Kemp, NMR in Chemistry: A Multinuclear Approach, MacMillan Press, Basingstoke, 1986.
5. J. Mason (Ed.), Multinuclear NMR, Plenum Press, New York, 1987.
6. D.H. Williams and I. Fleming, Spectroscopic Methods in Organic Chemistry, 4th edn, McGraw-Hill, London, 1989.
7. E.A.V. Ebsworth, D.W.H. Rankin and S. Cradock, Structural Methods in Inorganic Chemistry, 2nd edtion, Blackwell Scientific Co., 1991.
8. A.E. Derome, Modern NMR Techniques for Chemistry Research, Pergamon Press, Oxford, 1987.
9. M.L. Martin, G.J. Martin and J.J.Delpuech, Practical NMR Spectroscopy, Heyden and Son, London, 1986.
10. F.A.Bovey, Nuclear Magnetic Resonance Spectroscopy, 2nd edition, Academic Press, San Diego, 1988.
11. H. Gunther, NMR Spectroscopy, 2nd edition, Wiley, Chichester, 1980.
12. P.J. Hore, Nuclear Magnetic Resonance, Oxford University Press, Oxford, 1995.
13. J.K.M. Sanders and B.K. Hunter, Modern NMR Spectroscopy, 2nd edition, Oxford University Press, Oxford, 1993.
14. E. Breitmaier and W.Voelter, ^{13}C NMR Spectroscopy, 3rd edition., VCH Publishers, New York, 1987.
15. F.W. Wehrli, A.P. Marchand and T. Wirthlin, Interpretation of Carbon-13 NMR Spectra, 2nd edition, Wiley, Chichester, 1988.
16. L.D. Quin and J.G. Verkade (Eds.), Phosphorous-31 NMR Spectral Properties in Compound Characterization and Structural Analysis, VCH Publishers, New York, 1994.
17. D.G. Gorenstein, Phosphorous-31 NMR, Academic Press, London, 1984.
18. A-U-Rahman and M.I.Choudhary, Solving Problems with NMR Spectroscopy,Academic Press , London , 1996.

19. R.V. Parish, NMR, NQR, EPR and Mossbauer Spectroscopy in Inorganic Chemistry, Ellis Horwood, 1990.

20. R. Freeman, A Handbook of Nuclear Magnetic Resonance, Longman, Harlow, 1987.

21. E.D. Becker, High Resolution NMR, 2nd edition, Academic Press, New York, 1980.

22. R.R. Ernst, G. Bodenhausen and A. Wokaun, Principles of Nuclear Magnetic Resonance in One and Two Dimensions, Clarendon Press, Oxford, 1987.

23. J. Sandstrom, Dynamic NMR Spectroscopy, Academic Press, New York, 1982.

24. T.C. Ferrar and E.D. Becker, Pulse and Fourier Transform NMR– Introduction to Theory and Methods, Academic Press, New York, 1971.

25. G.N. La Mar, W. De. W. Horrocks and R.H. Holm, NMR of Paramagnetic molecules, Principles and Applications, Academic Press, London, 1973.

EXERCISES

1. Calculate the NMR frequency for the ^{31}P nucleus in a field of 2.5 T; $I = ½$ and $g = 2.261$. Calculate the corresponding energy in kJ mol^{-1}. Also calculate the g value for ^{31}P if μ(magnetic moment) is 1.1305 μ_N.

2. (a) Obtain the energy difference for protons oriented along and opposite to the direction of an applied magnetic field of 2.0 T. Find the frequency of the radiation that has photons of this energy, $g_N = 5.5857$ for protons.

   (b) ^{32}S has a nuclear spin of 3/2 and $g = 0.429$. Calculate the energies of the nuclear spin states in a magnetic field of B = 7.5 T.

3. (a) Find the ratio of the equilibrium population of the α and β spin states of protons at 20 °C in an external magnetic field of 4.69 T, given $\gamma = 26.752 \times 10^7$ rad T^{-1} s^{-1}.

   (b) Calculate the energy difference between the lowest and highest nuclear spin states of a ^{14}N nucleus in a magnetic field of 15.0 T.

4. Classify the following molecules according to the nuclear spin system (such as AB, AX, ABX, etc.):

   (a) $H_2C{=}CHCl$ (b) CH_3F (c) 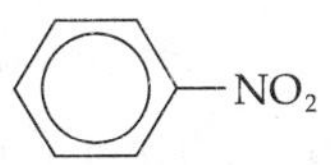 (d) $^{13}CH_3F$

   (e) 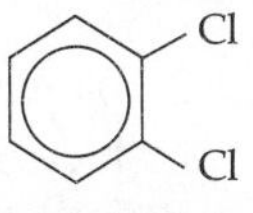 (f) PHF_2 (g) *anti* conformation of CH_2BrCH_2Cl

(h) $CH_2{=}CF_2$ (*i*) 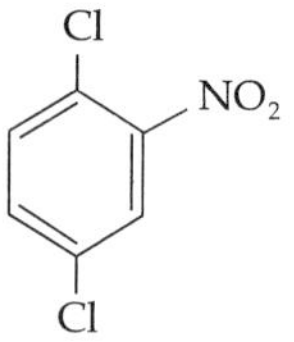

(j) 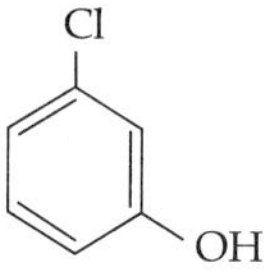

(k) $ClCH{=}CHCH_3$

(*l*) *cis* CHCl=CHCl (m) rapidly rotating CH_2BrCH_2Cl

(n) 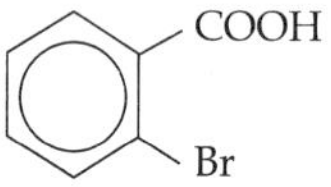

5. The difference between the ^1H chemical shifts of TMS and benzene is 505 Hz when the magnetic field is 1.5 T. What is the value of chemical shift of benzene?

6. The ^1H NMR spectrum of paraldehyde shows a quartet in the intensity ratio of 1:3:3:1 at δ 1.4 ppm of three times the total area. Arrive at the structure of paraldehyde.

7. A certain NMR spectrometer operates at 220 MHz. What magnetic fields are required to bring ^1H, ^{13}C and ^{19}F to resonance given : γ-value is 26.752, 6.728 and 25.18 $\times$ 10^7 rad $T^{-1}s^{-1}$, respectively?

8. Explain whether ^{14}N NMR spectral lines would be sharper in NH_3 or NH_4^+.

9. Predict the ^{31}P and ^{19}F NMR spectrum of PF_5.

10. Explain why the proton–proton coupling constant of CH_2BrCH_2Br varies considerably with temperature.

11. Predict the ^{31}P NMR spectrum of P_4S_3.

12. The spectral lines of an AB system are observed at δ5.9, 5.8, 1.2 and 1.1 ppm as measured from TMS with a spectrometer operating at 100 MHz. Find the chemical shifts of A and B nuclei and the coupling constant J_{AB} in Hz.

13. Explain the effect of substitution by deuterium the hydrogens of the CH_2 group of CH_3CH_2OH on the appearance of the ^1H spectrum.

14. In which of the following systems is the separation between the nuclear spin energy levels the largest?

    (a) protons in a 500 MHz NMR spectrometer and

    (b) deuterons in the same spectrometer

15. What are the relative population differences of ^{13}C spin states between 60 and 600 MHz spectrometer at 25°C?

16. The chemical shift of the CH_3 protons in acetaldehyde is δ2.20 ppm and that of the CHO proton is δ9.80 ppm. Find the difference in the local magnetic fields between the two types of protons of the molecule when the applied field is 1.5 T.

17. Two groups of protons of a molecule become equivalent by isomerisation. At low temperature, when the isomerization is slow, one group has the chemical shift δ 4.0 ppm and the other δ 5.2 ppm. At what rate of interconversion will the two signals overlap in a spectrometer operating at 250 MHz?
18. How can lines due to spin-spin splitting be differentiated from chemical shift lines?
19. What is the Larmor frequency for protons in magnetic fields of 7.05 T? Given $\mu = 2.793$ in units of μ_N.
20. Why is ^{13}C–^{13}C spin-spin splitting usually not observed in organic compounds?
21. Why liquid samples are spun while being examined in an NMR experiment?
22. Which of the following molecules have chemically equivalent sets of nuclei but not magnetically equivalent ones?

    (a) $^{13}CH_3{}^{13}CH_3$ (b) $CH_2{=}CH_2$ (c) $CH_2{=}C{=}CF_2$

    (d) $HC{\equiv}CH$ (e) CH_2FCH_2F (f) CH_2FCHF_2
23. Assume a pentavalent compound H_4B–CH_3 where BH_4 group forms a square pyramid with B at the apex and the –CH_3 group is free to rotate about the B–C bond, predict the first order NMR spectrum.
24. How does the magnetogyric ratio influence the sensitivity of a nucleus in the NMR experiment?
25. The ^{1}H NMR frequency at 1.0 T is 42.4 MHz. The magnetogyric ratio of ^{1}H and ^{13}C are 26.752×10^7 and 6.728×10^7 rad $T^{-1}s$. What will be the ^{13}C NMR frequency at 1.0 T ?
26. Find the number of ^{31}P NMR signals in $PF_2H(^{15}NH_2)_2$.
27. Broad band ^{13}C NMR spectrum of isobutylacetate $(CH_3)_2$–CH–CH_2–OCO-CH_3 shows bands at δ 20(*q*), δ 22(*q*), δ 28(*d*), δ 71(*t*) and δ 171 (s). Multiplicities (*s*, *d*, *t*, *q*) come from off-resonance spectra. Assign the bands.
28. In an NMR spectrometer which operates on 2.50 T magnetic field, Larmor frequency of ^{1}H is 100 MHz. The radiofrequency used has an associated magneic field of 2.50×10^{-4} T. Find the duration of the 90° pulse.
29. Predict the ^{19}F and ^{1}H NMR spectrum of CH_3CHF Br.
30. In a 600 MHz spectrometer, the two CH_3 resonances of N,N-dimethylnitrosamine are separated by 390 Hz. At what rate of interconversion will the two resonances merge into a single line?
31. Predict the number of peaks in the ^{13}C NMR spectrum (in the aromatic region) (a) for o, m and p isomers of dimethoxybenzene, (b) propan-1-ol and propan-2-ol.

32. If the ^1H NMR spectrum of $CHCl_3$ is recorded at high gain, a weak signal is observed on both sides of the ^1H resonance. Account for this observation.

33. The duration of a 90° or 180° pulse depends on the strength of the B_1 field. If a 90° pulse requires 10 μs, what is the strength of the B_1 field? How long would the corresponding 180° pulse require?

34. Describe the ^1H NMR you would expect for $[\eta^5\text{-}(C_5H_5)_2TiCl_2]$ and $[\eta^5\text{-}(C_5H_5)_2Fe(CO)_2]_2$.

35. For the molecule

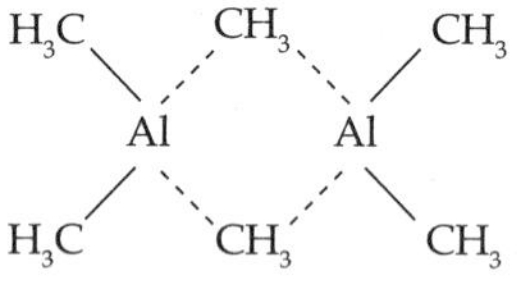

predict the ^1H NMR spectra you would expect over a temperature range.

36. (a) The NMR spectrum of an organic compound of empirical formula $C_5H_{10}O_2$ shows four types of peaks in the intensity ratio 3:2:2:3 protons. It shows a single peak at δ 3.6 and a triplet at δ 0.9 and a triplet at δ 2.2 and a multiplet at δ 1.7. Identify the compound.

    (b) The proton NMR spectrum of a compound of composition $C_4H_7BrO_2$ shows resonances at δ 1.1(t), δ 2.0(q), δ 4.2(t) and at δ 11.0(s) ppm. Identify the compound.

37. The following spectrum is that of an AB type of molecule

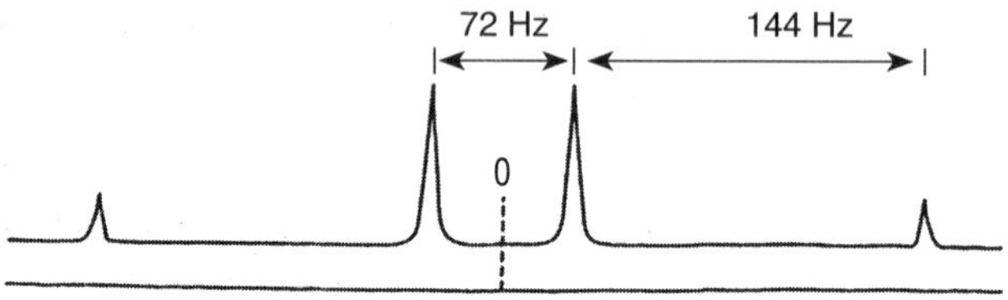

    (a) Obtain the J and $\Delta\upsilon$ values.

    (b) If a 40 MHz probe was used, find the difference in δ between the two peaks.

    (c) If the midpoint 0 is at a δ value of 2.8 ppm, find the δ values for A and B.

38. Predict the ^1H NMR spectrum of $Cd(CH_3)_2$.

39. Consider benzene and cyclohexane,

    (a) in which compound ^1H and ^{13}C chemical shift would be upfield?

    (b) in which $J(^{13}C\text{–}^1H)$ would be larger? Explain.

40. The ^{13}C NMR spectrum of a compound of molecular formula $C_7H_6O_2$ showed the signals at δ 117(d), δ 130(s), δ 133(d), δ 164(s) and δ 191(d). Deduce the structure.

41. Assuming J_{P-H} coupling to be unresolved, explain the expected low temperature and high temperature ^{31}P and ^{19}F NMR spectrum for $(H_3C)_2N-PF_4$.

42. Sketch the expected ^{31}P NMR spectrum of HPF_2 if

(a) $J_{P-F} > J_{P-H}$

(b) $J_{P-H} > J_{P-F}$

In most compounds, $^1J_{P-F}$ is about ~ 1500 Hz and $^1J_{P-H}$ is ~ 200 Hz.

43. 60 MHz proton NMR spectrum of an AB system is shown below. Find the chemical shifts and the coupling constant and the intensity of the lines.

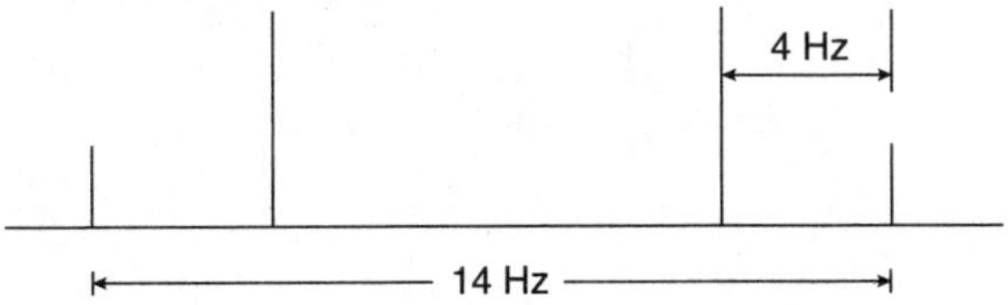

44. Give the spectral types which the protons in the following compounds exhibit.

(a) *cis*-cyclopropane-1,2-dicarboxylic acid,

(b) *trans*-cyclopropane-1,2-dicarboxylic acid

(c) 1,4-dichlorobenzene (d) 2-chlorophenol and (e) 4-chlorophenol

45. Classify the spectra of the ^{13}C satellites of the following molecules.

(a) 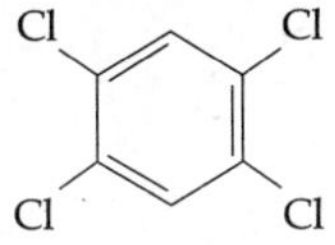

(b) 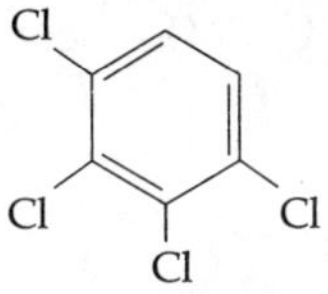

(c) $^{13}CHCl_2-^{12}CH_2Cl$ (d) $^{13}CH_2Cl-^{12}CH_2Cl$ (e) $H^{13}C\equiv{}^{13}CH$

46. The ^1H NMR spectrum of a mixture of acetone, chloroform and methylene chloride is integrated and results in step heights of 10, 18 and 36 mm for the signals at δ 7.27, δ 2.17 and δ 5.30, respectively. In what molar ratio are the three samples present?

47. Describe the expected proton NMR spectrum of Si_2H_6.

48. The ^1H NMR spectrum of an organic compound with molecular formula C_8H_{10} shows signals at δ 7 ppm (singlet), δ 2.4 ppm (quartet) and δ 1.2 ppm (triplet) with peak intensity ratio of 5:2:3. Identify the compound.

49. Predict the proton NMR spectrum of $Sn(CH_3)_4$.

50. At room temperature, the 200 MHz ^1H spectrum of N, N–dimethyl amide ($(CH_3)_2NCOR$) shows two sharp methyl signals at δ 3.05 and 2.85 ppm.

On heating the solution, the two signals broaden and merge at 84 °C, above which only a single signal is observed. Explain how would the spectrum appear if it was measured on a 500 MHz instrument at 84 °C.

51. The actual chemical shifts for the four lines of a 60 MHz spectrum are δ 6.13, 6.29, 7.28 and 7.74 pm. Calculate J, $v_0\delta/J$ and the chemical shifts and the intensity of the spectral lines. (b) How can one demonstrate that the NMR signals do not form a first order spectrum?

52. The proton NMR spectrum of HD shows lines separated by 45 Hz. Give the number of lines, and their relative intensities for the HD proton and the HD deutron NMR spectra. Obtain the proton–proton coupling constant.

53. The NMR spectrum of vinyl fluoride shows the following coupling constants. Predict the ^{19}F NMR spectrum.

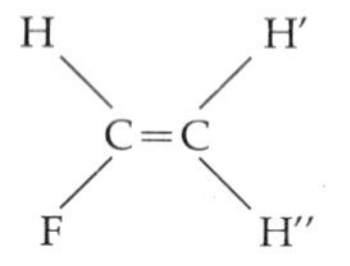

J_{HF} 85 Hz

$J_{H'F}$ 50 Hz

$J_{H''F}$ 20 Hz

54. Predict the ^{19}F NMR spectra one would observe for (a) GeF_4, (b) TeF_4, (c) NF_3, (d) ClF_3 and (e) SF_4.

55. The ^1H NMR spectrum taken on a 500 MHz spectrometer shows a quartet are 1759, 1753, 1747 and 1741 Hz. Find the coupling constant and the chemical shift.

56. The proton decoupled ^{13}C NMR spectrum of the compound

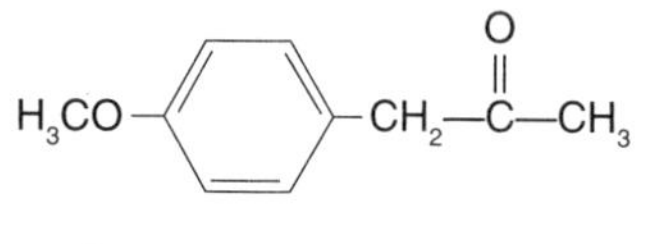

exhibits resonances at δ 29.1, 50.4, 55.0, 114.2, 126.3, 130.1, 159.2 and 207.0 ppm. The DEPT 135° spectrum shows positive signals at δ 29.1, 55.0, 114.2 and 130.1 ppm and a negative peak at δ 50.4 ppm. Assign the bands.

57. The ^{13}C NMR spectrum of a compound of molecular formula $C_7H_6O_2$ showed the following signals: δ 117(d), δ 130(s), δ 133(d), δ 164(s) and δ 191(d). The relative intensities observed were nearly 8:1:8:1:2 respectively. Deduce the structure of the compound.

58. Sketch the $^{13}C\{^1H\}$ NMR spectra of (a) $Fe(CO)_5$ and $Fe_2(CO)_9$ without considering the values of chemical shifts and coupling constant.

59. Predict the ^1H, ^{19}F and ^{31}P NMR spectra $Me_2P–CF_2Me$.

60. Predict the ^{19}F NMR spectrum of AsF_5Cl.

61. Two compounds of the molecular formula C_4H_8O exhibit the following proton decoupled ^{13}C spectral bands. Identify the compound.

Sample A: δ 13.3, 15.7, 45.7 and 201.6 ppm

Sample B: δ 7.9, 29.4, 36.8 and 209.5 ppm

62. A compound was known to be one of the following:

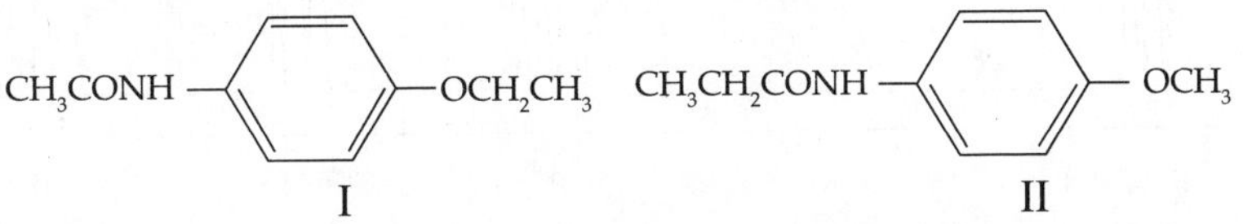

Its carbon–13 NMR spectrum showed peaks as listed below and the multiplicities obtained from off-resonance decoupled spectrum are given in the parenthesis: δ 15(q), 24(q), 64(t), 115(d), 123(s), 157(s) and 171(s). Which isomer is it?

63. Assign the following ^1H spectra and identify the compounds.

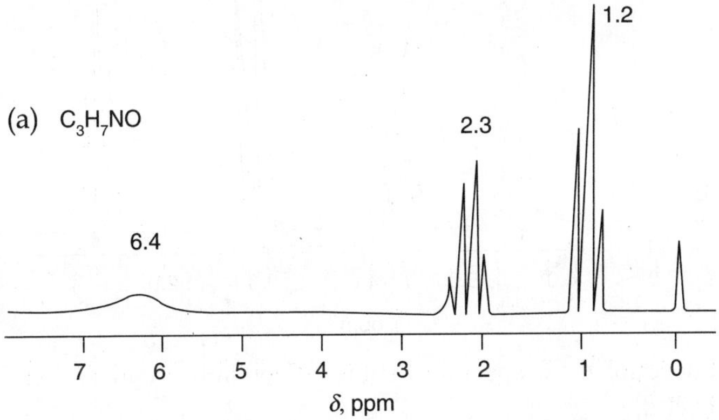

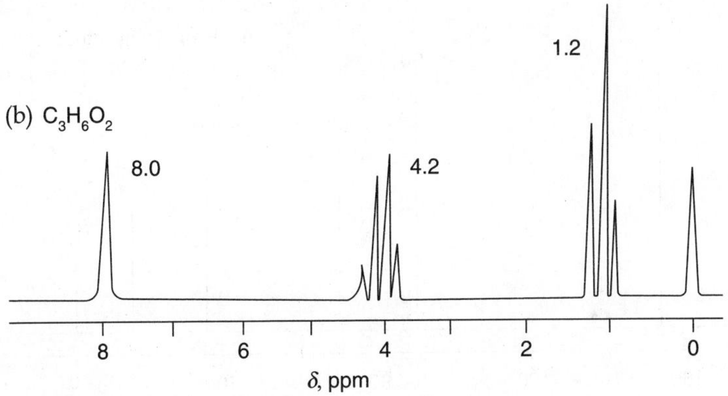

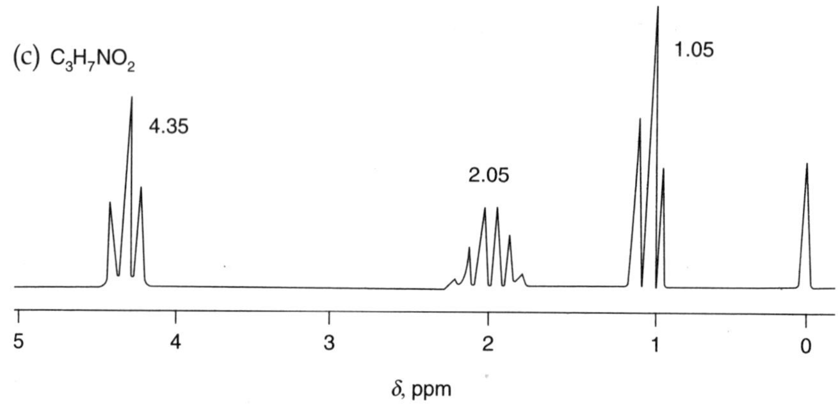

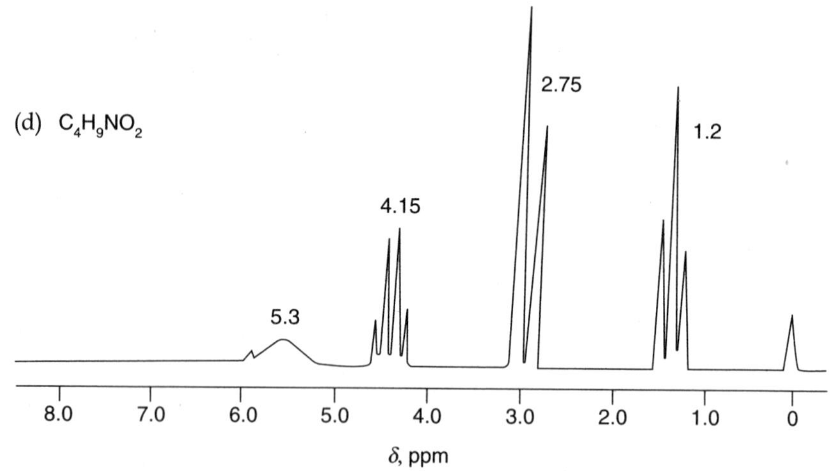

64. Proton decoupled ^{13}C spectra of some compounds are given below. Assign their spectra.

(a)

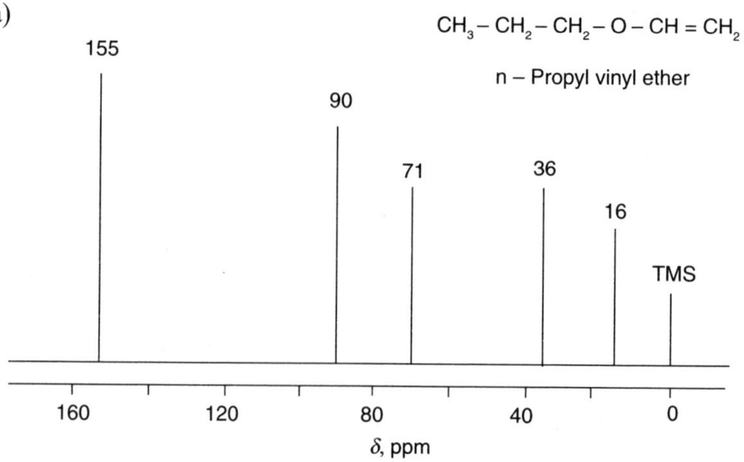

(b)

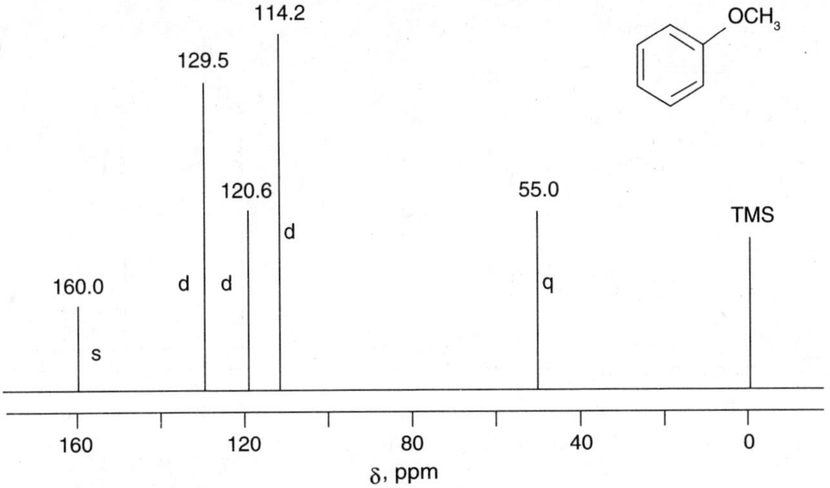
OCH$_3$
114.2
129.5
120.6
55.0
TMS
160.0
d
d
d
d
q
s
160
120
80
40
0
δ, ppm

(c)

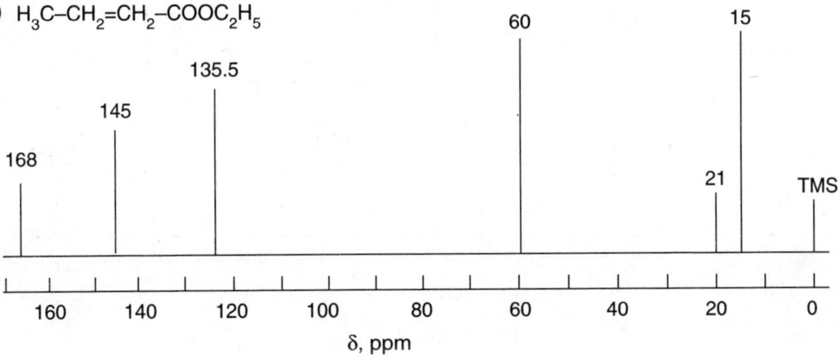
H$_3$C–CH$_2$=CH$_2$–COOC$_2$H$_5$
60
15
135.5
145
168
21
TMS
160
140
120
100
80
60
40
20
0
δ, ppm

(d)

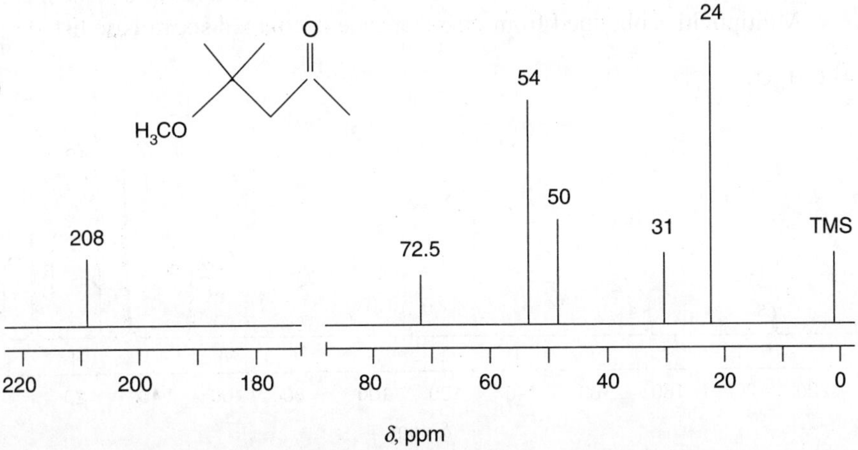
O
H$_3$CO
24
54
50
31
TMS
208
72.5
220
200
180
80
60
40
20
0
δ, ppm

(e)

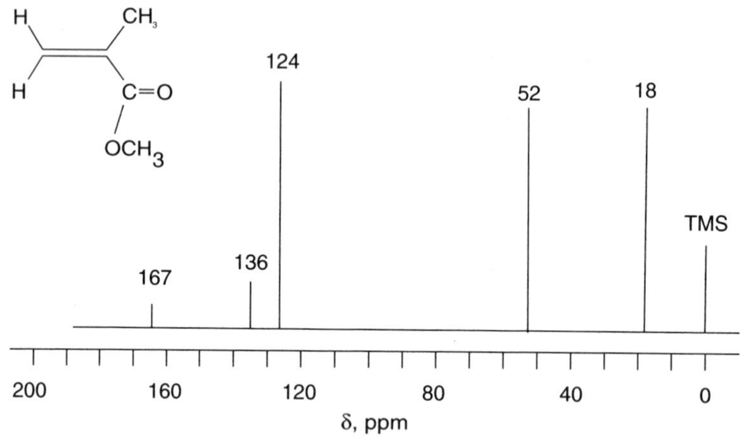

(f)

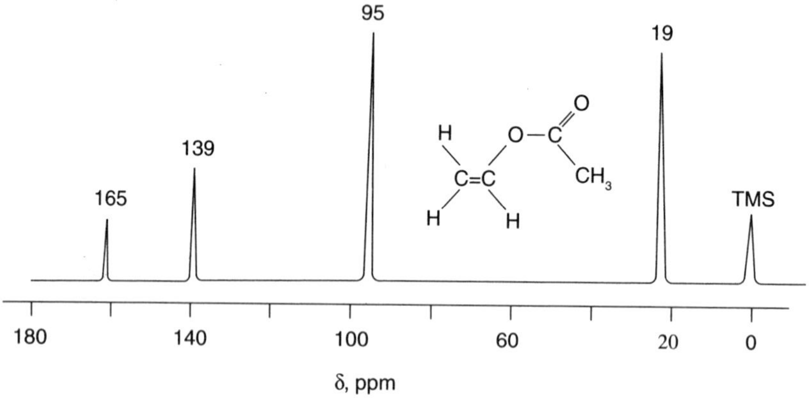

65. Identify the following compounds from the proton decoupled ^{13}C spectra. Multiplicities obtained from off-resonance decoupled spectra are listed.

(a) $C_7H_{12}O_3$

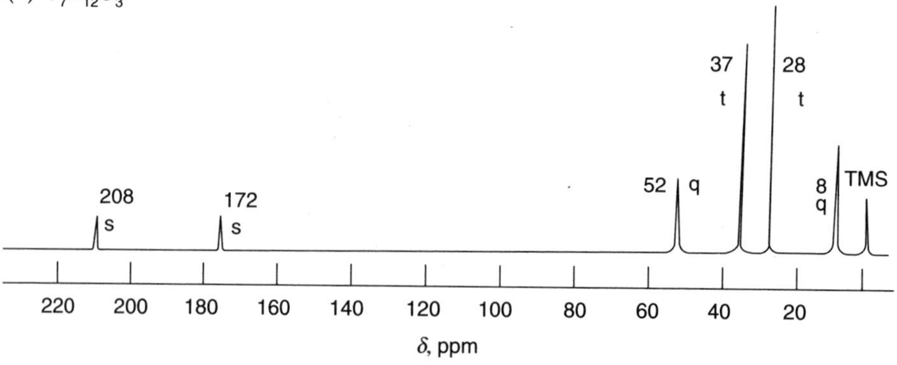

(b) $C_7H_{14}O_2$

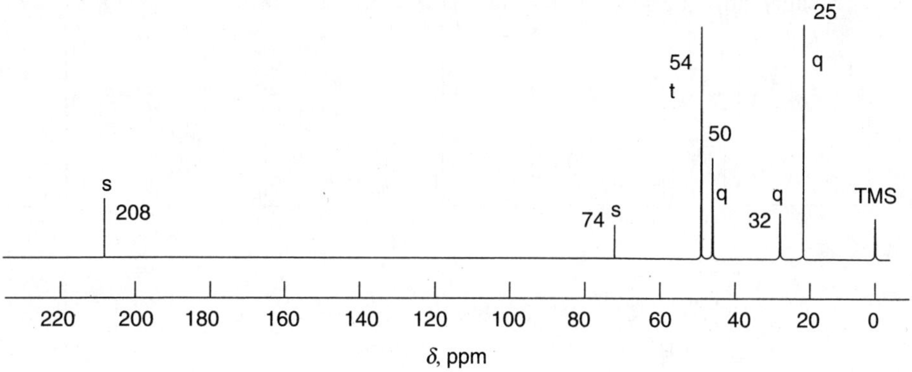

(c) $C_6H_{10}O_2$

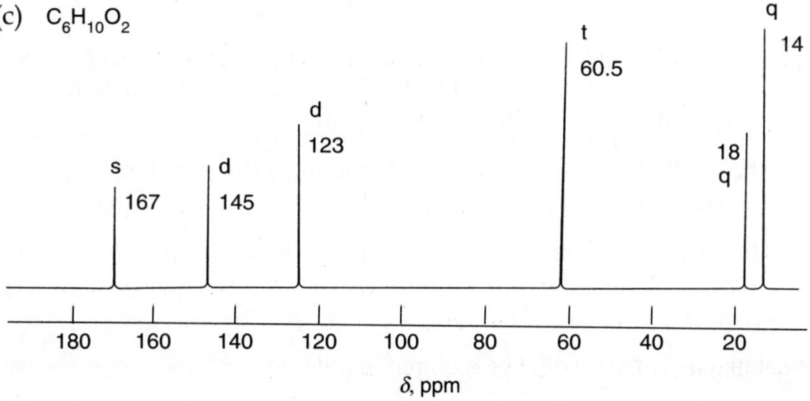

(d) $C_6H_{12}O_2$

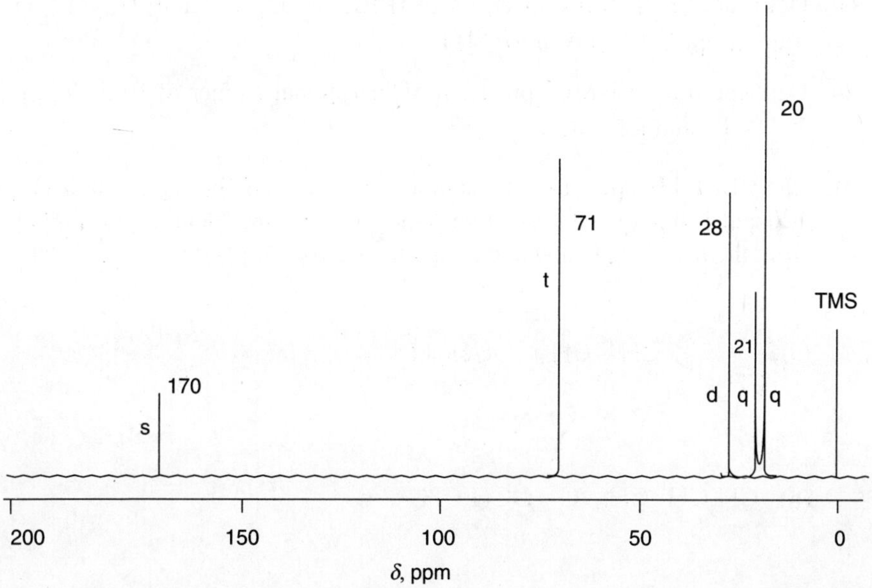

66. Assign the spectra shown below to three structural fragments given below by analysing the scalar coupling.

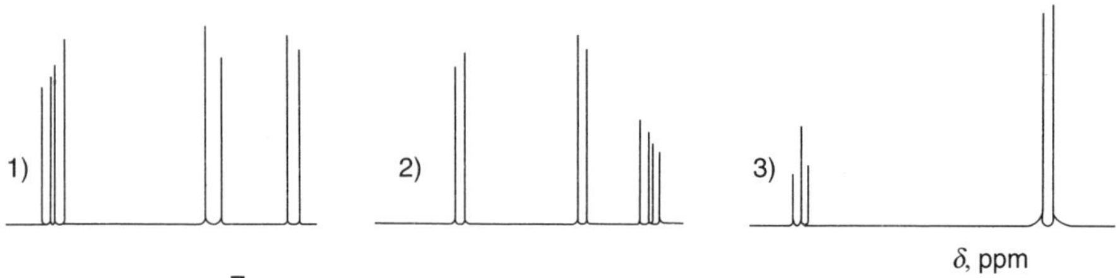

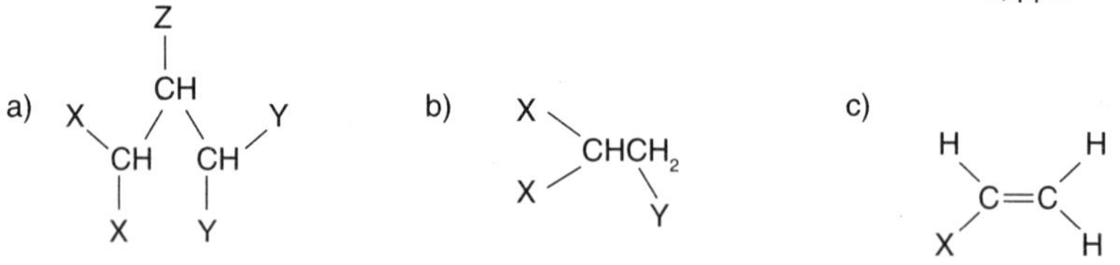

67. (a) The NMR spectrum of a compound $C_8H_{11}N$ showed resonances at δ 1.4 (d), δ 1.6 (s), δ 4.1 (s) and δ 7.3 (s). Derive the structure of the compound.

(b) The NMR spectrum of a compound of formula C_7H_8O exhibits signals at δ 2.6 (s), δ 4.7 (s) and δ 6.7 (m). Predict the structure of the compound.

(c) The NMR spectrum of a compound of molecular formula $C_9H_{10}O_2$ has signals at δ 2.6 (s, 3H), δ 3.9 (s, 3H), δ 6.8 (d, 2H) and δ 7.2 (d, 2H). Predict the structure of the compound.

68. What magnetic field would be required in order to use

(a) an ESR X-band (9 GHz) to observe ^{1}H NMR and

(b) 300 MHz NMR spectrometer to observe ESR transitions?

69. Describe the ^{31}P NMR spectrum of $H_3P.BCl_3$ ($J_{PH}>J_{PB}$) and ^{1}H NMR spectrum of Ph_3 $P.BH_3$ (given for ^{11}B I = 3/2).

70. Describe the ^{19}F NMR sprectrum of meridional isomer of RhF_3Cl_3, given $J_{Rh\text{-}F} > J_{F\text{-}F}$ and for ^{103}Rh I = 1/2.

71. The DEPT 135 spectrum of a compound shows positive signals at δ 11, 20, 120 and 140 ppm and a negative signal at δ 61 ppm. Identify the compound from the list given below and assign the bands.

(a) 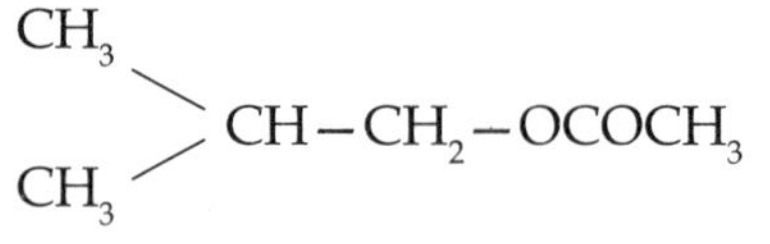

(b) 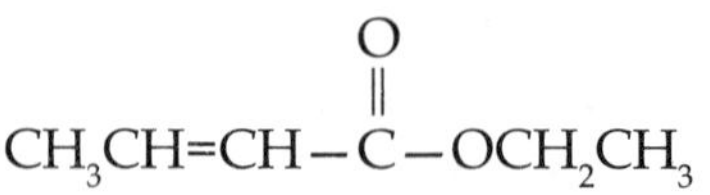

(c) $CH_3-CH_2-CH(CH_3)-OCOCH_3$

72. The DEPT 135 spectrum of a compound shows negative peaks at δ 30.2, 31.9, 61.8 and 114.7 ppm and a positive peak at δ 130 ppm. Identify the compound among the following:

(a) $H_2C{=}CH{-}CH_2{-}CH_2{-}CH_2{-}OH$

(b) $CH_2{=}CH-CH_2-O-CH_2-CH_3$

(c) $H_3C{-}CH_2{-}CH{=}CH{-}CH_2{-}OH$

(d) $CH_2-C(=CH_2)-CH_2CH_2-OH$

73. Which of the following compounds show AA′XX′ type of ^1H NMR spectrum?

(a) 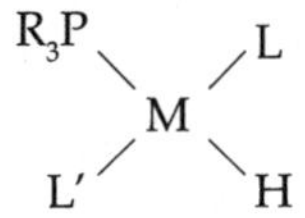

(b) 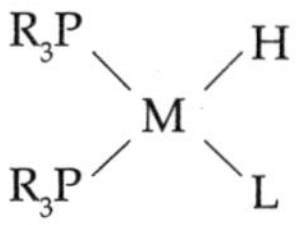

(c) 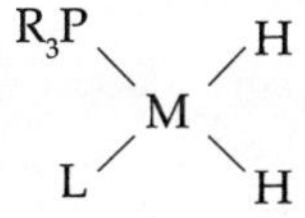

(d) 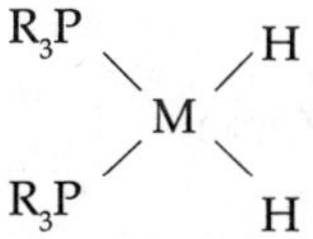

74. The ^1H NMR spectrum (400 MHz) shows resonances at δ 7.38 (d), 7.25 (d) and 1.23 (s) ppm. Select the compound among the choices given.

(a) 

(b) 

(c) 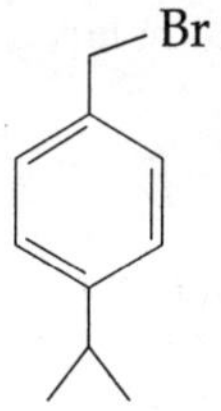

(d) 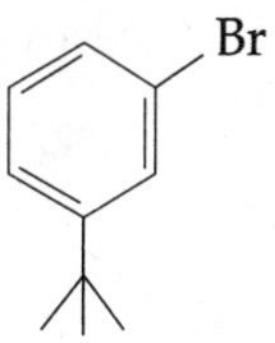

75. A compound shows the following ^1H and ^{13}C NMR spectral characteristics. Identify the compound among the choices given below.

(i) ^1HNMR : δ 2.5 (s, 3H), 3.8 (s, 3H), 6.9 (d, 2H J 8 Hz), 7.8 (d, 2H J 8 Hz), ppm

^{13}C NMR: δ 197, 165, 130, 129, 114, 56, 26 ppm.

(a)

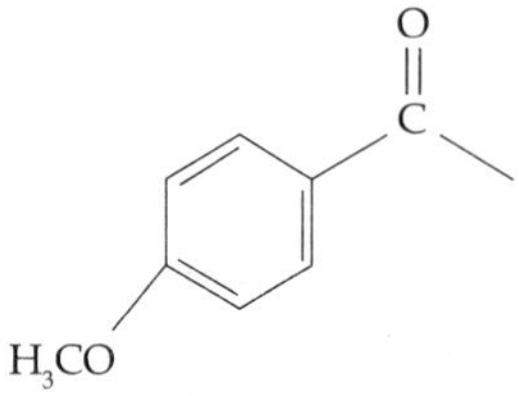

(b)

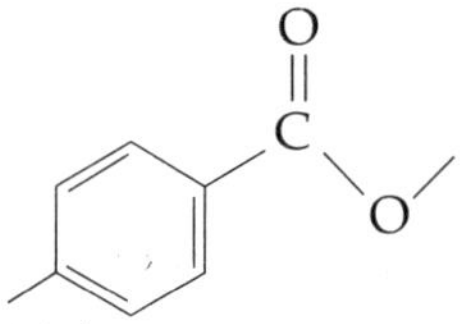

(c)

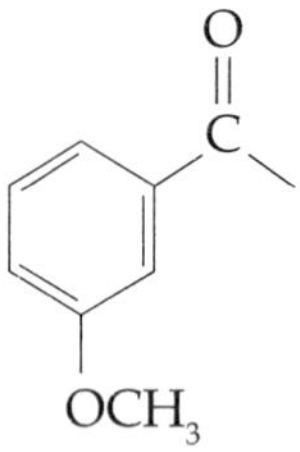

(d)

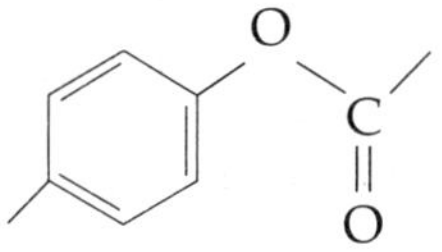

(ii) ^{1}H NMR: δ 3.59 (s, 3H), 3.32 (t, 2H), 2.25 (t, 2H) 1.85–1.75 (m, 2H), 1.73–1.62 (m, 2H) ppm

^{13}C NMR: δ 174.0, 51.0, 32.9, 32.8, 31.0, 23.0 ppm.

(a)

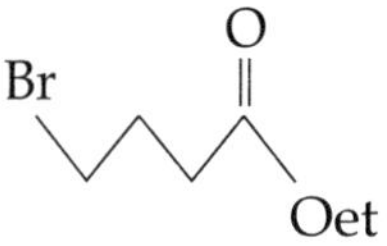

(b)

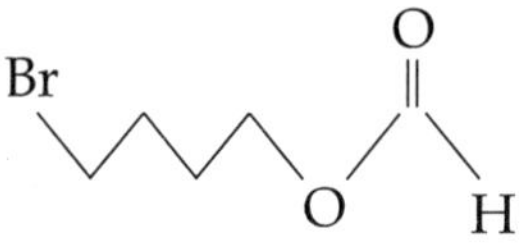

(c)

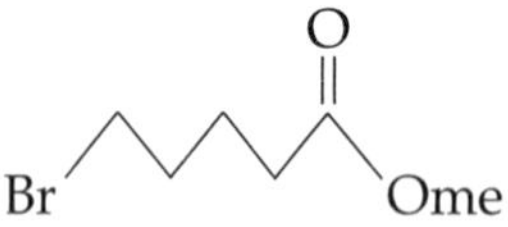

(d)

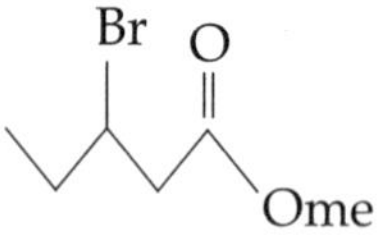

(iii) ^{1}H NMR: δ 1.30 (3H, t, J 7.2 Hz), 2.41 (2H, q, J 7.2 Hz), 2.32 (3H, s), 7.44 (1 H, t, J 7.0 Hz) 7.52 (1H, dt 7.0, J 3.0 Hz), 7.77 (1H, t, J 3.0 Hz), 7.9 (1H, dt 7.0, J 3.0 Hz) ppm.

^{13}C NMR: Selected bands: δ 9.0, 21.1, 35, 128.5, 136, 137.5, 198 ppm.

(a)

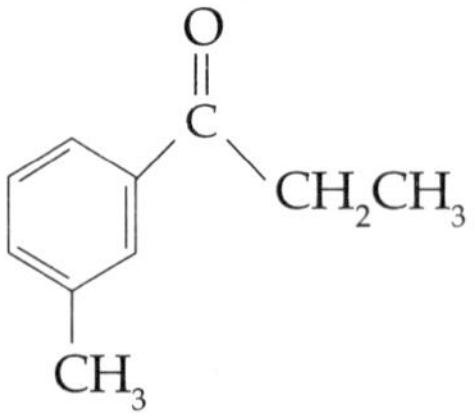

(b)

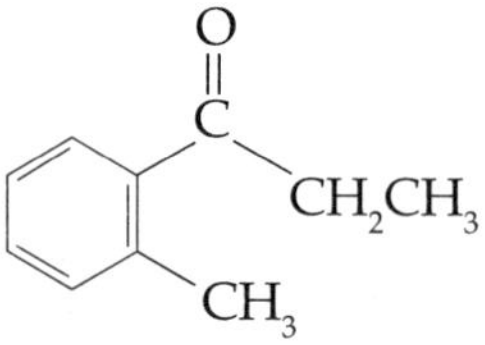

(c)

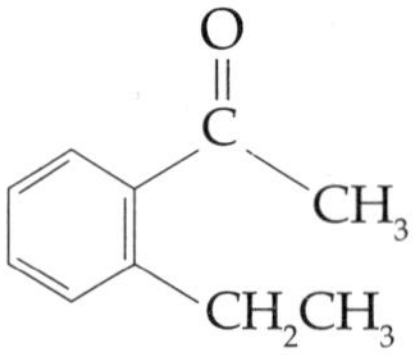

(d)

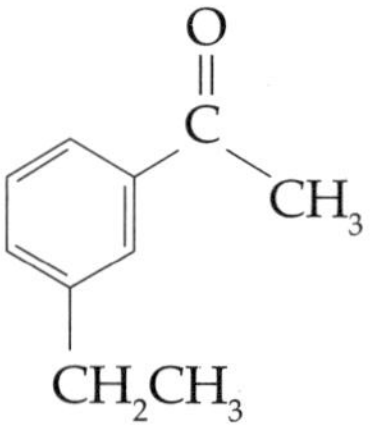

(iv) ^{1}H NMR: δ 8.0 (d, 1H, J 12.3 Hz), 7.7(d, 2H, J 8 Hz), 6.8 (d, 2H J 8.0 Hz), 5.8 (d, 1H, J 12.3 Hz), 3.8(s, 3H), 3.0 (s, 6H) ppm.

(a)

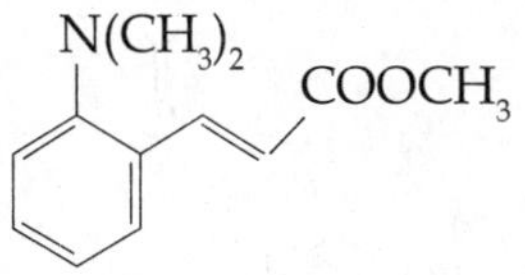

(b)

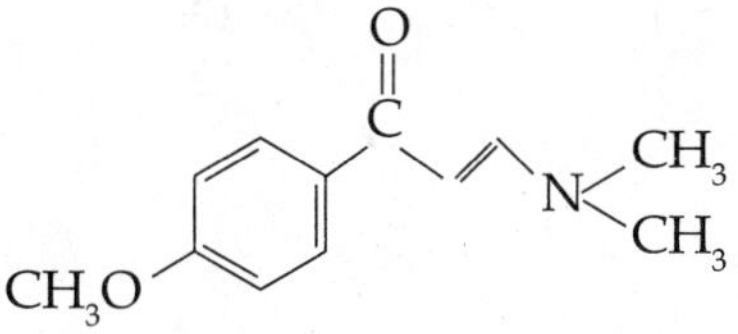

(c)

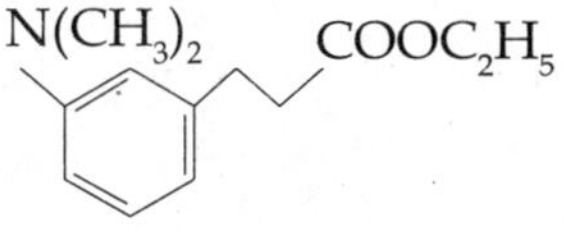

(d)

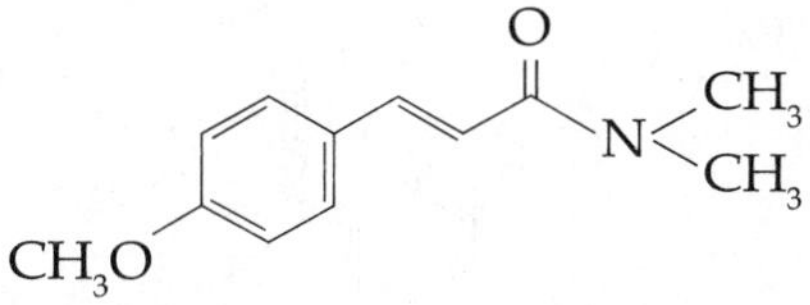

(v) ^{1}H NMR δ 7.9 (d, 2H, J 8 Hz), 6.6 (d, 2H, J 8 Hz), 4.3 (q, 2H, J 6 Hz), 4.0 (br, s, 2H, D_2O exchangeable) 1.4 (t, 3H, J 6 Hz) ppm.

(a)

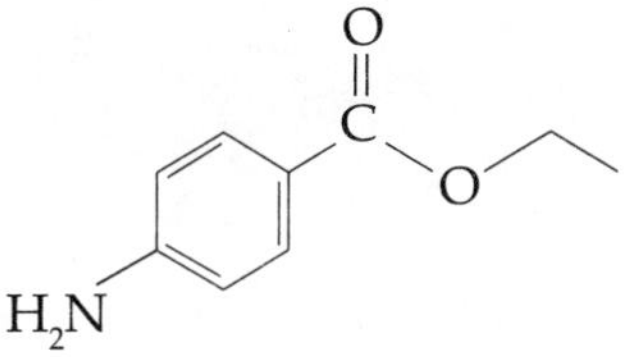

(b)

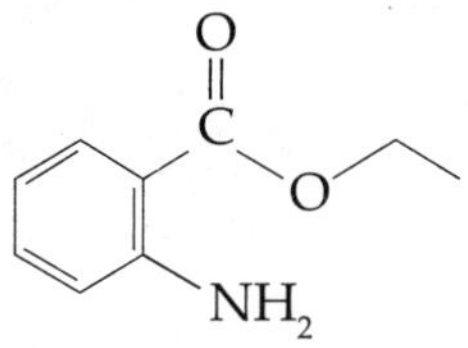

(c)

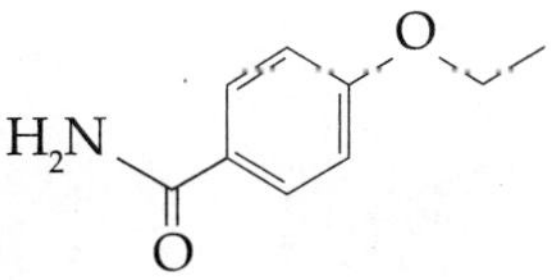

(d)

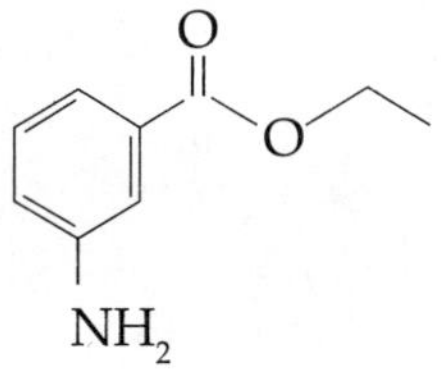

(vi) ^{1}H MNR : δ 7.8 (d, 2H, J 8 Hz), 6.8(d, 2H, J 8 Hz) 2.4 (s, 3H), 1.25 (t, 3H, J 7.2 Hz) 4.1(q, 2H, J 7.2 Hz) ppm.

(a)

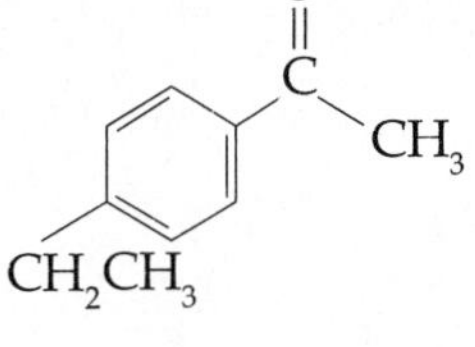

(b)

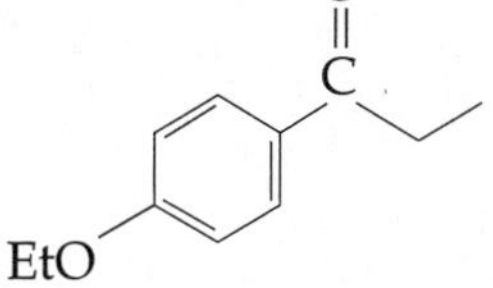

(c)

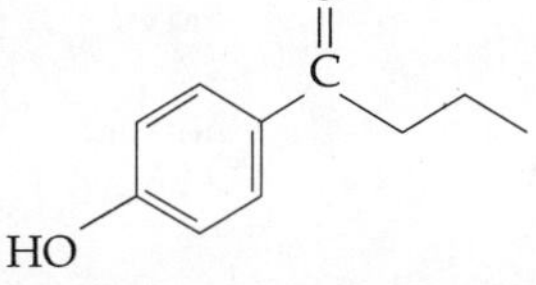

(d)

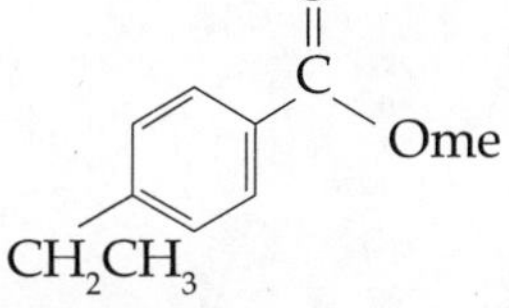

(vii) ^{1}H NMR: δ 7.8 (2H, J 8 Hz), 6.8 (2H, d, J 8 Hz), 4.1 (2H, q, J 7 Hz), 2.4 (3H, s), 1.25 (3H, t, J 7.2 Hz) ppm.

(a)

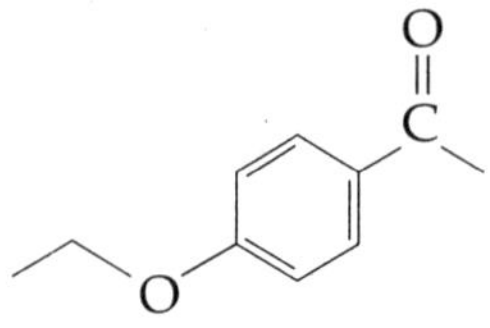

(b)

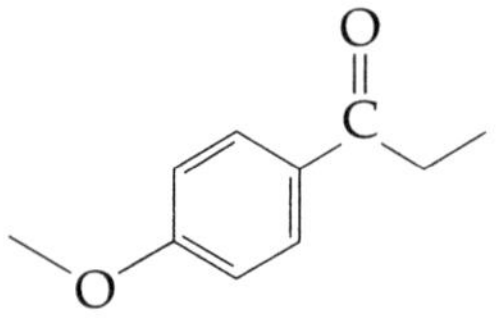

(c)

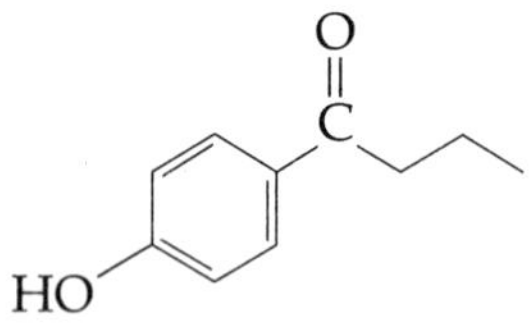

(d)

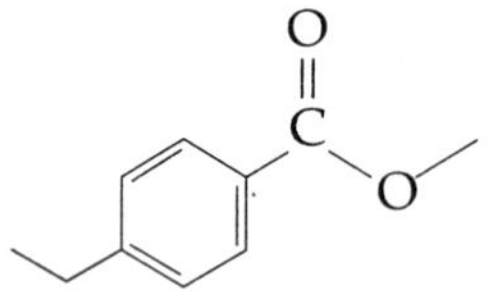

Part Three

Nuclear Quadrupole Resonance

NUCLEAR QUADRUPOLE RESONANCE SPECTROSCOPY 20

INTRODUCTION

Nuclear quadrupole resonance (NQR) spectroscopy is an analytical method which is useful to investigate solid materials. It is applicable to a wider range of nuclei. However, NQR spectroscopy is not routinely used because the range of problems to which it can be applied is limited and also because the necessary instrument is not easily available.

A dipole occurs when two different point charges are separated. An electric quadrupole consists of two dipoles with dipole moments that are equal in magnitude but opposite in direction. Thus a quadrupole has neither a net charge nor a dipole moment as for example CO_2.

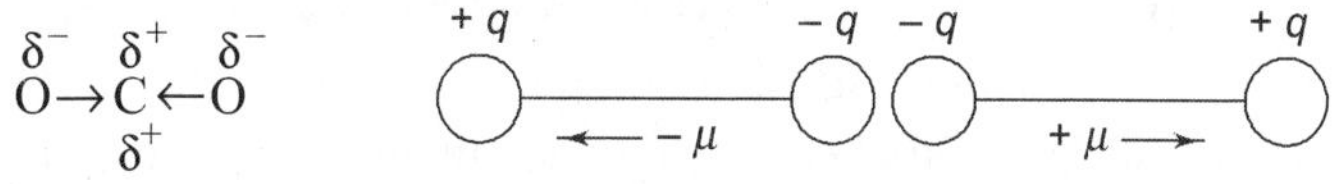

If the nuclear spin quantum number I is zero, the distribution of positive charges will be spherical, when the nucleus is said to possess a zero nuclear quadrupole moment eQ where e is the charge of the electron and Q, the quadrupole moment is a measure of the deviation of charge distribution from spherical symmetry. A nucleus with $I = 1/2$ is represented in Fig. 20.1a. Here too the charge distribution is spherical i.e. eQ is zero. All nuclei whose spin is greater than 1/2 possess an electric quadrupole moment. It arises from the nonspherical nature of the nuclei (Figs. 20.1b and c). Non-uniform charge distribution can be represented by an ellipsoid. A positive value of Q indicates that the protonic charge is oriented along the direction of the principal axis (prolate) while a negative value of Q indicates charge distribution perpendicular to the principal axis (oblate). Some common NQR isotopes are given in Table 20.1.

Table 20.1 Common NQR isotopes

Isotope	*Natural abundance*	*I*	*eQ**	*QCC/MHz*
^{14}N	99.6	1	1.6×10^{-2}	– 10
^{35}Cl	75.5	3/2	-1×10^{-1}	110
^{37}Cl	24.5	3/2	-7.9×10^{-2}	86.7
^{79}Br	50.5	3/2	3.7×10^{-1}	770
^{81}Br	49.5	3/2	3.1×10^{-1}	645
^{127}I	100.0	5/2	-7.9×10^{-1}	–2293

*Quadrupole moment 10^{-28} m^2; QCC: Quadrupole coupling constant

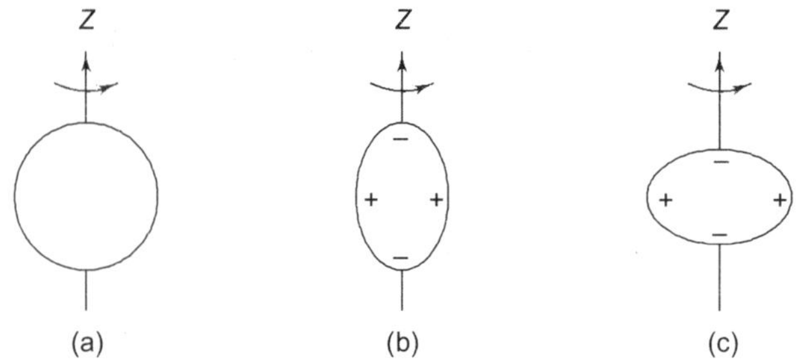

Fig. 20.1 Representation of nuclei: (a) $I = 1/2$, $eQ = 0$, (b) $I = 1$, $eQ > 0$ and (c) $I = 1$, $eQ < 0$.

A quadrupole moment will orient in a linear field gradient just as a dipole orients in a linear field. The electric field gradient (efg) is produced at the nucleus by the unsymmetrical distribution of charges arising from the extranuclear electrons, nonbonding or bonding electrons in the molecule of which the nuclei are a part. An electric field gradient (efg) is generated if for example the molecule is composed of atoms of varying electronegativity such as PCl_3, phosphorus and chloride carrying positive and negative charges respectively. The efg will interact with the nuclear quadrupole moment resulting in a splitting of the otherwise degenerate quadrupole states. The nuclear quadrupole energy levels are fixed molecular properties and NQR spectroscopy deals with the transitions between these energy levels. The magnitude of the transition energy is a measure of the electric field gradient. The transitions, which are magnetic dipole type, are induced by the interaction between the quadrupole moment of the nucleus and the magnetic vector of the incident electromagnetic radiation in the radiofrequency region.

The first evidence that certain nuclei possess electric quadrupole moment came from a study of the hyperfine structure in the atomic spectra of europium by Schuler and Schmidt. Transitions between nuclear quadrupole energy levels are generally observed in solids. However, molecular beam resonance experiments have revealed that pure quadrupole transitions are detectable in gases at low pressures, where the time between molecular collisions is of the order of the reciprocal of the natural line widths. The information on quadrupole energy levels may also be obtained from the fine structure in the pure rotational spectra of molecules in the gas phase. The different nuclear quadrupolar orientations give slightly different moment of inertia resulting in the fine structure in the microwave spectrum. For a detailed account of NQR spectroscopy, the standard books by Das and Hahn, and Lucken should be consulted.

NUCLEAR QUADRUPOLE MOMENT

When a nucleus with an electric quadrupole moment is surrounded by a low symmetry (lower than octahedral or tetrahedral) electric field resulting from the asymmetric electron distribution, the quadrupolar nucleus will interact with this electric field gradient and the extent of interaction will differ for different orientations of the elliptical quadrupolar nucleus. Each orientation corresponds to a different quantized energy. The total spread of energies is very small, considerably less than kT. The lowest energy quadrupole level corresponds to the orientation in which the greatest amount of positive nuclear charge is closest to the greatest density of negative charge in the electron environment. The energy differences between the different orientations are small and at room temperature in a group of molecules all the orientations are largely populated but to different extents. Transitions may therefore take place from any level subject to the selection rules. The intensity of the signal is dependent on the difference in population between the two levels involved in the transition which increases with the lowering of temperature.

Different values of the nuclear-spin quantum number m_I correspond to different orientations of the nuclear quadrupole moment in an inhomogeneous electric field of the electrons and other nuclei of the molecule and therefore correspond to different energies. In NQR spectroscopy, one observes transitions between these levels when appropriate radiation in the radiofrequency region 0.1 to 1000 MHz is employed but it is not necessary to scan the whole range. Unlike NMR and ESR, one does not apply an external magnetic field; the internal molecular electric field serves to produce the energy levels. The intensity of NQR spectral peak is maximum when the radiofrequency field is perpendicular to the symmetry axis and vanishes when it is parallel. Thus a study of the dependence of the intensity of the spectral line on the direction of the radiofrequency field with respect to the crystal axis helps to determine the axis of symmetry of the electric field gradient.

Direct measurement of the absorption of the radiofrequency radiation is made on solids because in a gas or liquid, molecular tumbling averages out the electric field gradient to zero and hence no splitting of the quadrupole energy levels occur.

A charged body such as a nucleus causes an electric field in the space around it. The electric field $\vec{E}$ at a particular nucleus is determined by the electric fields produced by the other nuclei. Electric potential V, a scalar quantity, is the potential energy per unit charge at any point in an electric field. It is usually simply called potential. Electric potential has meaning only for electric fields that are produced by static charges.

Potential V at any point in an electric field is defined as the potential energy per unit charge associated with a test charge q_0 at that point. The electric field and the potential are closely related. If we know the potential at various points, we can use it to determine $\vec{E}$. Regarding V as a function of the coordinates, (x, y, z) of a point in space we can show that the components of $\vec{E}$ are directly related to the partial derivatives of V with respect to x, y and z. That is, the component of $\vec{E}$ in any direction is the negative of the rate of change of the potential with distance in that direction. So we have,

$$E_x = -\frac{\partial V}{\partial x}, \quad E_y = -\frac{\partial V}{\partial y}, \quad E_z = -\frac{\partial V}{\partial z} \tag{20.1}$$

These are the components of $\vec{E}$ in terms of V. In terms of the unit vectors, we can write,

$$\vec{E} = -\left(\vec{i}\frac{\partial V}{\partial x} + \vec{j}\frac{\partial V}{\partial y} + \vec{k}\frac{\partial V}{\partial z}\right) \tag{20.2}$$

In vector notation, equation (20.2) is written compactly as

$$\vec{E} = -\text{ grad. V} \tag{20.3}$$

Thus $\vec{E}$ is the negative of the gradient V. At each point, the gradient of V points in the direction in which V increases rapidly with change in position. Hence at each point, the direction of $\vec{E}$ is the direction in which V decreases rapidly and is always perpendicular to the equipotential surface through the points. This is consistent with the observation that moving in the direction of the electric field means moving in the direction of decreasing potential.

Equation (20.2) does not depend on the particulars of the zero point for V. If we were to change the zero point, the effect would be to change V at every point by the same amount, the derivatives of V would be the same. We can compute the electric field caused by the charge distribution by first calculating the potential and then taking the gradient to find the field. This method is easier since the potential is a scalar quantity requiring the integration of a scalar function. If we know $\vec{E}$ as a function of position, we can calculate V (by integration) and if we know V as a function of position, we can calculate $\vec{E}$ (by differentiation).

ELECTRIC FIELD GRADIENT

A dipole in the field of an electric charge is shown in Fig. 20.2 (a). The energy associated with the orientation of the dipole depends upon how the electrostatic potential changes over the dipole, $\partial V/\partial Z$. Consider now the interaction of the quadrupole with an electric field along the z-axis as shown in Fig. 20.2(b). It may be viewed as an electric field of the electron interacting with two dipoles whose orientation relative to each other is fixed. The energy of interaction will depend upon the rate of change (i.e. gradient) of the electric field over the quadrupole. This is the second derivative of the electric potential V with respect to z, that is, $\partial^2 V/\partial z^2$ known as the electric field gradient denoted as q or eq. It denotes the change in the electric field component $\partial E_z/\partial z$. The asymmetry of the electric field is measured by the electric field gradient (efg) in the z-direction (where z has been chosen as the direction of maximum gradient); it is the negative of the second derivative of the potential:

$$\text{efg} = eq_{zz} = V_{zz} = -\frac{\partial^2 V}{\partial z^2} \tag{20.4}$$

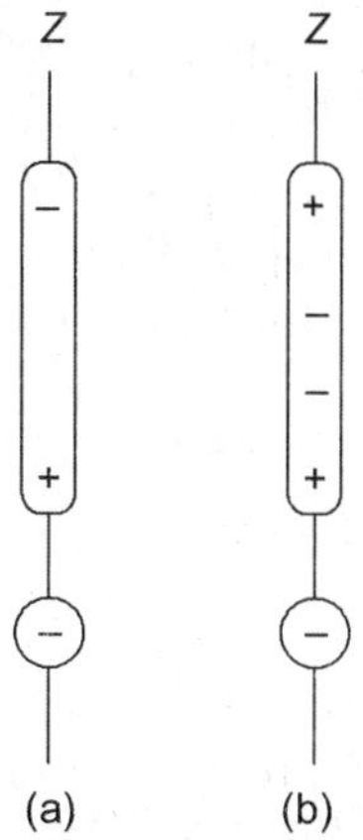

Fig. 20. 2 The interaction of (a) a dipole and (b) a quadrupole with the Z-component of the electric field of a unit negative charge.

In general, the efg for a system is described by a tensor having nine components like the *g* tensor in ESR spectroscopy. The components are defined as follows:

$$q_{xx} = \frac{-\partial E_x}{\partial x} = \frac{-\partial^2 V}{\partial x^2};\quad q_{xy} = \frac{-\partial E_x}{\partial y} = \frac{-\partial^2 V}{\partial x \partial y};\quad q_{zz} = \frac{-\partial E_x}{\partial z} = \frac{-\partial^2 V}{\partial z^2}, \text{ etc.} \tag{20.5}$$

It can be represented by a square symmetric matrix since the order of partial differentiation is immaterial, i.e. $q_{ij} = q_{ji}$. The negative sign comes from the definition of the electric field, which is the negative of the potential gradient (see equation (20.3)).

$$\text{efg} = -\begin{pmatrix} q_{xx} & q_{xy} & q_{xz} \\ q_{xy} & q_{yy} & q_{yz} \\ q_{xz} & q_{yz} & q_{zz} \end{pmatrix} \tag{20.6}$$

It is desirable to select a Cartesian coordinate system in which the nondiagonal terms q_{xy}, q_{xz}, etc. become zero. This type of coordinate system is known as the principal axes of the efg tensor and the principal components of efg then become q_{xx}, q_{yy} and q_{zz}. As noted from equation (20.7), then only two parameters are needed to specify the efg, since diagonal elements of the tensor will obey the Laplace rule, a square matrix can be diagonalized.

$$q_{xx} + q_{yy} + q_{zz} = 0 \tag{20.7}$$

By convention, q_{zz} refers to the largest field gradient, q_{yy} the next largest and q_{xx} to the smallest, that is, $|q_{zz}| \geq |q_{yy}| \geq |q_{xx}$. If $q_{zz} = q_{xx} = q_{yy}$, the field gradient is said to be spherical and this leads to degenerate quadrupole levels. When $q_{zz} \neq q_{xx} = q_{yy}$, the molecule is said to have axial symmetry about the z-axis. The NQR spectra of the molecules of this type are the easiest to interpret. By convention, for a molecule with axial symmetry q_{zz} is denoted as q and the maximum field gradient lies along the z-axis (i.e. the highest-fold symmetry axis). For example, the molecules Cl_2 and CH_3Cl are axially symmetric. The field gradient is said to be nonsymmetric if $q_{zz} \neq q_{yy} \neq q_{xx}$.

THE ASYMMETRY PARAMETER

As noted above when the electron density is the same in the x– and y– directions, it is commonly described as axial symmetry; it occurs when the principal axis of symmetry has an order greater than two. If this is not the case, an additional parameter known as the asymmetry parameter η is required. It is a measure the departure of the field gradient from axial symmetry when $q_{xx} \neq q_{yy} \neq q_{zz}$ and it is given by

$$\eta = \frac{q_{xx} - q_{yy}}{q_{zz}} \tag{20.8}$$

Since the sum of q_{xx}, q_{yy} and q_{zz} is zero, η is never greater than unity. Generally η varies from 0 to 1; $\eta = 0$ corresponds to axial symmetry along the z–axis, when $q_{xx} = q_{yy} = -\frac{1}{2}\, q_{zz}$ and $\eta = 1$ corresponds to the relation $q_{xx} = 0$ and $q_{yy} = -q_{zz}$.

The asymmetry of the electric field is measured by the efg in the z-direction, which is chosen as the direction of maximum field gradient. If the groups around the nucleus have cubic (tetrahedral or octahedral) symmetry, the charge cloud is spherical and the value of q is zero. If the molecule has a cylindrical symmetry i.e. if a three-fold or higher-fold rotation axis exists, the deviation from spherical symmetry is expressed by the magnitude of q. If the molecule has a still lower symmetry, two parameters q and η are usually needed. However, certain combinations of the angles and charges may cause fortuitous cancellation of the effects leading to $\eta = 0$.

The fundamental requirement for the occurrance of an efg is that the electron density in the z-direction be different from that in the x- and y- directions. Electron imbalance in the populations of either p or d-orbitals will give rise to an efg. Through bond formation, such p- and d- imbalance will occur if the local symmetry of the atom is lower than cubic (O_h or T_d). For transition metal compounds, there may be contribution to the efg from the nonbonding d electrons. This is often of opposite sign to that from the bonding electrons.

The contribution of an individual charge to V_{zz} defined by the equation (20.4) is given by $q\ (3 \cos^2\theta - 1)/r^3$ where q is the magnitude of the charge and θ and r are its polar coordinates. This relation shows that the valence electrons of a given atom make larger contribution than shared electron pairs in bonding orbitals which in turn have greater effect than the charges on surrounding atoms or ions. The core electrons of the atom have spherical symmetry and their contribution is negligible. Lone pair of electrons in the valence shell always produce a substantial negative contribution to the efg. When there are lone pairs, rather large efg values, either negative or positive, may be found when the molecule has a lower symmetry.

The deviation of the nucleus from spherical symmetry is measured by its quadrupole moment. The nuclear quadrupole moment eQ is defined by

$$eQ = \int \rho\,(x, y, z)\, r^2\,(3 \cos^2 \theta - 1)\, d\tau \tag{20.9}$$

where e is the elementary charge, $\rho\,(x, y, z)$ is the nuclear charge density at a point (x, y, z) and r is the distance of the volume element $d\tau$ from the nucleus and θ is the angle which the radius vector r makes with the nuclear spin axis. The z-axis is taken parallel to the nuclear spin vector and the integration is carried out over the nuclear volume.

The energy difference between the different levels and hence the frequency of the transition will depend upon both the electric field gradient, eq, produced by the valence electrons and the quadrupole moment eQ of the nucleus. For a given isotope, eQ is constant and the values for many isotopes are available in the literature (see Table 20.1). From equation (20.9), the dimension of eQ is the product of charge and the square of the distance. However, it is usual in practice to express the moment Q simply in units of cm^2 or m^2. For example, ^{37}Cl with a nuclear spin I = 3/2 has a quadrupole moment Q of $- 0.08 \times 10^{-28}$ m^2, the negative sign indicating that the charge distribution is flattened relative to the spin axis.

The ^{35}Cl and ^{14}N are the most commonly studied nuclei. The resonances of chlorine and bromine are in the accessible region. Even when all the atoms of chlorine or bromine occupy crystallographically identical sites, two resonances should be seen, one for each isotope (^{35}Cl, ^{37}Cl; ^{77}Br, ^{81}Br). They may be recognized from their relative intensities and frequencies which are in the ratio of their nuclear quadrupole moments eQ (see Table 20.1) (1.27:1 and 1.19:1 respectively for chlorine and bromine). In the literature, it is common to quote only for ^{35}Cl because of the low intensity of the signal for ^{37}Cl. The interaction between the electric feld gradient and the quadrupole is measured by their product, the quadrupole coupling constant (QCC) e^2qQ or e^2qQ/h where e denotes the absolute value of the electronic charge, i.e. a positive quantity (1.6022×10^{-19} C). From NQR spectroscopy, one obtains quadrupole coupling constants.

The nuclear quadrupole coupling energy E_Q is given by

$$E_Q = -\frac{e}{6}\, \boldsymbol{Q}_{ij}\, \boldsymbol{V}_{ij} \tag{20.10}$$

where $\boldsymbol{Q}_{ij}$ denotes the nuclear quadrupole moment tensor (second order) and $\boldsymbol{V}_{ij}$ is the electric field gradient tensor. The product will depend on the mutual orientation of the two axis systems. For Q, it is convenient to choose a coordinate system such that it coincides with that of the spin system. When this is followed, the tensor can be defined in terms of only one parameter, the nuclear quadrupole moment eQ. If the sign of this quantity is known, then the sign of the electric field gradient can be obtained.

When the principal coordinate axes of the molecule coincide with the principal axes of the electric field gradient tensor, the potential energy E_Q for the interaction of the quadrupole moment with the electric field gradient at the nucleus is given by

$$E_Q = \frac{e}{6}\, [V_{xx}\, Q_{xx} + V_{yy}\, Q_{yy} + V_{zz}\, Q_{zz}\} \tag{20.11}$$

The equation (20.11) can be rewritten using the definitions of V and η and the relation (20.7), when we get

$$E_Q = \frac{e^2Q}{6}\, [\tfrac{1}{2}(\eta - 1)\, Q_{xx} - \tfrac{1}{2}(\eta + 1)\, Q_{yy} + Q_{zz}] \tag{20.12}$$

When axial symmetry exists, η becomes zero and E_Q becomes equal to (1/4) e^2qQ_{zz}, since $q_{xx} = q_{yy} = -\tfrac{1}{2}\, q_{zz}$.

The three most important quantities in NQR spectroscopy are e^2Qq, eq and η. The unit of eQ is in multiples of 10^{-28} m^2 while that of eq is in multiples of 10^{-30} m^{-3}. The value of q (or eq) is generally reported with a negative sign. Let us calculate the value of e^2Qq/h for nitrogen nucleus (atomic number 7) in a molecule.

$$eQ = 1.6 \times 10^{-30}\ \text{m}^2 \text{ (see Table 20.1)}$$

$$\frac{1}{4\pi\varepsilon_0} = 9.0 \times 10^9\ \text{N m}^2\ \text{C}^{-2}\ (\varepsilon_0 \text{ denotes permittivity of vacuum})$$

$$e = 1.602 \times 10^{-19}\ \text{C} \qquad 1\ \text{J} = 1\ \text{Nm}$$

Since eq = (electronic charge)/ (average electronic distance)^3, we obtain,

$$\frac{e^2Qq}{h} = \frac{7 \times (1.602 \times 10^{-19} C)^2\ (1.6 \times 10^{-30} \text{m}^2) \times 9.0 \times 10^9 \text{Nm}^2\ \text{C}^{-2}}{(6.626 \times 10^{-34} \text{Js})\ (10^{-10} \text{m})^3}$$

$$= 3.90 \times 10^6\ \text{Hz} = 3.90\ \text{MHz}$$

This is consistent with the experimental value obtained for pyridine, which is discussed later.

NUCLEAR QUADRUPOLE TRANSITIONS

(a) Axially Symmetric Molecules

The various energy levels of a given quadrupolar nucleus in an axially symmetric field ($\eta = 0$) are given by the equation:

$$E_{m_I} = \frac{e^2Qq[3\,m_I^2 - I(I+1)]}{4I(2I-1)} \tag{20.13}$$

where eq is the magnitude of the electric field gradient in the direction of the axis of symmetry and m_I is the nuclear spin magnetic quantum number which can take $(2I + 1)$ values with $m_I = I, I - 1, 0 \ldots - I$. The energy levels and transitions for the quadrupolar nuclei with $I = 3/2$, $5/2$ and $7/2$ in axially symmetric electric fields derived from the equation (20.13) are shown in Fig. 20.3. The transition energies are obtained as a function of e^2Qq/h. Thus when the spin is half-integral, $I + ½$ energy levels are obtained. Transitions are allowed between adjacent levels, the selection rule being $\Delta m_I = \pm 1$; there are therefore $I - ½$ lines in the spectrum. For nuclei with integral spins, $I + 1$ energy levels and I transitions are possible. In either case, the magnitude of the quadrupole coupling constant can be obtained from the spectrum. The quadrupole coupling constant may be either positive or negative. When it is negative, the energy level pattern is reversed but the energies of the transitions remain the same; the sign of efg cannot be determined from the spectrum. For the nuclear spin $I = 3/2$ e.g. ^{11}B, ^{35}Cl, ^{79}Br, m_I can have the values of 3/2, 1/2, –1/2 and –3/2. Substitution of $m_I = 3/2$ in equation (20.13) yields

$$E_{3/2} = e^2Qq/4$$

Since m_I is squared, the quadrupole energy for $m_I = -3/2$ will be the same as that for $m_I = +3/2$ and thus a doubly degenerate energy level will result. Similarly the energy level for $m_I = \pm 1/2$ will also be doubly degenerate. The transition energy, ΔE, indicated by an arrow in Fig. 20.3(a) corresponds to

$$[(e^2Qq/4) - (-e^2Qq/4)] = e^2Qq/2.$$

Thus for a nucleus with spin $I = 3/2$, a single transition is expected and the nuclear quadrupole coupling constant e^2Qq can be determined directly from the frequency of absorption:

$$\Delta E = h\nu = (1/2)e^2Qq \text{ (Joules)}$$

For example, *t*–butyl chloride at 20 K shows two NQR signals, an intense band at 31.195 MHz and a weaker band at 24.586 MHz. The intense band is assigned to the more abundant ^{35}Cl isotope and the weak band to the less abundant ^{37}Cl isotope.

For both the nuclei $I = 3/2$ and since $\nu\,(\pm 1/2 \rightarrow \pm 3/2) = \dfrac{1}{2}\dfrac{e^2Qq}{h}$

$$\frac{e^2Qq}{h} = 2\nu = 62.39 \text{ MHz for } ^{35}\text{Cl and } 49.17 \text{ MHz for } ^{37}\text{Cl}.$$

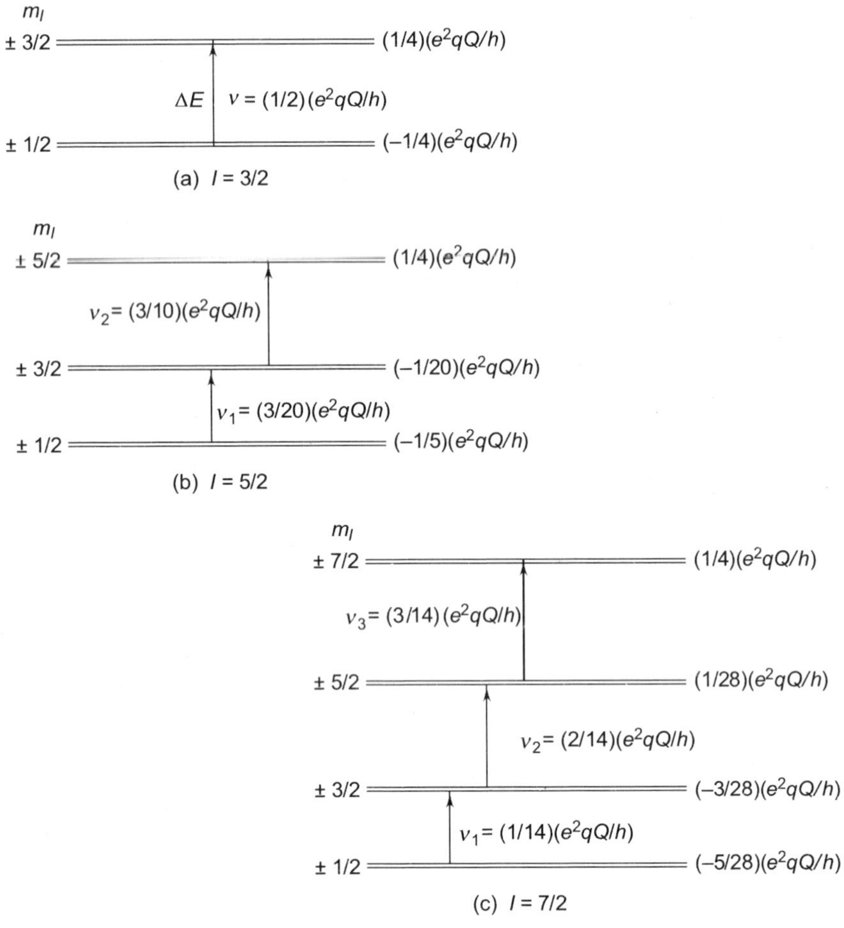

Fig. 20.3 Nuclear quadrupole energy levels and transitions in an axially symmetric field $\eta = 0$ for nuclei with spin (a) $I = 3/2$, (b) $I = 5/2$ and (c) $I = 7/2$. The energy levels are labelled with m_I values.

In the present case, e^2Qq is twice the frequency of the NQR transition (some authors use the symbol eQq to represent the same quantity as e^2Qq). The quantity e^2Qq is often expressed as a frequency in MHz although strictly speaking this should be e^2Qq/h. The spectrum of the γ phase of As_2S_3 is shown in Fig. 20.4. It is a typical NQR spectrum showing two resonances from two non-equivalent ^{75}As nuclei, $I = 3/2$.

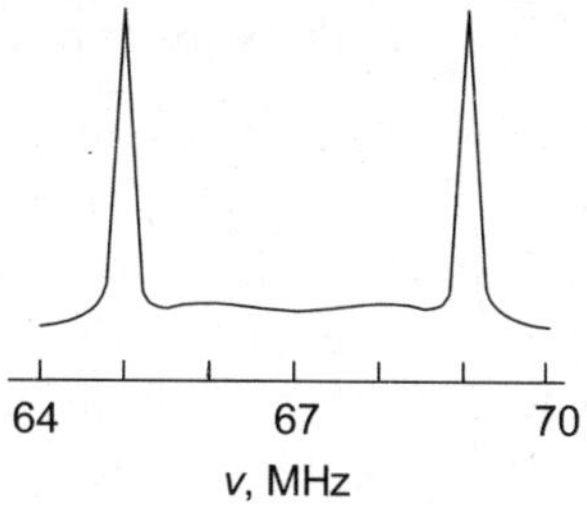

Fig. 20.4 NQR spectrum of the γ phase of As_2S_3.

The frequency of the transition $(m_1-1) \rightarrow m_I$ can be calculated using the equation (20.14),

$$\nu = \frac{3e^2qQ}{4I(2I-1)h}(2\,|m_I| - 1) \tag{20.14}$$

The relative intensities of the NQR resonances for the axially symmetric molecules are proportional to $[J(J+1) - M_I(M_I+1)]$. However, in NQR work, intensities are generally not considered because of the experimental and theoretical difficulties encountered in their interpretation.

For nuclei with $I = 5/2$, e.g. ^{27}Al and ^{127}I, three energy levels ($E_{\pm 1/2}$, $E_{\pm 3/2}$ and $E_{\pm 5/2}$) result and the two observable transitions are $E_{\pm 1/2} \rightarrow E_{\pm 3/2}$ and $E_{\pm 3/2} \rightarrow E_{\pm 5/2}$. Substitution of I and m_I into the equation (20.13) shows that the energy for the transition $E_{\pm 3/2} \rightarrow E_{\pm 5/2}$ is twice of that for the transition $E_{\pm 1/2} \rightarrow E_{\pm 3/2}$. In the experimental spectrum, deviations from the predicted frequencies may occur due to deviations from axial symmetry and it serves as a measure of asymmetry.

For $I = 7/2$ (e.g. ^{59}Co and ^{133}Cs), as shown in Fig. 20.3(c), from equation (20.13) four energy levels, $E_{\pm 7/2}$, $E_{\pm 5/2}$, $E_{\pm 3/2}$, and $E_{\pm 1/2}$, result Three frequencies ν_1, ν_2 and ν_3 in the ratio of 1:2:3 are expected. The effect of the asymmetry parameter η on the energies of the nuclei $I = 7/2$ is to diverge the energy levels arising from $m_I = +1/2$ and $m_I = 3/2$. While the energy of the former level is lowered, that of the latter level so also that of $m = 5/2$ level are raised although to a smaller extent.

For integral spins, the equation (20.13) yields one non-degenerate and $(I + 1)$ doubly degenerate levels. The nuclear quadrupole energy level diagram and the transitions for nuclei with $I = 1$ (e.g. ^{14}N, ^{6}Li) are shown in Fig. 20.5. A single resonance line is predicted with frequency ν given by

$$\nu = \frac{3e^2qQ}{4h} \tag{20.15}$$

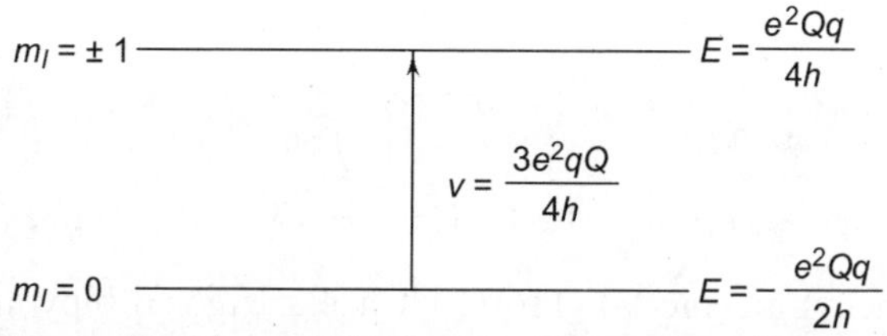

Fig. 20.5 Nuclear quadrupole energy levels and transition for nuclei with $I = 1$ in an axially symmetric electric field.

(b) Axially Non-Symmetric Molecules

For axially non-symmetric molecules, the spectra obtained are more complex than those obtained for axially symmetric molecules. For molecules belonging to non-axial symmetry, i.e. when $\eta \neq 0$, the energies of the various quadrupole levels are no longer given by the equation (20.13). In the case of half-integral spins, the energies are slightly affected. For $I = 3/2$ the following equations have been derived for the energies of the two states.

$$E_{\pm 3/2} = \frac{3e^2Qq\sqrt{1+\eta^{2/3}}}{4I(2I-I)} = \frac{e^2Qq}{4}\left(1+\frac{\eta^2}{3}\right)^{1/2} \quad (20.16a)$$

$$E_{\pm 1/2} = \frac{-3e^2Qq\sqrt{1+\eta^{2/3}}}{4I(2I-I)} = \frac{-e^2Qq}{4}\left(1+\frac{\eta^2}{3}\right)^{1/2} \quad (20.16b)$$

The nuclear quadrupole energy levels for $I = 3/2$ are shown in Fig. 20.6. A single transition ($\pm 1/2 \rightarrow \pm 3/2$) with frequency given by

$$\nu = (e^2Qq/2h)\left(\sqrt{1+\eta^2/3}\right). \quad (20.17)$$

is expected. Since there are two unknowns, the asymmetry parameter η and the field gradient q, the value of e^2Qq cannot be obtained from the measured single NQR frequency. From structural considerations, if the asymmetry parameter η is known to be very small, it can be neglected and e^2Qq can be determined with sufficient accuracy. However, this problem can be solved from NQR experiments on the sample in the presence of a weak external magnetic field as discussed later.

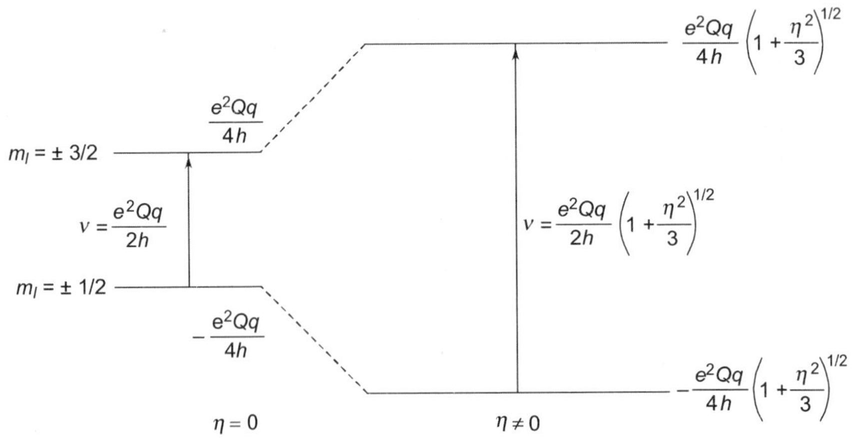

Fig. 20.6 Nuclear quadrupole energy levels and transitions for $I = 3/2$ when $\eta = 0$ and $\eta \neq 0$.

Tetrachlorophenyl arsenic(V) ($PhAsCl_4$) has a trigonal bipyramidal structure with the phenyl group in the equatorial position. The ^{35}Cl NQR resonances of this compound are found at 24.0, 25.0, 34.18 and 34.68 MHz. Since the axial chlorines are expected to be more ionic than the equatorial ones, their signals are assigned at lower frequencies. The two axial chlorines may be crystallographically inequivalent. The resonances at 24.0 and 25.0 MHz are assigned to the axial chlorine atoms and the remaining two resonances to the equatorial chlorines. The positions of ^{37}Cl resonances in the same compound are expected at 18.92, 19.71, 26.94 and 27.34 MHz. according to the relation $QCC(^{35}\text{Cl})/QCC(^{37}\text{Cl}) = 110/86.7 = 1.2686$ (see Table 20.1).

In the other half-integral cases, I = 5/2, 7/2, etc., the energy of each level shows a complex dependence on η. For small values of η, the energy can be expressed as a power series in η^2. In none of these cases is the number of transitions affected, only their energies change slightly. For integral spins however, the energies of the $\pm\, m_I$ substates become different. For example, for I = 1, the $m_I = \pm 1$ state splits as shown in Fig. 20.7 .The nuclear quadrupole energy levels and allowed transitions are shown in Fig. 20.7 for nuclei with I = 1 and $\eta \neq 0$. The energy levels are given by

$$E_0 = \frac{-2e^2Qq}{4I(2I-1)} = -\frac{1}{2}\,(e^2Qq) \tag{20.18a}$$

$$E_{\pm 1} = \frac{e^2Qq(1 \pm \eta)}{4I(2I-1)} = \frac{1}{4}\,(e^2Qq)\,(1 \pm \eta) \tag{20.18b}$$

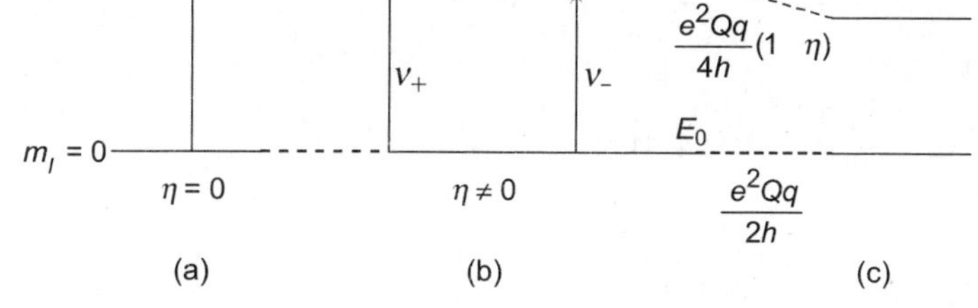

Fig. 20.7 Nuclear quadrupole energy levels and allowed transition for I = 1 when (a) η = 0, B = 0, (b) $\eta \neq 0$, B = 0 and (c) $\eta \neq 0$, $B \neq 0$ where B denotes constant external magnetic field.

As noted from Fig. 20.7(b), three transitions are expected for $I = 1$, $\eta \neq 0$, and hence the two unknowns, e^2Qq and η can be obtained directly from the experimental measurement of transition frequencies. An important system for $I = 1$ is that of ^{14}N. Since the quadrupole coupling constant for nitrogen is of the order of 4 MHz, the ν_0 transition is observed only when η is large. The perturbation of levels by the application of an external magnetic field is shown in Fig. 20.7(c). The effect will be discussed later.

The transition frequencies shown in Fig. 20.7 are given by

$$\nu_+ \,(0 \rightarrow 1) = \frac{3}{4}\left(\frac{e^2Qq}{h}\right)\left(1+\frac{\eta}{3}\right)$$

$$\nu_- \,(0 \rightarrow -1) = \frac{3}{4}\left(\frac{e^2Qq}{h}\right)\left(1-\frac{\eta}{3}\right)$$

These two transitions are separated by ν_o given by

$$\nu_o = \frac{e^2Qq}{2h}\,\eta$$

As an example, the ^{14}N NQR spectrum of pyridine is considered. It shows three peaks at 0.908, 2.984 and 3.892 MHz, assignable to ν_o, ν_- and ν_+, respectively. We have

$$\nu_+ = \frac{3}{4}\frac{e^2qQ}{h}\left(1+\frac{\eta}{3}\right) = 3.892 \text{ MHz}$$

$$\nu_- = \frac{3}{4}\frac{e^2qQ}{h}\left(1-\frac{\eta}{3}\right) = 2.984 \text{ MHz}$$

$$\nu_o = \frac{e^2qQ}{2h}\,\eta = 0.908 \text{ MHz}$$

$$\text{Sum of } \nu_- + \nu_+ = \frac{3}{2}\frac{(e^2qQ)}{h} = 6.876 \text{ MHz}$$

$$\text{giving, } \frac{e^2qQ}{h} = 4.584 \text{ MHz}$$

$$\text{Since } \eta = \left(\frac{h}{e^2Qq}\right) 2\nu_o$$

$$\text{we get, } \eta = \frac{2\,(0.908 \text{ MHz})}{4.584 \text{ MHz}} = 0.396$$

It is to be noted that when $\eta \approx 0$, the NQR frequency tends towards that given for the axially symmetric molecules. The magnitude of the quadrupole coupling constant represents chiefly the presence of a lone pair of electrons on pyridine nitrogen. The large value of η is due to the planarity of the ring showing that the components of efg are quite different in the plane of the ring and perpendicular to it.

The equations for the quadrupolar energy levels as a function of η have been tabulated for axially non-symmetric molecules with values of I other than 1 and 3/2. The tabulation permits the determination of η and e^2Qq from the spectral data for nuclei with $I = 5/2$, 7/2 and 9/2. When η is appreciable say $\eta > 0.3$, the expressions given for the energy levels of $I = 5/2$, 7/2 and 9/2 tend to be approximate. In several cases when more than one line is observed in the spectrum, both the asymmetry parameter, η and the quadrupole coupling constant e^2Qq can be obtained from the spectrum. When η is large, the selection rule $\Delta m_I = \pm 1$ breaks down and it is possible to observe some of the forbidden transitions with $\Delta m_I = \pm 2$.

EFFECT OF AN EXTERNAL MAGNETIC FIELD

The energies of nuclear spin states are affected by the presence of an external magnetic field. By recording the NQR spectrum of the sample in an applied magnetic field, additional information can be derived. Magnetic fields up to about 0.5 T are used.

The effect of an applied magnetic field is to lift the degeneracy of the $\pm m_I$ states and the Zeeman splitting that occurs is proportional to m_I. The selection rule is however unaltered. Therefore we would expect double the number of lines in the spectrum. However, the two new states produced by the external magnetic field for $m_I = \pm½$ are actually mixtures of $m_I = +½$ and $m_I = -½$ states so that the transitions can occur from both these levels to both of the $m_I = \pm 3/2$ levels as shown in Fig. 20.8. In a magnetic field therefore $m_I = ½ \rightarrow 3/2$ transition splits into four. The intensities of the spectral lines observed in the presence of the magnetic field depend on the relative orientations of the efg and the magnetic field direction. By a detailed study of the spectra for different orientations of the single crystal in a magnetic field, the asymmetry parameter and the direction of the field gradient q_{zz} relative to the crystal axes can be determined. In the external magnetic field, splitting of the NQR spectral lines does not occur for paramagnetic compounds although the unpaired electrons also produce a magnetic field. This is traced to the rapid fluctuations of the spins on individual atoms which averages to zero on the NQR time scale.

In a weak magnetic field, that is, when $g_N \beta_N B << e^2Qq$, the magnetic field acts as a perturbation on the nuclear quadrupole. As a result, generally shift in the energies of the nondegenerate quadrupole levels occur as shown in Fig. 20.7(c) for a nucleus with $I = 1$ and $\eta \neq 0$ and splitting of the doubly-degenerate levels as shown in Fig. 20.8 for $I = 3/2$ and $\eta \neq 0$. For nuclei with $I = 1$, a spectrum with two bands can arise from (i) $\eta \neq 0$ as discussed above (Fig. 20.7) or (ii) for nuclei with $\eta = 0$ situated in two inequivalent lattic e sites. A study of the spectrum of the sample in an external magnetic field will help to resolve between these two possibilities. In the first case ($\eta \neq 0$), two lines should be observed but with different energies than those obtained in the absence of the field. In the second case, each doubly degenerate level should split giving a four-line spectrum.

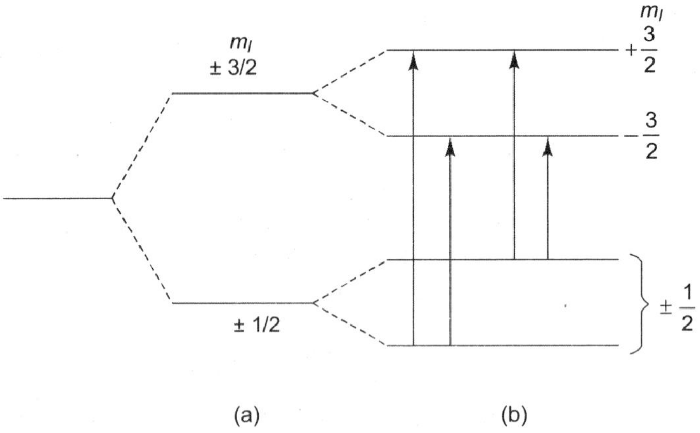

Fig. 20.8 Splitting of the energy levels for the nuclei with $I = 3/2$ in an external magnetic field: (a) $eq > 0$, $B = 0$ and (b) $eq > 0$, $B > 0$.

As discussed earlier, the energy levels $m_I = \pm 1/2$ and $\pm 3/2$ are each doubly degenerate for $I = 3/2$ in a non-symmetric field. This degeneracy is removed by an external magnetic field resulting in four levels. Theoretical consideration have shown that the $m_I = +3/2$ sublevel would be higher in energy than the $-3/2$ sublevel. Considerable mixing of states occurs for the $\pm 1/2$ sublevels so that the $+1/2$ sublevel has some character of the $-1/2$ sublevel and vice-versa as stated before.

The transition between $+1/2$ and $-1/2$ levels is not of interest since its frequency is much smaller than the NQR frequency. When the Zeeman field is of the order of 0.1 T as is the case with several nuclei of chemical interest, this frequency is of the order of 0.1 MHz or less. The transition between the mixed states $+1/2 \rightarrow -1/2$ which corresponds to the selection rule $\Delta|m_I| = 0$ is not of interest, so also the transition $+3/2 \leftrightarrow -3/2$. The following transitions between the states $\pm 1/2$ and $\pm 3/2$ obeying $\Delta|m_I| = \pm 1$ are of interest.

$$\nu_\alpha\ (+1/2 \rightarrow +3/2)$$
$$\nu'_\alpha\ (-1/2 \rightarrow -3/2)$$
$$\nu_\beta\ (-1/2 \rightarrow +3/2)$$
$$\nu'_\beta\ (+1/2 \rightarrow -3/2)$$

These four frequencies are symmetric with respect to the unperturbed NQR frequency ν_0. The inner pair of lines ν_α and ν'_α have equal intensity and this pair is more intense than the outer pair ν_β and ν'_β as shown in Fig. 20.9. For a given orientation of the magnetic field both the pairs ν_α, ν'_α and ν_β, ν'_β have maximum intensity when the radiofrequency field is perpendicular to the symmetry axis of the electric field gradient tensor. The intensity of the two pairs depends on the orientation of the magnetic field. The expressions for the four Zeeman– split energy levels and the frequencies need not concern us here (see Das and Hahn pp.8–10). The energy differences corresponding to these transitions are functions of B, e^2Qq and η. Therefore for this system, q and η can be determined from the spectrum of the sample obtained with and without the applied magnetic feld.

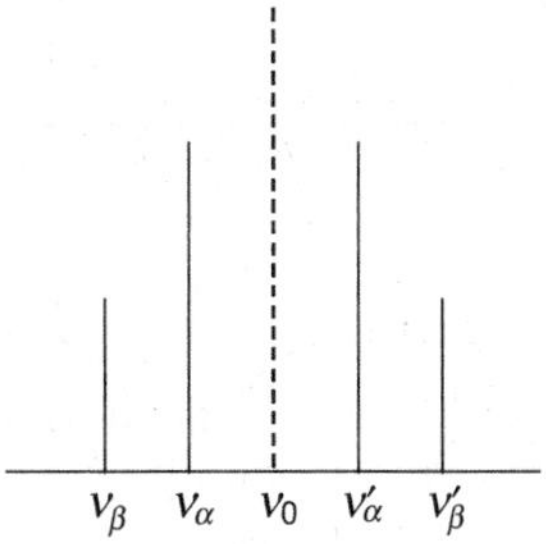

Fig. 20.9 Splitting of the NQR spectrum in an external magnetic field for the nuclei with $I = 3/2$.

APPLICATIONS

The chief information obtainable from NQR spectroscopy relates to the number of chemically distinct sites occupied by the quadrupolar nuclei in the sample and to the local electric field of the nuclei in each of these sites. A few examples given below serves to illustrate the applications of NQR spectroscopy to chemical and structural problems chiefly to inorganic systems.

(i) Chemical Bonding and Structure

NQR spectroscopy essentially uses the quadrupolar nucleus as a probe to estimate the electric field gradients in the crystal and the data are useful in the applications of crystal field theory. The NQR spectra are not easy to interpret however. The values of nuclear quadrupole moments eQ determined experimentally have been tabulated. Thus the electric field gradient eq can be determined from the known values of e^2Qq and eQ. For the first and second row elements, important contribution to field gradient comes from the valence electrons in the p-orbitals. Theoretical calculations show that $q_p >> q_d >> q_f$. The contribution to q from s-electrons is zero because of spherically symmetric charge distribution. According to the theory of Townes and Dailey, the efg in a molecule eq_{mol} is related to that in an atom eq_{at} by the relation

$$eq_{mol} = feq_{at} \tag{20.19}$$

where f is a quantity which can be found from chemical and physical considerations.

Table 20.2 Values of e^2Qq_{mol} for some diatomic halides (in MHz)

Molecule	e^2Qq	*Molecule*	e^2Qq
F Cl	–146	Br Cl	877
Br Cl	–103	F Br	1089
I Cl	–83	Li Br	37
K Cl	0.04	NaBr	58
Rb Cl	0.77	Na I	260
Cs Cl	3.0	Cl Cl	109

Values of e^2Qq_{at} for ^{35}Cl – 109.7, ^{79}Br, 769.8 and ^{127}I –2292.8; e^2Qq_{mol} refers to the underlined atom.

An important use of NQR spectroscopy is that the resonance frequency of a particular quadrupolar nucleus can be correlated to the ionic nature of bonding in the molecule being studied. The data in Table 20.2 show how the value of e^2Qq varies for some halogen chlorides and alkali metal halides. As the electronegativity difference between the elements increases, the bonding becomes more ionic and the values become smaller. In the free chloride ion all three p-orbitals will be equaly filled with two electrons each and hence the electron distribution is nearly symmetrical resulting in a small e^2Qq value.

When the atom being investigated is more electronegative than the atom to which it is bonded, the quadrupolar atom would have greater electron density around it in the molecule than in the isolated atom. A relationship exits between the electron "occupation" of the p-orbitals of a quadrupolar atom in a molecule and q_{mol}. The following equation relates the quantity e^2Qq_{mol}, which is determined from the NQR spectrum of the molecule and e^2Qq_{at}.

$$e^2Qq_{mol} = (1 - s + d)\,(1 - i)\,e^2Qq_{at} \tag{20.20}$$

where e^2Qq_{at} is the quadrupole coupling constant of the isolated atom for the occupancy of the p-orbital by a single electron, s and d are the fraction of the s and d character respectively used by the atom in its bond to its neighbour and i is the fraction of the ionic character of the bond. When the atom being studied is electropositive, i changes its sign. Recently a modified form of equation (20.20) has been reported.

$$e^2Qq_{mol} = (1 - s + d - i - \pi)\,e^2Qq_{at} \tag{20.21}$$

where π is the extent of π bonding and all the other quantities are as defined above. As noted from equation (20.21), the NQR spectrum of a molecule gives one measurable parameter but there are four unknowns, s, d, π and i. An unambiguous solution is hence not possible unless additional equations relating other measurable variables to these parameters are obtained.

Attempts have been made to relate the resonance frequencies to the net charges on the halogen atoms obtained from the molecular orbital calculations. This type of procedure has been used to show that chlorine atoms in *cis*-$PtCl_2(NH_3)_2$ have slightly higher negative charges than those in *trans* –Pt $Cl_2(NH_3)_2$ and hence Pt–Cl bonds in the *cis* isomer are slightly more ionic. For halogens a linear relationship exists between the resonance frequency and the ionic character of the bond to the halogen atom. Consequently the same halogen attached to different elements has similar frequencies and these may be distinctive. For example, ^{35}Cl resonances for Si-Cl bonds are 15–21 MHz, for S-Cl 30–40 MHz and for N-Cl 43–57 MHz. The highest values are observed for the most covalent compounds.

As more data become available, it may be possible to correlate certain ranges of quadrupole coupling constants with specific functional groups, like the group frequency in vibrational spectroscopy. For example, the e^2Qq values (in MHz) for nitrogen in nitriles RCN are in the range –4.2 to 4.6, while those in isonitriles RNC are about +0.5. The e^2Qq value of +1.2 MHz for nitrogen resonance supports HNCS structure and not HSCN. The magnitude of the quadrupole coupling constants could be used to distinguish between the geometrical isomers. For example, the e^2Qq value of *cis*-$[Co(en)_2Cl_2]Cl$ is 33.71 while that of the *trans*-isomer is 60.63 MHz.

In favourable cases, isomers can be distinguished solely on the basis of the number of bands. The compound $PFCl_4$ can exist in two different isomeric forms (Fig. 20.10). For the isomer I having the fluoride ligand in the axial position, there are two different chloride environments in the ratio of 1:3. For the isomer II, two signals are expected on the basis of the two different chloride environments in the ratio of 1:1. The ^{35}Cl NQR spectrum of $PFCl_4$ exhibits two resonances in the ratio of 1:3 indicating that structure I is found in the solid state.

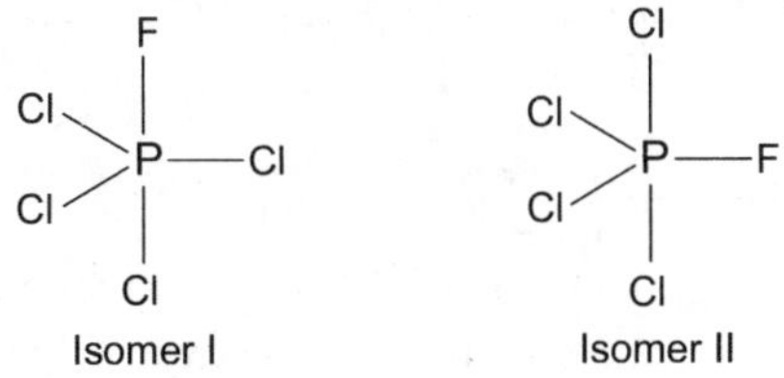

Fig. 20.10 Isomers of $PFCl_4$.

With increase in the ionic character of the bond, the electron environment approaches spherical symmetry (where $q_{mol} = 0$) and e^2Qq decreases. Mixing of the *p* orbital with *s* orbital decreases the field gradient. In a covalent molecule, *d* orbital contribution to the bonding increases the field gradient. The covalent nature of the metal-halide bond calculated from the NQR spectra for $TiCl_3$, VCl_3 and $CrBr_3$ were found to be consistent with those based on electronegativities. Halogen NQR frequencies were found to depend chiefly on the ionic character of the M–X bond (X=halide) where *M* is a metal such as Al, Ti, Cr, etc. Other substituents on

M will shift the NQR frequencies for a given M–X bond. Thus the NQR frequencies can be helpful for the identification of compounds.

Halogen nuclei have been most extensively investigated by NQR spectroscopy because of their large quadrupole moment. The NQR spectra of group III trihalides have been of interest. The ^{79}Br NQR spectrum of $AlBr_3$ exhibits two closely spaced resonances at 113.79 and 115.45 MHz (terminal bromine) and another resonance at 97.945 MHz (bridging bromine). The spectrum suggests a dimer structure for $AlBr_3$ with two halogen-bridged bonds as shown in Fig. 20.11.

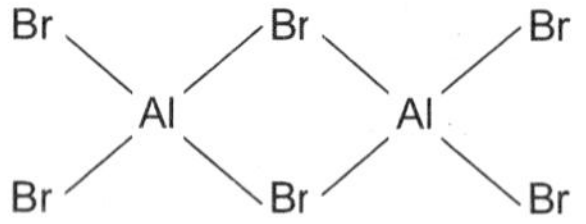

Fig. 20.11 Structure of $AlBr_3$ dimer.

Other trihalides such as $GaCl_3$, $AuCl_3$, etc. show similar spectra. The NQR frequencies are observed at 19.084, 20.225 and 14.667 MHz in $GaCl_3$. The low frequency signal is attributed as due to the bridging atoms and the high frequency peaks to the terminal halogens. From the NQR spectra, it is possible to determine the bond angles and the type of hybridization. From the results, it was inferred that the bonding of metal atoms with bridging halogens is weaker than that of the terminal bonds.

Assignment of signals for halogen containing materials is complicated since both chlorine and bromine have two isotopes. The isotopes of bromine have nearly equal abundance so that all the signals are doubled. The ratio of frequencies will be the same as that of the quadrupole moments and the appropriate signal pairs can be recognised since the nuclei must be in identical environments. Similar considerations apply to chlorine isotopes, which occur in the 3:1 natural abundance.

(ii) Solid State Effects

For complex molecules, NQR spectroscopy is the only source of nuclear quadrupolar information because of the complexity of the microwave spectrum. For solids, the complexities of lattice effects too must be considered. For those molecules which have been studied by both the methods (direct measurement and microwave), it was found that e^2Qq value is usually 10 to 15 per cent lower in the solid state than that in the gaseous state. This decrease has been attributed as due to increased ionicity of the bonds in the solid. The compound may be extensively associated in the solid but not so in the gas phase or in solution for example, I_2 and CNCl. The spectrum of solid iodine yielded a large value for the asymmetry parameter, η. Since the iodine molecule is axially symmetric, a large asymmetry parameter is taken as an evidence for intermolecular interaction in the solid state. A large value of η as noted from the iodine NQR spectrum of HIO_3 supports its structure as $IO_2(OH)$ instead of HIO_3. HIO_3 has a C_3 axis of symmetry so that

$q_{xx} = q_{yy}$. The NQR spectra (^{37}Cl, ^{35}Cl, ^{121}Sb, ^{127}I) of 2 ICl.$AlCl_3$ and 2 ICl.$SbCl_3$ indicate that these materials should be formulated as ICl_2^+ $AlCl_4^-$ and $ICl_2^+SbCl_6^-$.

The problem of distinguishing inequivalent positions in the lattice from chemical inequivalence in the molecular configuration is difficult in some cases. In the solid state, the molecule is subjected to different crystal fields resulting in crystallographically inequivalent nuclei. Such splittings are usually small, less than 500 kHz. Atoms which are chemically inequivalent give rise to large splittings.The inequivalent lattice positions occur in the bromine NQR spectrum of K_2SeBr_6 which exhibits a single line at room temperature but shows two lines at – 40 °C. This difference arises due to a change in the crystalline phase. Although K_2PtI_6, K_2SnBr_6 and K_2TeBr_6 contain 'octahedral' anions, the halide NQR spectra of all the compounds consist of three lines, indicating the existence of at least three different halide environments, showing that the halides are not equivalent in the solid lattice. Phase changes can be detected by making measurements over a range of temperature.

The ^{35}Cl NQR spectrum of $SiCl_4$ should show only a single resonance if all the four chlorine atoms are equivalent in the solid state. However, the NQR spectrum of $SiCl_4$ in the solid state at 77 K displays four peaks at 20.273, 20.408, 20.415 and 20.464 MHz, showing that the four chlorine atoms are not equivalent although the molecule has tetrahedral symmetry at room temperature. Similarly ^{35}Cl NQR spectrum of the solid $GeCl_4$ shows up to four resonances, one for each chloride because the low site symmetry of the molecule in the crystal results in every chloride ion becoming inequivalent. Four resonance lines were also found in the chlorine quadrupole spectra of $TiCl_4$ and $SnCl_4$. It was concluded hence that the crystal structures of $SiCl_4$, $TiCl_4$ and $SnCl_4$ are similar.

Another difficulty that arises in the interpretation of spectra is illustrated by the ^{35}Cl NQR spectrum of Nb_2Cl_{10}. Even though this molecule has both terminal and bridging chlorines, only a single chlorine resonance assigned to the bridging chlorine was observed in the spectrum. In Re_2Cl_{10} and W_2Cl_{10} however, a large number of resonances were found.

The number of resonances observed in a NQR spectrum can provide useful information about the species present in the solid. For example, the NQR spectrum of PCl_5 at lower temperatures shows ten resonances. A group of four resonances have been assigned to PCl_4^+ and the remaining six at lower frequencies to PCl_6^-. For PCl_4^+ PCl_6^- each chlorine gives a separate resonance. If PCl_5 vapour is rapidly cooled, a meta stable phase is formed and this gives only three resonances. Two of these have been assigned to non-equivalent chlorine nuclei in the axial positions of the trigonal bipyramidal PCl_5 and the remaining resonance to three equatorial chlorines, which are apparently equivalent.

The NQR studies of ^{14}N are difficult to carry out. However, nitrogen quadrupole resonance is of interest because of its different types of bonding and structure in

Table 20.3 ^{14}N Quadrupole transitions, quadrupole coupling constant and asymmetry parameter for pyridine and its complexes (at 77 K)

	MHz				
Compound	ν_+	ν_-	ν_0	e^2Qq/h	η
Pyridine	3.892	2.984	0.908	4.584	0.40
Py_2ZnCl_2	2.387	2.078	0.309	2.977	0.207
	2.332	2.038	0.294	2.913	0.202
$Py_2Zn(NO_3)_2$	2.124	1.884	0.240	2.672	0.580
	2.097	1.877	0.220	2.649	0.566

inorganic, organic and biological compounds. It occurs in the low frequency region (2–10 MHz). According to Townes and Dailey, for nuclei such as ^{14}N and halogens, the source of electric field gradient is mainly the valence shell p electrons. The ^{14}N quadrupole transitions in pyridine and two coordination compounds of pyridine with zinc salts are listed in Table 20.3. On coordination, the lone pair on pyridine is partially delocalized on to the metal leading to a decrease in the value of e^2Qq/h. The acceptor ability of zinc is affected slightly by the presence of other ligands.

$Co(NH_3)_6Cl_3$ has a low spin d^6 ($t_{2g}^{\ 6}$) configuration and symmetrical electron density around Co^{3+}. Therefore there is no efg at the cobalt nucleus. The nitrogen atom of ammonia has an asymmetric (local C_{3v}) environment. Therefore the NQR resonance for ^{14}N is observed but not for ^{59}Co. In the NH_3 molecule, the efg is chiefly due to the lone pair of electrons on nitrogen. When the lone pair coordinates to Co^{3+}, its contribution to efg decreases because the electrons have moved away from nitrogen.

In BrCN, the ^{14}N NQR resonance is a doublet. This could result from two inequivalent nitrogen atoms in the crystal lattice or from a splitting of the nitrogen resonance due to asymmetric field gradient. The structure of the solid BrCN was found by single crystal X-ray diffraction study to consist of linear chains of the type Br–C≡N … Br–C≡N…Br–C≡N… The nitrogen atom has an axial symmetry and hence only one line is expected. However the interaction between the chains possibly reduces the symmetry at the nitrogen atom and it leads to two lines in the spectrum. The e^2Qq values indicate that bromine in BrCN has a formal positive charge. In the solid state Br^+–CN^- is the predominant structure. An appreciable increase in e^2Qq value for bromine was observed in the solid state relative to that in the gaseous state, because of the stabilization of Br^+ by coordination. If the N…Br–C bond is described as a pd hybrid, the increased contribution in the bromine-carbon bond will also lead to an increase in e^2Qq value.

(iii) Hydrogen Bonding

NQR spectroscopy could be used for studying hydrogen bonding. In general, lowering of the quadrupole coupling constant is found on going from the vapour phase to the solid state. For example, the ^{35}Cl quadrupole coupling constant of HCl and likewise, the ^{14}N quadrupole coupling constant of NH_3 in the solid state is lower by over 20% than that in the vapour phase. This is traced to the reduction in the field gradient at the nucleus due to intermolecular hydrogen bonding in the solid state. When the temperature is increased the extent of hydrogen bonding decreases and consequently the NQR spectral line shifts.

EXPERIMENTAL ASPECTS

A schematic diagram of a NQR spectrometer is shown in Fig. 20.12. In a NQR experiment, radiation in the radiofrequency region is employed. The power from a radiofrequency oscillator is supplied to a previously balanced bridge circuit. At resonance, energy absorbed by the solid sample causes an imbalance in the circuit. The resultant output voltage as a function of radiofrequency is amplified and recorded. This is the quadrupole resonance spectrum.

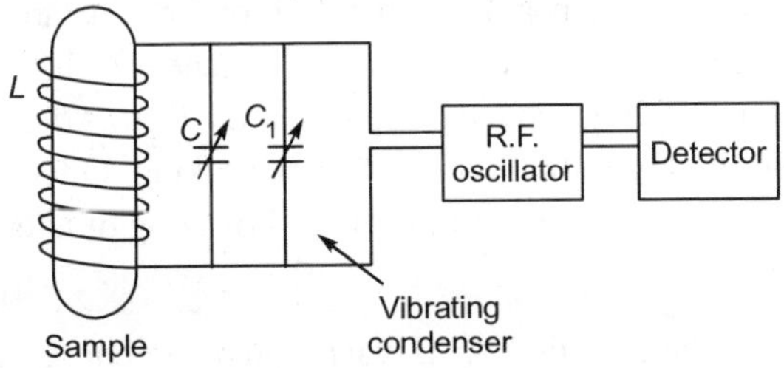

Fig. 20.12 Block diagram of a NQR spectrometer.

Two types of oscillators have been commonly employed, the super regenerative oscillator and the marginal oscillator both for excitation of the nuclei and as detector. The former oscillator is most common since it permits broad band scanning of the resonances and is easier to operate. However, it suffers from the limitation of producing a multiplet of lines for each resonance. The true resonance frequency is the centre line of the multiplet as shown in Fig. 20.13. The marginal oscillator gives a single peak for each absorption but it is rather difficult to operate. It is particularly useful at low frequencies. The quadrupole frequencies to be detected range from 0.1 to 1000 MHz. Therefore during the search for resonance absorption, the applied frequency must be varied continuously maintaining good stability and sensitivity. Because of the smaller spin lattice relaxation times of the quadrupolar nuclei, experiments require much larger *rf* power than that for the magnetic resonance experiments.

In the first two methods, the *rf* oscillator is used both for excitation of the nuclei and as detector, while in the third method a separate transmitter and a receiver carry out these functions. Spectra can also be measured by observing the free induction decay following a pulse excitation as in the Fourier transform NMR.

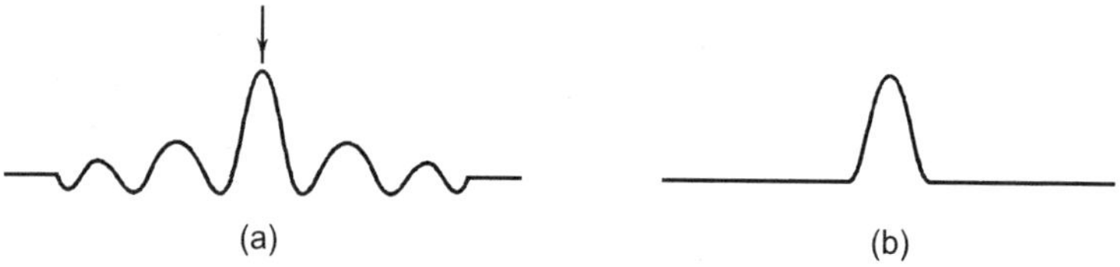

Fig. 20.13 For the same single resonance (a) multiplet peaks from super regenerative oscillator and (b) a single peak from marginal oscillator.

The spectrometer is a fairly simple device. The sample is placed in a coil which is a part of the radiofrequency circuit. The sample must always be a crystalline solid, usually in the form of a powder. Gaseous and liquid samples may be investigated by freezing. Glassy solids generally give very weak signals or no signal may be observed. Different orientations of the crystallites relative to the direction of the radiofrequency affect only the intensities of the transitions but not their energies. The sample size required depends on the strength of the resonance. The amount of the substance required for NQR measurement is generally high, typically a few grams for an isotope of high abundance. Essential factors for obtaining strong signals are high purity, crystallinity and efficient packing. Often it is necessary to record the spectra at liquid nitrogen temperature. Lowering of temperature increases the signal strength.

The frequency of the *rf* radiation is varied and the energy absorbed by the sample at resonance is recorded. The resonances are typically very weak and broad and may be found over a wide range of frequencies rendering the recording of the spectrum difficult. The NQR spectral half line widths are in the range 1–10 kHz. In order to observe the weak signals, the spectra are usually measured in the first or second derivative form (i.e. $dI/d\nu$ or $d^2I/d\nu^2$ respectively) and is plotted against ν. The Cl NQR spectrometer is standardized with substances such as $HgCl_2$, hexachlorophosphazene ($N_3P_3Cl_6$), *p*-dichloro benzene, chloropyridines and dichloropyridines.

GENERAL REFERENCES

1. H. Schuler and T. Schmidth, Z. Physik. **94**, 457 (1935).
2. T.P. Das and E.L. Hahn, Nuclear Quadrupole Resonance Spectroscopy, Academic Press, New York, 1958.
3. E.A.C. Lucken, Nuclear Quadrupole Coupling Constants, Academic Press, New York, 1969.
4. J.A.S. Smith (Ed.), Advances in Nuclear Quadrupole Resonance, Heyden and Sons, London, Vol. 1 (1974); Vol. 2 (1975); Vol. 3 (1977).
5. M. Kubo and D. Nakamura, Advances in Inorganic Chemistry and Radiochemistry, **8**, 257 (1966).
6. E. Schemp and P.B.Bray, Nuclear Quadrupole Resonance Spectroscopy, *in* Physical Chemistry – An Advanced Treatise, D. Henderson (ed.), Vol. 4, Academic Press, New York, 1970.
7. J.A.S. Smith, J. Chem. Educ. **48**, 39, A77, A147, and A243 (1969).
8. T. B. Brill, J. Chem. Educ. **58**, 519 (1981).
9. R.V. Parish, NMR, NQR, EPR and Mossbauer Spectroscopy in Inorganic Chemistry, Ellis Horwood, Chichester, England, 1990.

EXERCISES

1. Derive the quadrupole energy levels and the transition energies for nuclei with $I = 7/2$ with axial symmetry. Show the possible NQR transitions.
2. (a) Show by substitution in E_Q formula why the highest energy level for a quadrupolar nucleus in a positive EFG is always $e^2qQ/4$. What is the effect of changing the sign of EFG from positive to negative on (a) the energy levels and (b) the spectrum?
3. The e^2qQ values for the ^{35}Cl nucleus in the methyl and ethyl chlorides are given below. Account for the trend.

Compound	CH_3Cl	CH_3CH_2Cl	$(CH_3)_2CHCl$	$(CH_3)_3CCl$
$\frac{e^2qQ}{h}$ MHz	68.05	66.0	64.15	62.10

4. (a) Why the complex $[Co(NH_3)_6]^{3+}$ shows NQR resonance for ^{14}N but not for ^{59}Co for both isotopes $I > 1$?

   (b) Why is the ^{14}N quadrupole coupling constant larger for NH_3 than for amine complexes of transition metals?
5. The cation $[Co\{P(OCH_3\}_5]^+$ has D_{3h} symmetry. Explain why it shows a three-line spectrum? For ^{59}Co, abundance 100%, $I = 7/2$.
6. The e^2qQ/h value for KNCS is 2.43 MHz while for $Zn(NCS)_2Py_2$ (Py = pyridine) it is 2.89 MHz, $\eta = 0.92$. Explain.

7. The observed (QCC) quadrupole coupling constant e^2qQ/h (^{35}Cl) values for interhalogen compounds Cl_2, ClF, ClBr and ClI are 109, 146, 104 and 83.0 MHz, respectively. Calculate the bond ionicities. Given QCC for ^{35}Cl is 110 MHz.
8. For pyridine the NQR signals are observed at 0.908, 2.984 and 3.892 MHz. Calculate e^2qQ/h and η for pyridine. (Note that when $\eta \neq 0$, the selection rule $\Delta m_I = \pm 1$ no longer strictly applies).
9. The value of e^2qQ/h for pyridine is 4.584 MHz but in the complexes $ZnCl_2.Py_2$ and $ZnBr_2.Py_2$, the value decreases to 2.95 and 2.89 MHz, respectively. Account for this observation.
10. Obtain the NQR energy levels, transition energies and transitions for $I = 2$.
11. Define quadrupole coupling constant.
12. How is the nuclear quadrupole moment calculated?
13. Quenching of vapour of PCl_5 gives a new meta stable low temperature crystalline state. It shows two resonances at 29.26 and 33.75 MHz. Assign the bands.
14. The electric quadrupole moments of ^{35}Cl and ^{37}Cl are -0.1×10^{-28} m^2 and 0.079×10^{-28} m^2 respectively. At what frequencies would you expect to find ^{37}Cl resonances in $HgCl_2$ if ^{35}Cl lines in $HgCl_2$ occur at 22.05 and 22.25 MHz at 300 K?
15. How NQR spectroscopy could be used to distinguish between *cis*– and *trans*– isomers of $PtCl_2(NH_3)_2$?
16. (a) The ^{37}Cl resonances of tetrachlorophenyl arsenic(V), $C_6H_5AsCl_4$ occur at 18.92, 19.71, 26.94 and 27.34 MHz. The compound has a trigonal bipyramidal structure with phenyl group in the equatorial position. Assign the observed frequencies. (b) Find the frequencies of the ^{35}Cl resonances for the same compound.
17. The ^{35}Cl NQR spectra of CCl_4, $CHCl_3$, CH_2Cl_2 and CH_3Cl show resonances at 40.6, 38.3, 36.0 and 34.0 respectively. Account for this trend.
18. Explain the number of ^{35}Cl NQR signals expected for the following compounds.

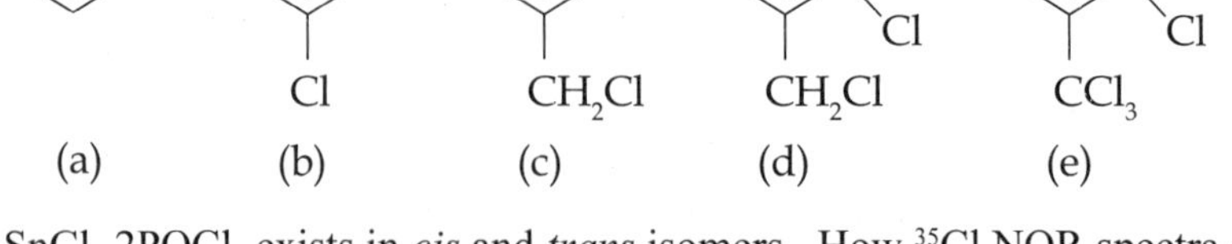

(a) (b) (c) (d) (e)

19. $SnCl_4.2POCl_3$ exists in *cis* and *trans* isomers. How ^{35}Cl NQR spectra could be useful to distinguish between the two?
20. Explain how NQR spectroscopy is useful to detect the explosives.

Appendices

Appendix I: General Data and Fundamental Constants

Quantity	Symbol	Value
Speed of light	c	2.997925×10^8 m s^{-1}
Boltzmann constant	k	1.38066×10^{-23} J K^{-1}
		8.6174×10^{-5} eV K^{-1}
Gas constant	R	8.31451 JK^{-1} mol^{-1}
		8.20578×10^{-2} dm^3 atm K^{-1} mol^{-1}
Planck's constant	h	6.62608×10^{-34} Js
Avogadro number	N_A	6.02214×10^{23} mol^{-1}
Elementary charge	e	1.602177×10^{-19} C
Atomic mass unit	u	1.66054×10^{-27} kg
Vacuum permittivity	ϵ_0	8.85419×10^{-12} J^{-1} C^2 m^{-1}
	$4\pi\epsilon_0$	1.11265×10^{-10} J^{-1}C^2 m^{-1}
Bohr magneton	μ_B	9.27402×10^{-24} J T^{-1}
Bohr radius	a_0	5.29177×10^{-11}m
Rydberg constant	R_∞	1.09737×10^5 cm^{-1}

Useful Relations

1 eV = 1.60218×10^{-19} J = 96.485 kJ mol^{-1} = 8065.5 cm^{-1}

1 cm^{-1} = 1.986×10^{-23} J = 11.96 J mol^{-1}

1 Cal = 4.184 J

1 D (debye) = 3.33564×10^{-30} C m

1 T (tesla) = 10^4 G

1 Å = 100 pm   10 mm = 30.0 GHz

Appendix II: Hückel Molecular Orbitals

A brief summary of the HMO approach to the calculation of orbital energies and unpaired-electron distributions in π-electron systems is given here. Because of the crude assumption of non-interaction among the electrons, we can treat all anions, neutral molecules and cations with the same theory. This approach has been widely described in textbooks and hence here we shall only map out successive steps.

1. We define the molecular orbitals to be linear combinations of n normalized atomic orbitals ϕ_n. The conjugated organic molecules are generally planar and the valance orbitals of carbon are sp^2 hybridized, the three hybridized σ orbitals are in the direction of neighbouring atoms in the plane while the unhybridized $2p_z$-orbital lies perpendicular to the molecular plane. Since σ and π electrons are contained in planes perpendicular to each other, the interaction between the two types of electrons is likely to be negligible in the planar conjugated molecules. Therefore, molecular orbitals for π-electrons in conjugated molecules may be treated independently of σ electrons. These orbitals are set up as linear combination of $2p_z$ atomic orbitals of each carbon atom in a conjugated molecule.

   Consider a system with n-carbon atoms each with one $2p_z$ orbital with one π-electron in it, then according to the linear combination atomic orbitals approximation, the molecular orbital is written as

$$\Psi = \sum_i C_i\phi_i \quad (1)$$

   where ϕ_i is a $2p_z$-atomic orbital of atom i and the summation is taken over all the $2p_z$ atomic orbitals. The coefficients and energies of these orbitals are obtained by solving the secular equation:

$$\sum_{i=1}^{n} C_i(H_{ij} - S_{ij}E) = 0 \quad j = 1, 2, 3 \quad (2)$$

where $H_{ij} = <\phi_i|H|\phi_j>$ (3)

$S_{ij} = <\phi_i|\phi_j>$ (4)

Hückel introduced a set of approximations. They are

1. The terms $H_{ii} = <\phi_i|H|\phi_i>$ called the Coulomb integrals represent approximately the energy of an electron in a $2p_z$ orbital on the ith carbon atom. All such integrals are assumed to be equal and denoted by α.

2. The terms H_{ij} ($<\phi_i|H|\phi_j>$) called the resonance integrals represent the energy of interaction between two neighbouring atomic orbitals. When atoms i and j are bonded, H_{ij} has a finite value and zero otherwise. All such integrals are assumed to be equal and denoted by the symbol β.

3. The integrals S_{ij} known as the overlap integrals are neglected. The integrals $S_{ii} = 1$ since they are normalisation integrals. This approximation is known as 'zero differential overlap'. With the introduction of these approximations, the secular equation for a noncyclic conjugated molecule becomes

$$C_1(\alpha - E) + C_2\beta = 0 \tag{5}$$
$$C_1\beta + C_2(\alpha - E) + C_3\beta = 0$$
$$C_2\beta + C_3(\alpha - E) + C_4\beta = 0$$

- - - - - - - - -

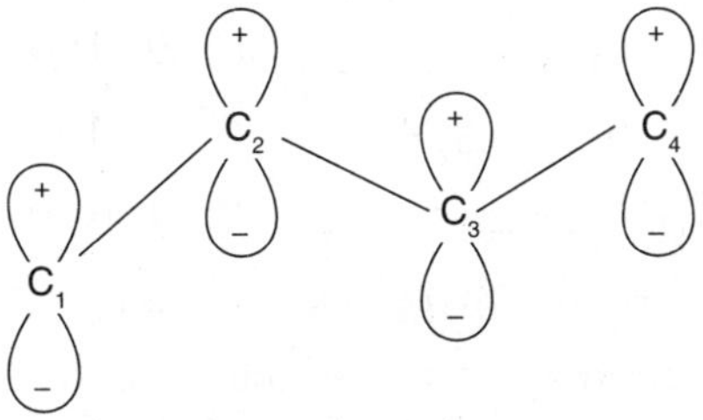

For a π electron system containing only (skeletal) carbon atoms, all $a_\mu s$ are given the same value α and all $\beta_{\mu\nu}$ are given the value β. The quantity α corresponds approximately to the negative of the ionization energy of an electron localized in the carbon atomic orbital and hence has a negative value. The energy β describes the additional stabilization due to the delocalization of the electron over the neighbouring atomic orbitals. Accordingly, β is also a negative quantity but of smaller magnitude.

In practice, the $H_{\mu\mu}$ and $H_{\mu\upsilon}$ are usually arranged as the elements of a square matrix of order n for a molecule with n $2p_z$-atomic orbitals. The Hückel determinant for example for 1,3–butadiene (H_2C–CH=CH=CH_2) is

$$\begin{vmatrix} H_{11} - E & H_{12} & H_{13} & H_{14} \\ H_{21} & H_{22} - E & H_{23} & H_{24} \\ H_{31} & H_{32} & H_{33} - E & H_{34} \\ H_{41} & H_{42} & H_{43} & H_{44} - E \end{vmatrix} = 0 \tag{6}$$

which simplifies to

$$\begin{vmatrix} a - E & \beta & 0 & 0 \\ \beta & a - E & \beta & 0 \\ 0 & \beta & a - E & \beta \\ 0 & 0 & \beta & a - E \end{vmatrix} = 0 \tag{7}$$

On setting $(\alpha - E)/\beta = x$, the determinant simplifies to

$$\begin{vmatrix} x & 1 & 0 & 0 \\ 1 & x & 1 & 0 \\ 0 & 1 & x & 1 \\ 0 & 0 & 1 & x \end{vmatrix} = 0 \tag{8}$$

The above determinantal equation can be written as a polynomial of fourth degree,

$$x^4 - 3x^2 + 1 = 0 \tag{9}$$

On solving the equation, we get four roots x_j. They are $x_1 = 1.618$, $x_2 = 0.618$, $x_3 = -0.618$ and $x_4 = -1.618$. The one electron HMO energies are obtained from the relationship, $E_j = \alpha_j + x_j\beta$ since $x_j = (\alpha - E)/\beta$. On substituting x_j values, the HMO energy levels are obtained as shown in Fig. 1.

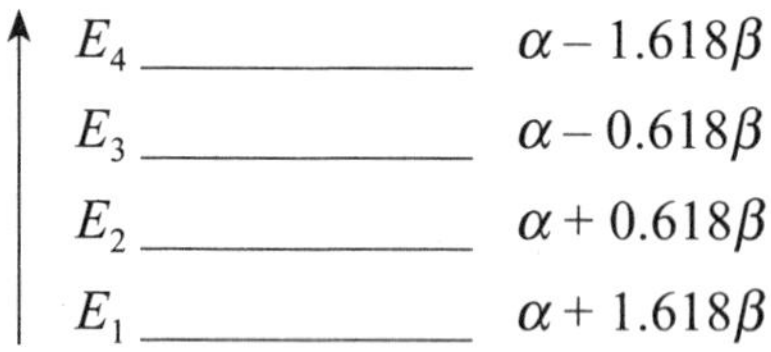

Fig. 1 HMO energy levels of butadiene

For butadiene, in the ground state, the four electrons are placed in the two lowest energy levels following aufbau principle. The two electrons occupy each molecular orbital in the antiparallel spin according to Pauli's exclusion principle. Since β is a negative quantity, a HMO with positive x_j has a value which is negative. An electron in such an HMO is more stable than isolated atomic orbitals; it is a bonding MO. When x_j is negative, the MOs are less stable than the isolated atomic orbitals and hence they are called antibonding HMOs. The HMOs are referred to as one-electron MOs since interelectron repulsions are not taken into account. The total π electron energy for butadiene is $2E_1 + 2E_2 = 2(\alpha + 1.618\beta) + 2(\beta + 0.618\beta) = 4\alpha + 4.47\beta$.

MO Wave functions

The coefficients of the LCAO giving the HMO ψ_j of energy E_j can be obtained from the secular determinant by inserting the E_j value. For butadiene, we obtain the following simultaneous equations,

$$\begin{aligned} &C_{j1}\,x_j + C_{j2} = 0 \\ &C_{j1} + C_{j2}\,x_j + C_{j3} = 0 \\ &C_{j2} + C_{j3}\,x_j + C_{j4} = 0 \\ &C_{j3} + C_{j4}\,x_j = 0 \end{aligned} \tag{10}$$

The coefficients $C_{j\mu}$ which are the eigenvectors of the Hückel matrix define the LCAO forming the MO ψ_j associated with HMO energy E_j

$$\psi_j = \sum_{\mu=1}^{n} C_{j\mu}\phi_\mu \quad (j = 1,2,3,\ldots) \tag{11}$$

The resulting MOs ψ_j are normalized, that is

$$\sum_{\mu=1}^{n} C_{j\mu}^2 = 1 \tag{12}$$

Further, any two (non-degenerate) HMOs say ψ_i and ψ_j are mutually orthogonal. That is

$$\sum C_{i\mu}C_{j\mu} = 0 \quad (13)$$

For butadiene, for the HMO with $x_1 = -1.618$ ($j = 1$) we get,

$$-1.618\ C_{11} + C_{12} = 0$$

$$C_{11} - 1.618\ C_{12} + C_{13} = 0$$

$$C_{12} - 1.618\ C_{13} + C_{14} = 0$$

$$C_{13} - 1.618\ C_{14} = 0$$

These equations yield,

$$\frac{C_{11}}{C_{12}} = \frac{1}{1.618}\ \frac{C_{13}}{C_{14}} = 1.618$$

$$\frac{C_{11}}{C_{13}} = \frac{1}{1.618}\ \frac{C_{12}}{C_{14}} = 1.618$$

On simplification and employing the normalization condition,

$$C^2_{11} + C^2_{12} + C^2_{13} + C^2_{14} = 1 \quad (14)$$

We obtain,

$$C_{11} = C_{14} = 0.372,\ C_{12} = C_{13} = 0.602$$

The coefficients for the other three MOs are obtained likewise for the energies E_j where $j = 2,3$ and 4. The following MOs are thus obtained.

$$\Psi_1 = 0.372\ (\phi_1 + \phi_4) + 0.602\ (\phi_2 + \phi_3)$$

$$\Psi_2 = 0.602\ (\phi_1 - \phi_4) + 0.372\ (\phi_2 - \phi_3)$$

$$\Psi_3 = 0.602\ (\phi_1 + \phi_4) - 0.372\ (\phi_2 + \phi_3)$$

$$\Psi_4 = 0.372\ (\phi_1 - \phi_4) - 0.602\ (\phi_2 - \phi_3)$$

The bonding nature of the MOs decreases with increase in the number of nodes.

Electron density

Integrating Ψ_1^2 over all space yields a value of 1.0 since the MO wave function is normalized. We have

$$\int \psi_1^2 d\tau = <\psi_1|\,\psi_1> = 1 \quad (15)$$

Substituting for Ψ_1, we get

$$\Psi_1^2 = [0.372\ \phi_1 + 0.602\ \phi_2 + 0.602\ \phi_3 + 0.372\ \phi_4]^2$$

$$= (0.372)^2\ \phi_1 + (0.602)^2\ \phi_2 + (0.602)^2\ \phi_3 + (0.372)^2\ \phi_4$$

$$= 0.138\ \phi_1 + 0.362\ \phi_2 + 0.362\ \phi_3 + 0.138\ \phi_4$$

because of zero differential overlap (ZDO). Within the ZDO approximation, the square of the HMO is particularly simple. The sum of the squares of the coefficients

$$C_{11}^2 + C_{12}^2 + C_{13}^2 + C_{14}^2 = 1$$

showing the presence of an electron in Ψ_1. It shows that an electron will spend $(0.372)^2$ of its time in the atomic orbital ϕ_1, a fraction of time equal to $(0.602)^2$ in the atomic orbital ϕ_2 and so on.

$$\psi_j^2 = \sum_{\mu=1}^{n} C_{j\mu}^2 \phi_\mu \tag{16}$$

The sum, $$\sum_{j=1} b_j C_{j\mu}^2 = q_\mu \tag{17}$$

is called the charge density (q_μ) and b_j is the number of electrons in the MO ψ_j. For butadiene, from the MOs we obtain the following electron density distribution.

For two electrons in Ψ_1, $0.276\,\phi_1 + 0.724\,\phi_2 + 0.724\,\phi_3 + 0.276\,\phi_4$

For two electrons in ψ_2 $0.724\,\phi_1 + 0.276\,\phi_2 - 0.276\,\phi_3 - 0.724\,\phi_4$

The sum of the electron densities $\sum_{\mu-1}^{n} q_\mu$ gives the total number of π electrons (Z_π) in the π-electron system. For butadiene $Z_\pi = 4$. A useful check is that the sum of all q is equal to the total number of π electrons in the conjugated system. The task of solving the secular determinant can be simplified if use is made of the symmetry properties of the molecule by straightforward methods of group theory.

Allyl System

This system contains three $2p_z$ orbitals. The determinantal form of the secular equation is

$$\begin{vmatrix} x & 1 & 0 \\ 1 & x & 1 \\ 0 & 0 & x \end{vmatrix} = 0$$

This determinant gives the polynomial

$x^3 - 2x = 0$ giving $x = 0, \pm\sqrt{2}$

The energy levels are given by

$E_1 = \alpha + \sqrt{2}\beta$ (bonding)

$E_2 = \alpha$ (non-bonding)

$E_3 = \alpha - \sqrt{2}\beta$ (antibonding)

The total π electron energies of the radical, cation and anion are given by

$E_\pi(\text{radical}) = 3\alpha + 2\sqrt{2}\beta$

$E_\pi(\text{cation}) = 2\alpha + 2\sqrt{2}\beta$

$E_\pi(\text{anion}) = 4\alpha + 2\sqrt{2}\,\beta$

The HMOs are obtained after substituting the values of x in any two of the following secular equations:

$$C_1x + C_2 = 0$$

$$C_1 + C_2x + C_3 = 0$$

$$C_2 + C_3x = 0$$

For $x = -\sqrt{2}$, we get

$$-\sqrt{2}C_1 + C_2 = 0$$

$$C_1 - \sqrt{2}C_2 + C_3 = 0$$

$$C_2 - \sqrt{2}C_3 = 0$$

The normalization condition gives the equation,

$$C_1^2 + C_2^2 + C_3^2 = 1$$

On solving, we get

$$C_1 = C_3 = \frac{1}{2} \quad C_2 = \frac{1}{\sqrt{2}}$$

Thus the molecular orbitals for the three values of x are

$$\Psi_1 = \frac{1}{2}(\phi_1 + \sqrt{2}\phi_2 + \phi_3)$$

$$\Psi_2 = \frac{1}{\sqrt{2}}(\phi_1 - \phi_3)$$

$$\Psi_3 = \frac{1}{2}\left(\phi_1 - \sqrt{2}\phi_2 + \phi_3\right)$$

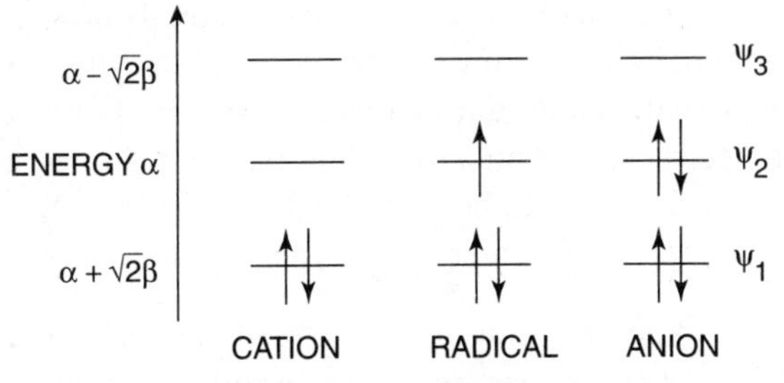

Fig. 2 The HMO energy levels of the allyl cation, radical and anion.

In the ground state, two π-electrons of allyl radical occupy the molecular orbital ψ_1 and the odd electron occupies the molecular orbital ψ_2 (Fig. 2). Thus the odd electron has an equal probability of being found on carbon atoms 1 and 3 with zero probability on the middle carbon atom. This result is in accord with the following resonance structures of allyl radical, where the position of the odd electron is indicated by a dot.

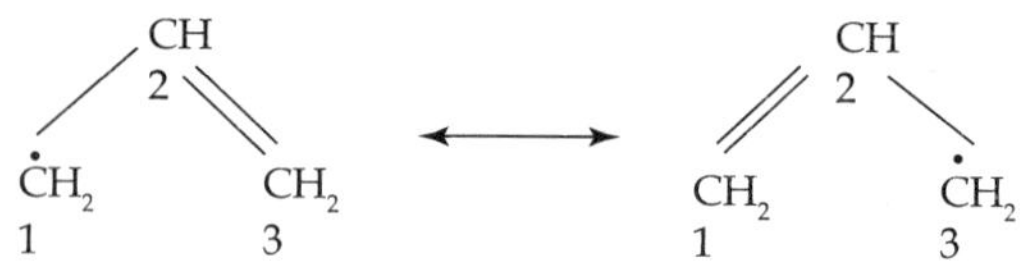

The same argument can be extended to the cation and the anion, a hole in one case and an extra electron in the other case has the same probability distribution as the odd electron in the radical. We can write the following resonance structures for the cation and anion.

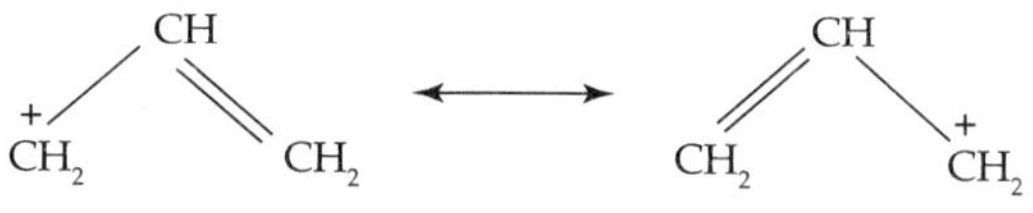

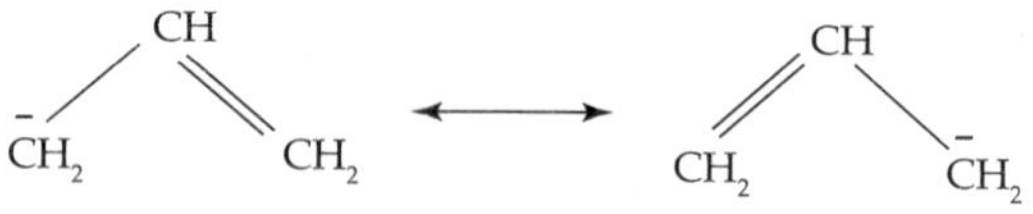

For allyl radical, the calculation of electron density follows the above procedure.

$$q_1 = q_3 = 2\times\frac{1}{4}+1\times\frac{1}{2} = 1$$

$$q_2 = 2\times\frac{2}{4}+1\times 0 = 1$$

For allyl cation: $q_1 = q_3 = 0.5$, $q_2 = 1.0$

For allyl anion: $q_1 = q_3 = 1.5$, $q_2 = 1.0$

Each unpaired electron of a π-type radical is expected to be distributed over the molecular framework. For example, in benzene anion the probability of finding its unpaired electron in the vicinity of any one carbon atom is 1/6 as required by symmetry. For radicals with lower symmetry, there is no such obvious guide to determine the unpaired-electron distribution. The HMO approach provides valuable guidance towards determining this distribution. For example, butadiene anion has five π electrons. The unpaired electron resides in ψ_3 and the unpaired electron populations are found to be $\rho_1 = \rho_4 = 0.36$ and $\rho_2 = \rho_3 = 0.14$. These are indeed the positions at which larger ESR hyperfine splittings are observed and that the ratio of the hyperfine splitting $a_1/a_2 = -0.762/-0.279 = 2.7$ which agrees satisfactorily with the ratio of the unpaired-electron populations, $\rho_1/\rho_2 = 2.6$.

In allyl radical, the HMOs show zero charge density at the central carbon atom. However, the electron spin resonance spectrum shows proton hyperfine coupling constants of 1.393 and 1.483 mT for the terminal carbon atoms and 0.406 mT at the central carbon atom. This is one of the limitations of HMO theory. More advanced molecular orbital methods place non-zero spin density on the central carbon atom.

Heteroatomic compounds: The simple HMO method may be relatively less satisfactory when applied to heteroatomic compounds. Since the coloumb integral represents the negative of the ionization potential, it may be expected to follow the electronegativity. Coulomb integrals of the heteroatoms (X) are expressed by adding β_{C-C} to α_C as an increment of certain energy. For example, we write $\alpha_N = \alpha_C + h\beta$ where h is a constant and β is the usual resonance integral. Resonance integrals of the heteroatom bonds, β_{C-X}, that is $\beta_{C-X} = k\beta_{C-C}$ where k is a constant, usually close to unity. The various values of h and k employed have been reviewed (E. Heilbronner and P.A. Straub, Hückel Molecular Orbitals, Springer, New York, 1966). Two sets of values for heteroatoms such as nitrogen and oxygen are used depending on whether the heteroatom contributes one or two electrons to the π system.

$\alpha_N(1)$	$\alpha + 0.5\beta$	β_{C-N}	0.80β
$\alpha_N(2)$	$\alpha + 1.5\beta$	$\beta_{C=N}$	1.0β
$\alpha_o(1)$	$\alpha + \beta$	$\beta_{C=O}$	0.80β
$\alpha_o(2)$	$\alpha + 2\beta$	$\beta_{C=O}$	β

The HMO method for heteroatomic compounds is illustrated taking the simplest system $H_2C{=}NH$ (methylene imine). The secular determinant is

$$\begin{vmatrix} \alpha - E & \beta_{C-N} \\ \beta_{C-N} & \alpha + (1) - E \end{vmatrix} = 0$$

Taking β_{C-N} as β and $\alpha_N(1) = \alpha + 0.5\beta$, the determinant becomes

$$\begin{vmatrix} \alpha - E & \beta \\ \beta & \alpha + 0.5\beta - E \end{vmatrix} = 0$$

Putting $(\alpha - E)/\beta = x$, we get

$$\begin{vmatrix} x & 1 \\ 1 & x + 0.5 \end{vmatrix} = 0$$

We obtain, $E_1 = \alpha + 1.28\beta$ and $E_2 = \alpha - 0.78\beta$. The total π electron energy for $H_2C{=}NH$ is $E_\pi = 2\alpha + 2.56\beta$, it differs from the value of ethylene by 0.56β. The HMO wavefunction and electron density are determined as before.

$$\Psi_1 = 0.62\phi_1 + 0.78\phi_2$$

$$\Psi_2 = 0.78\phi_1 - 0.62\phi_2$$

The electron population on C and N are calculated to be respectively 0.78 and 1.22 giving charge density of (C)$q_1 = 0.22$ and (N)$q_2 = -0.22$. The charge density for ethyleneimine is consistent with the structure such as $H_2\overset{+}{C}-\overset{-}{N}H$.

Appendix III: Solutions to Exercises

I hear and I forget, I read and I remember, I do and I understand

—*Confucius*

Chapter 1

1. Frequency $\upsilon = 2 \times 10^{13}\ s^{-1}$; wavenumber $\overline{\upsilon} = 6.667 \times 10^{6}\ cm^{-1}$

2. Frequency $\upsilon = 5.25 \times 10^{13}\ s^{-1}$

   Wavelength $\lambda = 5.7 \times 10^{8}$ cm

   Energy = 3.479×10^{-20} J/molecule

3. Frequency $\upsilon = \dfrac{c}{\lambda} = \dfrac{3\times10^{10}}{254\times10^{-7}\,\text{cm}} = 1.181\times10^{15}\,s^{-1}$

   Energy = 7.825×10^{-19} J per molecule

   $= 7.825 \times 10^{-19} \times 6.022 \times 10^{23} = 471$ kJ mol^{-1}

4. 380 kJ mol^{-1} = 63.10×10^{-20} J/molecule

   $$E = h\upsilon \qquad \upsilon = \frac{63.10\times10^{-20}\ \text{J}}{6.626\times10^{-34}\ \text{Js}} = 9.523\times10^{-14}\,s^{-1}$$

   $\lambda = 315$ nm; ultra-violet region

5. $\Delta E.\Delta t \leq h/2\pi$

   $\Delta E = 1.054 \times 10^{-30}$ J ; $\Delta\upsilon = 1.591 \times 10^{3}$ Hz

6. $$\frac{\text{Rate of spontaneous emission}}{\text{Rate of stimulated emission}} = \exp(h\upsilon/kT) - 1$$

   $$\frac{h\upsilon}{kT} = \frac{hc}{kT\lambda} = \frac{(6.626\times10^{-34}\ \text{Js})(3\times10^{8}\ \text{ms}^{-1})}{(1.38\times10^{-23}\ \text{JK}^{-1})(900\,\text{K})(5000\times10^{-10}\ \text{m})} = 32.0$$

   $e^{32.0} - 1 = 3.41 \times 10^{-13}$

7. 100 MHz = 100×10^{6} Hz = $1 \times 10^{8}\ s^{-1}$

   $$\lambda = \frac{c}{\upsilon} = \frac{3\times10^{10}\,\text{cm}\,s^{-1}}{1\times10^{8}\,s^{-1}} = 300\ \text{cm}$$

   E = 6.626×10^{-34} Js $\times 1 \times 10^{8}\ s^{-1}$

   $= 6.626 \times 10^{-26}$ J/molecule $\times 6.022 \times 10^{23}$ mol^{-1}

   $= 0.04 \times 10^{-3}$ kJ mol^{-1}

8. The changes in the magnetic states of nuclei are of lower energy than the ESR transitions and hence we expect to observe these changes as fine structure in the ESR spectrum.

9. The broadest bands will be observed for the quickest relaxation processes. It suggests ultra-violet visible spectroscopy. Time scale of 10^{-14} s^{-1}.

10. $kT = 1.38 \times 10^{-23}$ $JK^{-1} \times 298$ K = 4.11×10^{-21} J

 $kT = 1.38 \times 10^{-23} \times 4.2$ K = 5.796×10^{-23} J

 4.11×10^{-21}J = 6.21×10^6 MHz at 298 K = 207 cm^{-1} = 2.48 kJ mol^{-1}

 5.796×10^{-23} = 8.751×10^4 MHz at 4.2 K

 = 34.92 J mol^{-1}

 Since kT is much higher than the rotational energy, rotational excited states are significantly occupied while most molecules occupy the ground vibrational state so that the transitions from the first excited state (v = 1) are very weak or do not occur at all.

11. $N_1/N_0 = 0.999889$

12. $E = h\upsilon \quad \upsilon = \dfrac{2.14\times10^{-16}\,\text{J}}{6.626\times10^{-34}\,\text{Js}} = 0.323\times10^{8}\,\text{s}^{-1}$

 $$\lambda = \frac{c}{\upsilon} = \frac{3.0\times10^{10}\,\text{cm/s}}{0.323\times10^{18}\,\text{s}^{-1}} = 9.288\times10^{-6}\,\text{cm}$$

13.

(a) 1.033×10^4 cm^{-1}	(b) 66.67×10^6 cm^{-1}	(c) 0.3 cm^{-1}	(d) 8.0×10^4
Visible region Valence electron transitions	X-ray region. Changes involving the inner electrons; Ionization	Microwave region Electron spin transitions, Rotational transition	Ultraviolet region. Transitions of bonding electrons

14. 10.0 cm, E = 1.192 J/mol

15. (a) Mössbauer spectroscopy; γ-Ray region
 (b) Electronic spectroscopy; Ultra-violet visible region
 (c) Rotational spectroscopy; Microwave region
 (d) NMR spectroscopy, Radiofrequency region

16. (a) 500 nm, $\nu = 0.6 \times 10^{15} s^{-1}$

 $$\frac{A_{21}}{B_{21}} = \frac{8\pi h\nu^3}{c^3} = 1.333 \times 10^{-13}\ \text{J m}^{-3}\text{s}$$

 (b) 3000 cm^{-1}, $\nu = 9 \times 10^{13}$ s^{-1} $\dfrac{A_{21}}{B_{21}} = 4.5 \times 10^{-16}$ J m^{-3}s

 (c) 500 MHz, $\nu = 5 \times 10^8$ s^{-1} $\dfrac{A_{21}}{B_{21}} = 7.73 \times 10^{-32}$ J m^{-3}s

 (d) 3.0 cm, $\nu = 1 \times 10^{10}$ s^{-1} $\dfrac{A_{21}}{B_{21}} = 6.17 \times 10^{-28}$ Jm^{-3}s

 Spontaneous emission occurs predominantly at higher frequencies.

Chapter 2 ESR

1. (a) 1.25 T (b) 1.072 T

2. $N_\alpha / N_\beta = 1 - \frac{g\beta B}{kT} = 1 - (1.56 \times 10^{-3}) = 0.00944$

3. $\frac{\delta_N}{N} = \frac{N_\beta - N_\alpha}{N_\beta + N_\alpha} \approx \frac{\Delta E}{2kT} = 0.762 \times 10^{-3}$

4. 2.0017

5. 0.3318 T

6. a_H = 50.7 mT B = 0.339 T, a_D = 7.79 mT

7. N_α/N_β $\frac{\text{at liq. } N_2}{\text{at R.T}}$ $\frac{0.58}{0.149} = 3.9$

   Population difference, nearly 4 times higher

8. 0.1 mT = 2.8 MHz; 50.73 mT

   $hA_i/g\beta$ is the coupling constant in Tesla, A_i is the coupling constant in MHz

9. (a) 2.30 mT (b) 2.003, 64.4 MHz

10. 330.2 mT, 332.2 mT, 332.8 mT, 334.8 mT, 1:1:1:1

11. 2.20 mT, g = 1.992

12.

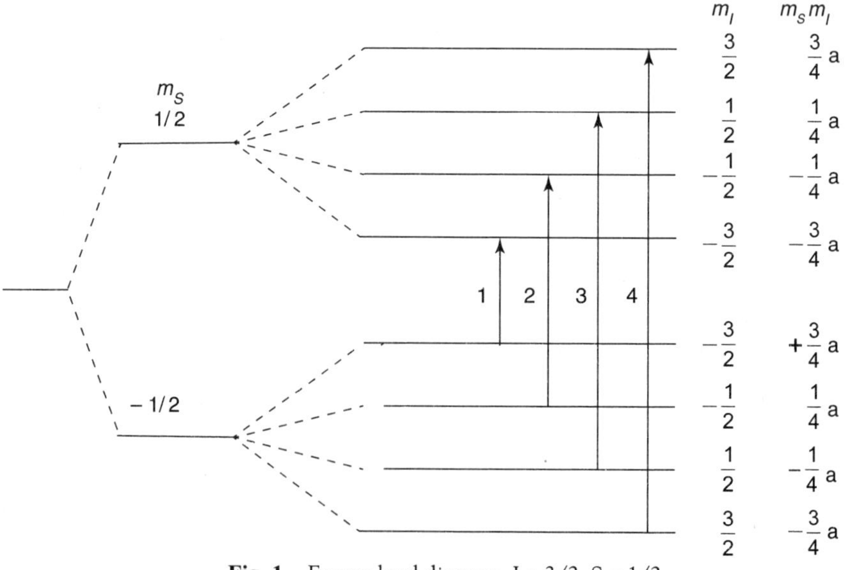

Fig. 1 Energy level diagram, I = 3/2, S = 1/2

13. (a) 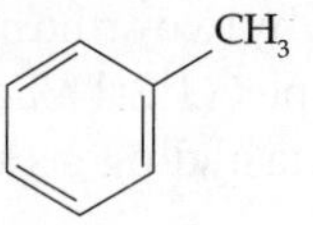 $3 \times 3 \times 2 \times 4 = 72$ lines

(b) 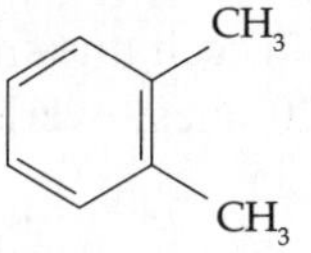 $3 \times 3 \times 7 = 63$ lines

(c) 1-hydroxyethyl radical $CH_3\dot{C}HOH$ (a_{CH_3} 2.2 a_{CH} 1.5 mT),
Quartet (1:3:3:1) due to CH_3, each line splits due to CH into a doublet (1:1)

(d) 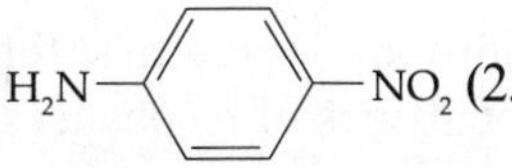 (2.1.1+1) (2.1.1+1) (2.2.1/2 + 1) (2.2.1/2 + 1) = 81 lines

H_2 of NH_2 do not give rise to splitting due to Drago's rule

(e) 9 lines (1:8:28:56:70:56:28:8:1); $a_H = 0.321$ mT

(f) 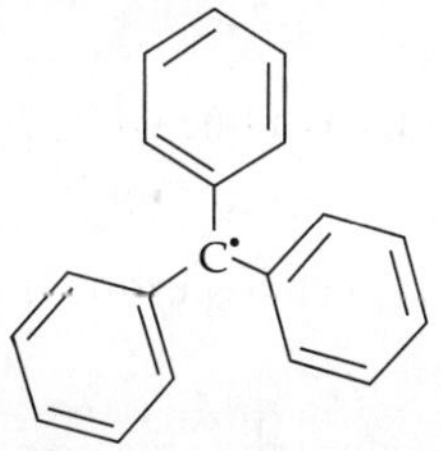

Six ortho hydrogens, six meta hydrogens and three para hydrogens
$(2 \times 6 \times 1/2 + 1)\ (2 \times 6 \times 1/2 + 1)$
$(2 \times 3 \times 1/2 + 1) = 7 \times 7 \times 4$
– 196 lines. All the 196 lines have been observed: a(o–) 0.253, a(m) 0.111 and a(p) 0.277 mT

14. (a) 3 triplets (1:2:1) due to H_2 of a triplet due to N

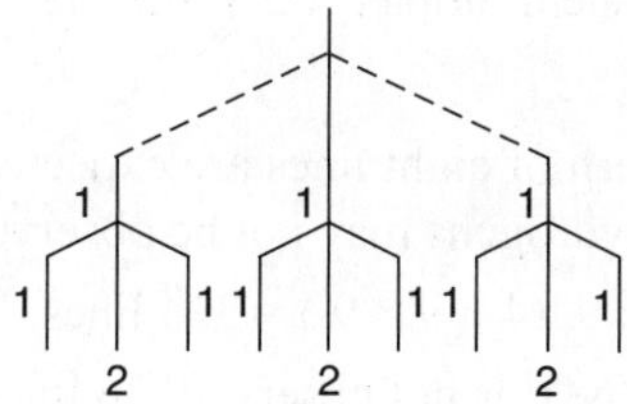

If $a_N = 2a_H$, then the intensity distribution will be 1:2:2:2:2:2:1

(b) 4 quartets (1:3:3:1) of a triplet (1:1:1) (due to N).

15. (a) Triplets (1:1:1) of a triplet (1:2:1) $a_H > a_D$

(b) Triplet (1:1:1) (due to N), each line splits into a heptet (1:6:15:20:15:6:1)

(c) A doublet (1:1) due to H, each line splits into a triplet (1:2:1)

(d) A doublet (1:1) due to CH, each line splits into a heptet (1:6:15:20:15:6:1)

(e) Quartet (^{11}B I = 3/2) (1:1:1:1), each of the lines splits into a quartet (1:3:3:1). Small value of the coupling constant shows BH_3 to be planar

(f) Triplet due to $\dot{C}H_2$ (1:2:1), each line splits into a quartet (1:3:3:1)

(g) Triplet due to N (1:1:1), (a_N 0.9 mT) each line splits by two ortho hydrogens of benzene nearer to N (0.31 mT) into another triplet (1:2:1). Each of the lines of the latter triplet splits again due to two meta hydrogens as a triplet (0.1 mT) (1:2:1). Total number of lines 27.

(h) A quintet (1:2:3:2:1) due to two N, (a_N = 0.15 mT) each line splits again into a quintet (1:2:3:2:1) (a_H = 0.11 mT); total 20 lines; overlapping of lines likely.

16. $g_{\parallel}$ 1.992 $g_{\perp}$ 2.0024

17. Eight lines, 332.8 ± 1.055 ± 1.435 ± 1.445, all of equal intensity

18. $(R_3Ge)_2 \xrightarrow{h\upsilon} 2R_3\dot{G}e$ (2 × 1 × 9/2 + 1) = 10 lines

19. $a(CD_3)$ = 0.35 mT = 2.3/6.51. Overall width for CH_3 6.9 mT, for CD_3 2.1 mT; CD_3: heptet (1:3:6:7:6:3:1). Relative intensities of resonances in NMR multiplets due to coupling with spins I > 1/2 are often referred to as non-Pascalian.

20. $CH_3(4)CD_2(5)$ = 20 lines

$a(CD_2)$ = 2.24/6.51 = 0.344 mT

The spectrum is a quartet due to CH_3 (1:3:3:1), each line splits by (CD_2) into a quintet (1:2:3:2:1).

21. $(CH_3)_2\dot{C}OH$, splitting due to $(CH_3)_2$ results in seven hyperfine lines, splitting due to OH not observed.

22. $CH_3\dot{C}HOH$; a quartet (1:3:3:1) due to CH_3, each line splits as a doublet due to CH (1:1), $a(CH_3) > a(CH)$. Free radical is $CH_3\dot{C}HOH$

23. $H_2\dot{A}$ type of spin system; triplet 1:2:1 centered at 0.3245 T, g = 2.048, a = 2.5 mT.

24. For V, I = 7/2, a pattern of eight lines are expected $\parallel$ and $\perp$ to V = O axis. In (b) coupling due to hydrogens may not be observed.

25. (a) Cu: 4; N: 9; CH_2; 9, (4 × 9 × 9) = 324 lines. For Cu, I = 3/2.

If coupling due to NH_2 is not observed, 36 lines. (Drago's rule).

26. $CH_3\dot{C}H_2$ $a(CH_3)$ 2.69 mT, $a(CH_2)$ 2.24 mT 4 × 3 = 12 lines. Considerable overlapping is expected and hence less number of lines are observed.

27. $O_2N.N_2O_4$ (2.1.1+1) (2.2.1+1) = 3 × 5 = 15 lines. NO_2 N_2O_4 is consistent with the spectrum.

28. SCl_2 (2.2.3/2+1) 7 lines (1:2:3:4:3:2:1); SCl_3 = 10 lines in the intensity ratio 1:3:6:10:12:12:10:6:3:1.

29. In McConnell relation, $a_H = Q\rho$

Q = –2.9 mT gives the best results

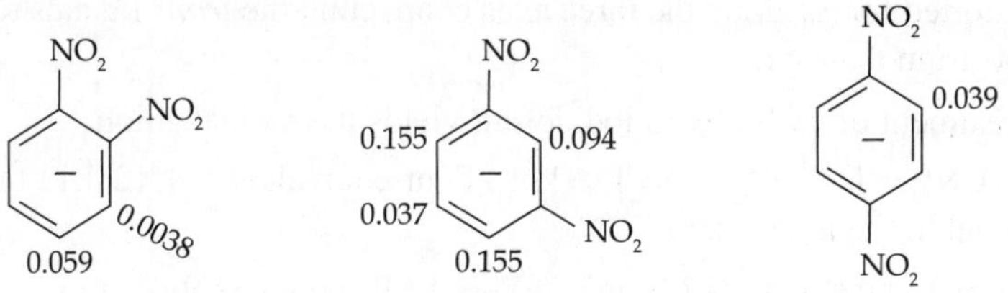

30. $\frac{\rho_1}{\rho_2} = 2 \quad \frac{\rho_9}{\rho_2} = \frac{0.194}{0.048} = 4.04$

HMO theory predicts ρ_2: ρ_1: ρ_9 that is, 0.048, 0.096, 0.194, that is 1:2:4 Observed coupling constants a_2:a_1:a_9 = 1:1.8:3.6. The agreement between the observed coupling constants and the HMO calculated charge densities is satisfactory.

31. (a) $0.425^2 = 0.181 \quad 0.263^2 = 0.069$

$\rho_1 = 0.181 \quad \rho_2 = 0.069$

$\rho_1/\rho_2 = 2.62 \quad \frac{a_1}{a_2} = \frac{0.495}{0.187} = 2.65$

32. (a) B_2 (b) by selective deuteration

(c) No. Sign of the coupling constants can be determined by studying ESR spectra of radicals trapped in solids.

33. $\dot{C}HCH_2$ a(CH) doublet (1:1), $a(CH_2)$ triplet (1:2:1), triplets of a doublet.

34. (a) $[Cu(NH_3)_4]^{2+}$ Cu I = 3/2, N I = 1, 4 × 9 = 36 lines

Hydrogens of NH_3 do not split the signal - Drago's rule.

(b) Due to Jahn-Teller distortion, two H_2O molecules would be at a longer distance from Cu than the other four.

D_{4h}: Cu(4), $4H_2O(9)$ $2H_2O(5)$ = 4 × 9 × 5 = 180 lines. For, Cu I = 3/2.

(c) Ti I = 5/2 (2.1 5/2 +1), $6H_2O$ (2.12.1/2+1) = 6 × 13 = 78 lines.

(d) Cr I = 3/2 4 × 9 × 5 = 180 lines.

Since four unpaired electrons are present in Cr^{2+}, the number of hyperfine lines expected is 270 × 4 = 1080. Hydrogens bonded to nitrogen in NH_3 do not split the ESR signal acc. to Drago's rule.

(e) $[Ti(H_2O)_4(NH_3)_2]^{2+}$ d^2 (2.5/2.1+1) (2.8.1/2+1) (2.2.1+1) = 6 × 9 × 5 = 270 lines. For ^{47}Ti I = 5/2.

Since two unpaired electrons are present, we have 2 × 270 = 540 lines

35. Three distorted forms (D_{4h}) with same energy exist; the spectrum becomes anisotropic. At 90 K an equilibrium of fast interconverting Jahn-Teller

distorted forms along the three axes connecting the *trans* ligands renders the spectrum isotropic.

36. Treatment of TCNE with iodide salt yields the radical anion, $C_2(CN)_4 + I^- \rightarrow [C_2(CN)_4]^- + ½.I_2$. Four equivalent ^{14}N, (2.4.1+1) splits the signal in the given ratio.

37. $a(CH_3)$ 2.43 mT (17/7 mT). The ESR signal splits into seven lines (1:6:15:20:15:6:1) due to $(CH_3)_2$ group each line being split into a doublet, a(CH) 0.26 mT. Usually the coupling constant varies as $sp^3 > sp^2 > sp$.

38. (a)

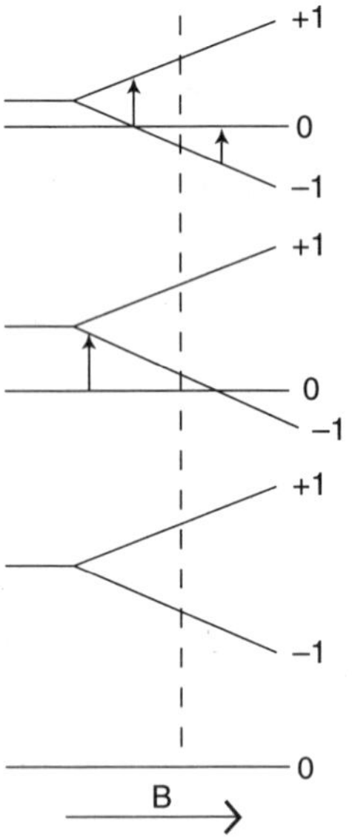

Fig. 2 Three different zero-field splittings. The dotted vertical line mark the position of the single resonance signal in the absence of zero-field splitting.

(b) Ni(II) is a d^8 system, with S = 1. As seen from Fig. 2(a), two transitions m_S −1 → 0 and 0 → +1 should be observed under the selection rule $\Delta m_S = \pm 1, 0$.

39. (a) Solution spectrum is isotropic, even though individual ions have non-cubic symmetry. A simple quartet spectrum is obtained. (b) The ions are immobilized when the solution is frozen. A random distribution of diffe rent orientations occur, resulting in an anisotropic spectrum; $g_x \neq g_y \neq g_z$.

40. (a) $Co(H_2O)_6^{2+}$ d^7, unpaired electrons 3 S = 3/2 ground state is a quartet. ZFS gives rise to the Kramers doublets ±1/2 and ±3/2, whose degeneracy is removed by the applied magnetic field. With $\Delta m_S = \pm 1$, three ESR lines are expected: −3/2 → −1/2, −1/2 → 1/2 and +1/2 → −1/2. Since I = 7/2 for Co, hyperfine splitting for each of the lines, though not all of them, may be observed.

(b) d^3 system, $S = 3/2$. The ground state is a quartet. If the ZFS is small, three ESR lines corresponding to $+1/2 \rightarrow 3/2, -1/2 \rightarrow +1/2$ and $-3/2 \rightarrow -1/2$ occur. If the ZFS is large, only the transition $-1/2 \rightarrow + 1/2$ will be observed.

41. (a) Much higher population difference between the ground and excited states. Boltzmann population N_{upper}/N_{lower} is about 200 times more than in NMR.

(b) Much faster relaxation

(c) Much larger transition energies/frequencies

42. (i) g-value anisotropy depends on the environment of the molecule in the solid/crystal and on the orientation of the radicals with respect to the external magnetic field in the crystal.

(ii) The values of the spin-orbit coupling (λ) and the crystal field 10 Dq and also the zero-field splitting determine the g-value to a large extent. These two have opposite effect on the orbital degeneracy. $g = 2 - \alpha k^2\lambda/10$ Dq where α is covalency factor, k is a constant. The value of g varies from 1.4 to 3.0 depending on the geometry of the complex. The g-value is also related to the electronic structure.

43. In O_h field (d^7), for CoF_6^{4-} due to zero-field splitting of $^4T_{1g}$, three Kramers doublets (±5/2, ±3/2, ±1/2) occur. The lowest doublet is separated from the next higher lying doublet by more than 200 cm^{-1} (kT). It behaves like a simple system with effective spin $S' = ½$. The g-value is anisotropic due to spin-orbit coupling and in distorted octahedral environments.

For CrF_6^{3-} (d^3), the ground state is $^4A_{2g}$. Small zero field splitting gives rise to two Kramers doublets (±3/2, ±1/2). The excited states are far removed from the ground state. Since no orbital angular momentum exists, there is no spin orbit coupling. The spectra are very nearly isotropic and can be obtained at room temperature.

44. Abundance of Mo ($I = 5/2$) is low, < 10% and hence only one line may be observed. When enriched with ^{13}C, 9 hyperfine lines are observed.

45. (a) Cu (2.1.3/2+1), N (2.2.1+1), H (2.2.1/2+1) = 4 × 5 × 3 = 60 lines. Experimental spectrum shows that the four signals due to Cu peaks split into 11 (not 15) lines due to overlapping. According to Drago's rule, hydrogen which is bonded to magnetically active atom (e.g., N) will not take part in hyperfine splitting. Deuteration of N–H produced an identical spectrum.

(b) Cu(4) x N(5) = 20 lines. Pentets (due to N) of a quartet (due to Cu, $I = 3/2$). The electron is mainly localized on Cu and hence shows large hyperfine splitting.

46. Lowest unoccupied HMO of butadiene

$\psi_3 = 0.602\ (\phi_1 + \phi_4) - 0.372\ (\phi_2 + \phi_3)$

$\rho_1 = \rho_4 = (0.602)^2 = 0.362 \quad \rho_2 = \rho_3 = (0.372)^2 = 0.138$

$\rho_1/\rho_4 = 2.62 \quad a_1/a_2 = 0.762/0.279 = 2.73$. The results are fairly consistent.

47. 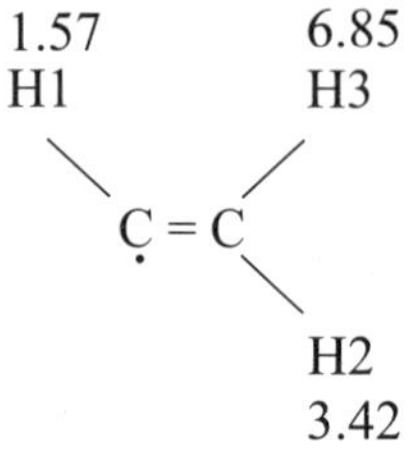

$a_1 = 1.57$ mT $a_2 = 6.85$ mT $a_3 = 3.42$ mT

A doublet due to a_2 (1:1), each line of a_2 split by a_3 into a doublet (1:1), which is split by a_1 into a doublet (1:1), total: $2 \times 2 \times 2 = 8$ lines (see Fig. 3). Eight lines are observed in frozen solution (see Fig.). In solution because of rapid interconversion due to resonance only two hyperfine splitting (1.57 and 10.28 mT) lines are found.

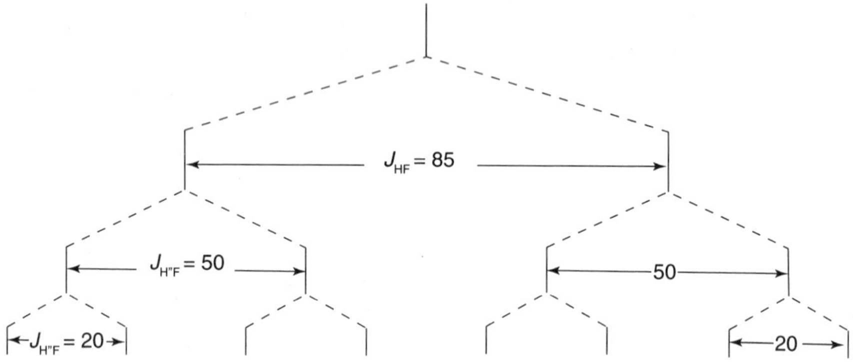

Fig. 3

48. (a) ESR spectra are simplified due to reduction in second order effects.
    (b) Increase in sensitivity in the study of disordered systems.
    (c) Precision and information obtainable from ENDOR increases.
    (d) Accessibility of spin systems with large zero-field splitting due to larger frequency/$h\upsilon$.
    (e) Study of the relaxation and dynamics of paramagnetic systems can be undertaken.
    (f) Higher spectral resolution over g factor. This is useful to study the structure and dynamics of spin-modified organic and biological systems.

49. In solid state $K_2[NO(SO_3)_2]$ exists as a spin-paired dimer which is diamagnetic. In aqueous solution, it dissociates into monomer radicals which show ESR spectrum, a sharp triplet, showing that the electron is mainly localized on N.

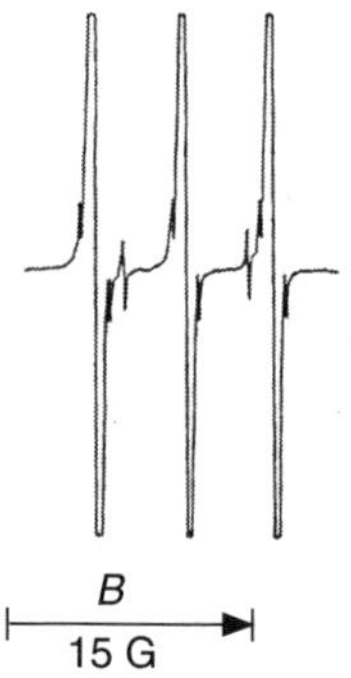

Fig. 4 X-band ESR spectrum of $K_2(SO_3)_2$ NO in aqueous solution. The smaller peaks arise from ^{15}N and ^{33}S hyperfine interaction.

50. If the complex is in a cubic environment (O_h or T_d), the g is isotropic, i.e., $g_x = g_y = g_z$. If the complex is in a tetragonal symmetry (i.e., trigonal bipyramid, square pyramid, square planar or elongated/compressed octahedral), $g_x = g_y \neq g_z$ (that is $g_{\|} \neq g_{\perp}$). If the complex is located in a rhombic symmetry (D_{2h}, C_{2v}, C_{2h}, etc.), then g is anisotropic, $g_x \neq g_y \neq g_z$. $g_{av} = (g_{11} + 2g_{\perp})$.

51. The ground state is $^3A_{2g}$. Here zero-field splitting can be ignored. The ground state is far removed from the excited states. In an applied magnetic field, $^3A_{2g}$ state splits into $M_s = +1, 0, -1$. The g-value is given by $g = g_e - 8\lambda/10$ Dq. The value of 10 Dq obtained from electronic spectral band due to $^3A_{2g} \rightarrow {}^3T_{2g}$ is 10,700 cm^{-1}. For the free ion, the value of spin-orbit coupling λ is -325 cm^{-1}; the calculated value of g is 2.245. The experimental value is 2.162. The decrease in g (also λ) also is attributed to the covalent bonding. Using the experimental g and 10 Dq, the value of λ is calculated to be -211 cm^{-1}.

52. The ground state of Mn^{2+}, ^6S becomes $^6A_{1g}$ in an O_h field. Small zero-field splitting produces three Kramers doublets, $M_s = \pm 5/2, \pm 3/2$ and $\pm 1/2$. These doublets split into two singlets by the applied magnetic field. Five transitions are expected as shown in the figure. Hyperfine splitting due to Mn nuclear spin $I = 5/2$ occurs giving totally 30 lines.

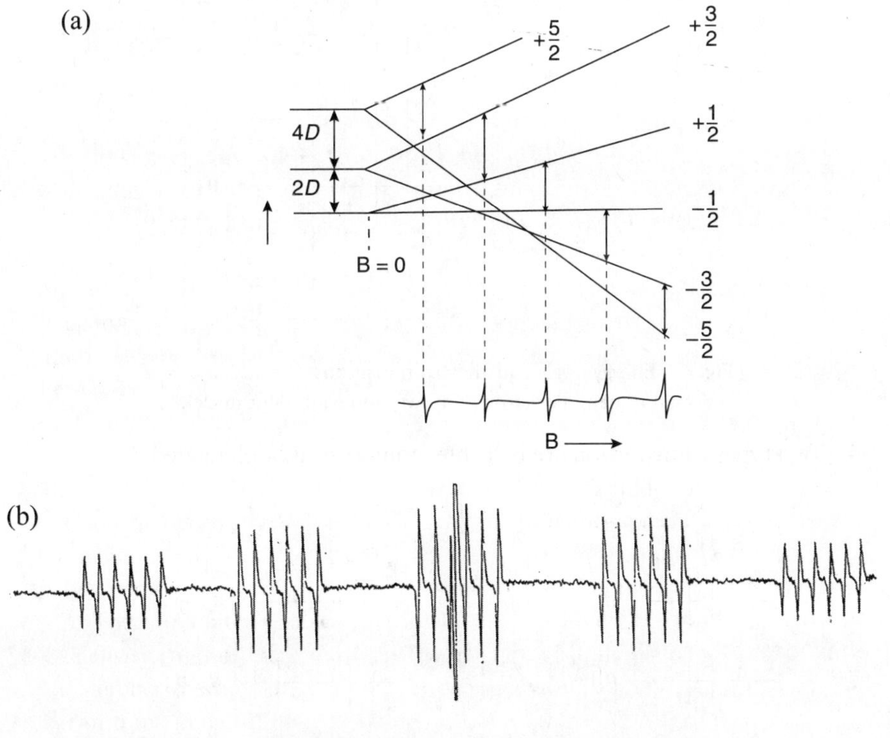

Fig. 5 (a) Energy levels and allowed transitions for a d^5 ion in a weak tetragonal field with B parallel to the tetragonal axis. (b) Hyperfine splitting due to Mn I = 5/2

53. The energy levels and the allowed ESR transitions are shown in the figure given below. ($I = 1/2$ $a_1 > a_2$). The four transitions are given by

Transition 1: $-g\beta B + \frac{1}{2}a_1 + \frac{1}{2}a_2$

2: $-g\beta B + \frac{1}{2}a_1 - \frac{1}{2}a_2$

3: $-g\beta B - \frac{1}{2}a_1 + \frac{1}{2}a_2$

4: $-g\beta B - \frac{1}{2}a_1 - \frac{1}{2}a_2$

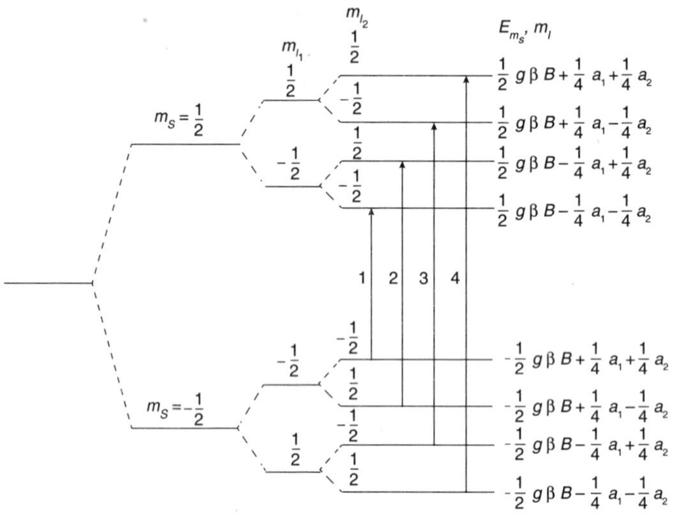

Fig. 6 Energy levels and allowed transitions, ($I = 1/2$, $a_1 > a_2$) for an unpaired electron with two non-equivalent nuclei.

54. Two types of distortion are possible, compressed or elongated.

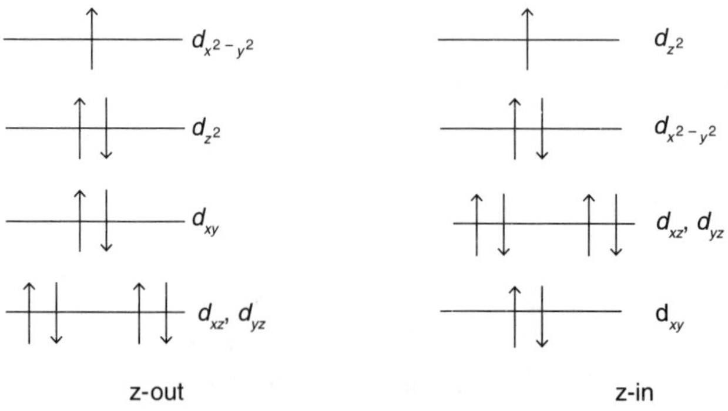

z-out: The electron density is concentrated along the z-axis.

z-in: The electron density is concentrated along the x-axis. Here, $g_{\perp} > g_{11}$.

Example: in $CuSO_4.5H_2O$, $g_{\parallel} > g_{\perp}$, OSO_3a weak ligand and H_2O, are coordinated along the z-axis. Here, electron density is higher in the z-direction.

55. Interaction energy: $\Delta E = g\beta B$

$= (2.0)(2.0\ T)(9.274 \times 10^{-24}\ JT^{-1})$

$= 0.371 \times 10^{-22}$ J

$$E = h\upsilon \qquad \upsilon = \frac{0.371 \times 10^{-22} J}{6.626 \times 10^{-34} Js} = 56.0\ \text{GHz}$$

56. (a) and (d)

57. $a_H = 50.8$ mT $\quad a_D = 7.8$ mT

Three lines will occur at 344.46, 352.26, 360.06 mT

58. The radical produced is $CH_3\dot{C}HOCH_2CH_3$. The ESR signal splits into a quartet (1:3:3:1) due to CH_3; each of the lines of the quartet splits into a doublet. Each of the lines of the doublet is split by CH_2 into a triplet giving altogether eight triplets, $a_{CH_3} > a_{CH} > a_{CH_2}$. The original positions of the quarter lines lie in between the triplets. The γ-CH_3 group is too distant to contribute to further splitting.

59. The hydrogens at 1, 4, 1′, 4′ give rise to a pentet (1 : 4 : 6 : 4 : 1) which again splits into a pentet giving pentets of a pentet a_0 2.89, a_m 0.1 G.

60. From the ESR spectrum of biphenyl, the coupling constants a_H(o) 2.63G, a_H(m) 0.41G and a_H(p) 5.26 G have been reported. Accordingly the spectrum should split into a triplet (1:2:1) (2.2. 1/2 +1) due to a_H(p). Each of the triplet lines should again split into a pentet giving three pentets (1 : 4 : 6 : 4 : 1) due to a_H(o) coupling. Each of the lines of the pentets should again split into a pentet (1 : 4 : 6 : 4 : 1) due to a_H(m) coupling giving totally 75 lines. Only nine pentets are seen due to overlapping. In benzyl radical, a similar trend is observed in the ESR coupling constants, a_H(o)5.1G, a_H(m) 1.8G and a_H(p) 6.1G.

Chapter 3 NMR

1. $$\upsilon = \frac{g_N \beta_N B}{h} = \frac{2.263 \times (5.051 \times 10^{-27}\ JT^{-1})(2.5\ T)}{6.626 \times 10^{-34}\ Js} = 43.12\ \text{MHz}$$

$$g = \frac{\mu}{I} = \frac{1.1305}{1/2} = 2.261$$

$E = h\upsilon = 6.626 \times 10^{-34}$ Js $\times 43.13 \times 10^{6}$ s^{-1}

$= 2.855 \times 10^{-29}$ kJ $= 1.72 \times 10^{-5}$ kJ mol^{-1}

2. (a) $\Delta E = h\upsilon = g_N\beta_N B = 5.5857 \times (5.051 \times 10^{-27}\ \text{JT}^{-1})\ (2.0\ \text{T})$
$= 5.641 \times 10^{-26}$ J

$$\upsilon = \frac{5.641 \times 10^{-26}\ \text{J}}{6.626 \times 10^{-34}\ \text{Js}} = 85.13\ \text{MHz}$$

(b) 1.625×10^{-26} J m_I

3. (a) $\frac{N_i}{N_0} = \exp(-\Delta E / kT) \approx 1 + \frac{g_N\beta_N B}{kT}$

Neglecting the second and higher powers

$$= 1 + \frac{(5.585)(5.051 \times 10^{-27}\ \text{JT}^{-1})(4.69\ \text{T})}{(1.38 \times 10^{-23}\ \text{JK}^{-1})(293\ \text{K})} \approx 1 + 3.27 \times 10^{-5}$$

$$\frac{N_i}{N_0} = \exp\left(\frac{-\Delta E}{kT}\right) = \frac{(2.68 \times 10^{8}\ \text{Ts}^{-1})(6.626 \times 10^{-34}\ \text{Js})(4.69\ \text{T})}{2(3.143)(1.38 \times 10^{-23}\ \text{JK}^{-1})(293\ \text{K})}$$

$= \exp\ (-3.28 \times 10^{-5}) = 0.999967$

$\frac{N_0}{N_i} = 1.000033$; 33 molecules in excess in the lower state for 10^6 molecules

(b) 6.116×10^{-26} J

4. (a) ABC (b) A_3X (c) AA′BB′C (d) A_3MX
(e) AA′BB′ (f) AM_2X (g) AA′BB′ (h) AA′XX′
(i) ABC (j) ABB′C (k) A_3XY (l) AB
(m) AA′BB′ (n) ABCD

5. B = 1.5 T, υ = 64 MHz, δ 7.89 ppm

6. Paraldehyde is a cyclic trimer of acetaldehyde. The triplet is due to -CH and the doublet arises from $-CH_3$.

7. ^{1}H: 5.167 T, ^{13}C: 20.54 T, ^{19}F: 5.49 T $\quad \nu = \frac{\gamma B_0}{2\pi};\ B_0 = \frac{2\pi\nu}{\gamma}$

8. Spectral line would be sharper in NH_4^+, nuclear quadrupole interaction is absent because of cubic symmetry (T_d).

9. The ^{31}P NMR of PF_5 is a sextet showing all the five fluorines to be equivalent. The axial and equatorial fluorines become chemically equivalent due to Berry pseudo rotation. At room temperature ^{19}F NMR shows only a single (1 : 1) doublet due to splitting from ^{31}P showing all the fluorines to be equivalent. At low temperature (–100 °C), ^{19}F NMR exhibits two signals in the ratio of 2 : 3 for axial and equatorial fluorines respectively.

10. The four protons in 1,2–dibromoethane appear to be chemically equivalent. However, the Newman diagrams (Fig. 1) show clearly that protons in different

conformations are not magnetically equivalent. If the NMR spectrum is recorded at low temperature, the rotation around the C–C bond are 'frozen' sufficiently for these differences to be detected giving rise to a spectrum with multiplet structure. At room temperature, rotation is so rapid that each proton experiences the same time-averaged environment giving only one sharp signal.

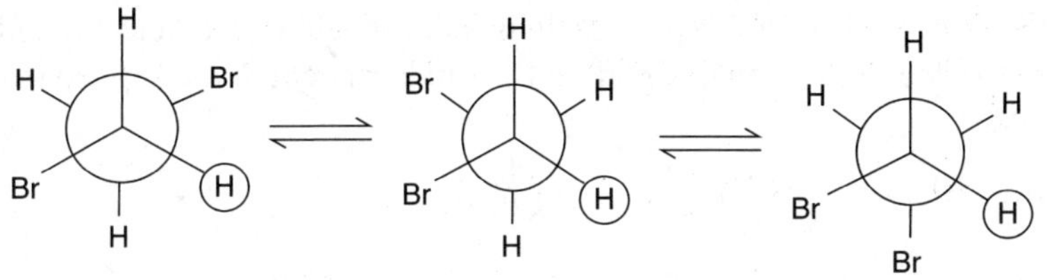

Fig. 1

11. Three of the four ^{31}P atoms are equivalent (Fig. 2). The spectrum would consist of a doublet (1:1) and a quartet (1:3:3:1) in the intensity ratio of 1 : 3.

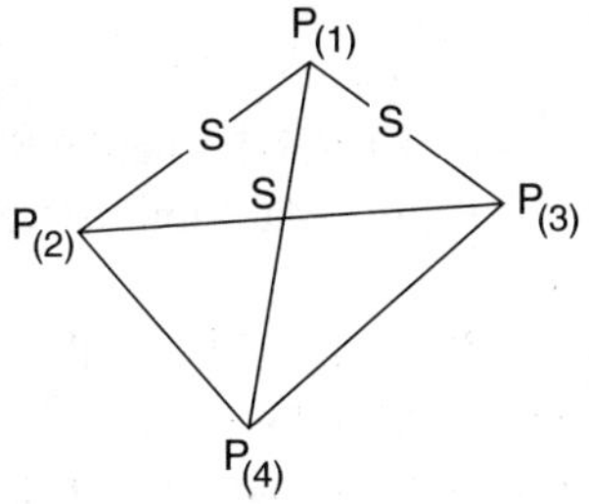

Fig. 2

12.

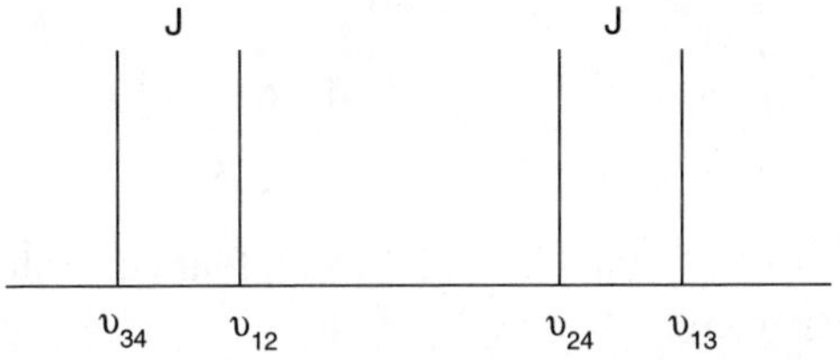

$$J_{AB} = \frac{0.1+0.1}{2} = 0.1 \qquad 0.1 \text{ in 100 MHz unit} = 10 \text{ Hz}$$

$$\upsilon_{34} - \upsilon_{24} = \upsilon_{12} - \upsilon_{13} = \delta\, 4.7 \quad \text{chemical shift} = \delta\, 4.7 \text{ ppm}$$

$$A = \delta\ 5.8 \quad B = \delta\, 1.15$$

13. CH_3CD_2OH
The CH_3 signal would split into a pentet (1:2:3:2:1) due to spin $I = 1$ of D. The CD_2 signal would be a quartet (1:3:3:1). If the sample is dry, splitting due to OH is observed. Then each of the lines of the quartet will split into a doublet and the OH signal would be a much weaker pentet (1:2:3:2:1).

14. (a)

15. 10

16. (a) 11×10^{-6} T (b) 0.110×10^{-3} T

17. $1.90 \times 10^{-3}\ s^{-1}$

18. Chemical shift in Hz is directly proportional to the applied magnetic field but the chemical shift value on δ scale is independent of the field strength. The magnitude of the coupling constant J is independent of the applied magnetic field.

19. 301 MHz; $h\nu = \mu B_0 / I$ or $\nu = g\mu_N B_0 / h$

20. Because of very low natural abundance (about 1.1%) of ^{13}C.

21. To reduce the applied magnetic field inhomogeneity by spinning the sample along its longitudinal axis.

22. (b) and (c)

23. The spectrum will show a quartet (1:3:3:1) arising from coupling with CH_3 and the CH_3 group will show a 1:4:6:4:1 pentet by coupling with ^{11}B, I = 3/2 (abundance 30%) with the same separation between the lines as found in the quartet.

24. The sensitivity of an NMR nucleus depends on the magnetic moment μ which determines the energy differences between nuclear spin states (Larmor frequency) and therefore the population excess in the lower energy state. At constant magnetic field, the signal strength is proportional to μ^3 and $\upsilon = \gamma B_0/2\pi$.

25. $\upsilon = \dfrac{\gamma B}{2\pi} \left(\dfrac{\nu_H}{\gamma_H} = \dfrac{\nu_C}{\gamma_C} \right) \quad \nu_C = \dfrac{\nu_H \gamma_C}{\gamma_H} = \dfrac{42.4 \times 6.73 \times 10^7}{26.75 \times 10^7} = 10.6 \text{ MHz}$

26. Two equivalent Fs (3), one ^1H(2), 2 ^{15}N(3), four equivalent H(5); total number of lines expected : $3 \times 2 \times 3 \times 5 = 90$.

27. The assignment of ^{13}C spectrum is straightforward : δ 171 OCO, δ 22 OCH_3, δ 20 $(CH_3)_2$, δ 28 CH and δ 71 CH_2. This is consistent with the off-resonance spectrum.

28. θ (radians) $= \gamma t_p B_1$ $\quad 90° = \pi/2$

$$\frac{3.143}{2} = 26.752 \times 10^7 \text{ rad T}^{-1}\text{s}^{-1} \times 2.5 \times 10^{-4} \text{ T} \times t_p$$

$$t_p = 23.5 \times 10^{-6} \text{ s}$$

29. The ^{19}F signal is split into a doublet by CH proton (J 50 Hz) and each line is further split into a quartet by CH_3 protons (J 22 Hz). The ^{19}F resonance is a doublet of quartets.

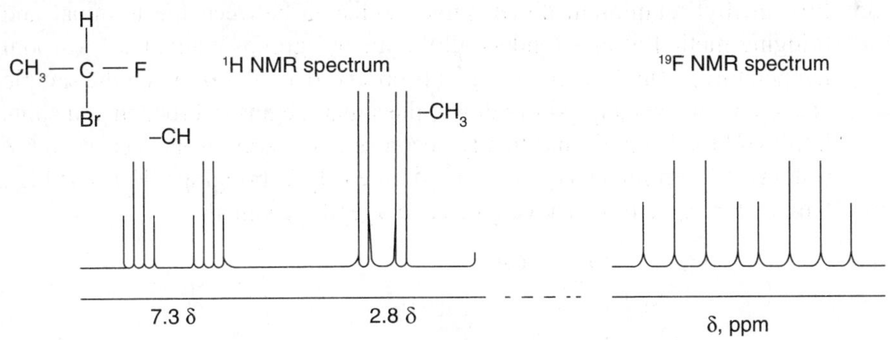

Fig. 3

In the proton spectrum, the CH proton is split (J 6 Hz) by CH_3 into a quartet and each line of the quartet is split by fluorine into a doublet (J_{HF} 50 Hz) giving a doublet of quartets. CH_3 resonance is split by CH (J 6 Hz) into a doublet and each line of doublet is split by ^{19}F vicinal coupling ($J = 22$) into doublet giving two doublets for CH_3 resonance.

30. $\delta\upsilon = 390\,\text{Hz} \quad \tau = \dfrac{\sqrt{2}}{\pi \times 390\,\text{s}^{-1}} = \dfrac{1.414}{1225.8\,\text{s}^{-1}} = 1.15 \times 10^{-3}\,\text{s}$

The signal will collapse into a single line when the rate of interconversion exceeds about 867 s^{-1} ($1/\tau$).

31. (a) ortho-3, meta 4, para 2
    (b) 3 and 2 respectively
32. An organic compound contains 1.1% of the isotope ^{13}C ($I = ½$) in natural abundance. These molecules show scalar coupling between ^{13}C and the proton that leads to doublet splitting in the ^1H NMR spectrum $J(^{13}C–^1H)$ is observed at 104 Hz on either side of the ^1H resonance. These are the so-called ^{13}C satellites.
33. 0.59 mT; For 180° pulse 20 μs
34. Both molecules have a plane of symmetry which bisects the five membered rings and we expect three ^1H resonances in the ratio 1:2:2. Each ring rotates rapidly in its own plane so that all protons have the same average environment and become indistinguishable.

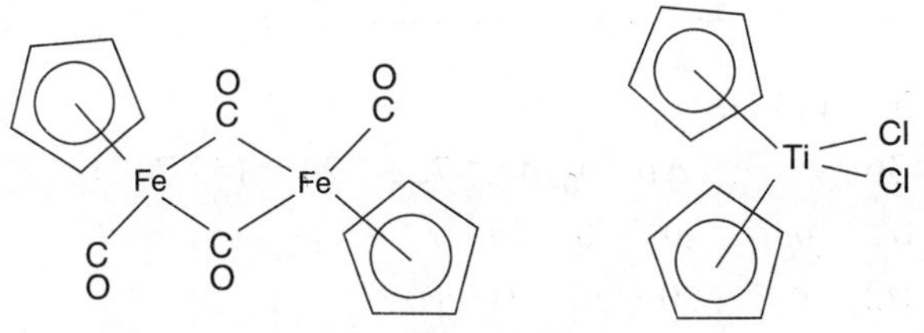

Fig. 4.

35. In trimethyl aluminium dimer, rapid exchange between the terminal and bridging methyl groups renders all the methyl groups equivalent. At room temperature, a single average signal is observed at –0.3 δ. When the sample is cooled, the exchange slows down, the signal begins to broaden and splits into two broad signals and finally becomes two sharp resonances at –0.6 δ (intensity 4, terminal CH_3) and +0.5 δ (intensity 2, bridging CH_3). The high temperature spectrum is a weighted average of the signals.

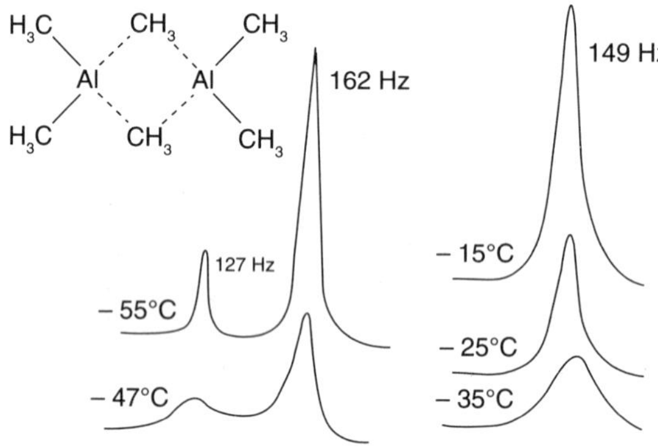

Fig. 5

36. (a) The single peak at δ 3.6 can be assigned to a –OCH_3 group. A triplet at δ 2.2 suggests a CH_3 group adjacent to a CH_2 group. Subtracting CH_3, OCH_3 and CH_2 groups from the molecular formula leaves C_2H_2O which is like to be CO–CH_2 and hence the compound is likely to be $CH_3COCH_2CH_2CH_3$. The central CH_2 gives rise to a multiplet signal at δ 1.7.

(b) The signal at δ 1.1 can be assigned to the CH_3 group adjacent to a CH_2 group, δ 2.0. The resonance at δ 4.2 is attributable to CH adjacent to CH_2 group. The singlet at δ 11.0 is due to -COOH group. The compound is

Br
|
CH_3CH_2CH—COOH

37.

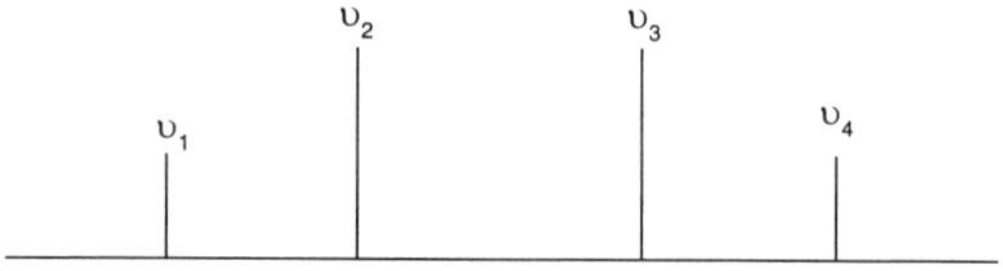

$J = \upsilon_1 - \upsilon_2 = 144$ Hz

$2C - J = 36 \qquad \Delta\upsilon = \upsilon_0\,\delta = \sqrt{72 \times 324} = 152.74$ Hz

$\upsilon_A = z - 1/2 \quad \upsilon_0\,\delta = 90 - 76 = 14$ Hz

$\upsilon_B = z + 1/2 \quad \upsilon_0\,\delta = 90 + 76 = 166$ Hz

$$\delta_A = \frac{14}{40} = 0.35 \quad \delta_B = \frac{166}{40} = 4.15 \quad \delta_B - \delta_A = 3.80$$

38. Naturally occurring cadmium is a mixture of ^{111}Cd (13%) and ^{113}Cd (12%) isotopes each with $I = 1/2$ and for the remaining ^{112}Cd $I = 0$. In the spectrum, satellites caused by spin–spin interaction with Cd are observed, which disappears when the individual methyl groups are exchanged between different cadmium atoms. With increasing temperature, the cadmium satellites initially broaden and finally disappear. However, one can still observe ^{13}C satellites after cadmium satellites disappear showing that the exchange involves entire methyl groups and not individual protons.

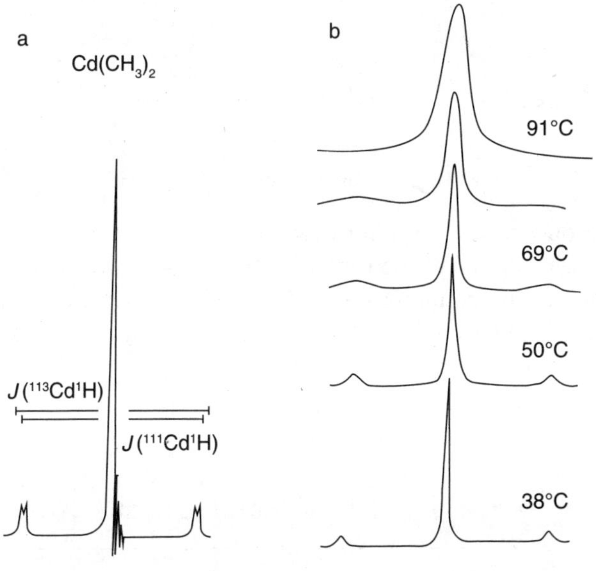

Fig. 6

39. In benzene, δ(^{1}H) and δ(^{13}C) occur at δ7.27 and δ128.5, respectively while in cyclohexane, δ(^{1}H) occurs at δ1.44 and δ(^{13}C) at δ26.3. The ^{13}C–^1H coupling constant in benzene and cyclohexane occurs at 159 Hz and 123 Hz, respectively. ^{13}C–^1H coupling constant increases with increasing *s* character of the ^{13}C–^1H bond. In benzene, δ(^{1}H) occurs at a high field due to ring current effect; proton is deshielded.

40. The signal at δ 191 can be assigned to –CHO since it is a doublet. The ^{13}C spectral region is typical of a phenyl group. Subtracting –CHO from $C_7H_6O_2$ leaves C_6H_5O. The compound is likely to be p–hydroxybenzaldehyde. The bands at δ 130 and δ 164 can be assigned (taking benzene δ 128 as reference) to carbons with substituent C_1(CHO) and C_4(OH), respectively. The signals at δ 133 and δ 130 can be assigned to C_2, C_6 and C_3, C_5 carbons, respectively.

41. ^{19}F NMR: At low temperature, two groups of resonance of equal intensity, each a doublet $^1J_{PF}$ of triplets ($^2J_{FF}$) are expected. All the lines broaden on warming and coalesce to give a simple doublet ($^1J_{PF}$).

^{31}P NMR: At low temperatures, the ^{31}P spectrum of $PF_4N(CH_3)_2$ is a triplet of triplets with coupling to two axial and two equatorial fluorine nuclei. On warming, the pattern changes and becomes a regular quintet due to fluorine atoms exchanging sites.

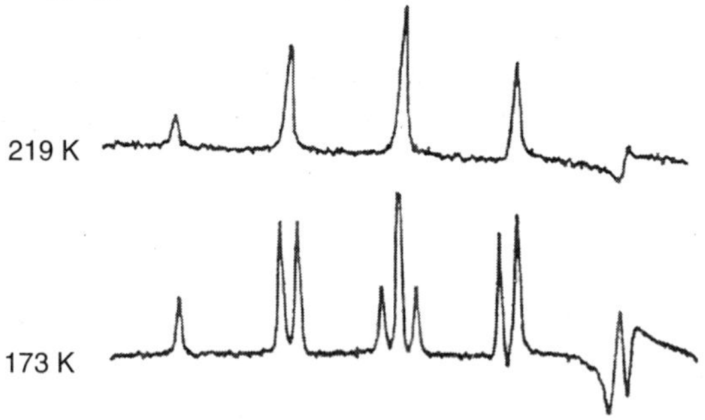

Fig. 7 ^{31}P NMR spectrum of $PF_4N(CH_3)_2$

42. (a) ^{31}P resonance will be split by two (equivalent) fluorine nuclei into a triplet (1:2:1). Each line of the triplet is further split into a doublet resulting in three doublets by coupling with ^{1}H. (b) ^{31}P resonance is split by ^{1}H into a doublet (1:1) and each line of the doublet is split into a triplet (1:2:1) by two equivalent fluorines.

43.

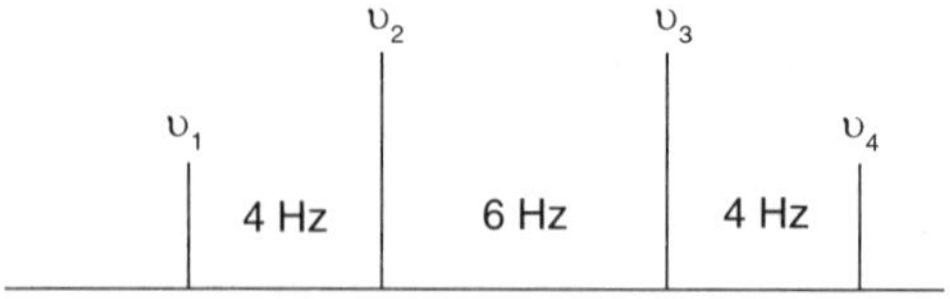

$J = \upsilon_1 - \upsilon_2 = \upsilon_3 - \upsilon_4 = 4$ Hz $\qquad 2C = \upsilon_1 - \upsilon_3 = 10$ Hz $= \upsilon_2 - \upsilon_4$

$\upsilon_0\delta = \sqrt{4C^2 - J^2} = \sqrt{(2C - J)(2C + J)} = \sqrt{(\upsilon_2 - \upsilon_3)(\upsilon_1 - \upsilon_4)} = \sqrt{6 \times 14} = 9.165$

z = centre of multiplet

$\upsilon_A = z - 1/2\ \upsilon_0\delta \qquad \upsilon_B = z + 1/2\ \upsilon_0\delta$

$\quad = 7 - 4.6 = 2.4 \qquad\qquad = 7 + 4.6 = 11.6$

$\delta_A = 0.04 \qquad \delta_B = 0.193$

$$\frac{I_2}{I_1} = \frac{I_3}{I_4} = \frac{(\upsilon_1 - \upsilon_4)}{(\upsilon_2 - \upsilon_3)} = \frac{14}{6} = 2.33$$

44. (a) ABX_2 (b) AA′ XX′ (c) A_4 (d) ABCD (e) AA'BB'

45. (a) AX (b) AX (c) AX_2 (d) AA′XX′ (e) A_2X_2

46. The signals at δ 7.27, δ 5.30 and δ 2.17 arise from chloroform, methylene chloride and acetone, respectively. Taking into account the number of protons in each compound, the integration indicates a molar ratio of 10:9:6, respectively.

47. Natural abundance of ^{29}Si ($I = 1/2$) is 4.7% and the remaining 95.3% ^{28}Si has $I = 0$. Therefore, 90.82% of the molecules will have two ^{28}Si and 8.96% will have one ^{29}Si and one ^{28}Si. The resonance due to molecules with two ^{29}Si atoms (i.e., $H_3{}^{29}Si\,{}^{29}SiH_3$, 0.22%) are too weak to be observed.

The resonance due to three protons bound to ^{29}Si will split into a wide doublet ($^1J_{SiH}$ splitting) while that due to the other three hydrogens will split into a narrow doublet. But the two sets of protons are no longer equivalent and further splitting ($^3J_{HH}$) of the satellites into quartets is observed. In Fig. 8, the outer quartets are clearly seen but the inner ones are only partially resolved from the intense central peak. $J_{Si\text{–}H}$ 118 Hz, $^2J_{Si\text{–}H}$ 6.6 Hz and Si_2H_6 $^3J_{HH}$ 50 Hz.

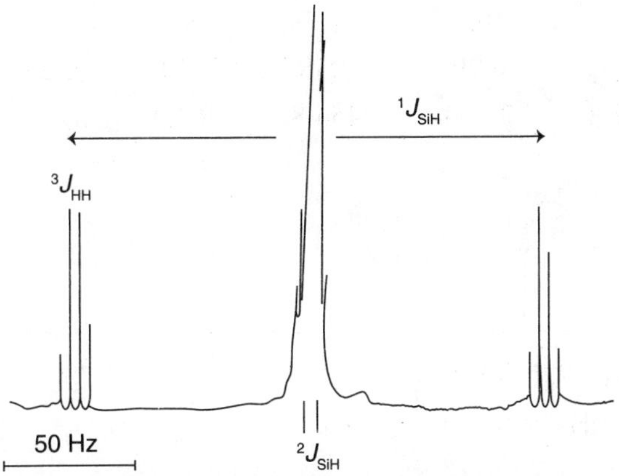

Fig. 8 ^{1}H NMR spectrum of Si_2H_6.

48. The signal at δ 7 ppm suggests the presence of a phenyl group. Subtracting C_6H_5 from C_8H_{10} leaves C_2H_5. The signal at δ 2.4 could be assigned to CH_2 group and that at δ 1.2 ppm to CH_3. These assignments are in accord with the given peak intensity ratio. The compound is ethylbenzene.

49. The ^{1}H NMR spectrum of $Sn(CH_3)_4$ shows a singlet (δ 0.075 ppm) arising from the 12-equivalent protons. Tin has two isotopes, ^{117}Sn (7.6%) and ^{119}Sn (8.6%) for both $I = 1/2$. Molecules which contain either of these spin-active nuclei will couple with protons to give two sets of weak doublets equally spaced about the central peak, J (^{117}Sn–H, 52 Hz) and J (^{119}Sn–H, 54 Hz). As shown in Fig. 9, coupling to the proton nuclei will occur resulting in a 13 line pattern (2 (12 × ½) + 1). The spectrum however does not show very weak peaks, only 9 lines may be seen.

50. A dynamic equilibrium exists due to barrier to rotation about the amide bond.

$$\text{Rate constant } k_c = \frac{\pi \delta \upsilon}{\sqrt{2}} \text{s}^{-1} = \frac{3.143 \times 0.2 \times 200}{\sqrt{2}} = 89.9\,\text{s}^{-1}$$

$$\tau = \frac{\sqrt{2}}{\pi \delta \upsilon} = \frac{1.414}{3.143 \times 40\,\mathrm{s}^{-1}} = 1.125 \times 10^{-2}\ \mathrm{s}$$

On a 500 MHz instrument

$$\tau = \frac{\sqrt{2}}{\pi \delta \upsilon} = \frac{1.414}{3.143 \times 100} = 4.5 \times 10^{-3}\ \mathrm{s}$$

$$k_c = \frac{\pi \delta \upsilon}{\sqrt{2}} = \frac{3.143 \times 100}{1.414} = 222.3\,\mathrm{s}^{-1}$$

51. (a)

	υ_1	υ_2	υ_3	υ_4	
δ	6.13	6.29	7.58	7.74	x 60
Hz	367.8	377.4	454.8	464.4	

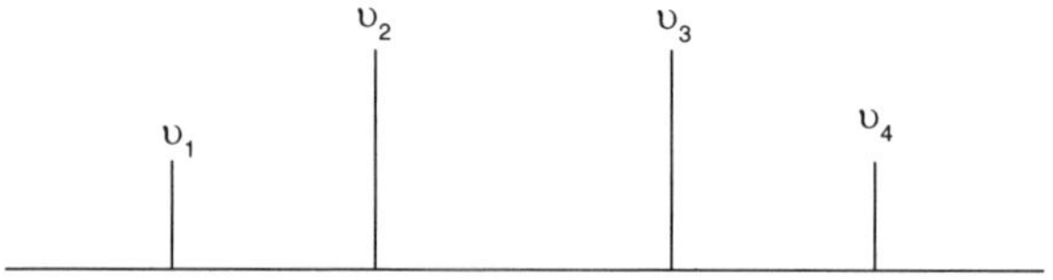

$\upsilon_1 - \upsilon_2 = \upsilon_3 - \upsilon_4 = 9.6$ Hz

$\upsilon_1 - \upsilon_3 = 87$ Hz $= \upsilon_2 - \upsilon_4$

$\upsilon_0 \delta = 86.5$

$\upsilon_A = 372.9$ Hz $\qquad \upsilon_B = 459.3$ Hz

δ_A 6.215 $\qquad \delta_B$ 7.655

Intensities $\dfrac{I_2}{I_1} = \dfrac{I_3}{I_4} = \dfrac{\upsilon_1 - \upsilon_4}{\upsilon_2 - \upsilon_3} = \dfrac{96.6}{77.4} = 1.25$

(b) When nuclei are strongly coupled,

(i) the centres of the multiplets do not reflect the true chemical shift of the resonance,

(ii) separation between the lines often do not represent the coupling constant

(iii) and intensities of lines become distorted.

This occurs when the chemical shift difference measured in Hz is less than about ten times the relevant coupling constant. The use of higher magnetic field will increase the probability of the first order approximation being valid.

52. $\dfrac{(\gamma H)\, 26.75}{(\gamma D)\, 4.11} = 6.51$ ^1H–^1H coupling constant $= 6.51 \times 44 = 286$ Hz

No. of lines in the proton spectrum, 3, intensity (1:1:1)

No. of lines in ^2D spectrum: 2, (1:1). Proton-proton coupling constant for H_2 molecule can be obtained from the spectrum of HD.

53. The J_{HF} coupling splits the signal into a doublet (1:1), each of the lines of the doublet is split by the coupling $J_{H'F}$ into another doublet (1:1), giving two doublets. The $J_{H''F}$ splits each of the lines of the latter two doublets into two doublets. Finally the spectrum would consist of four doublets. The proton coupling constants are not relevant to the ^{19}F NMR spectrum.

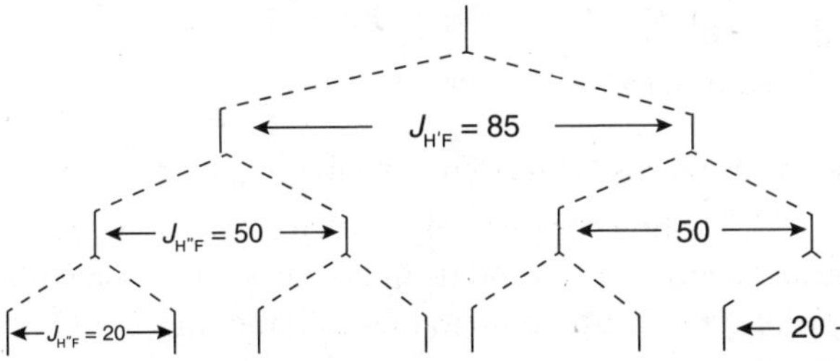

Fig. 9

54. (a) The spectrum of GeF_4 has a large central resonance from molecules containing non-spinning (I = 0) Ge isotopes with ten nearly equal satellites attributed to molecules containing ^{73}Ge (I = 9/2, natural abundance 7.8%) are found.

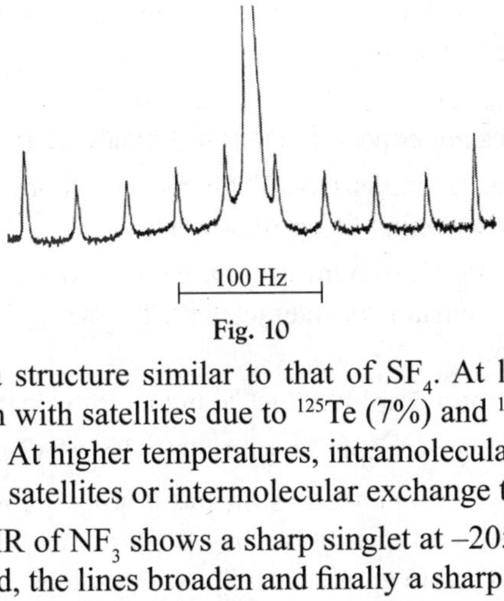

Fig. 10

(b) $TeCl_4$ has a structure similar to that of SF_4. At low temperature, two triplets each with satellites due to ^{125}Te (7%) and ^{123}Te (1%) both with I = ½ appear. At higher temperatures, intramolecular exchange leads to a singlet with satellites or intermolecular exchange to a singlet.

(c) The ^{19}F NMR of NF_3 shows a sharp singlet at –205 °C. As the temperature is raised, the lines broaden and finally a sharp triplet (I = 1, for ^{14}N) is observed at 20 °C. At low temperature, slow molecular motions are most effective for quadrupolar relaxation of ^{14}N and as a result, a singlet is observed. At higher temperatures relaxation is not as effective and the lifetime of a given ^{14}N is sufficient to cause spin-spin splitting.

(d) Axial fluorines of ClF_3 give rise to a doublet and the equatorial fluorine gives rise to a triplet.

(e) ^{19}F spectrum of SF_4 shows two triplets at –98 °C. At higher temperature, a sharp singlet is observed due to fluxional behaviour.

55. $\delta = (1741 + 1747 + 1753 + 1759)/4 = \dfrac{1750}{500\ \text{MHz}} \times 10^6 = 3.5$ ppm

$J = 1747 - 1741 = 6$ Hz

56. Consistent with DEPT spectra, the following assignments can be made.

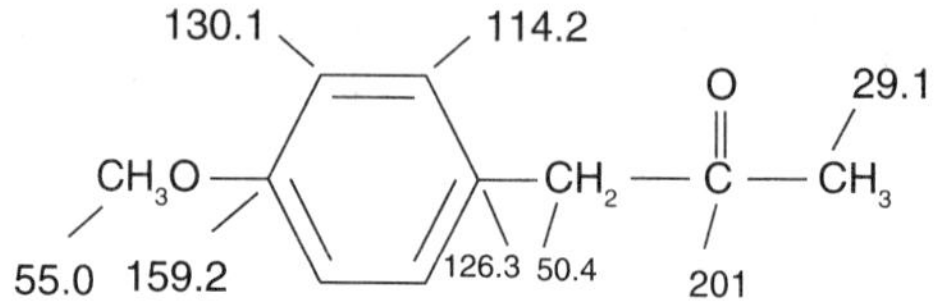

These assignments are also consistent with those made for benzene derivatives.

57. The signal at δ 191 shows the presence of HC=O group since it is a doublet in the off-resonance spectrum and its intensity is weak. The other signals are in the region of an aromatic compound. Subtracting –CHO and C_6H_4 from $C_7H_6O_2$ leaves OH. The compound is likely to be 4–hydroxybenzaldehyde. The signal δ 164(s) can be assigned to C–4 and that at δ 130(s) to C–1 consistent with the obsd intensity. The electronegative oxygen shifts C–4 signal upfield. The signals at δ 133(d) and δ 117(d) can be assigned to C_2, C_6 and C_3, C_5 respectively.

58. (a) The spectrum of $Fe(CO)_5$ should show two peaks in the intesity ratio 2 : 3 due to axial and equatorial groups each with weak ^{57}Fe satellites. But fluxional behaviour makes all the groups appear equivalent.

(b) In the spectrum of $Fe_2(CO)_9$ two peaks with intensity ratio 2 : 1 and with ^{57}Fe satellites are expected due to terminal and bridging groups.

In polynuclear metal carbonyls, the possibility of exchange between the terminal and bridging CO groups exists.

59. ^{1}H NMR: Coupling of ^{31}P with Me_2 gives rise to a doublet (J_{HP} 4 Hz). The Me_2 group is too distant to interact with the two ^{19}F nuclei as well as the single Me group.

The single Me group is split by the ^{19}F nuclei into a triplet (1:2:1, J_{HF} 20 Hz). Each of the lines of the triplet is split into a doublet by ^{31}P (J_{HP} 8 Hz).

^{19}F NMR: The ^{19}F resonance is split into two by ^{31}P (J_{PF} 70 Hz), each line being further split into quartet by the CH_3 protons (of the single Me group) (J_{HF} 20 Hz) resulting in approximately two quartets.

^{31}P NMR: The ^{31}P signal will split into a triplet by the two ^{19}F nuclei (J 70 Hz), each line would then be split into quartet by the single Me protons (J 8 Hz); finally each of these twelve lines would be split into septet by the Me_2 protons (J 4 Hz). Considerable overlapping of lines occurs and the spectrum would be difficult to analyze (not shown in Fig.11).

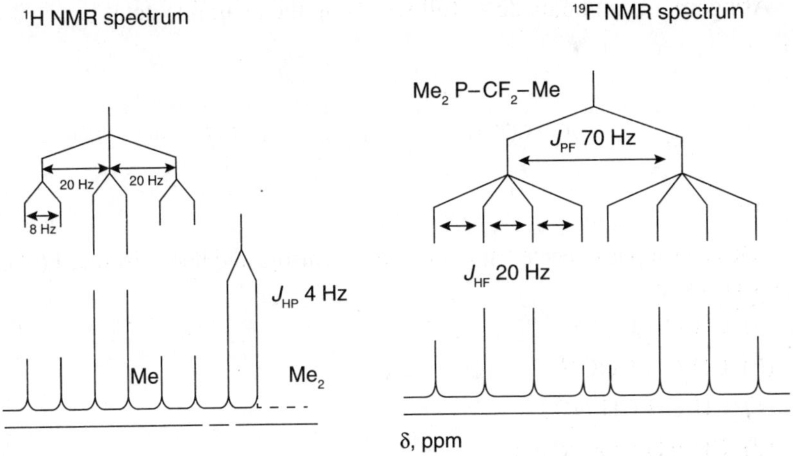

Fig. 11

60. An intense quartet appears due to ^{19}F bonded to As ($I = 3/2$). Each peak of this quartet is split into a doublet showing coupling to a single unique fluorine. The second quartet is composed of a symmetrical multiplet (1:4:6:4:1) indicating coupling to four ^{19}F. This suggests square pyramidal structure for AsF_5Cl.

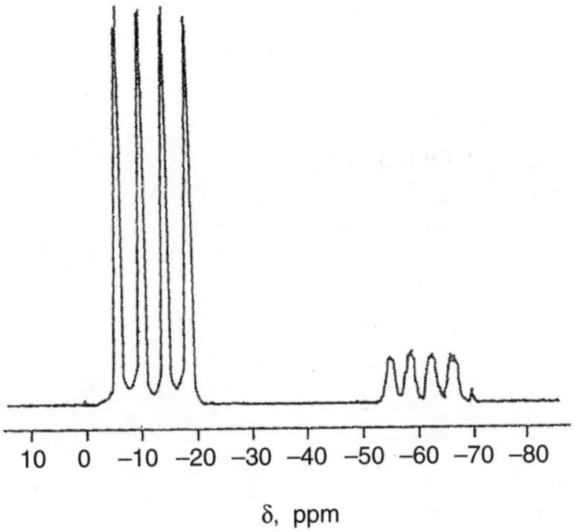

Fig. 12

61. Both samples have CO group ($\delta \sim 200$ ppm). Subtracting CO from C_4H_8O leaves C_3H_8. Aldehyde –CHO shows ^{13}C signal in the region (δ 200–210), the keto group (δ 205–220). This leaves $CH_3CH_2CH_2CHO$. The band at δ 45.7 can be assigned to CH_2 group next to CO. In sample A δ 13.3 and δ 15.7 can be assigned to CH_3 and CH_2 respectively. In sample B, the signals at δ 29.4 and δ 36.8 can be assigned to CH_3 and $-CH_2$ group next to –CO respectively. The signal at δ 7.9 can be assigned to CH_3 next to CH_2. The sample B is $CH_3CH_2COCH_3$.

62. Assignment can be made as follows using the available data

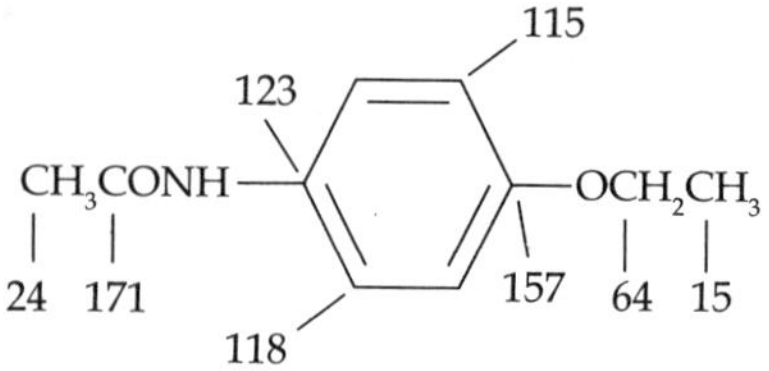

–OCH_3 group is expected at δ 50-55 as in anisole and hence isomer I (phenacetin) is likely.

63. (a) $CH_3CH_2CONH_2$
    (b) CH_3CH_2COOH
    (c) $CH_3CH_2CH_2NO_2$
    (d) $CH_3CH_2COONHCH_3$

64. (a) $CH_3–CH_2–CH_2–O–CH{=}CH_2$
    δ 16 36 71 155 90 ppm
    (b)

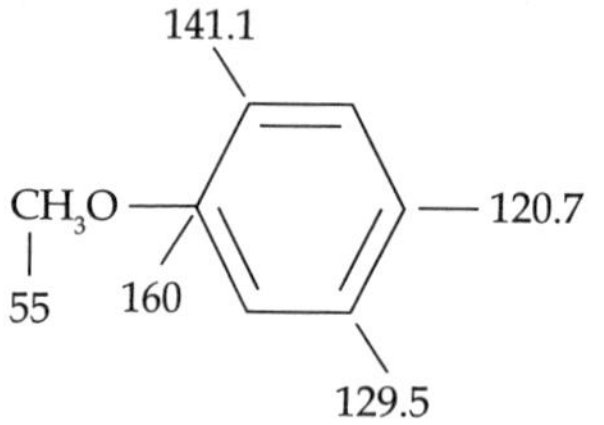

(c) $H_3C–CH_2{=}CH_2–COOCH_2CH_3$

δ 21 135.5 145 168 60 15

(d)

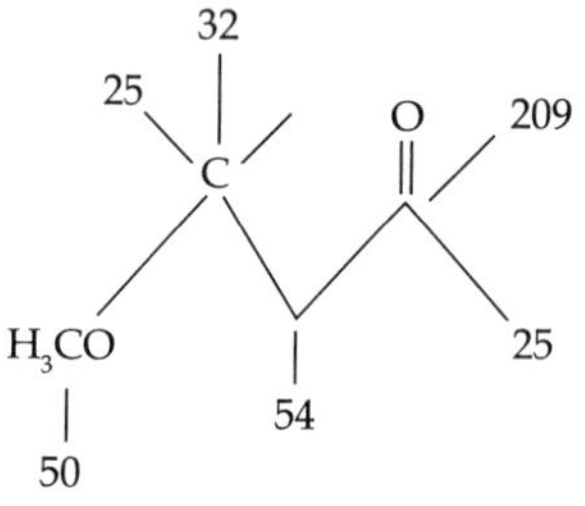

(e)

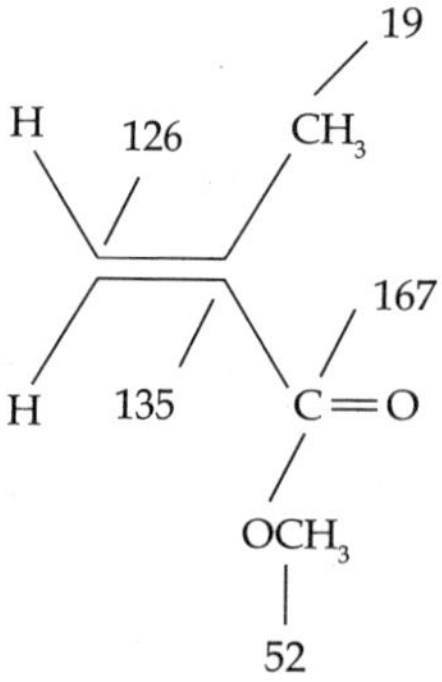

(f)

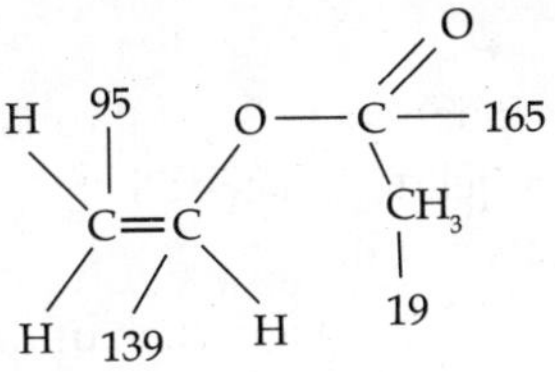

65. (a)

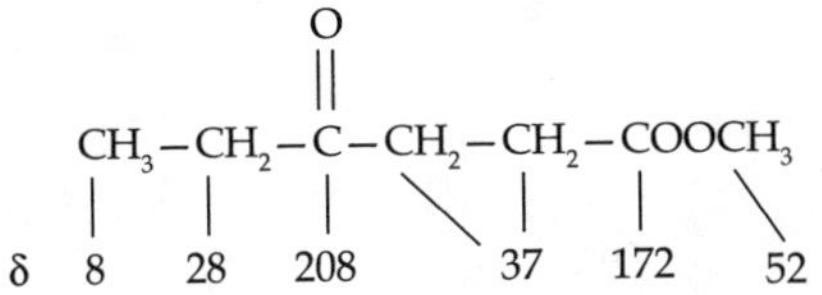

(b)

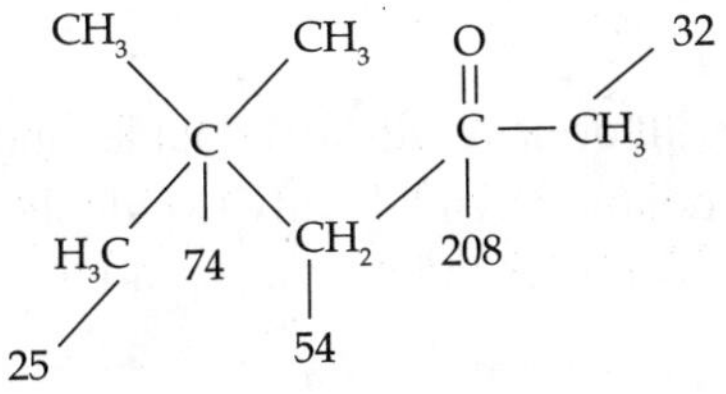

(c)

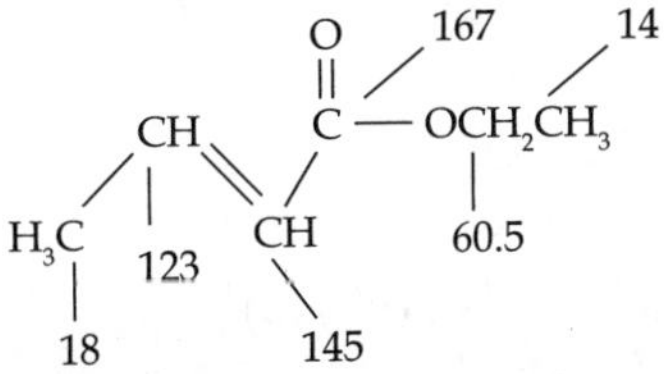

Methyl methacrylate

(d)

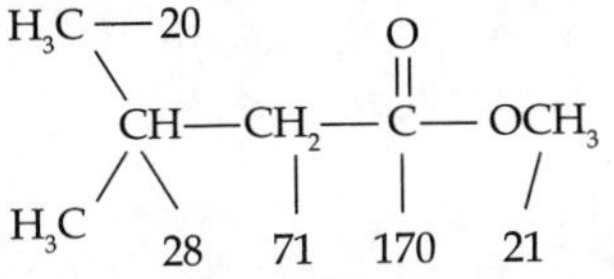

Isobutyl acetate

66. 2(a) The splitting of terminal CH groups by the central CH leads to a close double doublet. On the other hand, scalar coupling of terminal CH groups by the central CH leads to two doublets.

1(c) J_{trans} (13–18 Hz) > J_{cis} (7–12 Hz). Middle CH is coupled by the two *cis* protons, the splitting due to J_{trans} and J_{cis} results in two doublets with larger separation between them. Coupling by *cis* protons gives a closer double doublet.

3(b) CH_2 is split into a doublet by CH while CH is split by coupling with CH_2 into a triplet.

67. (a) N–methyl–p–toluidine
(b) benzyl alcohol
(c) p–methoxy acetophenone

68. (a) 211.4 T (b) 1.072×10^{-2} T
69. (a) Coupling of ^{31}P with three ^{1}H leads to a quartet (1 : 3 : 3 : 1) and coupling with ^{11}B splits each of the lines into a quartet (1 : 1 : 1 : 1); each line becoming a quartet, totally four quartets result.
   (b) Coupling with ^{11}B leads to a quartet (1 : 1 : 1: 1) and each of the lines of the quartet splits into a doublet giving totally 8 lines.

70.

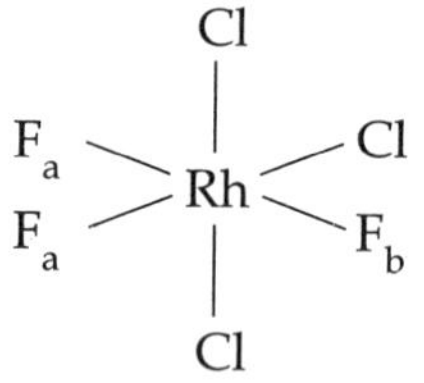

^{19}F NMR: Coupling of Rh with F_a fluorines leads to a doublet and coupling with F_b will split each line into a doublet giving totally two doublets. Coupling of Rh with F_b leads to a doublet and each line with split into a triplet (1 : 2 : 1) by coupling with F_a.

71. (b)
72. (a)
73. (d)
74. (a)
75. (i) (a), (ii) (c), (iii) (a), (iv) (b), (v) (a), (vi) (a), (vii) (a)

Chapter 4 NQR

1.

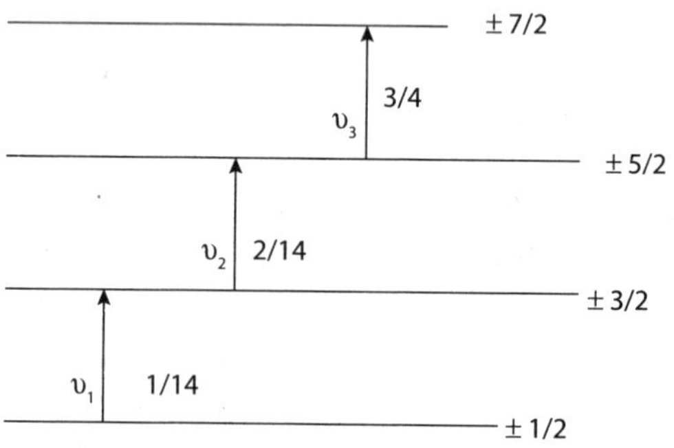

Fig. 1 Energy levels and NQR transitions

Transitions energies are

υ_1 ($\pm$ 1/2 $\rightarrow \pm$ 3/2) = 1/14 $(e^2qQ)/h$

υ_2 ($\pm$ 3/2 $\rightarrow \pm$ 5/2) = 2/14 $(e^2qQ)/h$

υ_3 ($\pm$ 5/2 $\rightarrow \pm$ 7/2) = 3/14 $(e^2qQ)/h$

We note that $\upsilon_1 : \upsilon_2 : \upsilon_3$ = 1:2:3

2. $E_Q = \dfrac{e^2qQ[3m_I^{\ 2} - I(I+1)]}{4I(2I-1)}$

   Substitute $m_I = 1$

   We get $E_Q = \dfrac{e^2qQ[3I^2 - I^2 - I]}{8I^2 - 4I} = \dfrac{e^2qQ}{4}$

   (b) The order of energy levels is reversed

   (c) No change occurs in the spectrum

3. e^2qQ represents the size of the 'hole' (i.e., vacancy) in the p-shell of Cl, which gets progressively filled as the organic group becomes less negative, that is, more electron donating.

4. (a) Cobalt(III) has a low spin d^6 configuration (t^6_{2g}) which provides spherical electron density. The complex has octahedral (O_h) symmetry. Therefore, there is no EFG at the cobalt nucleus. However, the nitrogen atom in NH_3 has an asymmetric environment (C_{3v}). The EFG represents the difference between the nature of N–H and N–Co bonds.

   (b) In NH_3, the EFG is largely due to the lone pair. When the lone pair becomes a bond pair, by bonding to cobalt, its contribution to the EFG decreases.

5. Cobalt(I) has a d^8 configuration and the complex has a D_{3h} symmetry. For both these reasons, the EFG is non-zero. Since $I = 7/2$, four energy levels, characterized by $m_I = \pm 7/2, \pm 5/2, \pm 3/2$ and $\pm 1/2$ occur. Three transitions under the selection rule, $\Delta m_I = \pm 1$ occur (See Fig. 1).

6. There is a substantial change between ionic KNCS and the thiocyanate complex, indicating that the thiocyanate is bound through nitrogen (isothiocyanato complex). In the $Zn(NCS)_2py_2$, the low QCC-value must be due to the nitrogen atom in the thiocyanate ligand.

7. $\text{Ionicity} = 1 - \dfrac{\text{QCC}}{\text{QCC}_o}$

   The value of ionicity for Cl_2, ClF, ClBr and ClI is 0, + 0.33, –0.95 and –0.25 respectively.

8. The frequencies of the transitions are $(e^2qQ/h)\ \eta/2$, $3/4\ (e^2qQ/h)\ (1 + \eta/3)$ and $3/4\ (e^2qQ)\ (1 - \eta/3)$. The sum of the last two is $3/2\ (e^2qQ/h) = 6.876$ and hence $e^2qQ/h = 4.584$ MHz and $\eta = 2(0.98 \text{ MHz})\ (4.584 \text{ MHz}) = 0.40$. The magnitude of e^2qQ/h represents chiefly the presence of a lone pair of electrons on the nitrogen. The high value of η is due to the planarity of the ring so that the components of EFG are very different in the plane of the ring and perpendicular to it.

9. On coordination, the lone pair is partially delocalized onto the metal giving a decrease in e^2qQ/h value. The acceptor ability of Zn^{2+} is affected only to a small extent by change from chloride to bromide.

10. 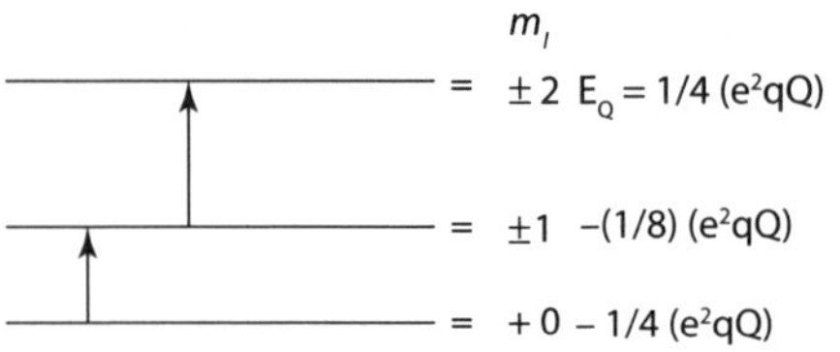

Transition $0 \rightarrow \pm 1$ $\quad \upsilon_1 = 1/8\ (e^2qQ)/h$

$\pm 1 \rightarrow \pm 2$ $\quad \upsilon_2 = 3/8\ (e^2qQ)/h$

Thus $\upsilon_1 : \upsilon_2 = 1{:}3$

11. The interaction of electric quadrupole moment with the electric field gradient is measured by their product known as the quadrupole coupling constant (e^2qQ).

12. If the charge distribution within the nucleus is known, the extent by which the shape of a nucleus is a prolate or oblate is determined by the scalar nuclear quadrupole moment Q, which can be calculated using

$$eQ = \int \rho(3z^2 - r^2)d\tau$$

where z-axis is along the direction of axial symmetry, e is the magnitude of the charge on the electron, and ρ is the nuclear charge density as a function of position.

13. The higher frequency signal at 33.75 MHz is assignable to axial chlorines and the lower frequency signal at 29.26 MHz to the equatorial chlorines of a trigonal bipyramidal structure. The axial chlorines are more ionic than the equatorial ones.

14. $\dfrac{-0.10}{0.079} = \dfrac{^{35}\text{Cl}}{^{37}\text{Cl}} = 1.266 \qquad \dfrac{22.05}{1.266} = 17.02 \quad \dfrac{22.25}{1.266} = 17.58$

15. The resonances at 22.05 and 22.25 occur at 17.02 and 17.58 MHz, respectively.

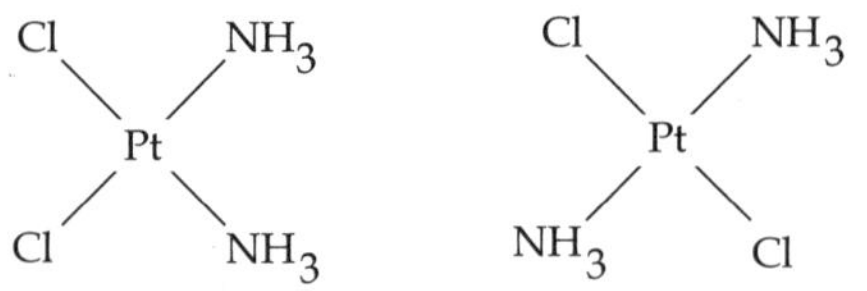

two bands one band

16. Two bands $= \dfrac{-0.10}{0.079} = 1.266$

(a) 18.92 19.71 26.94 27.34 MHz

(b) 23.95 24.95 34.11 34.61 MHz

$C_6H_5AsCl_4$ has a trigonal bipyramidal structure, with bulky phenyl group in the equatorial position.

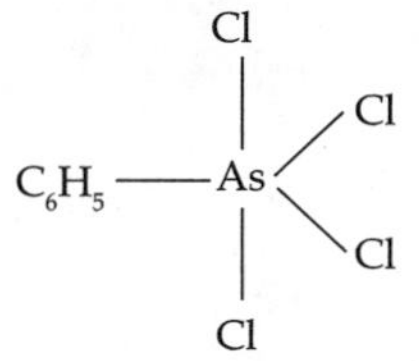

Since axial chlorines are expected to be more ionic than the equatorial ones, they are assigned at lower frequencies, 23.95 and 24.95 MHz and the equatorial chlorines are assigned at 34.11 and 34.61 MHz.

17. The ionic character of the bond decreases along the series.

18. (a) 1 (b) 1 (because of centre of symmetry) (c) 2 (d) 3 and (e) 3. The intensity of the bands are in the ratio 1:1:3. The band due to CCl_3 splits into a distinct doublet 2:1 due to the ortho-chloro group to it which interferes sterically with the free rotation of trichloromethyl group. As a result of this, the two chlorine atoms of the trichloromethyl group nearer the ortho group are different from the one which is farther.

19.

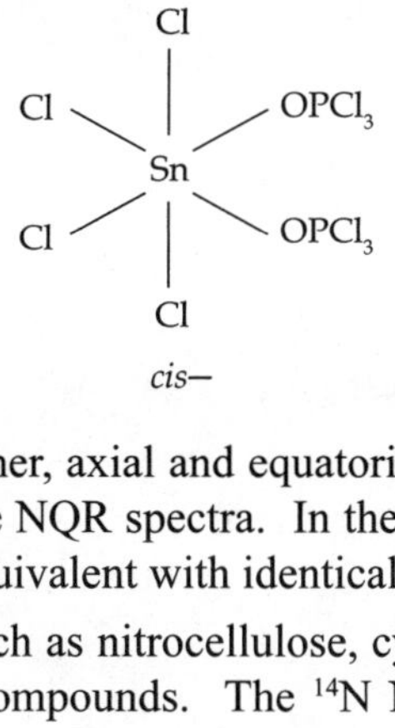

cis–

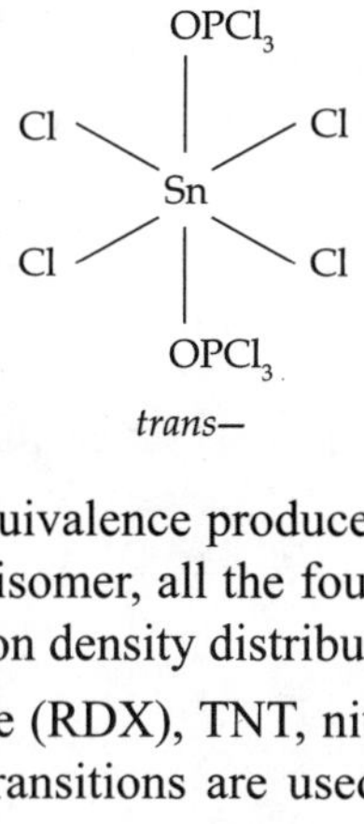

trans–

In the *cis* isomer, axial and equatorial inequivalence produces considerable splitting in the NQR spectra. In the *trans* isomer, all the four chlorines are chemically equivalent with identical electron density distribution.

20. Explosives such as nitrocellulose, cyclonite (RDX), TNT, nitroglycerin are nitrogenous compounds. The ^{14}N NQR transitions are used to detect the explosives.

INDEX

H

I

O

P

T

U

V

W

X

Z